ECONOMICS OF CLIMATE CHANGE AND ENVIRONMENTAL POLICY

Wherever possible, the articles in these volumes have been reproduced as originally published using facsimile reproduction, inclusive of footnotes and pagination to facilitate ease of reference.

For a list of all Edward Elgar published titles visit our website at www.e-elgar.com

Economics of Climate Change and Environmental Policy: Selected Papers of Robert N. Stavins, 2000–2011

Robert N. Stavins

Albert Pratt Professor of Business and Government
John F. Kennedy School of Government
Harvard University
and University Fellow, Resources for the Future
Research Associate, National Bureau of Economic Research, USA

Edward Elgar
Cheltenham, UK • Northampton, MA, USA

Published by
Edward Elgar Publishing Limited
The Lypiatts
15 Lansdown Road
Cheltenham
Glos GL50 2JA
UK

Edward Elgar Publishing, Inc.
William Pratt House
9 Dewey Court
Northampton
Massachusetts 01060
USA

A catalogue record for this book is available from the British Library

Library of Congress Control Number: 2012946676

ISBN 978 0 85793 737 7

Printed and bound by MPG Books Group, UK

Contents

PART VII INTERNATIONAL CLIMATE CHANGE POLICY

Acknowledgements

The editor and publishers wish to thank the authors and the following publishers who have kindly given permission for the use of copyright material.

American Economic Association for articles: Lori D. Snyder, Nolan H. Miller and Robert N. Stavins (2003), 'The Effects of Environmental Regulation on Technology Diffusion: The Case of Chlorine Manufacturing', *American Economic Review: Papers and Proceedings*, **93** (2), May, 431–5; Sheila M. Olmstead and Robert N. Stavins (2006), 'An International Policy Architecture for the Post-Kyoto Era', *American Economic Review: Papers and Proceedings*, **96** (2), May, 35–8; Robert N. Stavins (2011), 'The Problem of the Commons: Still Unsettled after 100 Years', *American Economic Review*, **101** (1), February, 81–108; Robin Cross, Andrew J. Plantinga and Robert N. Stavins (2011), 'What Is the Value of Terroir?', *American Economic Review: Papers and Proceedings*, **101** (3), May, 152–6; Lawrence H. Goulder and Robert N. Stavins (2011), 'Challenges from State-Federal Interactions in US Climate Change Policy', *American Economic Review: Papers and Proceedings*, **101** (3), May, 253–7.

Blackwell Publishing Ltd via the Copyright Clearance Center for articles: Adam B. Jaffe, Richard G. Newell and Robert N. Stavins (2002), 'Environmental Policy and Technological Change', *Environmental and Resource Economics*, **22**, 41–69; Richard G. Newell and Robert N. Stavins (2003), 'Cost Heterogeneity and the Potential Savings from Market-Based Policies', *Journal of Regulatory Economics*, **23** (1), January, 43–59; Lori S. Bennear, Robert N. Stavins and Alexander F. Wagner (2005), 'Using Revealed Preferences to Infer Environmental Benefits: Evidence from Recreational Fishing Licenses', *Journal of Regulatory Economics*, **28** (2), 157–79; Judson Jaffe and Robert N. Stavins (2007), 'On the Value of Formal Assessment of Uncertainty in Regulatory Analysis', *Regulation and Governance*, **1**, 154–71; Lori Snyder Bennear and Robert N. Stavins (2007), 'Second-Best Theory and the Use of Multiple Policy Instruments', *Environmental Resource Economics*, **37**, 111–29.

Elsevier Ltd for article: Adam B. Jaffe, Richard G. Newell and Robert N. Stavins (2005), 'A Tale of Two Market Failures: Technology and Environmental Policy', *Ecological Economics*, **54**, 164–74.

Elsevier Ltd via Copyright Clearance Center's Rightslink Service for articles: Richard G. Newell and Robert N. Stavins (2000), 'Climate Change and Forest Sinks: Factors Affecting the Costs of Carbon Sequestration', *Journal of Environmental Economics and Management*, **40**, 211–35; Robert N. Stavins, Alexander F. Wagner and Gernot Wagner (2003), 'Interpreting Sustainability in Economic Terms: Dynamic Efficiency Plus Intergenerational Equity', *Economic Letters*, **79**, 339–43; Ruben N. Lubowski, Andrew J. Plantinga and Robert N. Stavins (2006), 'Land-Use Change and Carbon Sinks: Econometric Estimation of the Carbon

Sequestration Supply Function', *Journal of Environmental Economics and Management*, **51**, 135–52; Richard G. Newell, Adam B. Jaffe and Robert N. Stavins (2006), 'The Effects of Economic and Policy Incentives on Carbon Mitigation Technologies', *Energy Economics*, **28** (5–6), November, 563–78; Sheila M. Olmstead, W. Michael Hanemann and Robert N. Stavins (2007), 'Water Demand under Alternative Price Structures', *Journal of Environmental Economics and Management*, **54**, 181–98.

Harvard Law School for article: Robert W. Hahn, Sheila M. Olmstead and Robert N. Stavins (2003), 'Environmental Regulation in the 1990s: A Retrospective Analysis', *Harvard Environmental Law Review*, 27 (2), 377–415.

Milken Institute Review for article: Robert N. Stavins (2005), 'Beyond Kyoto: Getting Serious about Climate Change', *Milken Institute Review*, **7** (1), 28–37.

Nature Publishing Group via the Copyright Clearance Center for article: Lawrence H. Goulder and Robert N. Stavins (2002), 'An Eye on the Future: How Economists' Controversial Practice of Discounting Really Affects the Evaluation of Environmental Policies', *Nature*, **419**, 17th October, 673–4.

Oxford University Press via the Copyright Clearance Center for article: Robert N. Stavins (2008), 'Addressing Climate Change with a Comprehensive US Cap-and-Trade System', *Oxford Review of Economic Policy*, **24** (2), 298–321.

Regents of the University of California for article: Judson Jaffe, Matthew Ranson and Robert N. Stavins (2009), 'Linking Tradable Permit Systems: A Key Element of Emerging International Climate Policy Architecture', *Ecology Law Quarterly*, **36**, 789–808.

Stanford Law School for article: Robert N. Stavins (2006), 'Vintage-Differentiated Environmental Regulation', *Stanford Environmental Law Journal*, **25** (1), Winter, 29–63.

Taylor and Francis Ltd (http://www.informaworld.com) via the Copyright Clearance Center for article: Joseph E. Aldy, Scott Barrett and Robert N. Stavins (2003), 'Thirteen Plus One: A Comparison of Global Climate Policy Architectures', *Climate Policy*, **3** (4), 373–97.

University of Chicago for article: Robert W. Hahn and Robert N. Stavins (2011), 'The Effect of Allowance Allocations on Cap-and-Trade System Performance', *Journal of Law and Economics*, **54** (4), November, S267–S294.

University of Wisconsin Press for articles: Forest L. Reinhardt, Robert N. Stavins and Richard H. K. Vietor (2008), 'Corporate Social Responsibility through an Economic Lens', *Review of Environmental Economics and Policy*, **2** (2), Summer, 219–39; Ruben N. Lubowski, Andrew J. Plantinga and Robert N. Stavins (2008), 'What Drives Land-Use Change in the United States? A National Analysis of Landowner Decisions', *Land Economics*, **84** (4), November, 529–50.

Introduction: Navigating a Two-Way Street Between Academia and the Policy World

Robert N. Stavins

Over the past decade, environmental and resource economics has continued to evolve from what was once a relatively obscure application of welfare economics to a prominent field of economics in its own right. The number of articles on the natural environment appearing in mainstream economics periodicals has continued to increase, as has the number of economics journals dedicated exclusively to environmental and resource topics. Likewise, the influence of environmental economics on public policy has increased significantly, particularly as greater use has been made of market-based instruments for environmental protection.

This introductory essay begins with some personal reflections on the professional path that has led to my research and writing, and concludes with a brief overview of the book.

A professional path

In retrospect, my professional path may now appear somewhat direct, if not altogether linear, but it hardly seemed so as I traveled along it. The path I describe took me back and forth across the United States and to several continents, and it took me from physics to philosophy, to agricultural extension, to international development studies, to agricultural economics, and eventually to environmental economics. It culminated in my receipt in 1988 of a Ph.D. degree in economics at Harvard University, where I have since been a faculty member at the John F. Kennedy School of Government. During this time, much has changed in the profession.

The early ascendancy of the field of environmental economics, during the period from 1970 to 1990, was centered within departments of agricultural and resource economics, mainly at U.S. universities, and at Resources for the Future (RFF), the Washington research institution. Within most economics departments, however, environmental studies remained a relatively minor area of applied welfare economics. So, when I enrolled in the Ph.D. program in Harvard's Department of Economics in 1983, and when I received my degree five years later, no field of study was offered in the field of environmental or resource economics.

Fortunately, Harvard permitted its graduate students to develop an optional, self-designed field as one of two fields on which they were to be examined orally before proceeding to dissertation research. Without a resident environmental economist in the Department of Economics (Martin Weitzman had yet to move to Harvard from the Massachusetts Institute of Technology), I developed an outline and reading list of the field through correspondence with leading scholars from other institutions, most prominently Kerry Smith, then at North Carolina State University. My proposal to prepare for and be examined in the special field of environmental and resource economics (along with econometrics) was approved by the Department's director of graduate study, Dale Jorgenson. So began my entry into the scholarly literature of the field.

But my interest in environmental economics pre-dated by a considerable number of years my matriculation at Harvard. Like many others before and since, I came to the field because of a personal interest in the natural environment (the origin of which I describe below). This personal interest evolved into a professional one while I was studying for an M.S. degree in agricultural economics at Cornell University in the late 1970s, where my thesis advisor and mentor was Kenneth Robinson. I had originally gone to Cornell to study for a professional degree in international development, but found agricultural economics more appealing, largely because of the opportunity to examine social questions with quantitative methods within a disciplinary framework.

The faculty at Cornell and the care given to graduate students (including masters students like me) were both outstanding. Ken Robinson, my first mentor within the economics profession, became my ongoing role model for intellectual integrity. It was a sad day just this past year when Professor Robinson passed away.

A course in linear algebra, brilliantly taught by S. R. Searle, inspired me to pursue quantitative methods of analysis, and I was fortunate to then have the opportunity to study econometrics with Tim Mount. One summer I had the great privilege of learning comparative economic systems in a small workshop setting from George Staller of the Cornell Department of Economics. Working with Bud Stanton, I had my first experience teaching at the university level, and with Olan Forker, I had my first try at serious writing. All of this led to research and writing of an M.S. thesis, 'Forecasting the Size Distribution of Farms: A Methodological Analysis of the Dairy Industry in New York State'. The methodology in question was a variable Markov transition probability matrix, the cells of which were estimated econometrically in a multinomial logit framework. Much to my surprise, this work subsequently received the Outstanding Master's Thesis Award in the national competition of the American Agricultural Economics Association.

Armed with my M.S. degree, I moved from Cornell to Berkeley, California, where I eventually met up with Phillip LeVeen, who had until shortly before that time been a faculty member in the Department of Agricultural and Resource Economics at the University of California, Berkeley. Phil was another superb mentor, and from him I learned the power of using simple models – by which I mean a set of supply and demand curves hastily drawn on a piece of scrap paper – to develop insights into real-world policy problems. He introduced me to a topic that was to occupy me for the next few years – California's perpetual concerns with water allocation. I remember many afternoons spent working with Phil at his dining room table on questions of water supply and demand.

This work with Phil LeVeen led to a consultancy and then a staff position with the Environmental Defense Fund (EDF), the national advocacy group consisting of lawyers, natural scientists, and – then almost unique among environmental advocacy organizations – economists. At EDF, I was able to experience for the first time the use of economic analysis in pursuit of better environmental policy. With W. R. Zach Willey, EDF's senior economist in California, as a role model, and Thomas Graff, EDF's senior attorney, as my mentor, I thrived in EDF's collegial atmosphere, while thoroughly enjoying life in Berkeley's 'gourmet ghetto', as my neighborhood was called. Sadly, Tom Graff – without whose mentorship I would not be where I am today – passed away last year after a heroic battle with cancer.

Although I found the work at EDF exceptionally rewarding, I worried that I would eventually be constrained – either within the organization or outside it – by my limited education. So, like many others in similar situations, I considered a law degree as the next logical step. In fact, I came very close to enrolling at Stanford Law School, but instead, in 1983, I accepted an offer of admission to the Department of Economics at Harvard, moved back east to Cambridge, Massachusetts, and began what has turned out to be a long-term relationship with the University.

But where did my interest in the natural environment begin? Not at Cornell; it was present long before those days. But it had not yet arisen when I was studying earlier at Northwestern University, from which I received a B.A. degree in philosophy, having departed from my first scholarly interest, astronomy and astrophysics.

Rather, the origins of my affinity for the natural environment and my interest in resource issues are to be found in the four years I spent in a small, remote village in Sierra Leone, West Africa, as a Peace Corps Volunteer, working in agricultural extension (in particular, paddy rice development). It was there that I was first exposed both to the qualities of a pristine natural environment and the trade-offs associated with economic development.

So, I had begun in astrophysics, moved to philosophy (both at Northwestern), then to agricultural extension in a developing country (Sierra Leone), then to international development studies and subsequently to agricultural economics (both at Cornell), then to environmental economics and policy (EDF), and eventually to graduate study in economics at Harvard.

My dissertation research at Harvard was directed by a committee of three faculty members: Joseph Kalt, Zvi Griliches, and Adam Jaffe. Joseph Kalt was the first faculty member at the Department of Economics to validate my interest in environmental and resource issues, and he was unfailingly generous to me and many other graduate students in making his office (and computer, then a rather scarce resource) available at all hours. Now a colleague at the Kennedy School, Joe provided examples never to be forgotten – that economics could be a meaningful and enjoyable pursuit, and that excellence in teaching was a laudable goal.

Zvi Griliches was not only my advisor and mentor, but my spiritual father as well. Generations of Harvard graduate students would offer similar testimony. My own father had died only a year before I entered Harvard, and Zvi soon filled for me many paternal needs. It is now more than a decade since Zvi himself passed away. I felt as if I had lost my father a second time.

If Zvi Griliches provided caring and inspiration, Adam Jaffe provided invaluable day-to-day guidance. It was Adam who convinced me not to go on the job market in my fourth year with what would have been a mediocre dissertation, but to put in another year and do it right. That turned out to be some of the best professional advice I have ever received. Our intensive faculty–student relationship from dissertation days subsequently evolved into a very productive professional (and personal) one that continues to this day. The name of Adam Jaffe appears frequently in my curriculum vitae as a co-author; he has been and continues to be much more than that.

Although they were not members of my thesis committee, I should acknowledge two other faculty members at the Harvard Department of Economics who played important roles in my education. I was fortunate to take two courses in economic history (a department requirement) from Jeffrey Williamson, who had recently arrived from the University of Wisconsin. Williamson's class sessions were as close as anything I have seen to being economic research laboratories. In class after class, we would carefully dissect one or more articles – examining hypothesis, theoretical model, data, estimation method, results, and conclusions. If there was any place where I actually learned how to carry out economic research, it was in those classes.

The other name that is important to highlight is that of Lawrence Goulder, then a faculty member at Harvard, and now a professor at Stanford. I say this not simply because he was willing to be my examiner in my chosen field of environmental and resource economics, nor because he subsequently became such a close friend. Rather, what is striking about my professional relationship with Larry is the degree to which he has been an unnamed collaborator on so many projects of mine. Although he and I have co-authored no more than a few articles, his name probably appears more frequently than anyone else's in the acknowledgments of papers I have written. There is no one whose overall judgement in matters of economics I trust more, and no one who has been more helpful.

When I began graduate school at Harvard in 1983, it was my intention to return to EDF as soon as I received my degree. But by my third year in the program, I had decided to pursue an academic career, although one that was heavily flavored with involvement in the real world of public policy. Within the context of this professional objective, it was not a difficult decision to accept the offer I received in February, 1988, to become an Assistant Professor at the Kennedy School. Although some of the other offers I received at that time were also very attractive, the choice for me was obvious, and I have never regretted it – not for a moment.

I remain at the Kennedy School today, where I was promoted to Associate Professor in 1992 (an untenured rank at Harvard), and to a tenured position as Professor of Public Policy in 1997. In 1998, I accepted an appointment as the Albert Pratt Professor of Business and Government.

Two years later, I launched the Harvard Environmental Economics Program, which today brings together – from across the University – 30 Faculty Fellows and 25 Pre-Doctoral Fellows, who are graduate students studying for the Ph.D. degree in economics, political economy and government, public policy, or health policy. The Program, which I continue to direct, forms links among faculty and graduate students engaged in research, teaching, and outreach in environmental, natural resource, and energy economics and related public policy, by sponsoring research projects, convening workshops, and supporting graduate (and undergraduate) education.

A key reason why the Program – and its various projects, including the Harvard Project on Climate Agreements – have been so successful is the marvelous administrative leadership and staff support it enjoys. Everyone who has been involved in virtually any way has come away impressed by our Executive Director, Robert Stowe, and Program Coordinator, Jason Chapman.

At the Kennedy School, I have had an excellent mentor, William Hogan, and a superb advisor and friend, Richard Zeckhauser. Over the years, five successive deans have provided leadership, guidance, and support (including abundant time for my research and writing) – Graham Allison, Robert Putnam, Albert Carnesale, Joseph Nye, and David Ellwood. At Harvard more broadly, I have benefitted from regular interactions with Daniel Schrag, director of the Harvard University Center for the Environment, and Martin Weitzman of the Department of Economics. For two decades, Marty and I have co-directed a bi-weekly Seminar in Environmental Economics and Policy, which has provided me with frequent opportunities to learn both from seminar speakers and from Marty's questions and comments. I will refrain from naming the many others at Harvard and elsewhere from whom I continue to learn – including my many co-authors – only because the list of such valued colleagues and friends is so long. Included have been a most remarkable set of Ph.D. students, many of whom have gone on to productive – indeed illustrious – careers.

Along the way, I have had my share of administrative responsibilities at Harvard, including serving as Director of Graduate Studies for the Doctoral Program in Public Policy and the Doctoral Program in Political Economy and Government, and Co-Chair of the Harvard Business

School–Harvard Kennedy School Joint Degree Programs. Outside of Harvard, I have had the privilege of being a University Fellow of Resources for the Future, a Research Associate of the National Bureau of Economic Research, and the founding Editor and now Co-Editor of the *Review of Environmental Economics and Policy*, as well as a member of the Board of Directors of Resources for the Future, the Scientific Advisory Board of the Fondazione Eni Enrico Mattei, and numerous editorial boards. I must also note that I serve as an editor of the *Journal of Wine Economics*. In 2009, I was elected a Fellow of the Association of Environmental and Resource Economists.

What originally attracted me to the Kennedy School was the possibility of combining an academic career with extensive involvement in the development of public policy. I have not been disappointed. Indeed, a theme that emerges from this book is the interplay between scholarly economic research and implementation in real-world political contexts. This is a two-way street. In some cases, my policy involvement has come from expertise I developed through research, following a path well worn by academics. But, in many other cases, my participation in policy matters has stimulated for me entirely new lines of research.

What I have characterized as involvement in policy matters is described at the Kennedy School as faculty outreach, recognized to be of great institutional and social value, along with the two other components of our three-legged professional stool – research and teaching. Because they relate to a number of the papers collected in this volume, I should note that my outreach efforts fall into five broad categories: advisory work with members of Congress and the White House (for example, Project 88, a bipartisan effort co-chaired by former Senator Timothy Wirth and the late Senator John Heinz, to develop innovative approaches to environmental and resource problems); service on federal government panels (for example, my role as Chairman of the Environmental Economics Advisory Committee of the U.S. Environmental Protection Agency Science Advisory Board); ongoing consulting – often on an informal basis – with environmental NGOs (most frequently, the Environmental Defense Fund) and private firms; advisory work with state governments; and professional interventions in the international sphere, such as service as a Lead Author for the Second and the Third Assessment Reports and a Coordinating Lead Author for the Fifth Assessment Report of the Intergovernmental Panel on Climate Change, professional roles with the World Bank and other international organizations, and advisory work with foreign governments.

Finally, because this book focuses on my articles, not my books, I should note that over the years I have been co-editor with Joseph Aldy of *Post-Kyoto International Climate Policy: Implementing Architectures for Agreement* (Cambridge University Press, 2010), *Post-Kyoto International Climate Policy: Summary for Policymakers* (Cambridge University Press, 2009), and *Architectures for Agreement: Addressing Global Climate Change in the Post-Kyoto World* (Cambridge University Press, 2007); editor of three editions of *Economics of the Environment* (W. W. Norton, 2000, 2005, 2012); co-editor with Bruce Hay and Richard Vietor of *Environmental Protection and the Social Responsibility of Firms: Perspectives from Law, Economics, and Business* (Resources for the Future, 2005); editor of *The Political Economy of Environmental Regulation* (Edward Elgar, 2004), co-editor with Paul Portney of *Public Policies for Environmental Protection* (Resources for the Future, 2000); and author of *Environmental Economics and Public Policy: Selected Papers of Robert N. Stavins, 1988–1999* (Edward Elgar, 2001).

The current volume

That last book is the predecessor of the current volume. Whereas the first volume (Stavins 2001) included selected papers from the period 1988 through 1999, this second volume covers the period from 2000 through 2011. To prepare this book, I have selected 26 articles from the many more (published and unpublished) papers I wrote – frequently with co-authors – over the past decade. Making this selection was not an easy task, but it was a rewarding one, because choosing the papers and organizing them has forced me to step back and reflect on the set of research endeavors in which I have been engaged over this decade, and thus to think more clearly about current and possible future directions.

The book is divided into seven parts. The papers in Part I provide an overview of environmental and resource economics, treating broadly several key topics, including economic views of: the problem of the commons (Stavins, *American Economic Review*, 2011); the history of U.S. environmental regulation (Hahn, Olmstead, and Stavins, *Harvard Environmental Law Review*, 2003); and corporate social responsibility (Reinhardt, Stavins, and Vietor, *Review of Environmental Economics and Policy*, 2008).

The articles in Part II deal with methods of environmental policy analysis, focusing, respectively, on: interpreting sustainability in economic terms (Stavins, Wagner, and Wagner, *Economic Letters*, 2003); the use of discounting in net present value analysis (Goulder and Stavins, *Nature*, 2002); the development of a new revealed-preference method for inferring environmental benefits (Bennear, Stavins, and Wagner, *Journal of Regulatory Economics*, 2005); and the value of formal assessment of uncertainty (that is, Monte Carlo analysis) in regulatory impact analysis (Jaffe and Stavins, *Regulation and Governance*, 2007).

Part III turns to economic analysis of alternative environmental policy instruments, with examinations of: vintage-differentiated environmental regulation (Stavins, *Stanford Environmental Law Journal*, 2006); cost heterogeneity and the potential savings from employing market-based environmental policies (Newell and Stavins, *Journal of Regulatory Economics*, 2003); the effects of allowance allocations on the performance of cap-and-trade systems (Hahn and Stavins, *Journal of Law and Economics*, 2011); and second-best theory and the use of multiple policy instruments (Bennear and Stavins, *Environmental and Resource Economics*, 2007).

Part IV focuses on a topic that also received considerable treatment in the predecessor to this volume, namely the economics of technological change. Here the articles include: a survey of the literature on environmental policy and technological change (Jaffe, Newell, and Stavins, *Environmental and Resource Economics*, 2002); an analysis of the interaction of environmental and technological market failures (Jaffe, Newell, and Stavins, *Ecological Economics*, 2005); an empirical assessment of the effect of environmental regulation on technology diffusion in the case of chlorine manufacturing (Snyder, Miller, and Stavins, *American Economic Review: Papers and Proceedings*, 2003); and the effects of economic and policy incentives on carbon mitigation technologies (Newell, Jaffe, and Stavins, *Energy Economics*, 2006).

Part V consists of three articles in the area of natural resource economics dealing with land and water resources: an analysis of the factors driving land-use change in the United States (Lubowski, Plantinga, and Stavins, *Land Economics*, 2008); an econometric examination of the significance of *terroir*, the notion that wine quality is primarily determined by location (Cross, Plantinga, and Stavins, *American Economic Review: Papers and Proceedings*, 2011); and an assessment of urban water demand under alternative pricing structures (Olmstead, Hanemann, and Stavins, *Journal of Environmental Economics and Management*, 2007).

Part VI consists of four articles on domestic (national and sub-national) climate change policy, beginning with a description and assessment of a comprehensive U.S. cap-and-trade system for carbon dioxide and other greenhouse gas emissions (Stavins, *Oxford Review of Economic Policy*, 2008), and followed by: an examination of the interactions of national and sub-national climate policies (Goulder and Stavins, *American Economic Review: Papers and Proceedings*, 2011); an econometric study of the carbon-sequestration supply function (Lubowski, Plantinga, and Stavins, *Journal of Environmental Economics and Management*, 2006); and an assessment of the factors that affect the costs of biological carbon sequestration (Newell and Stavins, *Journal of Environmental Economics and Management*, 2000).

Finally, Part VII focuses on the international dimensions of climate change policy, and consists of four articles: a comparison of alternative global climate change policy architectures (Aldy, Barrett, and Stavins, *Climate Policy*, 2003); an assessment of the Kyoto Protocol (Stavins, *Milken Institute Review*, 2005); an examination of a promising post-Kyoto international climate regime (Olmstead and Stavins, *American Economic Review: Papers and Proceedings*, 2006); and a detailed examination of a key element of emerging international climate policy architecture, namely the linkage of regional, national, and sub-national tradable permit systems (Jaffe, Ranson, and Stavins, *Ecology Law Quarterly*, 2009).

Final words

Selecting the essays for this second volume of my papers has permitted me to take note of some common themes that emerge from this decade of research and writing. First, there is the value – or at least, the potential value – of economic analysis of environmental policy. The cause of virtually all environmental problems in a market economy is economic behavior (that is, imperfect markets affected by externalities), and so economics offers a powerful lens through which to view environmental problems, and therefore a potentially effective set of analytical tools for designing and evaluating environmental policies.

A second message, connected with the first, is the specific value of benefit-cost analysis for helping to promote efficient policies. Economic efficiency ought to be one of the key criteria for evaluating proposed and existing environmental policies. Despite its limitations, benefit-cost analysis can be useful for consistently assimilating the disparate information that is pertinent to sound decision making. If properly done, it can be of considerable help to public officials when they seek to establish or assess environmental goals.

Third, the means governments use to achieve environmental objectives matter greatly, because different policy instruments have very different implications along a number of dimensions, including abatement costs in both the short and the long term. Market-based instruments are particularly attractive in this regard.

Fourth, an economic perspective is also of considerable value when reflecting on the use of natural resources, whether land, water, fisheries, or forests. Excessive rates of depletion of these resources are frequently due to the nature of the respective property-rights regimes, in particular, common property and open-access. Economic instruments – such as ITQ systems in the case of fisheries – can and have been employed to bring harvesting rates down to socially efficient levels.

Fifth and finally, policies for addressing global climate change – linked with emissions of carbon dioxide and other greenhouse gases – can benefit greatly from the application of economic thinking. On the one hand, the long time-horizon of climate change, the profound uncertainty in links between emissions and actual damages, and the possibility of catastrophic climate change present significant challenges to conventional economic analysis. But, at the same time, the ubiquity of energy generation and use in modern economies means that only market-based policies – essentially carbon pricing regimes – are feasible instruments for achieving truly meaningful emissions reductions. Hence, despite the challenges, an economic perspective on this grandest of environmental threats is essential.

On a personal level, the professional path I have taken offers some confirmation that research can influence public policy, and it also illustrates that involvement in public policy can stimulate new research. The quest – both professional and personal – that took me from Evanston, Illinois, to Sierra Leone, West Africa, to Ithaca, New York, to Berkeley, California, and finally to Cambridge, Massachusetts suggests some consistency of purpose and even function. I continue to find myself doing similar things, but in different contexts. It is fair to say that my professional life has taken me along a path that has brought me home (a home I am privileged to share with my wife, Joanna, and my children, Daniel and Julia). The words of T. S. Eliot (1943) ring true:

> We shall not cease from exploration
> And the end of all our exploring
> Will be to arrive where we started
> And know the place for the first time.

Selecting the papers for this volume forces me to reflect on the past and think more clearly about the future. The 26 articles that comprise this book and the 22 essays that comprised the predecessor volume are the product of 23 wonderful years on the faculty of the Harvard Kennedy School. During this time, I have continued to learn about environmental economics and related public policy from colleagues, collaborators, students, friends, and inhabitants of the 'real world' of public policy, individuals from government, private industry, advocacy groups, and the press. I hope and trust that the learning will continue.

References

Aldy, Joseph E., and Robert N. Stavins, eds. (2007). *Architectures for Agreement: Addressing Global Climate Change in the Post-Kyoto World.* New York: Cambridge University Press.

Aldy, Joseph E., and Robert N. Stavins (2009). *Post-Kyoto International Climate Policy: Summary for Policymakers.* New York: Cambridge University Press.

Aldy, Joseph E., and Robert N. Stavins, eds. (2010). *Post-Kyoto International Climate Policy: Implementing Architectures for Agreement.* New York: Cambridge University Press.

Eliot, T. S. (1943). 'Four Quartets: Little Gidding.' *Collected Poems, 1909–1962.* New York: Harcourt Brace, 1963.

Hahn, Robert W., and Robert N. Stavins (2011). 'The Effect of Allowance Allocations on Cap-and-Trade System Performance.' *Journal of Law and Economics*, 54(4), November 2011 (published August 2012): S267–S294.

Jaffe, Adam B., Richard G. Newell, and Robert N. Stavins (2006). 'The Effects of Economic and Policy Incentives on Carbon Mitigation Technologies.' *Energy Economics*, 28: 563–78.

Portney, Paul R., and Robert N. Stavins, eds. (2000). *Public Policies for Environmental Protection.* Second Edition. Washington, D.C.: Resources for the Future.

Stavins, Robert N. (2000). *Economics of the Environment: Selected Readings.* Fourth Edition. New York: W. W. Norton & Company.

Stavins, Robert N. (2001). *Environmental Economics and Public Policy: Selected Papers of Robert N. Stavins, 1988–1999.* Cheltenham, UK and Northampton, MA, USA: Edward Elgar.

Stavins, Robert N., ed. (2004). *The Political Economy of Environmental Regulation.* Cheltenham, UK and Northampton, MA, USA: Edward Elgar.

Stavins, Robert N., ed. (2005). *Economics of the Environment: Selected Readings.* Fifth Edition. New York: W. W. Norton & Company.

Stavins, Robert N. (2012). *Economics of the Environment: Selected Readings.* Sixth Edition. New York: W. W. Norton & Company.

To Joanna, Daniel, Julia, and Rudy,
with thanks for making it all worthwhile!

PART I

OVERVIEW

[1]

The Problem of the Commons: Still Unsettled after 100 Years

By ROBERT N. STAVINS*

The problem of the commons is more important to our lives and thus more central to economics than a century ago when Katharine Coman led off the first issue of the American Economic Review. *As the US and other economies have grown, the carrying capacity of the planet—in regard to natural resources and environmental quality—has become a greater concern, particularly for common-property and open-access resources. The focus of this article is on some important, unsettled problems of the commons. Within the realm of natural resources, there are special challenges associated with renewable resources, which are frequently characterized by open-access. An important example is the degradation of open-access fisheries. Critical commons problems are also associated with environmental quality. A key contribution of economics has been the development of market-based approaches to environmental protection. These instruments are key to addressing the ultimate commons problem of the twenty-first century—global climate change.* (*JEL* Q15, Q21, Q22, Q25, Q54)

As the first decade of the twenty-first century comes to a close, the problem of the commons is more central to economics and more important to our lives than a century ago when Katharine Coman led off the first issue of the *American Economic Review* with her examination of "Some Unsettled Problems of Irrigation" (Coman 1911). Since that time, 100 years of remarkable economic progress have accompanied 100 years of increasingly challenging problems.

As the US and other economies have grown, the carrying capacity of the planet—in regard to both natural resources and environmental quality—has become a greater concern. This is particularly true for common-property and open-access resources. While small communities frequently provide modes of oversight and methods for policing their citizens (Elinor Ostrom 2010), as the scale of society has grown, commons problems have spread across communities and even across nations. In some of these cases, no overarching authority can offer complete control, rendering

*Albert Pratt Professor of Business and Government, John F. Kennedy School of Government, Harvard University, Cambridge, MA 02138; University Fellow, Resources for the Future; and Research Associate, National Bureau of Economic Research. I am grateful to Lori Bennear, Maureen Cropper, Denny Ellerman, Lawrence Goulder, Robert Hahn, Geoffrey Heal, Suzi Kerr, Charles Kolstad, Gilbert Metcalf, William Nordhaus, Wallace Oates, Sheila Olmstead, Robert Pindyck, Andy Reisinger, James Sanchirico, Richard Schmalensee, Kerry Smith, Robert Stowe, Martin Weitzman, and Richard Zeckhauser for very helpful comments on a previous version of this article; and Jane Callahan and Ian Graham of the Wellesley College Archives for having provided inspiration by making available the archives of Katharine Coman (1857–1915), professor of history, then economics, and first chair of Wellesley College's Department of Economics. Any and all remaining errors are my own.

commons problems more severe. Although the type of water allocation problems of concern to Coman (1911) have frequently been addressed by common-property regimes of collective management (Ostrom 1990), less easily governed problems of open access are associated with growing concerns about air and water quality, hazardous waste, species extinction, maintenance of stratospheric ozone, and—most recently—the stability of the global climate in the face of the steady accumulation of greenhouse gases.

Whereas common-property resources are held as private property by some group, open-access resources are nonexcludable. This article focuses exclusively on the latter, and thereby reflects on some important, unsettled problems of the commons.[1] It identifies both the contributions made by economic analysis and the challenges facing public policy. Section I begins with natural resources, highlighting the difference between most nonrenewable natural resources, pure private goods that are both excludable and rival in consumption, and renewable natural resources, many of which are nonexcludable (Table 1). Some of these are rival in consumption but characterized by open access. An example is the degradation of ocean fisheries. An economic perspective on these resources helps identify the problems they present for management and provides guidance for sensible solutions.

Section II turns to a major set of commons problems that were not addressed until the last three decades of the twentieth century—environmental quality. Although frequently characterized as textbook examples of externalities, these problems can also be viewed as a particular category of commons problems: pure public goods, that are both nonexcludable and nonrival in consumption (Table 1). A key contribution of economics has been the development of market-based approaches to environmental protection, including emission taxes and tradable rights. These have potential to address the ultimate commons problem of the twenty-first century, global climate change. Section III concludes.

Several themes emerge. First, economic theory—by focusing on market failures linked with incomplete systems of property rights—has made major contributions to our understanding of commons problems and the development of prudent public policies. Second, as our understanding of the commons has become more complex, the design of economic policy instruments has become more sophisticated, enabling policy makers to address problems that are characterized by uncertainty, spatial and temporal heterogeneity, and long duration. Third, government policies that have not accounted for economic responses have been excessively costly, often ineffective, and sometimes counterproductive. Fourth, commons problems have not diminished. While some have been addressed successfully, others have emerged that are more important and more difficult. Fifth, environmental economics is well positioned to offer better understanding and better policies to address these ongoing challenges.

[1] Ostrom has made key contributions to our understanding of the role of collective action in common-property regimes, as she does in her article in this issue of the *Review* (Ostrom 2011). With her ably covering that territory, my focus is exclusively on situations of open access. As Daniel W. Bromley (1992) has noted, the better characterizations might be common property *regimes* and open-access *regimes*, because it is the respective institutional arrangements—as much as the resources themselves—that define the problems. However, I use the conventional characterizations because of their general use in the literature. Although my focus is on the natural resources and environmental realm, similar problems—and related public policies—arise in other areas, such as the allocation of the electromagnetic spectrum for uses in communication (Roberto E. Muñoz and Thomas W. Hazlett 2009).

TABLE 1—A TAXONOMY OF COMMON PROBLEMS IN THE NATURAL RESOURCE AND ENVIRONMENT REALM

	Excludable	Nonexcludable
Rival	*Pure private goods* Most nonrenewable natural resources (Fossil fuels & minerals) Some privatized renewable resources (Aquaculture)	*Renewable natural resources characterized by open access* (Ocean fishing) Some nonrenewable resources (Ogallala Aquifer)
Nonrival	*Club goods* (Water quality of municipal pond)	*Pure public goods* (Clean air, greenhouse gases and climate change)

I. The Problem of the Commons and the Economics of Natural Resources

Despite their finite supply in the earth's crust (and despite decades of doomsday predictions),[2] reserves of mineral and fossil fuel resources have not been exhausted. Price signals reflecting relative (economic) scarcity have stimulated exploration and discovery, technological progress, and supply substitution. Hence, the world of nonrenewable natural resources is characterized more by smooth transitions (Robert M. Solow 1991; William D. Nordhaus 1992) than by overshoot and collapse. Reserves have increased, demand has changed, substitution has occurred, and—in some cases—recycling has been stimulated. As a result, for much of the past century, the economic scarcity of natural resources had not been increasing, but decreasing (Harold J. Barnett and Chandler Morse 1963). Late in the twentieth century, increasing scarcity may have set in for a subset of nonrenewable resources, although the time trends are far from clear (V. Kerry Smith 1980; Junsoo Lee, John A. List, and Mark C. Strazicich 2006; John Livernois 2009).

The picture is quite different if we turn from nonrenewable natural resources—minerals and fossil fuels—to renewable natural resources (including many forests and most fisheries), which have exhibited *monotonically increasing scarcity*. The irony is obvious: many nonrenewable natural resources, which are in finite supply, have not become more scarce over time, and *none* has been exhausted; but renewable natural resources, which have the capacity to regenerate themselves, have in many cases become more scarce, and in some cases have indeed been exhausted, that is, become extinct.

This irony can be explained by the fact that while most nonrenewable natural resources are characterized by well-defined, enforceable property rights, many renewable resources are held as common property or open access (Table 1). Whereas scarcity is therefore well reflected by markets for nonrenewable natural resources (in the form of "scarcity rent," the difference between price and marginal extraction cost, originally characterized by Harold Hotelling in 1931 as "net price"),[3] such

[2] See, for example, Donella H. Meadows et al. (1972).

[3] This is not to suggest that the market rate of extraction of nonrenewable natural resources always matches the dynamically efficient rate. Under any one of a number of conditions, markets may lead to inefficient rates of extraction: imperfect information; noncompetitive market structure (the international petroleum cartel); poorly defined property rights (ground water); externalities in production or consumption (coal mining and combustion); or differences in market and social discount rates.

rents are dissipated for open-access resources, a reality well illustrated by the bio-economics of open-access fisheries.

A. *Biology*

Since the middle of the nineteenth century, open-access fishery stocks of numerous species have been depleted beyond sustainable levels, sometimes close to the edge of extinction. The basic biology and economics of fisheries—descendent from the Gordon-Schaefer model (H. Scott Gordon 1954; Anthony Scott 1955; M. B. Schaefer 1957; Colin W. Clark 1990)—makes clear why this has happened.

In the upper panel of Figure 1, a logistical growth function plots the time rate of change of the fishery stock (dS/dt) on the vertical axis against the stock's mass (S) on the horizontal axis:

$$F(S_t) = \delta S_t \left[1 - \frac{S_t}{K}\right], \tag{1}$$

where δ is the intrinsic growth rate of the stock, and K is the carrying capacity of the environment. As the size of the stock increases, its rate of growth increases until scarce food supplies and other consequences of crowding lead to decreasing growth rates. The maximum growth rate is achieved at S_{MSY}, where the "maximum sustainable yield" (MSY) occurs. A stable equilibrium is found where the rate of growth transitions from positive to negative, a level of the stock described by biologists as the "carrying capacity" or "natural equilibrium" of the fishery. Another stable equilibrium is found at the origin—exhaustion (extinction).

The likelihood of extinction is particularly acute when the natural growth function of a species exhibits "critical depensation," illustrated in the lower panel of Figure 1:

$$F(S_t) = \delta S_t \left[1 - \frac{S_t}{K}\right] \left[\frac{S_t}{K_0} - 1\right], \tag{2}$$

where K_0 is the minimum viable population level. Below this critical level of the stock, the natural rate of growth is negative. Hence there are three equilibria: extinction (the origin); the carrying capacity; and the minimum viable population. This reflects the reality that the large habitat ranges that exist for some species, such as whales and some species of birds, means that relatively small numbers are insufficient for mating pairs to yield birth rates that exceed the natural rate of loss to predators and disease.

This third equilibrium is unstable. Once the population falls below this critical level, it will proceed inevitably to extinction (unless "artificial" actions are taken, such as confined breeding of the California condor, man-made habitats for the whooping crane, or "zoos in the wild" for giant pandas in China). In the nineteenth century, hunters did not shoot down each and every passenger pigeon, but nevertheless, the species was driven to extinction. A similar pattern has doomed other species. A contemporary case in point could be the blue whale (Michael A. Spence 1974), the largest animal known to have existed. Harvesting has been prohibited

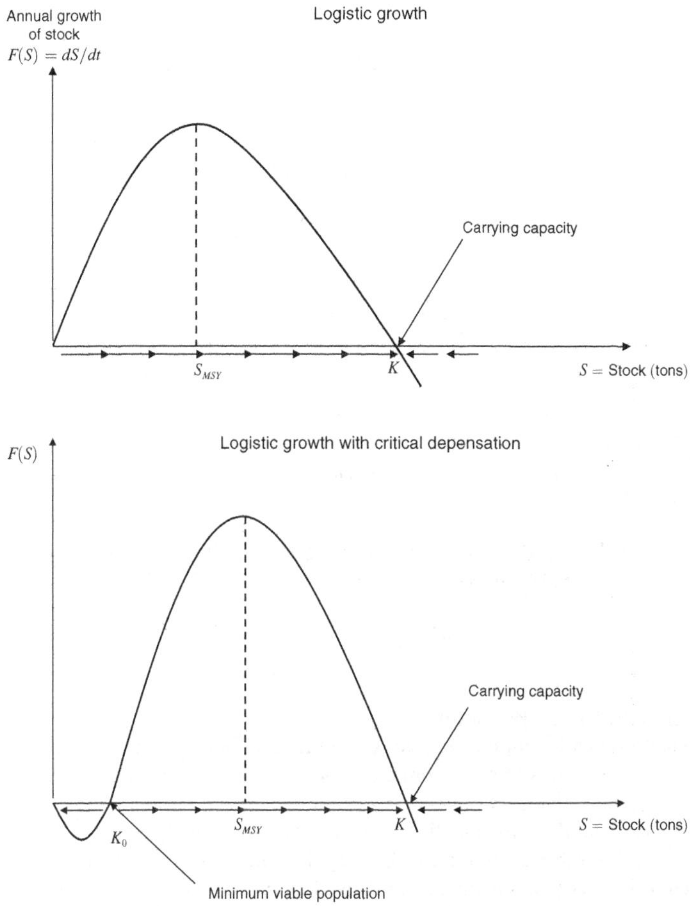

FIGURE 1. A SIMPLE MODEL OF THE FISHERY: THE BIOLOGICAL DIMENSION

under international agreements since 1965, but it is unclear whether stocks have rebounded, although numbers have been increasing in one region. Across species, there is a mixed picture. Stocks of some whale species are believed to be above and others below their respective minimum viable population (International Whaling Commission 2010).

B. *Bioeconomics*

A much greater threat to renewable natural resources than this unusual and unstable biological growth function is the way many of these resources are managed: as common property or open access. To see this, we add some basic economics to the

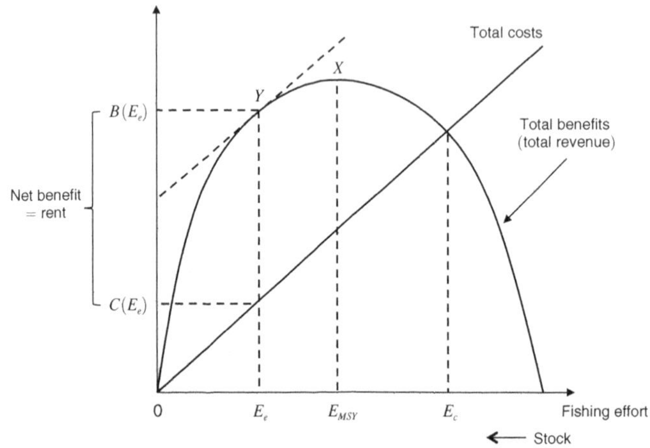

FIGURE 2. A SIMPLE MODEL OF THE FISHERY: THE ECONOMIC DIMENSION

biology of the fishery (Figure 2). First, a change in the stock of a fishery can be due not only to its biological fundamentals, but to harvests, that is, fishing:

$$\frac{dS}{dt} = \dot{S}_t = F(S_t) - q_t, \tag{3}$$

where q_t is the harvest rate at time t.[4]

The harvest is a function of the stock and the level of effort, E_t, by firms (fishing boats and crews). Abstracting from dynamics, we can identify the static efficient sustainable yield (that is, we ignore discounting over time, which is all that distinguishes this from the dynamically efficient sustainable yield), without loss of key insights. To keep things simple for the graphics, three assumptions are employed: (a) there is perfectly elastic demand, that is, the price of fish is constant, not a function of the quantity sold; (b) the marginal cost of a unit of fishing effort is constant; and (c) the quantity of fish caught per unit of effort is proportional to the size of the stock. With these assumptions, the relationship between effort and harvest is:

$$q_t(S_t, E_t) = \alpha_t S_t E_t, \tag{4}$$

where α_t is a proportional "catchability coefficient." And profits, π_t, are given by:

$$\pi_t = p_t q_t(S_t, E_t) - c_t E_t, \tag{5}$$

[4]If equation (3) is replaced by a stochastic differential equation—to characterize uncertainty inherent in the biological growth function—then even lower harvest rates than otherwise can lead to extinction (Robert S. Pindyck 1984).

where p_t is the market price of fish and c is the marginal cost of fishing effort. In the steady state, harvest is equal to growth:

$$F(S_t) = q_t \tag{6}$$

and so from equations (1) and (2):

$$\delta S_t \left[1 - \frac{S_t}{K}\right] = \alpha_t \, S_t \, E_t. \tag{7}$$

Solving for S and substituting into equation (4) yields steady-state harvest as a function of effort:

$$q_{SS} = \alpha_t \, E_t \, K \left[1 - \frac{\alpha_t E_t}{\delta}\right]. \tag{8}$$

With this, total revenue at the steady-state (or sustainable) level, equivalent to pq, is indicated in Figure 2 as the total benefits of fishing as a function of effort level. When effort exceeds level E_{MSY}, total fish catch and revenues decline. Total cost is equivalent to the constant marginal cost of effort, c, multiplied by the effort level. Hence, the efficient level of effort, E_e, is where net benefits—the difference between total revenue and total cost—are maximized, namely where marginal benefits equal marginal costs. Clearly the maximum sustainable yield is not the efficient harvest level (but would be if fishing were costless).

C. *The Consequences of Open Access*

What happens in actual markets with open access, which historically has characterized much of commercial fishing around the world (as well as markets for a number of other renewable natural resources)? At the efficient level of effort, E_e, each boat would make profits equal to its share of scarcity rent, $B(E_e)$ minus $C(E_e)$, but with open access these profits become a stimulus for more capital and labor to enter the fishery. Each fisherman considers his marginal revenue and marginal extraction cost, but—without firm property rights—scarcity rent is ignored, and each has an incentive to expend further effort (including more entry) until profits in the fishery are driven to zero: effort level E_c in Figure 2, where marginal cost is equal to *average* revenue rather than equal to *marginal* revenue. Thus, with open access, it is rational for each fisherman to ignore the asset value of the fishery, because he cannot appropriate it; all scarcity rent is dissipated (Scott 1955).

Because no one holds title to fish stocks in the open ocean, for example, everyone races to catch as much as possible. Each fisherman receives the full benefit of aggressive fishing—a larger catch—but none pays the full cost, an imperiled fishery for everyone. One fisherman's choices have an effect on other fishermen (of this generation and the next), but in an open access fishery—unlike a privately held copper mine—these impacts are not taken into account.

These consequences of open access—predicted by theory—have been validated repeatedly with empirical data. A study of the Pacific halibut fishery in the Bering Sea estimated that the efficient number of ships was nine, while the actual number was 140 (Daniel D. Huppert 1990).[5] An examination of the New England lobster fishery found that in 1966 the efficient number of traps set would have been about 450,000, while the actual number was nearly one million. Likewise, an analysis of the North Atlantic stock of minke whale found that the efficient stock size was about 67,000 adult males, whereas the open-access stock had been depleted to 25,000 (Erik S. Amundsen, Trond Bjørndal, and Jon M. Conrad 1995). In terms of social costs, an analysis of two lobster fisheries in eastern Canada found that losses due to unrestricted entry amounted to about 25 percent of market value of harvests, due mainly to excess deployment of resources for harvest, with fishery effort exceeding the efficient level by some 350 percent (J. V. Henderson and M. Tugwell 1979).

Under conditions of open access, two externalities may be said to be present. One is a contemporaneous externality (as with any public good) in which there is over-commitment of resources: too many boats, too many fishermen, and too much effort as everyone rushes to harvest before others. The other is an intertemporal externality in which overfishing reduces the stock and hence lowers future profits from fishing.

A classic time path of open-access fisheries has been repeated around the world. First, a newly discovered resource is open to all comers; eventually, large harvests and profits attract more entry to the fishery; boats work harder to maintain their harvest; despite increased efforts, the harvests decline; and this leads to greater increases in effort, resulting in even greater declines in harvest, resulting in essential collapse of the fishery. This pattern has been documented for numerous species, including the North Pacific fur seal (Wilen 1976) and the Northern anchovy fishery (Jean-Didier Opsomer and Conrad 1994), as well as Atlantic cod harvested by US and Canadian fishing fleets in the second half of the twentieth century (Figure 3).

Although open access drives the stock below its efficient level, it normally does not lead to the stock being exhausted (except possibly under critical depensation, as explained above), because below a certain stock level, the benefits of additional harvest are simply less than the additional costs. This is at the heart of a fundamental error in what is probably the most frequently cited article on common-property and open-access resources, Garrett Hardin's "The Tragedy of the Commons" (1968):

> Picture a pasture open to all … A rational herdsman concludes that the only sensible course for him to pursue is to add another animal to his herd. And another; and another … Each man is locked into a system that compels him to increase his herd without limit—in a world that is limited. Ruin is the destination toward which all men rush, each pursuing his own best interest in a society that believes in the freedom of the commons. Freedom in the commons brings ruin to all (Hardin 1968, 1244).

As Partha Dasgupta (1982) subsequently wrote: "It would be difficult to locate another passage of comparable length and fame containing as many errors as the one above." Ruin is not the outcome of the commons, but rather excessive employment

[5] These fisheries are actually "regulated open-access fisheries," because they are subject to restrictions (James N. Sanchirico and James E. Wilen 2007), as explained below.

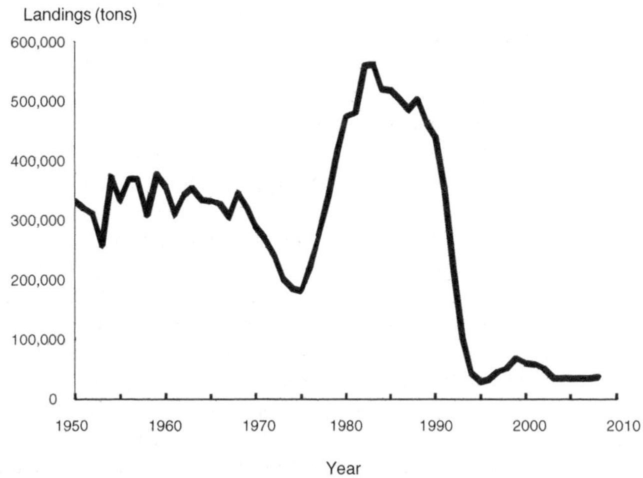

Figure 3. Annual Harvest of Atlantic Cod, 1950–2008

Source: United Nations Food and Agriculture Organization 2010.

of capital and labor, small profits for participants, and an excessively depleted resource stock.[6] Those are bad enough.

D. *Alternative Policies for the Commons Problem*

The most obvious solution to a commons problem—in principle—may be to enclose it, that is, put in place fee-simple or other well-defined property rights to limit access.[7] In the case of a natural fishery, this is typically not feasible, but it is if species are immobile (oysters, clams, mussels), can be confined by barriers (shrimp, carp, catfish), or instinctively return to their place of birth to spawn (salmon, ocean trout). Such fish farming (aquaculture) is feasible and profitable with a limited but important set of commercial species (Table 1). Presently, approximately one-third of global fisheries production is supplied by commercial aquaculture, much of it in Asia (United Nations Food and Agriculture Organization 2007).

Because aquaculture remains confined to a limited set of commercial species (and environmental concerns may preclude expansion), there has been a history of government attempts to regulate open-access fisheries through other means. The most frequent regulatory approach has been to limit annual catches (with the target typically being the maximum sustainable yield, not the efficient level of effort) through restrictions on allowed technologies, closure of particular areas, or

[6]The annual loss due to rent dissipation in global fisheries has been estimated to be on the order of $90 billion (Sanchirico and Wilen 2007).

[7]Recall that my focus is on open access, not common property. Arrangements of various kinds can and do serve to limit access to common-property resources (Ostrom 2010). An example in the fisheries realm would be the informal groups of lobster harvesters ("gangs") in coastal Maine that seek to restrict access to identified areas (James M. Acheson 2003).

imposition of limited seasons. These regulatory approaches have the effect of raising the marginal cost of fishing effort, in effect pivoting up the total cost function in Figure 2 until it intersects the benefit function at point X, thereby achieving the fishing effort associated with maximum sustainable yield, E_{MSY}, or potentially to point Y, thereby achieving the efficient effort level, E_e.

Marginal costs increase because each new constraint causes fishermen to reoptimize. In response to constraints on technology, areas, or season, fishermen employ excessively expensive methods (overcapitalization) to catch a given quantity of fish. Technology constraints can lead to the employment of more labor; area closures can lead to the adoption of more sophisticated technologies; and reduced seasons result in the use of more boats. Although the harvest may be curtailed as desired, the net benefits to the fishery are essentially zero. Costs go up for fishermen (as resources are squandered). Social efficiency is not achieved, nor is it approached.

A dramatic example is provided by New York City's once thriving oyster fishery. In 1860, 12 million oysters were sold in New York City markets. By 1880, production was up to 700 million oysters per year. "New Yorkers rich and poor were slurping the creatures in oyster cellars, saloons, stands, houses, cafes, and restaurants..." (Elizabeth Royte 2006). It became clear that the oyster beds were being depleted. First the city restricted who could harvest oysters, then when they were permitted to do so. Eventually, the city limited the use of dredges and steam power. Nevertheless, in 1927, the last of the city's oyster beds closed (a casualty not only of open access to the oyster habitats, but also of the use of the city's harbors as another sort of commons, namely as a depository for the city's sewage).

The economic implications of conventionally regulated open-access fisheries are typically worse than those that occur under unregulated open-access conditions (Frances R. Homans and Wilen 1997; Martin D. Smith and Wilen 2003). Overcapitalization is greater, as is the consequent welfare loss. Such situations with conventional open-access fisheries regulation are commonplace: overfishing occurs, the fishery stock is depleted, the government responds by regulating the catch, thereby driving up the cost of fishing, fishermen complain that they cannot make a profit, and harvests continue to fall. Is there a better way?

From an economic perspective, the most obvious way of assuring that harvest levels are maintained at an efficient level while providing incentives for cost reductions is a tax on fish harvests. Such an efficient tax, which increases marginal costs, rotates the total costs line in Figure 2 until it intersects total benefits at point Y, and thereby brings about E_e, similar to conventional regulation. The tax that would accomplish this would be equal to the difference between $B(E_e)$ and $C(E_e)$. Despite the apparent graphical similarity with the conventional regulatory outcome, this approach is efficient, because rather than destroying the rents through higher resource costs, the tax transfers the rents from the private to the public sector. Hence, the social net benefits of the tax approach are identical to those under the efficient outcome.

There is a problem, however. For the fishermen, these transfers are very real costs. The rent that would be received by a sole owner is received by the government instead. Any fishermen who might want the fishery to be managed efficiently will surely object to this particular approach. So, is there some way that the catch can be restricted to the efficient level, with real resource costs minimized, but without transferring the rents from fishermen to the government?

One answer is a system of individual transferable quotas (ITQs), by which the government sets the overall, annual allowable catch (equal to the efficient catch for the fishery), allocates this catch to fishermen in the form of quotas that entitle holders to catch a specified quantity of fish per year, and allows the fishermen to transfer (buy and sell) the quotas.[8] As we examine in more detail below in the context of tradable "pollution allowances," the quotas in the fishery will flow to those that gain the most net benefit from them due to lower costs. Hence, cost-reducing technologies and management are encouraged (as with a tax), but rents are retained by the fishing industry.[9]

Such ITQ systems have been used successfully in some 150 major fisheries of 170 species in seventeen countries—some with very significant fishing industries, such as Australia, Canada, Iceland, and New Zealand (Richard G. Newell, Sanchirico, and Suzi Kerr 2005). In fact, New Zealand regulates virtually its entire commercial fishery this way. Since 1986, the system has been effective, largely eliminating overfishing, restoring stocks to sustainable levels, and increasing fishermen's profits. Several ITQ systems are in operation in the United States, including ones for Alaska's pacific halibut and Virginia's striped-bass fisheries.

In addition to reducing catches in an efficient manner, these systems have been found to improve safety by reducing incentives for fishermen to go out (or stay out) when weather conditions are dangerous. Further, because ITQ systems eliminate the motivation for government to limit the duration of the fishing season, supplies available to consumers improve in quality. Since fishermen own shares of the assets under an ITQ system, the total allowable catch is self-enforcing, in the sense that all participants have incentives to report anyone not complying with the rules.

The example of the Pacific halibut fishery is illuminating (Huppert 1990, 2005; Homans and Wilen 1997). Open access had led to a gradually diminishing stock throughout the 1970s. In an effort to reduce the harvest, the season was first reduced from 125 days in 1975 to 25 days in 1980, and then to just 2 days in 1994. The result, of course, was more effort expended in a shorter time. Overcapitalization of the fishery was rampant. By 1994, crews remained out for the entire 48 hours of the season, leading to high rates of injury—and even mortality. Due to the rushed fishing, the by-catch (of other species) was exceedingly high, as was so-called "ghost fishing" from abandoned nets. Fresh halibut became a rarity, because nearly all of the catch had to be frozen; much of it decayed on docks due to insufficient processing capacity. Furthermore, the regulatory approach failed even to limit the catch, with the targeted total allowable catch exceeded in two out of three years (Homans and Wilen 1997).

An ITQ system was established in 1995. The season length increased from two days to more than 200 days (United States National Research Council 1999). Safety problems were diminished, by-catch was reduced by 80 percent, ghost fishing losses fell by 77 percent, and the quality of fish in the market increased. From 1994 to 1999, the number of fishing vessels decreased by 10 percent, while the value of the

[8] Martin L. Weitzman (2002) has shown that an optimal "landing fee" (tax on fish caught) can be superior to an optimal ITQ system when particular forms of biological (stock) uncertainty are present.

[9] If the government chose to auction the quotas, rather than distribute them freely, the distributional result would be the same as with the tax.

harvest increased by 34 percent. Total allowable catch has not been exceeded since the inception of the program.

In 2006, a group of scientists projected that at existing rates of ocean fisheries depletion, all commercial fisheries would collapse by the middle of the century (Boris Worm, et al. 2006). Two years later, Christopher Costello, Steven D. Gaines, and John Lynham (2008) compiled a global database of fisheries management and catch statistics for more than 11 thousand fisheries from 1950 to 2003. They concluded that, where implemented, ITQ systems had halted and reversed trends toward collapse.[10]

All of the approaches described above can be used in the context of inland bodies of water and within countries' exclusive economic zones—200 miles from coastlines, where the richest and most important ocean fisheries are located. But in the open ocean, beyond the 200-mile limit, international negotiation and regulation is required (as with whaling), and the challenges become much greater (Bromley 1992), a generic problem that is taken up below.

II. Environmental Quality as a Problem of the Commons

In the late twentieth century, concerns about the commons expanded well beyond renewable, open-access resources to include environmental degradation, that is, the use of common airsheds, watersheds, and land masses as repositories of pollution and waste. For much of the past 90 years, economists have thought of environmental pollution as a classic—indeed, textbook—example of a negative externality, that is, an unintentional consequence of production or consumption that reduces another agent's profits or utility (Arthur C. Pigou 1920). A separate but related strand of literature—stemming from Ronald Coase's work (1960)—has identified environmental pollution essentially as a public-good problem, that is, a problem of incomplete property rights.

Both perspectives identify the problem as one of the commons, but they lead to somewhat different policy prescriptions: emission taxes versus tradable emission rights. Both prescriptions can facilitate cost-effective environmental protection, and the respective economic literatures together constitute what is arguably the most important contribution of environmental economics to public policy—the notion and means of getting the prices right, that is, the development of market-based approaches to environmental protection.

A. *Cost Effectiveness*

Whereas much of normative economics has focused on questions of efficiency (maximizing net benefits), discussions in the environmental policy realm have tended to employ a more modest criterion—cost effectiveness (minimizing costs of achieving some given objective)—largely because of the difficulty of measuring

[10]To be precise, Costello, Gaines, and Lynham (2008) compared systems with caps to systems without; they were unable to distinguish between cap and trade (ITQ) and caps without trade; see Geoffrey Heal and Wolfram Schlenker (2008). As economic understanding of fisheries has advanced, more sophisticated policy instruments have been developed, such as those intended to address the spatial and dynamic features of these resources (Smith, Sanchirico, and Wilen 2009).

the benefits of environmental protection.[11] To be more precise, by cost effectiveness I mean that allocation of control efforts among pollution sources that results in an aggregate abatement target being achieved at the lowest possible cost, that is, the allocation that satisfies the following cost-minimization problem:

$$\min_{r_i} C = \sum_{i=1}^{N} c_i(r_i), \tag{9}$$

$$\text{s.t.} \sum_{i=1}^{N} [u_i - r_i] \leq E, \tag{10}$$

$$\text{and} \quad 0 \leq r_i \leq u_i, \tag{11}$$

where r_i is reductions in emissions (abatement or control) by source i ($i = 1$ to N); $c_i(r_i)$ is the cost function for source i; C is aggregate cost of control; u_i is uncontrolled emissions by source i; and E is the aggregate emissions target imposed by a regulatory authority.

If the cost functions are convex, then necessary and sufficient conditions for satisfaction of the constrained optimization problem posed by equations (9) through (11) are:

$$\frac{\partial c_i(r_i)}{\partial r_i} - \lambda \geq 0, \tag{12}$$

$$r_i \left[\frac{\partial c_i(r_i)}{\partial r_i} - \lambda \right] = 0, \tag{13}$$

where λ is a Lagrange multiplier that reflects the shadow price of emissions. Equations (12) and (13) together imply the crucial condition for cost effectiveness: that all sources that exercise some degree of control experience the *same marginal abatement cost* (William J. Baumol and Wallace E. Oates 1988). Thus, when considering alternative environmental policy instruments, a key question is whether instruments are likely to result in marginal abatement costs being equated across sources.[12]

Conventional approaches to regulating the environment—frequently characterized as command and control—allow relatively little flexibility in the means of achieving goals. Such policy instruments tend to force firms to take on similar shares of

[11] Despite the tendency for environmental standards to be set based on political and other considerations, economists have sought to inform policy through the comparison of benefits and costs. Considerable progress has been made in developing better methods for estimating the benefits of environmental protection (A. Myrick Freeman, III 2003) for use in normative analysis (Nicholas Kaldor 1939; John R. Hicks 1939), sometimes under conditions of uncertainty (Weitzman 1974). A concise survey is provided by Richard L. Revesz and Stavins (2007).

[12] The model of cost effectiveness I employ is of a uniformly mixed flow pollutant, that is, a pollutant for which the location of emissions has no effect on the location of damages and which does not accumulate in the environment. Little additional insight is gained but much is sacrificed in terms of transparency and tractability by modeling a more complex nonuniformly mixed stock pollutant.

the pollution-control burden, regardless of the cost, sometimes by setting uniform standards for firms, the most prevalent of which are technology- and performance-based standards.

Where there is significant heterogeneity of costs—which is a common feature of pollution abatement—command-and-control methods will not be cost effective. In reality, costs can vary enormously due to production design, physical configuration, age of assets, and other factors. Holding all firms to the same target will be unduly expensive.

In principle, governments could employ nonuniform performance standards to bring about the cost-effective allocation of control responsibility, but to develop such a set of source-specific standards, the government would need to know the marginal abatement cost functions of all sources within its jurisdiction, information that is generally not available to governments. Is there a means by which the government can achieve the cost-effective allocation of control responsibility among pollution sources, but without needing to have information about source-level control costs?

B. *Pigou and Environmental Taxation*

For some 40 years prior to Coase (1960), the sole economic response to the problem of externalities was that the externality in question should be taxed. In principle, a regulator could ensure that emitters would internalize the damages they caused by charging a tax on each unit of pollution equal to the marginal social damages at the efficient level of pollution (Pigou 1920). Such a system makes it worthwhile for firms to reduce emissions to the point where their marginal abatement costs are equal to the common tax rate. Hence, marginal abatement costs will be equated across sources, satisfying the condition for cost effectiveness (equations 12 and 13). Whenever abatement costs differ across emitters, conventional policies would not be cost-effective, but a uniform Pigouvian tax would be. This is true both in the short term, and in the long term by providing incentives for the innovation (Newell, Adam B. Jaffe, and Stavins 1999) and diffusion (Jaffe and Stavins 1995) of low-cost abatement technologies.

The conventional wisdom is that pollution taxes have been rarely, if ever, employed. This is not strictly correct if one defines environmental tax systems broadly (Stavins 2003). In this spirit, these systems can be divided into those for which behavioral impacts are central to their design and performance; and those for which anticipated behavioral impacts are secondary, at best. Within the first set, effluent charges have been employed—typically for water pollution and at low levels with minimal behavioral effects—in a number of European and other nations. Beyond that, deposit-refund systems—in which front-end charges are combined with refunds payable when particular behavior is carried out—have been used in 10 US states for beverage containers and 11 states for motor vehicle batteries, as well as in dozens of other countries for these and other products. In addition, various forms of tax differentiation—tax cuts, credits, and subsidies—intended to encourage environmentally desirable behavior are common in Europe, the United States, and many other countries (Stavins 2003).

Why have true Pigouvian taxes been used infrequently, despite their theoretical advantages (Allen V. Kneese and Charles L. Schultze 1975; Don Fullerton 1996)?

First, it is difficult to identify the appropriate tax rate. For social efficiency, it should be set equal to the marginal benefits of cleanup at the efficient level of cleanup, but policy makers are more likely to think in terms of a desired level of cleanup, and they do not know beforehand how firms will respond to a given level of taxation. A more important political problem posed by pollution taxes is associated with their distributional consequences for regulated sources. Despite the fact that such systems minimize aggregate social costs, these systems are likely to be *more* costly than comparable command-and-control instruments *for regulated firms*, because firms both incur their abatement costs and pay taxes on their residual emissions (James M. Buchanan and Gordon Tullock 1975). In practice, some of these costs will be passed on to consumers, but many firms may still be worse off under a tax.

Is there a way the government can achieve its pollution-control targets cost effectively, but eliminate the abatement uncertainty inherent in the Pigouvian tax approach, and—more important, politically—eliminate the distributional impacts on regulated firms?

C. *Coase and Tradable Rights*

Following Coase (1960), it became possible to think about solving the problem of pollution as one of clarifying poorly defined property rights. If resources such as clean air and water could be recognized as a form of property, whose corresponding rights could be traded in a market, private actors could allocate the use of this property in a cost-effective way. Some 40 years ago, Thomas D. Crocker (1966) and J. H. Dales (1968) each proposed a system of transferable discharge permits that could provide such a market solution: the regulator need only designate the total quantity of emissions allowed (the cap), distribute rights corresponding to this total, and allow individual sources of emissions to trade the permits until an optimal allocation had been reached. This was the fundamental thinking behind what has come to be known as "cap and trade."[13]

Under this approach, an allowable overall level of pollution is established by the government (not necessarily at the efficient level), and allocated among firms in the form of allowances. Firms that keep their emissions below their allotted level may sell their surplus allowances to other firms or use them to offset excess emissions in other parts of their operations. Under these conditions, it is in the interest of each source to carry out abatement up to the point where its marginal control costs are equal to the market-determined price of tradable allowances. Hence, the environmental constraint is satisfied, and marginal abatement costs are equated across sources, satisfying the condition for cost effectiveness (equations 12 and 13).[14]

[13] Cap-and-trade systems should not be confused with "emission reduction–credit" or "offset" systems, whereby permits are assigned when a source reduces emissions below some baseline, which may or may not be readily observable.

[14] In theory, a number of factors can adversely affect the performance of a cap-and-trade system, including: concentration in the permit market (Robert W. Hahn 1984), concentration in the product market (David A. Malueg 1990), transaction costs (Stavins 1995), nonprofit maximizing behavior, such as sales or staff maximization (John T. Tschirhart 1984), the preexisting regulatory environment (Douglas Bohi and Dallas Burtraw 1992), and the degree of monitoring and enforcement (Juan-Pablo Montero 2007). Some of these also affect the performance of pollution taxes.

The unique cost-effective equilibrium is achieved independent of the initial allocation of allowances (W. David Montgomery 1972).[15] This independence property is of central political importance and is the primary reason why cap-and-trade systems have been employed in representative democracies, where distributional issues are of paramount importance in mustering support for a policy. In principle, the government can set the overall emissions cap—whether on the basis of economic efficiency or, more likely, some other grounds—and then leave it up to the legislature to allocate the available number of allowances among sources to build a constituency of support for the initiative *without* reducing the system's environmental performance or driving up its cost.[16] This should be contrasted with most public policy proposals—environmental or otherwise—for which the normal course of events is that the political machinations that are necessary to develop sufficient legislative support reduce the effectiveness of the policy and/or drive up its costs.

Cap and trade has been used in the United States and Europe, as well as other countries (Stavins 2003). In the 1980s, leaded gasoline was phased out of the US market with a program similar to cap and trade among refineries, saving about $250 million per year compared with a program without trading (US Environmental Protection Agency 1985), and providing measurable incentives for cost-saving technological change (Kerr and Newell 2003). Since 1995, under the Clean Air Act amendments of 1990, a sulfur dioxide (SO_2) allowance trading program has reduced emissions by half, saving $1 billion per year compared with a conventional approach (Stavins 1998; Curtis Carlson et. al 2000).[17] Most recently, Australia, Canada, Europe, Japan, New Zealand, and the United States have turned their attention to employing cap and trade to address the ultimate problem of the commons.

D. *The Ultimate Commons Problem: Global Climate Change*

Anthropogenic emissions of greenhouse gases[18]—including carbon dioxide (CO_2) from fossil-fuel combustion and land-use changes (Stavins 1999)—are likely to change the earth's climate in ways that will have serious environmental, economic, and social consequences (Martin Parry et al. 2007). The atmospheric concentration of CO_2 increased from the preindustrial value of about 280 parts per million (ppm) to 379 ppm by 2005 (Rajendra K. Pachauri and Andy Reisinger 2008). These increased concentrations of CO_2 and other greenhouse gases (GHGs) have been accompanied by increases in global mean temperatures measured over land and oceans (Figure 4).

Although the 2008–2009 global recession slowed emissions growth significantly, global atmospheric GHG concentrations (in CO_2-equivalent terms) are likely to

[15] This property is likely to be violated under specific, but relatively infrequent conditions (Hahn and Stavins 2010).

[16] Experience has validated the political importance of this property. For example, in the Senate debate over the Clean Air Act Amendments of 1990, "bonus allowances" were awarded to electricity generators in Ohio, which were going to incur particularly high costs because of their reliance on high-sulfur coal (Paul L. Joskow and Richard Schmalensee 1998); the result was the key support of Senator John Glenn (D-Ohio) for the legislation.

[17] As with fisheries, greater economic understanding of the complexities of environmental problems has led to the development of more sophisticated policy instruments, such as in recognizing the spatial heterogeneity of some pollution problems (Nicholas Z. Muller and Robert Mendelsohn 2009).

[18] The major anthropogenic greenhouse gases in the atmosphere are carbon dioxide (CO_2), methane (CH_4), nitrous oxide (N_2O), and various halocarbons.

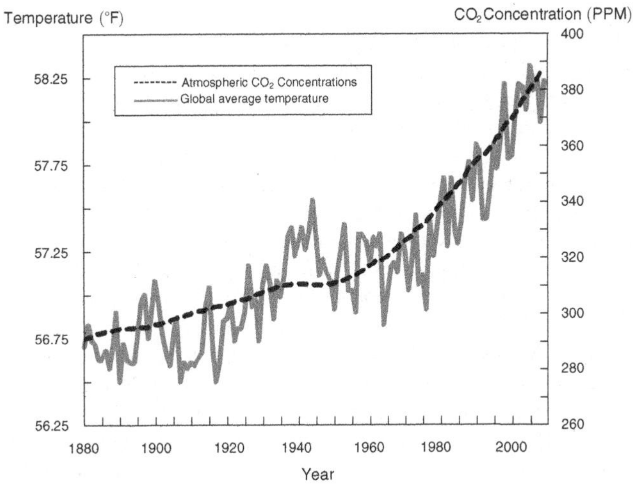

FIGURE 4. GLOBAL TEMPERATURE AND CARBON DIOXIDE CONCENTRATIONS, 1880–2010

Notes: Global annual average temperature measured over land and oceans. The dark dashed line shows atmospheric CO_2 concentrations in parts per million (PPM).

Source: US National Aeronautics and Space Administration 2010.

double well before the end of the century, leading to an average global temperature increase of 1.8 to 4.0°C (3.2 to 7.2° F), relative to 1980–1990 levels, depending upon the quantity of future emissions and the relationship between GHG concentrations and warming (Pachauri and Reisinger 2008). But increased temperatures—which might be welcome in some places—are only part of the story.

The most important anticipated consequences of climate change are changes in precipitation, decreased snowpack, glacier melting, droughts in mid to low latitudes, decreased cereal crop productivity at lower latitudes, increased sea level, loss of islands and coastal wetlands, increased flooding, greater storm intensity, species extinction, and spread of infectious disease. Climate change will also bring longer growing seasons to higher latitudes and some health benefits to temperate areas, such as fewer deaths from cold exposure. However, it is anticipated that such benefits will be greatly outweighed by negative impacts (Parry et al. 2007).

These biophysical impacts will have significant economic, social, and political consequences. Estimates of economic impacts of unrestrained climate change vary, with most falling in the range of 1 to 3 percent of world GDP per year by the middle of the current century (with large regional differences), assuming 4°C warming (Parry et al. 2007; Nordhaus 2010a). The best estimates of marginal damages of emissions (by midcentury) are in the range of \$75 to \$175 per ton of CO_2, in today's dollars (Nordhaus 2008; United States Environmental Protection Agency 2008). In order to have a 0.50 probability of keeping temperature increases below 2°C (a long-term goal acknowledged by most national governments), it would be necessary to stabilize atmospheric concentrations at 450 ppm, which in principle could be achieved by cutting global emissions 60 to 80 percent below 2005 levels by 2050 (Metz et al. 2007), a task that currently appears to be politically impossible, despite its apparent economic and technological feasibility. Of course, economic feasibility does not necessarily imply economic desirability, normally thought of as requiring a comparison of anticipated benefits and costs.[19]

Climate change is a commons problem of unparalleled magnitude along two key dimensions: temporal and spatial (Christopher Robert and Richard Zeckhauser 2010). In the temporal domain, it is a stock, not a flow problem, with greenhouse gases remaining in the atmosphere for decades to centuries. In the spatial domain, greenhouse gases uniformly mix in the atmosphere, and so the nature, magnitude, and location of damages are independent of the location of emissions. Hence, for any individual political jurisdiction, the direct benefits of taking action will inevitably be less than the costs, producing a free-rider problem, and thereby suggesting the importance of international—if not global—cooperation (Nordhaus 2010b).

Despite the apparent necessity of international cooperation for the achievement of meaningful GHG targets, the key political unit of implementation—and decision making—for any international climate policy will be the sovereign state, that is, the nations of the world. Therefore, before turning to the topic of international cooperation, it is important to ask what economics has to say about the best instruments for national action. In both cases, I limit my attention to the means—the instruments—of climate policy, although economists have and will continue to make important contributions to analyses of the ends—the goals—of climate policy.

There is widespread agreement among economists (and a diverse set of other policy analysts) that economy-wide carbon pricing will be an essential ingredient of any policy that can achieve meaningful reductions of CO_2 emissions cost effectively, at least in the United States and other industrialized countries (Gilbert E. Metcalf 2009; Louis Kaplow 2010).[20] The ubiquitous nature of energy generation and use and the diversity of CO_2 sources in a modern economy mean that conventional technology and performance standards would be infeasible and—in any event—excessively costly (Newell and Stavins 2003). There is somewhat less agreement among economists regarding the choice of specific carbon-pricing policy instruments, with some tending to support carbon taxes (N. Gregory Mankiw 2006; Nordhaus 2007;

[19] Such a comparison of anticipated benefits and costs presents challenges to economics. It has been argued that long duration, great uncertainty, and potentially unlimited liability to the planet characterize anthropogenic climate change. Hence, it is important to consider the risk of extreme outcomes, that is, catastrophic consequences with small but nonnegligible probabilities, in addition to—and some would argue, instead of—conventionally defined expected values of abatement costs and benefits (Weitzman 2009). At the heart of this concern is the possibility that the economic consequences of fat-tailed structural uncertainty will outweigh the economic effects of temporal discounting, because (under specific conditions regarding the structure of uncertainty and preferences) there will be an infinitely large expected loss from low-probability, high-consequence events (Weitzman 2010). Although it cannot be said that there is agreement regarding the specific analysis that is appropriate (Stern 2007; Lawrence Summers and Zeckhauser 2008; Nordhaus forthcoming; Pindyck forthcoming; Weitzman forthcoming), there is considerable agreement that benefit-cost analysis based exclusively on conventional expected values is of less use in this realm than in others and that the primary (economic) argument for limiting increases in GHG concentrations is to provide insurance against catastrophic climate risks.

[20] Carbon pricing is necessary but not sufficient, because other market failures limit the impacts of price signals (Jaffe, Newell, and Stavins 2005). One example is the well-known principal-agent problem that constrains incentives for energy-efficiency investments by either landlords or tenants in renter-occupied properties. Another is the public-good nature of research and development, whereby firms capture only a share of the benefits of the information their research produces. Both argue for specific public policies that would complement a carbon-pricing regime.

Metcalf 2007) and others cap-and-trade mechanisms (A. Denny Ellerman, Joskow, and David Harrison 2003; Stavins 2007; Nathaniel Keohane 2009).

These two instruments—carbon taxes and carbon cap and trade—may be said to derive respectively from the externality (Pigou 1920) and property-rights (Coase 1960) perspectives. But, in truth, the two approaches are more similar than different. A carbon tax would directly place a price on carbon (most likely upstream, where fossil fuels—coal, petroleum, and natural gas—enter the economy), with quantities of carbon use and CO_2 emissions adjusting in response. An upstream carbon cap-and-trade system would constrain the quantity of carbon entering the economy through allowances on the carbon content of the three fossil fuels, with prices emerging indirectly from the market for allowances. Either instrument can be designed—in principle—to be equivalent to the other in distributional terms. If allowances are auctioned, a cap-and-trade system looks much like a carbon tax from the perspective of regulated firms. Likewise, if tax revenues are refunded in particular ways, a carbon tax can resemble cap and trade with free allowances.

What may appear to be key differences between the two instruments fade on closer inspection. First, a carbon tax would raise revenues that can be used for beneficial public purposes, such as for cutting distortionary taxes, thereby lowering the social cost of the overall policy (A. Lans Bovenberg and Lawrence H. Goulder 1996). Given the need for government revenues, this is an important attribute of taxes. But, an auction mechanism under cap-and-trade can do precisely the same. Second, it might appear that cap-and-trade systems offer greater opportunities to protect the profits of regulated firms (through free allocation of allowances). However, the same can be accomplished under a tax regime through inframarginal tax exemptions.

Third, an important question is the relative effect of the two approaches on technological innovation. A series of theoretical explorations have found that a tax and a cap-and-trade system with auctioned allowances are equivalent in their incentives for carbon-saving innovation (Scott R. Milliman and Raymond Prince 1989; Chulho Jung, Kerry Krutilla, and Roy Boyd 1996), or at least that neither system dominates (Carolyn Fischer, Ian W. H. Parry, and William A. Pizer 2003).[21]

Fourth, there is the simplicity of a carbon tax, in which firms would not need to manage and trade allowances, and the government would not need to track allowance transactions and ownership. However, experience with cap-and-trade systems indicates that the actual costs of trading institutions have not been significant. And whether a policy as important as a national carbon tax would turn out to be "simple" in its design and implementation is at least open to question.

Fifth, there is resistance to new taxes. In their simplest respective forms (a carbon tax without revenue recycling, and a cap-and-trade system without auctions), a carbon tax is more costly than a cap-and-trade system to the regulated sector, because with the former firms incur both abatement costs and the cost of tax payments to the government. This might argue politically against the tax approach, but now that cap and trade has been demonized—in US politics, at least—as "cap and tax," this difference has surely diminished.

[21] The optimal system in this regard is a hybrid instrument, that is, a cap-and-trade system with a price collar (Thomas A. Weber and Karsten Neuhoff 2010).

That said, there are some real differences between these two approaches. First, there is the reality of abatement cost uncertainty under a cap-and-trade system versus emissions uncertainty under a tax regime. From an economic perspective, it makes sense to allow emissions (of a stock pollutant) to vary from year to year with economic conditions that affect aggregate abatement costs.[22] This happens automatically with a carbon tax. With a cap-and-trade system, such temporal flexibility needs to be built in through provisions for banking and borrowing of allowances. Furthermore, a tax approach eliminates the potential for short-term price volatility, which can exist under a cap-and-trade system.

Second, there are fears of market manipulation, a relevant argument against the use of cap-and-trade systems in a developing-country context. In industrialized countries, however, appropriate regulatory oversight can address such concerns. Third, there has been considerable experience with the use of cap-and-trade systems, as noted above, but virtually no experience with Pigouvian taxes for pollution control. It should be noted, however, that there is abundant experience with a wide variety of taxes to accomplish a diverse set of social objectives.

Fourth, cap-and-trade systems generate a natural unit of exchange for international harmonization and linkage: allowances denominated in units of carbon content of fossil fuels (or CO_2 emissions). Hence, it is easier to harmonize with other countries' carbon mitigation programs, which are more likely to employ cap-and-trade than tax approaches (Judson Jaffe, Matthew Ranson, and Stavins 2009). However, through appropriate mechanisms, international linkage can include carbon tax systems (Metcalf and David Weisbach 2010).

Fifth and finally, there is a fundamental political-economy difference. Cap and trade leaves distributional issues up to politicians, and thereby provides a straightforward means to compensate burdened sectors. Of course, this political advantage is also an economic disadvantage in that it invites rent-seeking behavior. In any event, the compensation associated with free distribution of allowances based on historical activities can be mimicked under a tax regime through the assignment of specific tax exemptions. A real difference is that the cap-and-trade approach avoids likely battles over tax exemptions among vulnerable industries and sectors that would drive up the costs of the program, as more and more sources (emission-reduction opportunities) are exempted from the program, thereby simultaneously compromising environmental performance. Instead, a cap-and-trade system leads to battles over the allowance allocation, but these do not raise the overall cost of the program nor affect its climate impacts (Montgomery 1972).

Remaining differences between carbon taxes and cap and trade can diminish with implementation. Hybrid schemes that include features of taxes and cap-and-trade systems blur distinctions. The government can auction allowances in a cap-and-trade system, thereby reproducing many of the properties of a tax approach. Mechanisms that deal with uncertainty in a cap-and-trade system also bring it close to a tax approach, including a cost containment mechanism that places a cap or collar on

[22] Because climate change is a function of the accumulated GHGs in the atmosphere, it is reasonable to anticipate that the marginal damage (benefit) function has a smaller slope (in absolute value) than the marginal cost function, and that the more efficient instrument under conditions of uncertainty about abatement costs will be a price instrument, rather than a quantity instrument, such as cap and trade (Weitzman 1974; Newell and Pizer 2003).

allowance prices, banking that creates a floor under prices, and borrowing that provides flexibility similar to a tax. To some degree, the dichotomous choice between taxes and cap and trade can be a choice of design elements along a policy continuum (Weisbach 2010).

Because of the similarity between the two approaches to carbon pricing, it has been argued that the key questions that should be used to decide between these two policy approaches are: which is more politically feasible; and which is more likely to be well designed (Jason Furman, et. al 2007). To some degree, responses to these questions have been provided by the political revealed preference of individual countries, with the world's most significant climate policy employing a cap-and-trade system to constrain Europe's CO_2 emissions—the European Union Emission Trading Scheme (EU ETS).[23] Although the system had its share of problems in its pilot phase, it has functioned as anticipated since then (Ellerman, Frank J. Convery, and Christian de Perthuis 2010). This is despite the fact that the 2008–2009 recession led to significantly lower allowance prices and hence fewer emission reductions than anticipated (Richard N. Cooper 2010b).[24] In addition, New Zealand has launched a GHG cap-and-trade system, and Australia and Japan have considered doing likewise for CO_2. Canada has indicated that it will launch a domestic system when and if the United States does so, but domestic US politics slowed developments in 2010.[25]

Even as domestic climate policies move forward in some countries but not others, it is clear that due to the global commons nature of the problem, meaningful international cooperation will eventually be necessary. The Kyoto Protocol (1997) to the United Nations Framework Convention on Climate Change (1992) will expire in 2012, and is, in any event, insufficient to the long-term task, due to the exclusion of developing countries from responsibility. Although the industrialized countries accounted for the majority of annual CO_2 emissions until 2004 (and the majority of annual GHG emissions until 2000), that is no longer the case. China has surpassed the United States as the world's largest emitter, and most growth in CO_2 emissions in the coming decades will come from countries outside of the Organization of Economic Cooperation and Development (OECD), with emissions in nearly all OECD countries close to stable or falling (Figure 5).

A wide range of potential paths forward are possible (Joseph E. Aldy, Scott Barrett, and Stavins 2003), including: top-down international agreements involving targets and timetables that involve more countries as they become more wealthy (Jeffrey Frankel 2010), harmonized national policies, such as domestic carbon taxes (Cooper 2010a), and bottom-up, loosely coordinated national policies, such as the linkage of regional and national cap-and-trade systems through bilateral

[23] One factor which is said to have influenced the choice of cap and trade over a tax approach was the fact that fiscal measures—such as a carbon tax—require unanimity in the Council of the European Union, whereas most other measures require only a majority.

[24] Note that the lower allowance prices and fewer reductions that occur with cap and trade during a recession are an economic virtue, i.e., countercyclicality.

[25] With political stalemate in Washington, attention has turned to subnational policies, including the Regional Greenhouse Gas Initiative in the Northeast and California's Global Warming Solutions Act (Assembly Bill 32). These subnational policies will interact in a variety of ways with federal policy when and if a federal policy is enacted. Some of these interactions would be problematic, some benign, and some could be positive (Goulder and Stavins 2010).

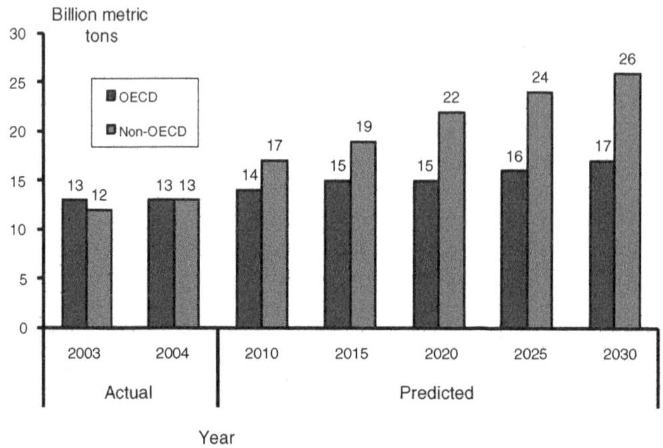

Figure 5. Energy-Related World Carbon Dioxide Emissions by Region

Source: US Energy Information Administration 2010.

arrangements (Jaffe, Ranson, and Stavins 2009). The most promising alternatives can—in principle—achieve reasonable environmental performance cost effectively by including not only the currently industrialized nations, but also the key emerging economies (Aldy and Stavins 2009).

Political feasibility is another matter, partly due to countries' asymmetric situations (Edward A. Parson and Zeckhauser 1995). The United States and other industrialized countries have come to insist that the large, emerging economies—China, India, Brazil, Korea, Mexico, and South Africa—begin to take on proportionate shares of the mitigation burden, while some of those emerging economies plus most developing countries insist that the rich countries go first, and possibly compensate developing countries for climate damages (due to cumulative emissions from the industrialized world). At a minimum, developing countries want their mitigation and adaptation to be financed by the wealthier countries, but such large financial transfers are unlikely.

Given the spatial and temporal nature of this global commons problem, political incentives around the world are to rely upon others to take action. Since sovereign nations cannot be compelled to act against their wishes, successful cooperation—whether in the form of international treaties or less formal mechanisms—must create internal incentives for compliance, along with external incentives for participation. Because no single approach guarantees a sure path to ultimate success, the best strategy to address this ultimate commons problem may be to pursue a variety of approaches simultaneously. The difficulties inherent in addressing the climate problem cannot be overstated.[26] As Coman (1911) observed a century ago, appropriate

[26] It might be asked whether a useful precedent is provided by the successful regime put in place under the Montreal Protocol on Substances That Deplete the Ozone Layer (1987) to address another global commons problem—stratospheric ozone depletion linked with emissions of chlorofluorocarbons (CFCs) and hydrochlorofluorocarbons

regimes to govern the commons—though sometimes theoretically clear—may nevertheless be very difficult to achieve in practice.

III. Conclusions

Problems of the commons are both more widespread and more important today than when Coman wrote about unsettled problems in the first issue of the *Review* 100 years ago. A century of economic growth and globalization have brought unparalleled improvements in societal well-being, but also unprecedented challenges to the carrying capacity of the planet. What would have been in 1911 inconceivable increases in income and population have come about and have greatly heightened pressures on the commons, particularly where there has been open access to it.

The stocks of a variety of renewable natural resources—including water, forests, fisheries, and numerous other species of plant and animal—have been depleted below socially efficient levels, principally because of poorly defined property-right regimes. Likewise, the same market failures of open access—whether characterized as externalities, following Pigou, or public goods, following Coase—have led to the degradation of air and water quality, inappropriate disposal of hazardous waste, depletion of stratospheric ozone, and the atmospheric accumulation of greenhouse gases linked with global climate change.

Over this same century, economics—as a discipline—has gradually come to focus more and more attention on these commons problems, first with regard to natural resources, and more recently with regard to environmental quality. Economic research within academia and think tanks has improved our understanding of the causes and consequences of excessive resource depletion and inefficient environmental degradation, and thereby has helped identify sensible policy solutions. Conventional regulatory policies, which have not accounted for economic responses, have been excessively costly, ineffective, or even counterproductive. The problems behind what Hardin (1968) characterized as the "tragedy of the commons" might better be described as the "failure of commons regulation." As our understanding of the commons has become more complex, the design of economic policy instruments has become more sophisticated.

Problems of the commons have not diminished, and the lag between understanding and action can be long. While some commons problems have been addressed successfully, others continue to emerge. Some—such as the threat of global climate change—are both more important and more difficult than problems of the past. Fortunately, economics is well positioned to offer better understanding and better policies to address these ongoing challenges. As the first decade of the twenty-first century comes to a close, natural resource and environmental economics has emerged as a productive field of our discipline and one that shows even greater promise for the future.

(HCFCs) (Barrett 2003, 2007). It is expected that the ozone layer (which protects the planet from harmful ultraviolet radiation) will recover by 2050 (World Meteorological Organization 2007). However, the differences between stratospheric ozone depletion and global climate change explain why this success provides little precedent for addressing global climate change: (1) the cost of reducing CFC and HCFC production and use was extremely small relative to the cost of weaning economies away from fossil-fuel use; (2) substitutes for CFCs and HCFCs were readily available; and (3) production of these compounds was limited to a small set of countries.

REFERENCES

Acheson, James M. 2003. *Capturing the Commons: Devising Institutions to Manage the Maine Lobster Industry.* Lebanon, NH: University Press of New England.

Aldy, Joseph E., and Robert N. Stavins. 2009. *Post-Kyoto International Climate Policy: Summary for Policymakers.* New York: Cambridge University Press.

Aldy, Joseph E., Scott Barrett, and Robert N. Stavins. 2003. "Thirteen Plus One: A Comparison of Global Climate Policy Architectures." *Climate Policy,* 3(4): 373–97.

Amundsen, Eirik S., Trond Bjørndal, and Jon M. Conrad. 1995. "Open Access Harvesting of the Northeast Atlantic Minke Whale." *Environmental and Resource Economics,* 6(2): 167–85.

Barnett, Harold J., and Chandler Morse. 1963. *Scarcity and Growth: The Economics of Natural Resource Availability.* Baltimore, MD: Johns Hopkins University Press.

Barrett, Scott. 2003. *Environment and Statecraft: The Strategy of Environmental Treaty-Making.* New York: Oxford University Press.

Barrett, Scott. 2007. *Why Cooperate? The Incentive to Supply Global Public Goods.* New York: Oxford University Press.

Baumol, William J., and Wallace E. Oates. 1988. *The Theory of Environmental Policy.* 2nd Ed. New York: Cambridge University Press.

Bohi, Douglas, and Dallas Burtraw. 1992. "Utility Investment Behavior and the Emission Trading Market." *Resources and Energy,* 14(1–2): 129–53.

Bovenberg, A. Lans, and Lawrence H. Goulder. 1996. "Optimal Environmental Taxation in the Presence of Other Taxes: General-Equilibrium Analyses." *American Economic Review,* 86(4): 985–1000.

Bromley, Daniel W. 1992. "The Commons, Common Property, and Environmental Policy." *Environmental and Resource Economics,* 2(1): 1–17.

Buchanan, James M., and Gordon Tullock. 1975. "Polluters' Profits and Political Response: Direct Controls versus Taxes." *American Economic Review,* 65(1): 139–47.

Carlson, Curtis, Dallas Burtraw, Maureen Cropper, and Karen L. Palmer. 2000. "Sulfur Dioxide Control by Electric Utilities: What Are the Gains from Trade?" *Journal of Political Economy,* 108(6): 1292–1326.

Clark, Colin W. 1990. *Mathematical Bioeconomics: The Optimal Management of Renewable Resources.* 2nd Ed. New York: Wiley.

Coase, Ronald. 1960. "The Problem of Social Cost." *Journal of Law and Economics,* 3(10): 1–44.

Coman, Katharine. 1911. "Some Unsettled Problems of Irrigation." *American Economic Review,* 1(1): 1–19.

Cooper, Richard N. 2010a. "The Case for Charges on Greenhouse Gas Emissions." In *Post-Kyoto International Climate Policy: Implementing Architectures for Agreement,* ed. Joseph E. Aldy, and Robert N. Stavins, 151–78. New York: Cambridge University Press.

Cooper, Richard N. 2010b. "Europe's Emissions Trading System." Harvard Project on International Climate Agreements Discussion Paper 10–40.

Costello, Christopher, Steven D. Gaines, and John Lynham. 2008. "Can Catch Shares Prevent Fisheries Collapse?" *Science,* 321(5896): 1678–81.

Crocker, Thomas D. 1966. "The Structuring of Atmospheric Pollution Control Systems." In *The Economics of Air Pollution,* ed. Harold Wolozin, 61–86. New York: W. W. Norton and Co.

Dales, J. H. 1968. "Land, Water, and Ownership." *Canadian Journal of Economics,* 1(4): 791–804.

Dasgupta, Partha. 1982. *The Control of Resources.* Cambridge, MA: Harvard University Press.

Ellerman, A. Denny, Frank J. Convery, and Christian de Perthuis, ed. 2010. *Pricing Carbon: The European Union Emissions Trading Scheme.* New York: Cambridge University Press.

Ellerman, A. Denny., Paul L. Joskow, and David Harrison. 2003. *Emissions Trading in the US: Experience, Lessons, and Considerations for Greenhouse Gases.* Arlington, VA: Pew Center on Global Climate Change.

Fischer, Carolyn, Ian W. H. Parry, and William A. Pizer. 2003. "Instrument Choice for Environmental Protection When Technological Innovation Is Endogenous." *Journal of Environmental Economics and Management,* 45(3): 523–45.

Frankel, Jeffrey. 2010. "An Elaborate Proposal for a Global Climate Policy Architecture: Specific Formulas and Emission Targets for All Countries in All Decades." In *Post-Kyoto International Climate Policy: Implementing Architectures for Agreement,* ed. Joseph E. Aldy, and Robert N. Stavins, 31–87. New York: Cambridge University Press.

Freeman, A. Myrick, III. 2003. *The Measurement of Environmental and Resource Values: Theory and Methods.* 2nd Ed. Washington, DC: Resources for the Future.

Fullerton, Don. 1996. "Why Have Separate Environmental Taxes?" In *Tax Policy and the Economy,* Vol. 10, ed. James M. Poterba, 33–70. Cambridge, MA: MIT Press.

Furman, Jason, Jason E. Bordoff, Manasi A. Deshpande, and Pascal J. Noel. 2007. *An Economic Strategy to Address Climate Change and Promote Energy Security.* Hamilton Project Strategy Paper. Washington, DC: Brookings Institution.

Gordon, H. Scott. 1954. "The Economic Theory of a Common-Property Resource: The Fishery." *Journal of Political Economy,* 62(2): 124–42.

Goulder, Lawrence H., and Robert N. Stavins. 2010. "Interactions between State and Federal Climate Change Policies." National Bureau of Economic Research Working Paper 16123.

Hahn, Robert W. 1984. "Market Power and Transferable Property Rights." *Quarterly Journal of Economics,* 99(4): 753–65.

Hahn, Robert W., and Robert N. Stavins. 2010. "The Effect of Allowance Allocations on Cap-and-Trade System Performance." National Bureau of Economic Research Working Paper 15854.

Hardin, Garrett. 1968. "The Tragedy of the Commons." *Science,* 162(3859): 1243–48.

Heal, Geoffrey, and Wolfram Schlenker. 2008. "Sustainable Fisheries." *Nature,* 455(7216): 1044–45.

Henderson, J. V., and M. Tugwell. 1979. "Exploitation of the Lobster Fishery: Some Empirical Results." *Journal of Environmental Economics and Management,* 6(4): 287–96.

Hicks, John R. 1939. "The Foundations of Welfare Economics." *Economic Journal,* 49(196): 696–712.

Homans, Frances R., and James E. Wilen. 1997. "A Model of Regulated Open Access Resource Use." *Journal of Environmental Economics and Management,* 32(1): 1–21.

Hotelling, Harold. 1931. "The Economics of Exhaustible Resources." *Journal of Political Economy,* 39(2): 137–75.

Huppert, Daniel D. 1990. "Managing Alaska's Groundfish Fisheries: History and Prospects." Unpublished.

Huppert, Daniel D. 2005. "An Overview of Fishing Rights." *Reviews in Fish Biology and Fisheries,* 15(3): 201–15.

International Whaling Commission. 2010. "Status of Whales." International Whaling Commission. http://iwcoffice.org/conservation/status.htm. (accessed September 15, 2010).

Jaffe, Adam B., and Robert N. Stavins. 1995. "Dynamic Incentives of Environmental Regulations: The Effects of Alternative Policy Instruments on Technology Diffusion." *Journal of Environmental Economics and Management,* 29(3): S43–63.

Jaffe, Adam B., Richard G. Newell, and Robert N. Stavins. 2005. "A Tale of Two Market Failures: Technology and Environmental Policy." *Ecological Economics,* 54(2–3): 164–74.

Jaffe, Judson, Matthew Ranson, and Robert N. Stavins. 2009. "Linking Tradable Permit Systems: A Key Element of Emerging International Climate Policy Architecture." *Ecology Law Quarterly,* 36(4): 789–808.

Joskow, Paul L., and Richard Schmalensee. 1998. "The Political Economy of Market-Based Environmental Policy: The U.S. Acid Rain Program." *Journal of Law and Economics,* 41(1): 37–83.

Jung, Chulho, Kerry Krutilla, and Roy Boyd. 1996. "Incentives for Advanced Pollution Abatement Technology at the Industry Level: An Evaluation of Policy Alternatives." *Journal of Environmental Economics and Management,* 30(1): 95–111.

Kaldor, Nicholas. 1939. "Welfare Propositions of Economics and Interpersonal Comparisons of Utility." *Economic Journal,* 49(195): 549–52.

Kaplow, Louis. 2010. "Taxes, Permits, and Climate Change." National Bureau of Economic Research Working Paper 16268.

Keohane, Nathaniel O. 2009. "Cap and Trade, Rehabilitated: Using Tradable Permits to Control U.S. Greenhouse Gases." *Review of Environmental Economics and Policy,* 3(1): 42–62.

Kerr, Suzi, and Richard G. Newell. 2003. "Policy-Induced Technology Adoption: Evidence from the U.S. Lead Phasedown." *Journal of Industrial Economics,* 51(3): 317–43.

Kneese, Allen V., and Charles L. Schultze. 1975. *Pollution, Prices and Public Policy.* Washington, DC: Brookings Institution Press.

Lee, Junsoo, John A. List, and Mark C. Strazicich. 2006. "Non-Renewable Resource Prices: Deterministic or Stochastic Trends?" *Journal of Environmental Economics and Management,* 51(3): 354–70.

Livernois, John. 2009. "On the Empirical Significance of the Hotelling Rule." *Review of Environmental Economics and Policy,* 3(1): 22–41.

Malueg, David A. 1990. "Welfare Consequences of Emission Credit Trading Programs." *Journal of Environmental Economics and Management,* 18(1): 66–77.

Mankiw, N. Gregory. 2006. "The Pigou Club Manifesto." http://gregmankiw.blogspot.com/2006/10/pigou-club-manifesto.html.

Meadows, Donella H., Dennis L. Meadows, Jørgen Randers, and William W. Behrens, III. 1972. *The Limits to Growth.* New York: Universe Books.

Metcalf, Gilbert E. 2007. "A Proposal for a U.S. Carbon Tax Swap: An Equitable Reform to Address Global Climate Change." Brookings Institution Hamilton Project Discussion Paper 2007-12.

Metcalf, Gilbert E. 2009. "Market-Based Policy Options to Control U.S. Greenhouse Gas Emissions." *Journal of Economic Perspectives*, 23(2): 5–27.

Metcalf, Gilbert E., and David Weisbach. 2010. "Linking Policies when Tastes Differ: Global Climate Policy in a Heterogeneous World." Harvard Project on International Climate Agreements Discussion Paper 2010-038.

Metz, Bert, Ogunlade Davidson, Peter Bosch, Rutu Dave, and Leo Meyer, ed. 2007. *Climate Change 2007: Mitigation of Climate Change. Working Group III Contribution to the Fourth Assessment Report of the Intergovernmental Panel on Climate Change.* New York: Cambridge University Press.

Milliman, Scott R., and Raymond Prince. 1989. "Firm Incentives to Promote Technological Change in Pollution Control." *Journal of Environmental Economics and Management*, 17(3): 247–65.

Montero, Juan-Pablo. 2007. "Tradable Permits with Incomplete Monitoring: Evidence from Santiago's Particulate Permits Program." In *Moving to Markets in Environmental Regulation: Lessons from Twenty Years of Experience,* ed. Jody Freeman, and Charles D. Kolstad, 147–70. New York: Oxford University Press.

Montgomery, W. David. 1972. "Markets in Licenses and Efficient Pollution Control Programs." *Journal of Economic Theory*, 5(3): 395–418.

Muller, Nicholas Z., and Robert Mendelsohn. 2009. "Efficient Pollution Regulation: Getting the Prices Right." *American Economic Review*, 99(5): 1714–39.

Muñoz, Roberto E., and Thomas W. Hazlett. 2009. "A Welfare Analysis of Spectrum Allocation Policies." *RAND Journal of Economics,* 40(3): 424–54.

Newell, Richard G., and William A. Pizer. 2003. "Regulating Stock Externalities under Uncertainty." *Journal of Environmental Economics and Management*, 45(2S): 416–32.

Newell, Richard G., and Robert N. Stavins. 2003. "Cost Heterogeneity and the Potential Savings from Market-Based Policies." *Journal of Regulatory Economics*, 23(1): 43–59.

Newell, Richard G., Adam B. Jaffe, and Robert N. Stavins. 1999. "The Induced Innovation Hypothesis and Energy-Saving Technological Change." *Quarterly Journal of Economics*, 114(3): 941–75.

Newell, Richard G., James N. Sanchirico, and Suzi Kerr. 2005. "Fishing Quota Markets." *Journal of Environmental Economics and Management*, 49(3): 437–62.

Nordhaus, William D. Forthcoming. "Tail Events and Economic Analysis." *Review of Environmental Economics and Policy.*

Nordhaus, William D. 1992. "Lethal Model 2: The Limits to Growth Revisited." *Brookings Papers on Economic Activity,* 23(2): 1–43.

Nordhaus, William D. 2007. "To Tax or Not to Tax: Alternative Approaches to Slowing Global Warming." *Review of Environmental Economics and Policy*, 1(1): 26–44.

Nordhaus, William D. 2008. *A Question of Balance: Weighing the Options on Global Warming Policies.* New Haven, CT: Yale University Press.

Nordhaus, William D. 2010a. "Economic Aspects of Global Warming in a Post-Copenhagen Environment." *Proceedings of the National Academy of Sciences,* 107(26): 11721–26.

Nordhaus, William D. 2010b. "Some Foundational and Transformative Grand Challenges for the Social and Behavioral Sciences: The Problem of Global Public Goods." Unpublished.

Opsomer, Jean-Didier, and Jon M. Conrad. 1994. "An Open-Access Analysis of the Northern Anchovy Fishery." *Journal of Environmental Economics and Management*, 27(1): 21–37.

Ostrom, Elinor. 1990. *Governing the Commons: The Evolution of Institutions for Collective Action.* Political Economy of Institutions and Decisions series. New York: Cambridge University Press.

Ostrom, Elinor. 2010. "Beyond Markets and States: Polycentric Governance of Complex Economic Systems." *American Economic Review*, 100(3): 641–72.

Ostrom, Elinor. 2011. "Reflections on 'Some Unsettled Problems of Irrigation.'" *American Economic Review,* 101(1): 49–63.

Rajendra K. Pachauri, and Andy Reisinger, ed. 2008. *Climate Change 2007: Synthesis Report. Contribution of Working Groups I, II and III to the Fourth Assessment Report of the Intergovernmental Panel on Climate Change.* Geneva: Intergovernmental Panel on Climate Change.

Parry, Martin, Osvaldo Canziani, Jean Palutikof, Paul van der Linden, and Clair Hanson, ed. 2007. *Climate Change 2007: Impacts, Adaptation and Vulnerability. Contribution of Working Group II to the Fourth Assessment Report of the Intergovernmental Panel on Climate Change.* New York: Cambridge University Press.

Parson, Edward A., and Richard J. Zeckhauser. 1995. "Equal Measures or Fair Burdens: Negotiating Environmental Treaties in an Unequal World." In *Shaping National Responses to Climate Change: A Post-Rio Guide,* ed. Henry Lee, 81–114. Washington, DC: Island Press.

Pigou, Alfred C. 1920. *The Economics of Welfare.* London: MacMillan.

Pindyck, Robert S. Forthcoming. "Fat Tails, Thin Tails, and Climate Change Policy." *Review of Environmental Economics and Policy.*

Pindyck, Robert S. 1984. "Uncertainty in the Theory of Renewable Resource Markets." *Review of Economic Studies*, 51(2): 289–303.

Revesz, Richard L., and Robert N. Stavins. 2007. "Environmental Law." In *Handbook of Law and Economics, Volume 1.* Handbooks in Economics, Vol. 27, ed. A. Mitchell Polinsky, and Steven Shavell, 499–589. New York: Elsevier Science.

Robert, Christopher, and Richard Zeckhauser. 2010. "The Methodology of Positive Policy Analysis." Harvard Kennedy School Faculty Research Working Paper 10–041.

Royte, Elizabeth. 2006. "The Mollusk that Made Manhattan." *New York Times, Sunday Book Review,* March 5.

Sanchirico, James N., and James E. Wilen. 2007. "Global Marine Fisheries Resources: Status and Prospects." *International Journal of Global Environmental Issues,* 7(2–3): 106–18.

Schaefer, M. B. 1957. "Some Considerations of Population Dynamics and Economics in Relation to the Management of the Commercial Marine Fisheries." *Journal of the Fisheries Research Board of Canada,* 14(5): 669–81.

Scott, Anthony. 1955. "The Fishery: The Objectives of Sole Ownership." *Journal of Political Economy,* 63(2): 116–24.

Smith, Martin D., and James E. Wilen. 2003. "Economic Impacts of Marine Reserves: The Importance of Spatial Behavior." *Journal of Environmental Economics and Management*, 46(2): 183–206.

Smith, Martin D., James N. Sanchirico, and James E. Wilen. 2009. "The Economics of Spatial-Dynamic Processes: Applications to Renewable Resources." *Journal of Environmental Economics and Management*, 57(1): 104–21.

Smith, V. Kerry. 1980. "The Evaluation of Natural Resource Adequacy: Elusive Quest or Frontier of Economic Analysis?" *Land Economics,* 56(3): 257–98.

Solow, Robert M. 1991. "Sustainability: An Economist's Perspective." The Eighteenth J. Seward Johnson Lecture to the Marine Policy Center, Woods Hole Oceanographic Institution, Woods Hole, MA, June 14.

Spence, A. Michael. 1974. "Blue Whales and Applied Control Theory." In *Systems Approaches and Environmental Problems,* ed. Hans W. Gottinger, 97–174. Göttingen, Germany: Vandenhoeck & Ruprecht.

Stavins, Robert N. 1995. "Transaction Costs and Tradeable Permits." *Journal of Environmental Economics and Management*, 29(2): 133–48.

Stavins, Robert N. 1998. "What Can We Learn from the Grand Policy Experiment? Lessons from SO2 Allowance Trading." *Journal of Economic Perspectives*, 12(3): 69–88.

Stavins, Robert N. 1999. "The Costs of Carbon Sequestration: A Revealed-Preference Approach." *American Economic Review*, 89(4): 994–1009.

Stavins, Robert N. 2003. "Experience with Market-Based Environmental Policy Instruments." In *Handbook of Environmental Economics, Volume 1: Environmental Degradation and Institutional Responses*. Handbooks in Economics, Vol. 20, ed. Karl-Göran Mäler and Jeffrey R. Vincent, 355–435. Amsterdam: North-Holland.

Stavins, Robert N. 2007. "A U.S. Cap-and-Trade System to Address Global Climate Change." Brookings Institution Hamilton Project Discussion Paper 2007-13.

Stavins, Robert N. 2008. "Addressing Climate Change with a Comprehensive US Cap-and-Trade System." *Oxford Review of Economic Policy*, 24(2): 298–321.

Stern, Nicholas. 2007. *The Economics of Climate Change: The Stern Review.* New York: Cambridge University Press.

Summers, Lawrence, and Richard Zeckhauser. 2008. "Policymaking for Posterity." *Journal of Risk and Uncertainty*, 37(2–3): 115–40.

Tschirhart, John T. 1984. "Transferable Discharge Permits and Profit-Maximizing Behavior." In *Economic Perspectives on Acid Deposition Control,* ed. Thomas D. Crocker, 157–72. Boston: Butterworth.

United Nations Food and Agriculture Organization. 2007. *The State of World Fisheries and Aquaculture 2006.* Rome: FAO Fisheries and Aquaculture Department.

United Nations Food and Agriculture Organization. 2010. Global Capture Production: US and Canadian Atlantic Cod. http://www.fao.org/fishery/statistics/global-capture-production/en. (accessed September 14, 2010).

United States Energy Information Administration. 2009. *International Energy Outlook 2009.* Washington, DC: United States Department of Energy.

United States Environmental Protection Agency. 1985. *Costs and Benefits of Reducing Lead in Gasoline: Final Regulatory Impact Analysis.* Schwartz, Joel, Hugh Pitcher, Ronnie Levin, Bart Ostro, and Albert L. Nichols. Washington, DC: United States Environmental Protection Agency.

United States Environmental Protection Agency. 2008. "Technical Support Document on Benefits of Reducing GHG Emissions." http://www.eenews.net/public/25/10084/features/documents/2009/03/11/document_gw_04.pdf.

United States National Aeronautics and Space Administration. 2010. Global Land-Ocean Temperature Index. http://data.giss.nasa.gov/gistemp/tabledata/GLB.Ts+dSST.txt. (accessed September 14, 2010).

United States National Research Council. 1999. *Sharing the Fish: Toward a National Policy on Individual Fishing Quotas.* Washington: DC: National Academy Press.

Weber, Thomas A., and Karsten Neuhoff. 2010. Carbon Markets and Technological Innovation. *Journal of Environmental Economics and Management,* 60(2): 115–32.

Weisbach, David. 2010. "Instrument Choice is Instrument Design." In *U.S. Energy Tax Policy,* ed. Gilbert Metcalf, 113–58. New York: Cambridge University Press.

Weitzman, Martin L. Forthcoming. "Fat-Tailed Uncertainty and the Economics of Climate Change." *Review of Environmental Economics and Policy.*

Weitzman, Martin L. 1974. "Prices vs. Quantities." *Review of Economic Studies*, 41(4): 477–91.

Weitzman, Martin L. 2002. "Landing Fees vs. Harvest Quotas with Uncertain Fish Stocks." *Journal of Environmental Economics and Management*, 43(2): 325–38.

Weitzman, Martin L. 2009. "On Modeling and Interpreting the Economics of Catastrophic Climate Change." *Review of Economics and Statistics*, 91(1): 1–19.

Weitzman, Martin L. 2010. "GHG Targets as Insurance against Catastrophic Climate Damages." Harvard Project on International Climate Agreements Discussion Paper 10-42.

Wilen, James E. 1976. "Common Property Resources and the Dynamics of Overexploitation: The Case of the North Pacific Fur Seal." University of British Columbia Programme in Natural Resource Economics Working Paper 3.

World Meteorological Organization. 2007. *Scientific Assessment of Ozone Depletion, 2006.* Geneva, Switzerland: World Meteorological Organization.

Worm, Boris, Edward B. Barbier, Nicola Beaumont, J. Emmett Duffy, Carl Folke, Benjamin S. Halpern, Jeremy B. C. Jackson, et al. 2006. "Impacts of Biodiversity Loss on Ocean Ecosystem Services." *Science,* 314(5800): 787–90.

[2]

ENVIRONMENTAL REGULATION IN THE 1990S: A RETROSPECTIVE ANALYSIS

*Robert W. Hahn**

*Sheila M. Olmstead***

*Robert N. Stavins****

I. Introduction

This Article addresses the influence of economics on environmental and resource policy-making during the 1990s. We focus on the Clinton administration and highlight important trends and changes in the impacts of economic concepts such as efficiency, cost-effectiveness and distributional equity.[1] The continuing controversy over the appropriate role for economics in environmental policy design makes this a particularly good time to analyze environmental policy during the 1990s from an economic perspective.

We note that the role of efficiency as a criterion for assessing environmental and natural resource rules and regulations was very controver-

---

* Director, AEI-Brookings Joint Center for Regulatory Studies, and Resident Scholar, American Enterprise Institute.

** Assistant Professor of Environmental Economics, School of Forestry and Environmental Studies, Yale University.

*** Albert Pratt Professor of Business and Government, John F. Kennedy School of Government, Harvard University, and University Fellow of Resources for the Future. Helpful comments on a previous version of this Article were provided by: Arthur Fraas, George Frampton, Myrick Freeman, José Gómez-Ibáñez, Alan Krupnick, Randall Lutter, Albert McGartland, Richard Morgenstern, Paul Portney, Richard Schmalensee, Jason Shogren, and Murray Weidenbaum. Research assistance was provided by Simone Berkowitz, and financial support was provided by the Savitz Family Fund for Environment and Natural Resource Policy and the Ford Fund at Harvard University. The authors alone are responsible for any remaining errors. A longer, related paper by Sheila M. Cavanagh et al. includes comprehensive tables describing specific environmental and resource statutes and regulations. *See* Sheila M. Cavanagh (Olmstead) et al. National Environmental Policy During the Clinton Years (John F. Kennedy School of Government, Center for Business and Government, Regulatory Policy Program Working Paper RPP-2001-10, 2001). For surveys of environmental and resource policy in the 1980s, see Paul R. Portney, Natural Resources and the Environment: The Reagan Approach (1984) and W. Kip Viscusi, *Health and Safety Regulation*, *in* American Economic Policy in the 1980s 453 (Martin Feldstein ed., 1994).

[1] We follow the standard definition of an "efficient" environmental policy as being one which involves a target—such as a fifty percent reduction in sulfur dioxide ("SO_2") emissions—that maximizes the difference between social benefits and social costs (i.e., a target level at which marginal benefits and marginal costs are equated). By "cost-effective" policies, we refer to those which take (possibly inefficient) targets as given by the political process, but achieve those targets with policy instruments—such as a tradeable permit system in the SO_2 case—that minimize aggregate costs. Assessments of the "distributional" implications of environmental policies include analyses of the distributions of costs and benefits.

sial in the Clinton administration, while efficiency emerged as a central goal of the regulatory reform movement in Congress. Cost-effectiveness was embraced by both the Administration and Congress in the 1990s as a criterion for adopting specific policy instruments. In addition, the decade witnessed an increasing role for equity concerns as a consideration in environmental policy-making.

The attention given to environmental and natural resource issues in the United States has grown over the past several decades, a period during which greater consideration has been given to economic analysis of laws and regulations intended to protect the environment or improve natural resource management. Although several of the major environmental statutes are ambivalent about the role of economic analysis, in some cases prescribing it, in others proscribing it, a series of Presidential executive orders has called for a larger role for economic analysis.

Administrations can have substantial influence over the application of economics to environmental policy through a variety of mechanisms. The conventional wisdom in the United States is that Democratic administrations are predisposed toward more active environmental regulation, and less inclined toward economic analysis of environmental policy than their Republican counterparts. The Clinton administration, for example, is widely perceived to have been predisposed to environmental quality and resource preservation, and less supportive of economic analysis of such issues, in comparison with its Republican predecessor and successor (the administrations of George H. W. Bush and George W. Bush, respectively).

In fact, environmental and natural resource policy in the 1990s was characterized by continuity and by change. Two important trends that began in the 1970s continued through the 1990s—environmental quality improved, and environmental targets were made more stringent. In some cases, these improvements can be linked directly to federal policies and regulations; in others, such linkage has yet to be established.[2]

Trends in emissions of Clean Air Act criteria air pollutants are described in Table 1 (see Appendix). Emissions of some of these pollutants decreased significantly during the decade.[3] Although a number of studies show continued improvements in water quality during the 1990s,[4] following

---

[2] In order to attribute environmental quality improvements to specific policies, we must compare actual emissions to what they would have been in the absence of policies.

[3] *See* U.S. EPA Pub. No. 454/R-00-002, National Air Pollutant Emission Trends 1900-1998 (2000) [hereinafter EPA, 1900-1998 Trends Report]; U.S. EPA Pub. No. 454/R-00-003, National Air Quality and Emission Trends Report, 1998 (2000). Real improvements in environmental quality would be measured by changes in exposure and resulting changes in human morbidity and mortality, ecosystem health, etc. Improvements in emissions are not, themselves, measures of environmental quality improvements, although they may be highly correlated with such improvements.

[4] *See* Tayler H. Bingham et al., A Benefits Assessment of Water Pollution Control Programs Since 1972 (U.S. EPA, revised draft report, 1998); Myrick A. Freeman, *Water Pollution Policy*, *in* Public Policies for Environmental Protection 169 (Paul R. Portney and Robert N. Stavins eds., 2000); Myrick A. Freeman, *Environmental Policy*

the pattern of thirty-year trends, improvements in water quality during the 1990s were both less dramatic and more difficult to measure than improvements in air quality.[5]

Emissions of many air and water pollutants declined dramatically from 1970 to 1990, when the "low-hanging fruit" among air and water quality problems were being addressed.[6] For example, air emissions of lead, which declined significantly due to the shift to unleaded gasoline (completed in 1987), saw little further improvement during the 1990s.[7] Pollutant emissions to water declined dramatically during the 1970s and 1980s due to expanded municipal sewage treatment, a shift that was largely completed before 1990.[8]

In addition to environmental quality, the stringency of environmental targets continued to increase during the 1990s. An important example was the Clinton administration's 1997 National Ambient Air Quality Standards ("NAAQS") for ambient ozone and particulate matter. The new NAAQS were far stricter than previous standards, carrying substantial potential benefits and costs.

Public policy affecting natural resource management during the Clinton years was heavily weighted toward environmental protection. The Administration proposed initiatives to reduce subsidies for private resource extraction on public lands, but Congress was not receptive. The Administration did, however, shift U.S. Forest Service ("USFS") priorities away from timber production to resource protection, placing some sixty million acres of federal forests off limits to road building. President Clinton also designated more than twenty new national monuments, thereby restricting the use of six million additional acres of federal lands.[9]

Our ability to offer sound judgments about the influence of Clinton-era policies on environmental quality improvements is restricted by two problems. First, the fact that quality improvements occurred contempora-

---

Since Earth Day I—What Have We Gained?, 16 J. Econ. Persp., Winter 2002, at 125.

[5] Improvements in water quality have been achieved largely through point source regulation. James Boyd, The New Face of the Clean Water Act: A Critical Review of the EPA's Proposed TMDL Rules 4 (Resources for the Future, Discussion Paper 00-12, Mar. 2000). Non-point source pollution in the form of runoff from cities and agricultural areas may actually have increased during the 1990s. Freeman, *supra* note 4, at 137.

[6] Important exceptions are emissions of toxic substances to air and water. Unlike conventional pollutants, decreases in air and water toxics emissions during the 1990s were likely greater than decreases in previous decades. The Toxics Release Inventory ("TRI") data show a decrease in toxic discharges to air of forty percent, and a decrease in toxic discharges to surface water of sixty-seven percent, between 1990 and 1994. Environmental Defense, *Toxics Release Inventory Data Summary*, *at* http://www.scorecard.org/env-releases/us.tcl#data_summary (last visited Apr. 25, 2003) (on file with the Harvard Environmental Law Review).

[7] *See* EPA, 1900–1998 Trends Report, *supra* note 3.

[8] Boyd, *supra* note 5, at 3. The percentage of the U.S. population connected to wastewater treatment systems increased from forty-two percent to seventy-four percent between 1970 and 1985.

[9] Reed McManus, *Six Million Sweet Acres*, Sierra, Sept.-Oct. 2001.

neously with the term of a particular administration or legislature is not proof that policies promulgated during this term actually caused those quality improvements. With the exception of reduced emissions of criteria air pollutants in the 1990s, we find no studies that establish such a causal relationship between 1990s policies and environmental quality changes.[10]

Second, a fundamental issue that would confront any assessment of policy initiatives associated with a particular administration is the choice of an appropriate basis of comparison for evaluating policy initiatives—a counterfactual. It might appear reasonable to contrast first-term Clinton administration initiatives with what might have been anticipated from a hypothetical second-term administration of George H. W. Bush. But what would be the appropriate counterfactual for the second Clinton term?

For these reasons, establishing a causal relationship between improvements in environmental quality or resource management and the policies of any particular administration or Congress is difficult, if not impossible, and is not attempted here. Instead, we apply economic criteria for policy assessment—principally efficiency, cost-effectiveness and distributional equity.

The combined trends of more stringent standards for air and water quality, and increased private land-use restrictions and protections for public lands, have brought both increased benefits and an increasing price tag. As a result, economic concepts like benefit-cost analysis and the selection of least-cost environmental and natural resource regulations have received more attention since the late 1980s than they did in the early years of U.S. federal environmental regulation.

We note in this Article that, rather than a simple split along party lines, politicians in the 1990s endorsed the use of the efficiency criterion where its results were likely to coincide with their own ideological agendas. For example, Congress during the 1990s supported improvements in the efficiency of pollution control standards, which would have lightened regulatory burdens on some industries, and did not support increased efficiency in natural resource management, where subsidy reduction would have hurt communities dependent on resource extraction in the conservative West. The Administration, likewise, promoted the reduction of natural resource extraction subsidies, but was unsupportive of benefit-cost analysis of pollution control regulations; both viewpoints were consistent with those of supporters in the environmental community. We analyze these issues in light of the increased focus on the distribution of benefits and costs of environmental and natural resource regulation.

---

[10] Based on EPA modeling of trends in emissions within and without the Clean Air Act, the observed decreases in emissions of criteria air pollutants between 1990 and 2000 can be attributed to the Clean Air Act and its amendments. Freeman, *supra* note 4, at 127–28.

Our analysis is primarily qualitative, although in cases in which quantitative economic analyses of environmental policies have been produced, we discuss those results. The analysis is not exhaustive, but we do our best to consider the most important and most prominent intersections of economics and environmental regulation over the decade.

In Part II, we highlight the ways in which the role of efficiency as a criterion for assessing environmental and natural resource rules and regulations was very controversial in the Clinton administration, while economic efficiency emerged as a central goal of the regulatory reform movement in Congress. In Part III, we examine how cost-effectiveness was embraced by both the Administration and Congress in the 1990s as a criterion for adopting specific policy instruments. In Part IV, we examine how and why the decade witnessed an increasing role for equity concerns as a consideration in environmental policy-making. In Part V, we conclude.

II. Efficiency as a Criterion for Assessing Rules and Regulations

The primary economic criterion for the analysis of environmental and natural resource regulation is efficiency. An efficient policy enacts a level of pollution control or rate of resource extraction that maximizes the difference between social benefits and social costs.[11] Assessing the efficiency of policies requires benefit-cost analysis.

The Clinton administration established a framework for benefit-cost analysis of major regulations that was very similar to those of previous administrations, but the influence of economic thinking in analyzing environmental rules and regulations within EPA declined significantly during the 1990s. While economists in other parts of the Administration strongly pressed for efficiency in natural resource management, a negligible portion of their initiatives became policy. Congress did not support the Administration's proposals for efficiency in natural resource management, but did embrace efficiency as a criterion for environmental policy as part of its overarching regulatory reform agenda, and succeeded in making substantive, efficiency-related changes to a handful of existing environmental statutes.

---

[11] In a dynamic context, the efficient rate of resource extraction or pollution control maximizes the present value of net social benefits.

A. Role and Acceptance of the Efficiency Criterion in the Clinton Administration

1. Executive Order on Regulatory Impact Analysis

The Clinton administration, like its two immediate predecessors, issued an Executive Order ("EO") requiring benefit-cost analysis of all federal regulations with expected annual costs greater than $100 million.[12] Throughout the Reagan and Bush administrations, these Regulatory Impact Analyses ("RIAs") were required under Reagan EOs 12,291 and 12,498.[13] President George H. W. Bush created a Council on Competitiveness, chaired by Vice President Dan Quayle, which reviewed the impact on industry of selected regulations.

Shortly after taking office in 1993, Clinton abolished the Council on Competitiveness and revoked both of the Reagan orders, replacing them with EO 12,866, "Regulatory Planning and Review."[14] The Clinton EO was substantively and administratively similar to the Reagan orders. It was qualitatively different in tone, however, signaling a less strict efficiency test. While the Reagan orders required that benefits *outweigh* costs, the Clinton order required only that benefits *justify* costs. The Clinton EO allowed that: (1) not all regulatory benefits and costs can be monetized; and (2) nonmonetary consequences should be influential in regulatory analysis.[15]

The requirements for RIA, however, have not necessarily improved the efficiency of individual federal environmental rules. In the first fifteen years of the review process, under both Republican and Democratic administrations, about two-thirds of the federal government's approved environmental quality regulations failed benefit-cost analyses using the gov-

---

[12] Exec. Order No. 12,866, 58 Fed. Reg. 51,735 (Sept. 30, 1993). The threshold is not indexed for inflation and has not been modified over time. Elsewhere in this Article, we refer to year 2000 dollars, unless we indicate otherwise.

[13] Exec. Order No. 12,291, 46 Fed. Reg. 13,193 (Feb. 17, 1981) required agencies to conduct a RIA for all proposed and final rules that were anticipated to have an effect on the national economy in excess of $100 million. EO 12,291 has been called the "foremost development in administrative law of the 1980s." *See* Richard D. Morgenstern, *The Legal and Institutional Setting for Economic Analysis at EPA*, *in* ECONOMIC ANALYSES AT EPA: ASSESSING REGULATORY IMPACT 5–23 (Richard D. Morgenstern ed., 1997). But, the Reagan EOs were not the first presidential effort at regulatory efficiency. Nixon required a "Quality of Life" review of selected regulations in 1971. Robert W. Hahn, *The Impact of Economics on Environmental Policy*, 39 J. ENVTL. ECON. & MGMT. 375, 385 (2000). Ford formalized this process in 1974 with Exec. Order 11,821, 39 Fed. Reg. 41,501 (November 29, 1974). Carter's EO 12,044 required analysis of proposed rules and centralized review by the Regulatory Analysis Review Group. Hahn, *supra*. The Administration of President George W. Bush has continued to enforce the RIA requirements of Clinton's EO 12,866 rather than issuing a new EO. *See* John D. Graham, *Presidential Review of Agency Rulemaking by OIRA*, Memorandum for the President's Management Council (2001), available at www.whitehouse.gov/omb/inforeg/oira_review-process.html, (last visited Apr. 25, 2003) (on file with the Harvard Environmental Law Review).

[14] Exec. Order 12,866, *supra* note 12.

[15] W. Kip Viscusi, *Regulating the Regulators*, 63 U. CHI. L. REV. 1423, 1430 (1996).

ernment's own numbers.[16] A good example during the Clinton years is the 1997 NAAQS for ozone, for which EPA submitted a RIA that listed $2.0 to $11.2 billion in monetized benefits and $12.7 billion in costs through 2010, assuming full attainment.[17]

Regulatory impact analysis is required only for major rules,[18] a small fraction of all rules issued by EPA and other agencies. Rules that do not meet this threshold pass under the efficiency radar, as do EOs such as those Clinton used to designate twenty new national monuments comprising six million additional acres, restricting natural resource extraction and other commercial activities therein.

2. *Diminished Role of Economic Analysis at EPA*[19]

Given the increase in requirements for and attention to benefit-cost analysis by Congress during the 1990s, discussed below, EPA probably was required to do more applied economic analysis during the 1990s than at any other time in its thirty year history. Perhaps in response to this workload, the share of EPA employees with graduate degrees in Economics grew during the 1990s.[20] However, the influence of economists and the acceptance of economic analysis at EPA were almost certainly lowered during the Clinton years.[21]

The mixed record of political and administrative integration of economic analysis within EPA during the Clinton years reflects the ambiva-

---

[16] Hahn, *supra* note 13.

[17] U.S. OFF. OF MGMT. & BUDGET, REPORT TO CONGRESS ON THE COSTS AND BENEFITS OF FEDERAL REGULATIONS 55 (1998). In other cases, issuing agencies do not provide enough information to assess the benefits and costs of rules. During the Clinton Administration, a good example is the RIA for the U.S. Forest Service's Roadless Areas proposal, which discusses benefits and costs in general and qualitative terms but does not offer the information necessary to make a direct, quantitative comparison of costs and benefits. *See* U.S. FOREST SERVICE, REGULATORY IMPACT ANALYSIS FOR THE ROADLESS AREA CONSERVATION RULE (2001).

[18] Exec. Order 12,866, *supra* note 12.

[19] We discuss at length the use and acceptance of economics at the EPA, since rules promulgated by EPA comprise a substantial majority of total costs and benefits of all federal environmental regulation. Fifty-four percent of total annual regulatory benefits and fifty percent of total annual regulatory costs identified by the Office of Management and the Budget ("OMB") in 1997 were attributed to environmental regulations. Susan Dudley and Angela Antonelli, *Shining a Bright Light on Regulators: Tracking the Costs and Benefits of Federal Regulation*, *in* THE HERITAGE FOUNDATION BACKGROUNDER (1997);OFF. OF MGMT. & BUDGET, *supra* note 17. Discussion of similar issues at the Departments of Energy, Agriculture, the Interior and other agencies is beyond the scope of this study.

[20] Between 1996 and 2000, the percentage of EPA employees with graduate degrees who held either masters or doctoral degrees in economics increased by fifteen percent, compared to a 7.7% overall increase in EPA employees with graduate degrees. Richard D. Morgenstern, *Decision making at EPA: Economics, Incentives and Efficiency*, Draft conference paper, "EPA at Thirty: Evaluating and Improving the Environmental Protection Agency," Duke University, 36–38 (2000).

[21] *See id.*

lence of the major environmental statutes with respect to the role of economic analysis.[22] EPA is not an economic agency. It has a mandate to protect human health and the environment through the Administration of the major statutes.[23] Many of those statutes constrain economic analysis, and the representation of economists within most EPA offices is relatively thin, particularly at the level of the Senior Executive Service.[24] However, there is a good deal of flexibility in the extent to which economic analysis influences EPA processes and decisions. As a result, the use and role of economic analysis at EPA has varied substantially from one administration to another.

a. Organizational Location of Core Economics Staff

During the Clinton administration, economics staff at the agency were marginalized. When Clinton took office in 1992, the core economics staff at EPA were located within the Office of Policy, Planning and Evaluation (OPPE), as they had been since before 1980. OPPE reviewed all draft regulations and provided the Administrator with an independent economic perspective, which could be quite different from program office analyses. Within weeks of the Clinton inauguration, however, this role was eliminated.

The substantive role of economic analysis in the development and review of EPA regulations was abandoned by EPA in 1995, when the program offices, rather than the Administrator, became the official recipients of these analyses.[25] In 1999, OPPE was eliminated, shifting the core economics staff to a new Office of Policy and Reinvention. The shifts in organizational location of the core economics staff at EPA are documented in Table 2.

Administrator Browner was openly dismissive of economics as an appropriate framework for environmental decisions. In her remarks in honor of the thirtieth anniversary of the first Earth Day, she commented on the establishment of EPA, and recalled that "the nation committed itself to the task of eliminating pollution, to restoring our lands and waters to their uses, and to protecting public health without regard to cost. Let me repeat

---

[22] U.S. environmental laws alternately "forbid, inhibit, tolerate, allow, invite, or require the use of economic analysis in environmental decision making." *Id.* at 20.

[23] The term "major environmental statutes" in this Article refers to the following federal laws (and all amendments thereto): the Clean Air Act ("CAA"); Federal Water Pollution Control Act (Clean Water Act, "CWA"); Toxic Substances Control Act ("TSCA"); Federal Insecticide, Fungicide and Rodenticide Act ("FIFRA"); Comprehensive Environmental Response, Compensation and Liability Act ("CERCLA"); Resource Conservation and Recovery Act ("RCRA"); and Safe Drinking Water Act ("SDWA").

[24] *See* Morgenstern, *supra* note 13, at 16. Of the 193 EPA Senior Executive Service members with graduate degrees in 1996, only four (two percent) held graduate Economics degrees; in contrast, almost one-third held law degrees, and one-fifth held graduate science degrees. Despite their minority status relative to lawyers, scientists and engineers, EPA probably employs more economists working on environmental issues than any other single institution. *Id.* at 14.

[25] Morgenstern, *supra* note 20, at 39.

those last four words—without regard to cost."[26] The Administrator referred to the introduction of benefit-cost analysis into EPA regulations intended to protect public health as "poison[ing] the well."[27] The reduction in acceptance of economic analysis at EPA was likely influenced by Vice President Al Gore, who was known to be skeptical about the application of benefit-cost analysis to environmental policy.[28]

b. Role of the Environmental Economics Advisory Committee

Despite the reduced role of economists within EPA, policy advising by government economists outside of EPA occurred throughout the 1990s. Deputy Administrator Fred Hansen worked closely with the Environmental Economics Advisory Committee ("EEAC") within EPA's Science Advisory Board to develop an aggressive mission statement for EEAC that focused on giving expert advice on broad issues of importance to the Agency, rather than simply carrying out end-of-pipe reviews of agency RIAs.[29] During the 1990s, the EEAC conducted the first comprehensive review and revision in fifteen years of EPA's Economic Analysis Guidelines.[30] They also thoroughly reviewed EPA's methodology for valuing reductions in cancer-induced mortality.[31] External economists also served on the Advisory Council on Clean Air Act Compliance, required under the 1990 CAA Amendments to provide technical and economic input on EPA's benefit-cost analyses of CAA impacts.[32] The Council had a major impact on the identification of key research issues and the treatment of uncertainty in these analyses.[33]

---

[26] Carol M. Browner, Speech marking the 30th anniversary of Earth Day, John F. Kennedy School of Government, Harvard University (Apr. 17, 2000) (transcript available at http://www.epa.gov/history/topics/epa/30a.htm) (last visited Apr. 25, 2003) (on file with Harvard Environmental Law Review).

[27] *Id.* Although she referred to benefit-cost analysis, what Administrator Browner described was more like a strict benefit-cost test that would disallow rules unless quantified benefits outweighed costs.

[28] *See generally* AL GORE, EARTH IN THE BALANCE: ECOLOGY AND THE HUMAN SPIRIT (1992).

[29] The EEAC was established by the Science Advisory Board in 1990.

[30] *See* U.S. EPA, GUIDELINES FOR PREPARING ECONOMIC ANALYSES, 240-R-00-003 (Sept. 2000); SCIENCE ADVISORY BOARD, U.S. EPA, AN SAB REPORT ON THE EPA GUIDELINES FOR PREPARING ECONOMIC ANALYSES, EPA-SAB-EEAC-99-020 (Sept. 1999).

[31] SCIENCE ADVISORY BOARD, U.S. EPA, AN SAB REPORT ON EPA'S WHITE PAPER "VALUING THE BENEFITS OF FATAL CANCER RISK REDUCTION," EPA-SAB-EEAC-00-013 (July 2000).

[32] 42 U.S.C. § 7612 (2000).

[33] *See* Morgenstern, *supra* note 20.

3. Role of Other Executive Branch Economists in Natural Resource Policy

Having noted the diminished role of economics at EPA during the Clinton years, it is also important to recognize economists external to EPA. In particular within the Council of Economic Advisors ("CEA"), OMB, and the Treasury Department, economists did have some influence over the Administration's policy proposals regarding efficiency in natural resource management.[34]

The most important artifact of the White House economic agencies' influence in emphasizing efficiency in environmental and natural resource policy is the Clinton administration's 1993 economic stimulus and deficit reduction proposal.[35] The Administration proposed a variety of policies related to natural resource subsidy reduction. First, it proposed increasing the baseline federal grazing fee on public lands by almost 200%. The baseline federal grazing fee had been calculated at only fifty-six to eighty-three percent of federal costs per animal unit month in 1990 and was a much smaller percentage (perhaps eighteen to forty percent) of private market rates.[36] In theory, below-market fees for grazing livestock on public lands cause (economic) over-grazing. In practice, low fees have been criticized from a budgetary perspective, since current fees do not cover the costs of federal public range management.[37]

Similarly, below-cost timber sales from federal lands theoretically lead to logging at faster-than-efficient rates, and where revenues do not cover costs, they also contribute to budget deficits. The Administration's 1993 budget proposal sought to phase out below-cost timber sales. By USFS estimates, 77 of the 120 national forests showed net losses from timber sales over the period FY 1989-FY 1993, and sixty reported losses in every year over this period.[38]

---

[34] *See* Jonathan Orszag et al., *The Process of Economic Policy-Making During the Clinton Administration*, *in* AMERICAN ECONOMIC POLICY IN THE 1990S 983, 994 (Jeffrey A. Frankel & Peter R. Orszag eds., 2002).

[35] Richard L. Berke, *Clinton Backs Off From Policy Shift on Federal Lands*, N.Y. TIMES, Mar. 31, 1993, at A1; *Last Round Up for the Old West*, ECONOMIST, Mar. 6, 1993, at 23.

[36] Betsy A. Cody, *Grazing Fees: An Overview* (Congressional Research Service Report for Congress 96–450 ENR, 1996), *available at* http://www.ncseonline.org/NLE/ CRSreports/ Agriculture/ag-5.cfm (on file with the Harvard Environmental Law Review).

[37] *Id.* The baseline grazing fee for federal lands in 1990 was $1.81 per animal unit month (AUM), while the various livestock grazing programs' cost to government ranged from $2.18 to $3.24 per AUM. The fair market value of grazing on federal land was last updated in 1986 and ranged from $4.68 to $10.26 per AUM for cattle and horses, varying by region. (These figures have not been converted to constant dollars.) The Administration continued to lobby for fee increases, and the 104th Congress established a new fee formula that resulted in a small increase in the baseline fee, still many times lower than the average private market rate.

[38] *See* Ross W. Gorte, Below-Cost Timber Sales: Overview (Congressional Research Service Report for Congress 95-15 ENR, 1994).

Neither subsidy reduction proposal—the grazing fee increase nor the below-cost timber sales phase-out—became law, however. The grazing fee proposal led to a Senate filibuster on FY 1994 Interior Appropriations during the 103d Congress, and was taken up again in the 104th Congress, resulting in a negligible price increase, leaving rates still many times lower than the average private market rate. The plan to reduce below-cost timber sales was eliminated from Clinton's final budget proposal, and a USFS draft plan to phase out below-cost sales on one-half of forest service lands over four years was not adopted by the Administration.[39]

The Administration's attempt to reduce natural resource subsidies in the 1993 budget proposal also included introduction of royalties for hardrock mining on public lands governed under the 1872 General Mining Law,[40] increased fees for recreational use of federal public lands, and a British Thermal Unit ("BTU") tax, which would have taxed essentially all fuels. The BTU tax proposal faced stiff opposition in the first session of the 103d Congress, narrowly passing the House of Representatives. Recognizing that the proposal did not have enough votes in the Senate, the Administration removed the BTU tax from its budget proposal.[41]

During the 1990s, economists at the U.S. Department of Commerce ("DOC") began work on the issue of "green accounting." Incorporating natural resource depletion and other non-market activity within the National Income and Product Accounts ("NIPA") has been a longstanding recommendation of economists.[42] In 1993 the Clinton administration ordered the Bureau of Economic Analysis ("BEA") at the DOC to begin working on this process.[43] The BEA produced the first official U.S. Integrated Environmental and Economic Satellite Accounts in 1994, accounting only for selected subsoil minerals. Shortly afterward, Congress suspended BEA's work on environmental accounting, pending external review by a blue-ribbon panel convened by the National Research Council's Committee on National Statistics. Though the panel's review, released in 1999, strongly supported BEA's efforts and endorsed further efforts to

---

[39] *Id.*

[40] 30 U.S.C §§ 22–54 (2000).

[41] The Senate later passed a much more modest Transportation Fuels Tax in 1993, with revenues flowing to the General Fund. This was a retail tax on commercial gasoline sales of less than five cents per gallon. The BTU tax would have been imposed on coal, natural gas, liquid petroleum gases, nuclear electricity, hydroelectricity, and all imported electricity ($0.0257/million BTU); a higher tax ($0.0599/million BTU) would have been imposed on refined petroleum products. *See* FEDERAL BUDGET ISSUE: DO WE NEED AN ENERGY TAX?, National Center for Policy Analysis Policy Backgrounder No. 127 (June 4, 1993).

[42] *See, e.g.*, ARTHUR C. PIGOU, THE ECONOMICS OF WELFARE (1920); Martin L. Weitzman, *On the Welfare Significance of National Product in a Dynamic Economy*, 90 Q. J. ECON. 156 (1976); Robert Solow, "An Almost Practical Step Toward Sustainability," Invited Lecture on the Occasion of the Fortieth Anniversary of Resources for the Future (October 1992); NATURE'S NUMBERS: EXPANDING THE NATIONAL ECONOMIC ACCOUNTS TO INCLUDE THE ENVIRONMENT (William D. Nordhaus & Edward C. Kokkelenberg eds., 1999).

[43] Nordhaus & Kokkelenberg, *supra* note 42, at 154.

extend the NIPA,[44] Congress did not fund additional work on green accounting.

B. Role and Acceptance of the Efficiency Criterion in Congress

While Congress was unsupportive of efficiency as a criterion for natural resource management, benefit-cost analysis of environmental regulation emerged as a major goal of Congressional regulatory reform efforts of the 1990s. We examine general and specific regulatory reform proposals considered by the 103d through 106th Congresses, as well as changes to individual environmental statutes.[45]

1. Cross-cutting Regulatory Reform Proposals

The 103d Congress (1993–1995), the Clinton administration's first legislative "partner," actively debated benefit-cost analysis and risk analysis as methods for informing environmental protection decisions.[46] Three of the lightning rods for regulatory relief interests were "takings" issues or private property rights, unfunded mandates, and risk analysis. With Democratic majorities in both houses, none of the Republican minority's initiatives were enacted into law during the 103d Congress, or even offered for Presidential signature.

The regulatory reform movement gained momentum when members of the 104th Congress (1995–1997) took their seats after the 1994 midterm election, in which Republicans gained control of both the Senate and the House of Representatives. Reform-oriented bills in 1995-1996 included mandates for benefit-cost analysis, maximum likelihood risk assessments (rather than upper bounds), and regulatory process reforms.[47]

a. General Regulatory Reform: The Contract with America

Most of the 104th Congress' general regulatory reform proposals either failed to pass both Houses or were vetoed by President Clinton. Item 8 of the 1994 Contract with America, the "Job Creation and Wage

---

[44] *Id.* at 155.

[45] A comprehensive summary of successful and unsuccessful regulatory reform initiatives of the Congresses of the 1990s that would have influenced the application of efficiency, risk analysis, or cost-effectiveness criteria to environmental regulation is found in Table 2 of SHEILA M. CAVANAGH (OLMSTEAD) ET AL., NATIONAL ENVIRONMENTAL POLICY DURING THE CLINTON YEARS (Regulatory Policy Program Working Paper RPP-2001-10, Center for Business and Government, John F. Kennedy School of Government, 2001).

[46] *See* John E. Blodgett, Environmental Policy and the Economy: Conflicts and Concordances (Congressional Research Service Report for Congress 95-147 ENR, 1995), *available at* http://www.ncseonline.org/NLE/CRS.../econ-1.cfm; Martin R. Lee, Environmental Protection: From the 103rd to the 104th Congress (Congressional Research Service Report for Congress 95-58 ENR, 1995).

[47] *See* Viscusi, *supra* note 15.

Enhancement Act of 1995,"[48] did not reach the President's desk. It would have made Reagan's EO 12,291 statutory, superseding the Clinton EO—as well as the language in several other important statutes—and would have required that the benefits of regulations outweigh their costs.[49] Although this component of the Contract with America did not become law, it did lead to a prominent public debate over regulatory reform, in which benefit-cost analysis was a central issue.

b. Specific Regulatory Reform Proposals

The Small Business Regulatory Enforcement Fairness Act[50] ("SBREFA") amended the 1980 Regulatory Flexibility Act. As one of the affected agencies, EPA must prepare a regulatory flexibility analysis of all rules with "significant economic impact" on a "substantial number" of small entities (businesses, non-profits, and small government organizations).[51] Embedded within SBREFA, but for the most part unrelated to its other provisions, was the Congressional Review Act (CRA),[52] which established a process of Congressional review and possible rejection of agency rules on efficiency grounds.[53]

In late 1996, in another attempt to emphasize efficiency in regulation, the 104th Congress attached a benefit-cost requirement to Section 645(a) of the Treasury, Postal Services and General Government Appropriations Act of 1997.[54] To meet this requirement, the OMB is required to

---

[48] H.R. 9, 104th Cong. (1995).

[49] Item 8 also focused on the reduction of so-called "unfunded mandates," and on strengthening the Regulatory Flexibility Act of 1980, 5 U.S.C. §§ 601–612 (2000), resulting in the Small Business Regulatory Enforcement Fairness Act of 1996 (SBREFA), 5 U.S.C. §§ 801–808 (2000) and the Unfunded Mandates Reform Act of 1995, 2 U.S.C. §§ 658, 1501–1571 (2000). There were many other unsuccessful attempts at regulatory reform legislation during the 104th Congress, including: "Risk Assessment and Cost-Benefit Act of 1995," H.R. 1022, 104th Cong. (1995); H.R.J. Res. 27 & 54, 104th Cong. (1995), proposing a Constitutional amendment to ban unfunded mandates; "Regulatory Relief and Reform Act," H.R. 47, 104th Cong. (1995); and H.R. 122, 104th Cong. (1995) to establish a Regulatory Sunset Commission. Detailed discussion of these is beyond the scope of this study. We mention them only to emphasize the scope and depth of the 104th Congress' focus on regulatory reform.

[50] 5 U.S.C. §§ 801–808.

[51] *Id.* These analyses, which are reviewed by Congress, examine the type and number of small entities potentially subject to the rule, record-keeping and compliance requirements, and significant regulatory alternatives. The statute does not require formal benefit-cost analysis beyond that already required by environmental regulations and EO; rather, it requires that EPA submit to Congress "a complete copy of the benefit-cost analysis of the rule, if any," along with the regulatory flexibility analysis. *Id.* From an economic efficiency perspective, the focus on small entities makes little, if any sense.

[52] *Id.* at §§ 801–802.

[53] The CRA was the basis for the George W. Bush Administration's overturning of the Occupational Safety and Health Administration's ergonomics rule in March 2001. Pub. L. 107-5, 115 Stat. 7 (2001). The CRA has not been used to overturn any environmental regulations.

[54] Pub. L. No. 104-208, 110 Stat. 3009 (1997). This provision was typically referred to

submit to Congress a report estimating the "total annual costs and benefits of federal regulatory programs, including quantitative and non-quantitative measures."[55] The legislation also requires OMB to estimate individually the benefits and costs of rules with annual costs to the economy of $100 million or more. Importantly, OMB also is required to recommend the reform or elimination of any regulation that appears to be inefficient. Under this requirement, reports were submitted yearly, 1997 through 2000.[56] The requirement has further centralized regulatory oversight in the hands of OMB, which already had been charged with reviewing the RIAs required by EOs since 1981.

Congressional regulatory reform efforts continued through the end of the Clinton administration. The 105th and 106th Congresses considered establishing further checks on agency regulation. The Regulatory Improvement Act of 1999 (also known as the Thompson-Levin bill) would have allowed courts to remand or invalidate rules formulated by an agency that fails to perform sufficient benefit-cost analyses.[57] While this bill never became law, the 106th Congress did pass a major piece of regulatory reform legislation, the Truth in Regulating Act ("TIRA")[58], which was signed into law by President Clinton in October 2000. The TIRA established a three-year pilot project beginning in early 2001, which required the Government Accounting Office ("GAO") to review RIAs to evaluate agencies' benefit estimates, cost estimates, and analysis of alternative approaches, upon request by Congress. Because funding was never provided, TIRA was not implemented. If TIRA had been implemented, it likely would have increased the importance of economic analysis in regulatory decision making.

2. *Successful Changes to Individual Statutes*

In addition to these attempts at cross-cutting regulatory reform, the Congresses of the Clinton years pursued efficiency within environmental statutes themselves.[59] In general, Congress was more successful during

---

as "regulatory accounting."

[55] *See* U.S. OMB, *supra* note 17.

[56] The continuation of this provision was proposed by the Regulatory Right-to-Know Act of 1999, S. 59, 106th Cong. (1999). Introduced as H.R. 1074, 106th Cong. (1999) in the House, the bill would have required much more stringent analysis by OMB: an annual accounting statement of total costs and benefits of federal regulations, including direct and indirect impacts on federal, state, local and tribal government; the private sector; small business; wages; and economic growth.

[57] The Regulatory Improvement Act was first proposed as S. 981, 105th Cong. (1997) in 1997 and continued with the same title into 1998. It was introduced in various versions in both Houses of Congress throughout 1997–1999, and took on the Thompson-Levin moniker in May 1999.

[58] Pub. L. No. 106-312, 114 Stat. 1248 (2000).

[59] During the 1990s, Congress also pursued reforms of non-environmental statutes that affected environmental regulation. For example, the Accountable Pipeline Safety and Part-

the 1990s at passing cross-cutting regulatory reform bills than it was at reforming individual environmental statutes, although important exceptions were the 1996 SDWA amendments[60] and the partial reform of pesticide permitting under the Federal Food, Drug and Cosmetic Act ("FFDCA").

a. SDWA Amendments of 1996

The 1996 SDWA amendments[61] include the most far-reaching requirement for benefit-cost analysis in any environmental statute. The amendments focus EPA regulatory efforts on contaminants that pose the greatest health risks by: (1) requiring benefit-cost analysis of new rules; (2) removing the mandate that EPA regulate twenty-five new contaminants every three years; (3) allowing EPA to use cost information to adjust its "feasibility standards" for water system reduction of contaminants; and (4) requiring the Administrator to balance risks among contaminants to minimize the overall risk of adverse health effects.[62] While the Amendments require EPA to determine whether the benefits of each new drinking water maximum contaminant level ("MCL") regulation justify the costs, they also allow the Agency to adopt more stringent standards than those that maximize net benefits, explaining the reasons for not selecting the efficient standard.[63]

b. Food Quality Protection Act of 1996

The Food Quality Protection Act of 1996 ("FQPA")[64] amends both FIFRA[65] and FFDCA,[66] removing pesticide residues on processed food from the group of Delaney "zero-risk standard" substances. The Delaney standard has long been a target of economic criticism on the grounds that it specifies an often unachievable regulatory standard for the benefits of regulation, and hence leads to associated costs that may greatly exceed benefits. While the standard continues to apply to non-pesticide food ad-

---

nership Act of 1996, 49 U.S.C. § 60102(b)(5) (2000), requires the Secretary of Transportation to issue pipeline safety regulations only upon justification that benefits exceed costs. *See* John E. Blodgett, *Environmental Reauthorizations and Regulatory Reform: From the 104th Congress to the 105th* (Congressional Research Service Report for Congress 96-949 ENR, 1998), *available at* http://www.ncseonline.org/NLE/CRSreports/legislative/leg-22.cfm (last visited Apr. 25 2003).

[60] 42 U.S.C. § 300 (2000), *amended by* Pub. L. No. 104-182, 110 Stat. 1613 (1996).

[61] 110 Stat. 1613.

[62] Mary Tiemann, *Safe Drinking Water Act Amendments of 1996: Overview of P.L. 104-182* (Congressional Research Service Report for Congress 96-722, 1999), *available at* http://www.ncseonline.org/nle/CRSreports/water/h2o-17.cfm (last visited Apr. 25 2003).

[63] *See* 42 U.S.C. § 300g-1(a). The amendments do not allow standards published before the SDWA to be subjected to an *ex-post* benefit-cost analysis.

[64] Pub. L. No. 104-170, 110 Stat. 1489 (1996).

[65] 7 U.S.C. § 136 (2000).

[66] 21 U.S.C. §§ 301–397 (2000).

ditives, the FQPA eliminated the distinction between pesticide residues on raw foods (which had been regulated under FFDCA section 408[67]) and processed foods (which had been regulated under FFDCA section 409—the Delaney Clause).[68]

c. Failed Attempts at Changes to Individual Statutes

Two of the environmental statutes most frequently criticized on efficiency grounds—CERCLA (Superfund)[69] and the CWA[70]—remained relatively untouched by Congress in the 1990s, despite its focus on regulatory reform. Superfund's critics have focused on the low benefits and high costs of achieving the statute's standards.[71] Reauthorization and reform were considered during the 105th Congress, but no legislation was passed. Rather than efficiency, liability issues and questions of how to finance Superfund were the major foci of legislative discussions. The taxes that support the Superfund trust fund (primarily excise taxes on petroleum and specified chemical feedstocks and a corporate environmental income tax) expired in 1995 and have not been reinstated.[72]

The 104th Congress also pursued efficiency-oriented reform of the CWA through the reauthorization process, but the effort failed in the Senate. During the 104th Congress, the House passed a comprehensive CWA reauthorization[73] that would have been more flexible and less prescriptive than the current statute, but the Senate did not take up the bill.[74] No reauthorization legislation was considered in the 105th or 106th Congress.

---

[67] 21 U.S.C. § 346a (1994).

[68] *Id.* at § 348. The FQPA also mandates that EPA coordinate pesticide regulation under FIFRA and FFDCA. For example, once a pesticide registration is canceled under FIFRA, the food-use tolerance under FFDCA must be revoked within 180 days, rather than the average six year time frame noted in a 1994 GAO report. *See* Linda Jo Schierow, *Pesticide Legislation: Food Quality Protection Act of 1996* (Congressional Research Service Report for Congress 96-759 ENR, 1996), *available at* http://www.ncseonline.org/nle/crsreports/pesticides/pest-8.cfm; U.S. Gen. Accounting Office, Pesticides: Reducing Exposure to Residues of Canceled Pesticides, GAO/RCED-95-23 (1994).

[69] 42 U.S.C. §§ 9601–9675 (2000).

[70] 33 U.S.C. §§ 1251–1387 (2000).

[71] *See, e.g.*, W. Kip Viscusi, Fatal Tradeoffs: Public and Private Responsibilities for Risk (1992); Stephen Breyer, Breaking the Vicious Circle: Toward Effective Risk Regulation (1993); James T. Hamilton & Kip W. Viscusi, *How Costly is Clean?: An Analysis of the Benefits and Costs of Superfund Site Remediations*, 18 J. Pol'y Anal. & Mgmt. 2 (1999).

[72] The revenue now flowing into the trust fund comes from so-called "potentially responsible parties," interest on the fund's investments, fines, and penalties. Then-Chairman of the House Ways and Means Committee, Bill Archer (R-Tex.), made it known that no reinstatement of the Superfund taxes would be considered without major reform of the statute's liability provisions and other features. Mark Reisch, Superfund Reauthorization Issues in the 106th Congress (Congressional Research Service Issue Brief for Congress IB10011, 2000).

[73] H.R. 961, 104th Cong. (1995).

[74] The 103d Congress had considered similar legislation H.R. 3948, S. 2093, 103d Cong. (1994). However, no floor action on CWA reauthorization was taken in either house.

C. Limited Effect of Regulatory Reform Legislation and Changes to Statutes

The cross-cutting legislative regulatory reform measures passed in the 1990s and the efficiency-related changes to specific environmental statutes had limited effects on regulation during the decade. This is in part due to differences between the Administration and Congress in the acceptance of efficiency as an appropriate criterion for managing the environment and natural resources. An additional explanation is the existing statutory bias against benefit-cost analysis in some cases, particularly under the CAA. In such cases, substantial movement toward efficiency in regulation cannot be expected without substantial changes in the authorizing legislation.

The SDWA Amendments of 1996 incorporated a strong benefit-cost criterion, in comparison to other environmental statutes. However, the decisions made on MCLs since the SDWA Amendments have not placed great weight on the results of required benefit-cost analyses. Two major rules proposed since the 1996 Amendments were those regulating allowable levels of arsenic and radon in drinking water.[75] EPA's benefit-cost analyses for the radon and arsenic MCLs can be interpreted as indicating that monetized costs exceed monetized benefits for both rules (by more than $50 million annually for radon, and $30 million annually for arsenic). The Agency maintained, however, that benefits of both rules justify their costs when unquantified benefits are included.[76]

Importantly, the regulatory reform initiatives passed by Congress in the 1990s apparently did not influence EPA's issuance of NAAQS for ozone and particulate matter in July 1997. Due to their high potential compliance costs, the revised standards were immediately controversial; both the decision to tighten the standards and the quality of the research used

---

[75] The arsenic rule was finalized on January 22, 2001, but implementation was delayed while the rule was taken under review by the George W. Bush Administration, citing concerns about the rule's costs and benefits. After an expedited review by the National Academy of Sciences, in October, 2001, EPA Administrator Christine Whitman announced the Agency's intention to enforce the Clinton arsenic standard. *See* Press Release, EPA, EPA Announces Arsenic Standard For Drinking Water of 10 Parts Per Billion (Oct. 31, 2001) (on file with Harvard Environmental Law Review). No final action has been taken on radon.

[76] *See* U.S. EPA, PROPOSED ARSENIC IN DRINKING WATER RULE: REGULATORY IMPACT ANALYSIS (2000), U.S. EPA, HEALTH RISK REDUCTION AND COST ANALYSIS FOR RADON IN DRINKING WATER (2000). EPA's cost and benefit figures for these rules were presented as annualized 1999 dollar values using a seven percent discount rate. The AEI-Brookings Joint Center for Regulatory Analysis performed its own benefit-cost analysis of the arsenic rule, and concluded that in all likely scenarios the cost per life saved by the rule would never be less than $6.6 million, and that in its "most likely" scenario, cost per life saved was approximately $67 million. *See* Jason K. Burnett & Robert W. Hahn, EPA's Arsenic Rule: The Benefits of the Standard Do Not Justify the Costs (AEI-Brookings Joint Center for Regulatory Studies, Regulatory Analysis 01-02, Jan. 2001). For a critical review of the EPA analysis and Burnett & Hahn, see Cass Sunstein, *The Arithmetic of Arsenic*, 90 GEO. L.J. 2255 (2002).

to support the new standards came under fire. EPA's cost estimates for the ozone standard were singled out for criticism; some analysts found them to be too low by a considerable margin.[77] On the other hand, the particulate standard exhibited expected benefits that could well exceed costs by a considerable margin. Table 3 provides EPA's estimated benefits and costs for both standards.

The regulated community challenged the new NAAQS in court, and the case reached the U.S. Supreme Court in October, 2000.[78] Under the CAA, EPA is required to set health-based standards for specified pollutants without consideration of costs. In February 2001, the Supreme Court ruled unanimously that the CAA does not allow EPA to consider costs in setting NAAQS for the criteria air pollutants, and that the statute's mandate that the NAAQS protect the public health with "an adequate margin of safety" allows an acceptable scope of discretion to EPA.[79]

Given that the ozone standard's estimated costs appear to outweigh its benefits by a significant margin, EPA has been under considerable pressure to revise the standard, despite the Supreme Court's decision.[80] The situation is very different, of course, for particulate matter, for which estimated benefits appear to outweigh estimated costs. If the courts continue to uphold the standards and if the statutes preventing cost considerations remain unchanged, the stricter NAAQS for ozone and particulate matter may be one of the Clinton administration's most enduring environmental legacies, in terms of both potential benefits and potential costs.[81]

The differences in opinion between Congress and the executive branch (especially EPA) on the usefulness of efficiency analysis resulted

---

[77] *See* Jason F. Shogren, *A Political Economy in an Ecological Web*, 11 ENVTL. & RESOURCE ECON. 557; Randall Lutter, *Is EPA's Ozone Standard Feasible?*, (REGULATORY ANALYSIS 99-6, AEI-Brookings Joint Center for Regulatory Studies) (1999).

[78] *See* Whitman v. Am. Trucking Ass'ns, Inc., 531 U.S. 457 (2001). A group of forty economists filed a brief *amici curiae* in the Supreme Court, suggesting that benefit-cost analysis should be considered in the setting of ambient air quality standards. *See* AEI-Brookings Joint Center *et al.* Brief Amici Curiae in the Supreme Court of the United States, American Trucking Ass'ns v. Browner, 530 U.S. 1202 (2000) (No. 99-1426).

[79] *See Am. Trucking Ass'ns*, 531 U.S. at 457. The Supreme Court decision was greeted positively by EPA Administrator Whitman: ". . . Congress delegated to EPA the standard-setting function, and EPA carried it out appropriately." *See* Press Release, EPA, Supreme Court Upholds EPA Position on Smog, Particulate Rules (Feb. 27, 2001) (on file with Harvard Environmental Law Review). The Court acknowledged that EPA and the states could continue to take costs into account in implementing the standards, which may serve as an impetus for the adoption of cost-effective policy instruments.

[80] EPA has agreed to reconsider its analysis of ozone NAAQS benefits in at least one respect. The agency's initial analysis did not consider the possible damages associated with *decreases* in ground-level ozone, which leads to increases in some ultraviolet radiation ("UV-B") exposure. *See* Randall Lutter & Christopher Wolz, *UV-B Screening by Tropospheric Ozone: Implications for the National Ambient Air Quality Standard*, 31 ENVTL. SCI. & TECH. 142A (1997).

[81] It remains to be seen whether some urban areas will be able to comply with the new ozone standards. One analyst estimates the costs to Los Angeles of meeting the ozone standard in 2010 will be about $15 billion in constant 2000 dollars, assuming a five percent decrease in current abatement costs due to technological change. Lutter, *supra* note 77, at 7.

in an effective stalemate. Even where statutes were explicitly altered to require benefit-cost analysis, as was the case for the setting of MCLs under the SDWA, rules promulgated during the 1990s do not appear to be any more or less efficient than rules promulgated during earlier decades.

III. Cost-Effectiveness as a Criterion for Assessing Public Policies

Many or most environmental laws and regulations are not cost-effective, typically specifying technologies or uniform emissions limits, despite tremendous variation in abatement costs among sources.[82] While uniform standards may effectively limit emissions of pollutants, they typically exact relatively high costs in the process, by forcing some firms to resort to unduly expensive means of controlling pollution. For example, under current regulations, the marginal cost of abating lead emissions ranges from $13 per ton in the non-metal products sector to $56,000 per ton in the food sector.[83]

Market-based approaches to environmental protection can be used to achieve the least-cost allocation of pollution reduction, even if the aggregate target is not efficient. Thus, cost-effectiveness is a criterion quite separate and distinct from efficiency.[84] A cost-effective regulatory policy takes environmental quality or natural resource extraction targets as given by the political process, but achieves those targets at minimum aggregate cost. Since the 1970s, the advantages of market-based (or economic-incentive) approaches in reducing the costs of environmental regulation have received serious political attention, and there have been increasing numbers of applications in the United States and other countries.[85] Both the Clinton Administration and Congress embraced cost-effectiveness as a criterion for adopting environmental and natural resource policies during the 1990s.

A. Support for the Cost-Effectiveness Criterion Within the Clinton Administration

The Clinton administration's support for the use of a cost-effectiveness criterion in choosing environmental policies was demonstrated in a

---

[82] *See* Richard G. Newell & Robert N. Stavins (2003), *Cost Heterogeneity and the Potential Savings from Market-Based Policies*, 23 J. Reg. Econ. 43 (2003); T.H. Tietenberg, *Economic Instruments for Environmental Regulation*, 6 Oxford Rev. Econ. Pol'y 17 (1990).

[83] *See* Raymond S. Hartman et al., The Cost of Air Pollution Abatement (World Bank Policy Research Working Paper #1398, Dec. 1994); Morgenstern, *supra* note 20, at 17–18.

[84] William J. Baumol & Wallace E. Oates, *The Use of Standards and Prices for Protection of the Environment*, 73 Swed. J. Econ. 42 (1971).

[85] Robert N. Stavins, *Experience with Market-Based Environmental Policy Instruments*, *in* The Handbook of Environmental Economics (Karl-Göran Mäler & Jeffrey Vincent eds., forthcoming 2003).

variety of contexts. The Administration included selection of cost-effective regulatory alternatives within Clinton EO 12,866, requiring regulatory impact analysis. And in the same Earth Day speech that was so critical of benefit-cost analysis, EPA Administrator Browner highlighted EPA's cost-effective regulatory measures and flexible approaches to pollution reduction.[86] During the Clinton years, EPA continued to emphasize cost-effective approaches to pollution control, including the use of information disclosure and voluntary programs, and the Administration aggressively promoted international market-based policy instruments for greenhouse gas emissions control (specifically, emissions trading).

1. Reinventing EPA

Administrator Browner announced the creation of EPA's Office of Reinvention in 1997, although it is fair to say that reform efforts at EPA had been underway since the mid-1980s. Vice President Gore's National Performance Review Report and the Government Performance and Results Act of 1993[87] brought increased attention to such efforts at EPA, and the Agency launched the centerpiece of its "reinvention" program, the Common Sense Initiative ("CSI") in 1994.[88]

Although the CSI can be considered within the umbrella of policies intended to foster greater cost-effectiveness, it is unclear whether the CSI improved the cost-effectiveness of environmental regulation in the 1990s. The CSI engaged six major industries in dialogue with EPA with the purpose of reducing compliance costs, introducing flexibility by moving toward regulation by industry rather than by pollutant, and reducing costly litigation through stakeholder participation.[89] But in 1997, two GAO reports found that too many CSI resources had been spent on process, and too few on substance and results. In addition, progress had been limited by the lack of consensus among industry workgroups on the most important issues, and the effort lacked results-oriented measures to assess progress.[90]

---

[86] Browner, *supra* note 26.

[87] Pub. L 103-62, 107 Stat. 285 (1993).

[88] Other organizations and institutions may also have played a role in EPA's focus on reinvention. A 1995 National Academy of Public Administration report suggested reforms at EPA, including better use of risk and cost information to rank priorities. In 1996, the Center for Strategic and International Studies launched "Enterprise for the Environment," an effort to build consensus for systematic environmental management reform. And the regulatory reform focus of the 104th Congress may also have prompted EPA to attempt to carry out reform efforts, in part to forestall Congressionally mandated changes. *See* Claudia Copeland, *Reinventing the Environmental Protection Agency and EPA's Water Programs* (Congressional Research Service Report to Congress 96-283 ENR, Mar. 1996), *available at* http://www.nsceonline.org/NLE/CRSreports/water/h2o-20.cfm.

[89] The participating industries were auto manufacturing, computers and electronics, iron and steel, metal finishing, petroleum refining, and printing.

[90] *See* U.S. GEN. ACCOUNTING OFFICE, GAO/RCED-97-155, ENVIRONMENTAL PROTECTION: CHALLENGES FACING EPA'S EFFORTS TO REINVENT ENVIRONMENTAL REGULATION (1997); U.S. GEN. ACCOUNTING OFFICE, GAO/RCED-97-164, REGULATORY REINVEN-

In 1995, Vice President Gore and Administrator Browner announced a set of twenty-five specific reinvention reforms at EPA, in addition to the CSI. One of these new programs was Project XL ("Excellence and Leadership"), which set a goal of fifty pilot projects allowing regulated firms to propose alternatives to existing command-and-control regulations that would attain higher levels of pollution control at lower cost.[91] The National Environmental Performance Partnership System sought to give states greater flexibility in achieving environmental goals by allowing them to convert some types of categorical federal grants into more flexible block grants.

In its assessment of EPA's reinvention program, GAO noted that EPA's efforts could have only limited success in introducing cost-effective changes, because significant progress would require reform of the legislative framework for environmental protection, rather than process reforms within EPA.[92]

2. *Information Disclosure and Voluntary Programs*

In addition to its reinvention efforts, EPA significantly increased use of information disclosure regulations during the 1990s. TRI was initiated in 1988 under the Emergency Planning and Community Right-to-Know Act Section 313[93] and requires firms to report on use, storage and release of hazardous chemicals. A 1993 Clinton EO required TRI reporting by federal facilities.[94] In 1994, EPA added 286 new chemicals to the list requiring TRI reporting, an eighty percent increase in the number of listed chemicals.[95] Further, EPA lowered reporting thresholds in 1999 for many persistent bioaccumulative toxic chemicals and added more of these chemicals to the TRI list.[96] The Clinton administration announced another expansion of TRI on January 17, 2001, considerably lowering the threshold for reporting lead emissions.[97]

TION: EPA'S COMMON SENSE INITIATIVE NEEDS AN IMPROVED OPERATING FRAMEWORK AND PROGRESS MEASURES (1997).

[91] Lisa C. Lund, *Project XL: Good for the Environment, Good for Business, Good for Communities*, 30 Envtl. L. Rep. (Envtl. L. Inst.) 10,140 (2000).

[92] U.S. GAO, GAO/RCED-97-155, *supra* note 90.

[93] P.L. 99-499, Title III, § 313, 100 Stat. 1741 (1986) (codified as amended at 42 U.S.C. § 11023 (2000)).

[94] Exec. Order 12,856, 58 Fed. Reg. 41981 (August 6, 1993).

[95] Linda Jo Schierow, Toxics Release Inventory: Do Communities Have a Right to Know More?, (Congressional Research Service Report for Congress 97-970 ENR, 1997).

[96] The EPA under Clinton also continued the 33/50 program, started under the Bush Administration, which engaged TRI-reporting industries in achieving voluntary accelerated emissions reduction targets in exchange for public "certification."

[97] 40 C.F.R. § 372.28 (2000). The previous standard required reporting by facilities that manufacture or process more than 25,000 pounds of lead annually, or that use more than 10,000 pounds annually. The newer standard required reporting by any facility that manufactures, processes, or uses more than 100 pounds annually. The Bush Administration announced its intention to uphold the new threshold on April 17, 2001.

Releases reported under TRI declined by forty-five percent from 1988 to 1998, but no analysis has yet been able to attribute that reduction to the policy itself. Limited evidence exists that publicly available information about firms' TRI emissions (either in absolute terms or relative to some benchmarks) negatively affects stock prices.[98] Other possible avenues through which the TRI may influence emissions are green consumerism, redirection of firms' attention toward measures that increase environmental performance while saving costs, and community pressure, but there is little solid evidence that any of these forces are at work.[99]

In addition to the TRI, EPA established new and expanded existing information programs during the 1990s. In 1997, EPA expanded the existing Energy Star Buildings program, consolidating it with the Green Lights program, both of which are information disclosure programs related to energy efficiency. In 1998, the Agency began requiring public water systems to issue annual Drinking Water Consumer Confidence Reports.[100] In 2000, it posted automobile "pollution rankings" on the EPA Web site, ranking vehicles based on hydrocarbon and NO_x tailpipe emissions. While these programs could, in theory, provide cost-effective ways of reaching environmental objectives, there is no solid evidence of their actual effects.

3. *Cost-Effectiveness and Climate Change Policy*

In October 1993, the Administration released its Climate Change Action Plan, which recommended fifty-two voluntary measures to meet greenhouse-gas emissions goals.[101] The nature of the initiatives in the plan is not unlike those that might have been expected from a second-term Bush administration, with their emphasis on voluntary programs, government-industry cooperation, cost-effectiveness, use of market incentives, and minimal mandatory government intervention.[102] But, even if not different

---

[98] *See* James T. Hamilton, *Pollution as News: Media and Stock Market Reactions to the Toxics Release Inventory Data*, 28 J. ENVTL. ECON. & MGMT. 98 (1995); Shameek Konar & Mark A. Cohen, *Information as Regulation: The Effect of Community Right to Know Laws on Toxic Emissions*, 32 J. ENVTL. ECON. & MGMT. 109 (1997); Madhu Khanna et al., *Toxics Release Information: A Policy Tool for Environmental Protection*, 36 J. ENVTL. ECON. & MGMT. 243 (1998).

[99] *See* Lori D. Snyder, Regulating Pollution Through Information Disclosure: Modeling Firm Response to the Toxics Release Inventory (Kennedy School of Government, Draft Working Paper, May 2001).

[100] U.S. EPA, Pub. No. 240/R-01-001, THE UNITED STATES EXPERIENCE WITH ECONOMIC INCENTIVES FOR PROTECTING THE ENVIRONMENT 161 (2001).

[101] Climate Change Action Policy, *available at* http://gcrio.gcrio.org/USCCAP/toc.html (Oct. 1993) (last visited Apr. 25, 2003) (on file with the Harvard Environmental Law Review).

[102] In 1993, the Administration also established the U.S. Initiative on Joint Implementation under the Climate Change Action Plan. Joint implementation arrangements allow firms or other entities in one country to meet part of their greenhouse gas reduction commitments by financing mitigation in another country. The U.S. Initiative through 1997 had approved twenty-two arrangements whereby U.S. firms agreed to finance projects in eleven

in substance, the Clinton administration's Climate Action Plan differed greatly in tone from what had been Bush administration policy. Whereas the Bush administration was moderate in its characterization of the climate change problem, the Clinton administration characterized the challenge in much more dramatic terms. Not surprisingly, this complex set of voluntary initiatives had relatively little effect. By 1995, the U.S. acknowledged that it would fall short of its goals by at least fifty percent.

A key component of the Clinton administration's climate change policy was its strong and unwavering support for cost-effective approaches, including market-based instruments, and in particular, tradeable permit mechanisms.[103] The Administration's formal proposal released in preparation for the Third Conference of the Parties of the Framework Convention on Climate Change, held in Kyoto, Japan in November 1997, called for domestic and international emissions trading.[104] In fact, it was largely because of the efforts of the U.S. negotiating team that the Kyoto Protocol included significant provisions for international emissions trading among industrialized nations, as well as what came to be known as the Clean Development Mechanism for offsets in developing countries.

Subsequently the United States proposed rules for international emissions trading in 1998, at preparatory talks for the Fourth Conference of the Parties. The U.S. proposal faced substantial opposition, most significantly from the European Union. No agreement was reached on emissions trading at the Fourth (1998), Fifth (1999), or Sixth (2000) Conferences of the Parties. Indeed, at the Sixth Conference of the Parties, which met in The Hague in November 2000, disagreements between the United States and the European Union over the role of carbon sequestration and emissions trading led to the ultimate breakdown of the talks.[105]

Economic considerations appear to have played a much more substantial role in the development of the Administration's international negotiating position on climate change than they did in the development of domestic regulatory policies with substantial economic costs, such as the NAAQS for ozone and particulate matter. Within the White House, weekly (and even more frequent) meetings on climate change leading up to the Kyoto conference were chaired by the National Economic Council ("NEC"),

---

other countries. WORLD BANK, ENVIRONMENTALLY SUSTAINABLE DEVELOPMENT STUDIES AND MONOGRAPHS SERIES NO. 18, FIVE YEARS AFTER RIO: INNOVATIONS IN ENVIRONMENTAL POLICY 40 (1997).

[103] The prior Bush Administration had taken a similar though less aggressive position. *See, e.g.*, Richard B. Stewart & Jonathan B. Wiener, *The Comprehensive Approach to Global Climate Policy: Issues of Design and Practicality*, 9 ARIZ. J. INT'L. & COMP. L. 83 (1992).

[104] *See* Press Release, White House Office of the Press Secretary, President Clinton to Participate in White House Conference on Climate Change (Oct. 2, 1997) (on file with the Harvard Environmental Law Review).

[105] Andrew C. Revkin, *Odd Culprits in Collapse of Climate Talks*, N.Y. TIMES, Nov. 28, 2000, at F1.

the coordinating body for economic policy during the Clinton years.[106] In contrast, EPA was relatively disengaged on this issue.

The NEC was created by Clinton to coordinate the development and implementation of the Administration's major domestic and international economic policies. During the Clinton years, the Council of Economic Advisers ("CEA") continued to provide economic analysis, forecasting, and advice on the topics of regulatory reform and the environment, as well its traditional areas of expertise. The NEC acted for the White House as a coordinating filter and organizer of information from agencies engaging in economic policy throughout the administration, including the CEA.[107]

CEA testimony on this and many other occasions emphasized the enormous cost savings that could be achieved through emissions trading and through participation by developing countries, possibly contributing to the passage of Senate Resolution 98.[108] In addition, in its 1998 report on the costs of complying with the Kyoto Protocol, the CEA resisted pressure to adopt overly optimistic assumptions about technological change and energy efficiency advanced by the so-called "DOE Five Lab study" and by the Interagency Analytical Team study on the economic effects of global climate change policies.

B. Support for the Cost-Effectiveness Criterion from Congress

In 1995, the 104th Congress enacted the Unfunded Mandates Reform Act.[109] The main purpose of the Act was to require quantitative assessment of benefits, and comparison of benefits with costs for proposed and final rules with expected costs of $100 million or more to state, local, and tribal governments or to the private sector. The Act also mandated that agencies choose the least-cost regulatory alternative, or explain why they have not done so.[110]

C. Mixed Results on Cost-Effectiveness of Specific Policies

Integration of the cost-effectiveness criterion into environmental policymaking made more progress than the efficiency criterion in the 1990s.

---

[106] The major role of the economic agencies in developing U.S. climate change policy began at least as early as July 1997, when then-Chair of the CEA, Janet Yellen, testified before the House Commerce Committee, Subcommittee on Energy and Power. Statement Before the Senate Committee on Environment and Public Works (July 17, 1997), *available at* www.senate.gov/~epw/105th/yell7-17.htm (last visited Apr. 25, 2003) (on file with the Harvard Environmental Law Review).

[107] Orszag et al., *supra* note 34, at 995.

[108] S. Res. 98, 105th Cong. (1997). The "Byrd-Hagel resolution" stated that the United States should not approve any agreement at the Third Conference of the Parties in Kyoto, that did not impose binding emission reduction targets on major developing countries as well as industrialized nations.

[109] Pub. L. No. 104-4, 109 Stat. 48 (codified in scattered sections of 2 U.S.C.).

[110] 2 U.S.C. § 1535 (2000).

We consider implementation of the 1990 CAA Amendments during the decade as a case study.

1. Implementation of the 1990 CAA Amendments

While the judiciary in the 1990s upheld CAA provisions preventing EPA from taking costs into account when setting the NAAQS, the 1990 Amendments provided the basis for implementation of cost-effective regulation. Under Title IV of the amendments, Congress directed EPA not to mandate specific pollution control technologies for sulfur dioxide ("SO_2") emissions from power plants, but set up instead a permit trading system.[111] Not all regulations promulgated under the 1990 CAA Amendments were equally as cost-effective, however. The Amendments explicitly required EPA to issue technology standards for 188 toxic air pollutants, perhaps one of the most expensive and least cost-effective components of the statute.[112]

a. Market-Based Instruments in CAA Amendment Implementation

EPA provided averaging, banking, and trading opportunities for most of the new standards promulgated under the 1990 CAA Amendments, including those aimed at mobile sources. EPA's implementation of the reformulated gasoline provisions of Title II of the Amendments allowed refinery-level trading of oxygen, aromatics, and benzene content.[113] Title II also authorized EPA to regulate particulate matter, NO_x, and other emissions from heavy-duty trucks. The resulting regulations were promulgated at the vehicle engine-manufacturing level, and allow averaging, banking, and trading.[114] The Tier 2 emissions standards for cars and light-duty trucks, issued in February 2000, allow vehicle manufacturers to average NO_x emissions throughout their fleets to meet the new national tailpipe standards. They also allow refiners and gasoline importers to average, bank, and trade gasoline sulfur content to meet new Tier 2 standards.[115]

With respect to stationary sources, the regional NO_x cap-and-trade program in the Northeast is another significant market-based policy in-

---

[111] Paul R. Portney, *Air Pollution Policy*, *in* PUBLIC POLICIES FOR ENVIRONMENTAL PROTECTION 77, 89 (Paul R. Portney & Robert N. Stavins eds., 2000).

[112] Paul R. Portney, *Policy Watch: Economics and the Clean Air Act*, J. ECON. PERSP., Fall 1990, at 173, 178.

[113] U.S. EPA, *supra* note 100, at 88. The initial guidance for the reformulated gasoline trading programs was issued in October 1992, during the Bush Administration. Trading at the refinery level has been very active.

[114] *Id.* at 89. While a great deal of averaging and banking has taken place, only one trade was completed through 2000.

[115] *Id.* The average sulfur content cap drops annually between 2004 and 2006, and credits produced within that time frame have a limited life, while credits produced after the introduction of the strictest standard (2006) have unlimited life.

strument developed and implemented under the 1990 CAA Amendments. Although the SO_2 allowance trading program was created under the Bush administration, implementation of Phase I and Phase II occurred during the 1990s. These two programs are described below, as are two significant rulemakings that have been more heavily criticized from an economic perspective: the revised NAAQS for ozone and particulate matter; and new regulations on toxic air pollutants.

b. SO_2 Allowance Trading

The tradeable permit system that regulates SO_2 emissions, the primary precursor of acid rain, was established under Title IV of the CAA Amendments of 1990. The statute is intended to reduce SO_2 and NO_x emissions from 1980 levels by ten million tons and two million tons, respectively.[116] The first phase of SO_2 emissions reductions was started in 1995, with a second phase of reduction initiated in the year 2000.[117]

A robust market of bilateral SO_2 permit trading emerged in the 1990s, resulting in cost savings on the order of $1 billion annually, compared with the costs under some command-and-control regulatory alternatives.[118] Although the program had low levels of trading in its early years,[119] trading levels increased significantly over time.[120]

c. Regional NO_x Budget Program

Under EPA guidance, twelve northeastern states and the District of Columbia implemented a regional NO_x cap-and-trade system in 1999 to reduce compliance costs associated with the Ozone Transport Commission ("OTC") regulations of the 1990 CAA Amendments.[121] Required reductions

[116] *See* Brian L. Ferrall, *The Clean Air Act Amendments of 1990 and the Use of Market Forces to Control Sulfur Dioxide Emissions*, 28 HARV. J. ON LEGIS. 235, 241 (1991).

[117] In Phase I, individual emissions limits were assigned to 110 plants, located largely at coal-fired power plants east of the Mississippi River. Under Phase II of the program, beginning January 1, 2000, all electric power generating units greater than 25 MW burning fossil fuels were brought within the system. Dallas Burtraw, *The SO_2 Emissions Trading Program: Cost Savings Without Allowance Trades*, 14 CONTEMP. ECON. POL'Y, at 79, 82 (1996).

[118] Curtis Carlson et al., *Sulfur Dioxide Control by Electric Utilities: What Are the Gains from Trade?*, 108 J. POL. ECON. 1292 (2000).

[119] *See* Burtraw, *supra* note 117, at 82.

[120] *See* R. Schmalensee et al., *An Interim Evaluation of Sulfur Dioxide Emissions Trading*, J. ECON. PERSP., Summer 1998, at 53; Robert N. Stavins, *What Can We Learn from the Grand Policy Experiment? Lessons from SO_2 Allowance Trading*, J. ECON. PERSP., Summer 1998, at 69; Dallas Burtraw & Erin Mansur, *Environmental Effects of SO_2 Trading and Banking*, 33 ENVTL. SCI. & TECH. 3489 (1999).

[121] 42 U.S.C. §§ 7401–7671 (1970), *amended by* Pub. L. No. 101-549 (1990). Seven OTC states have also implemented state-level NO_x trading programs: New Jersey, Connecticut, Delaware, New York, Massachusetts, New Hampshire, and Maine. *See* Barry D. Solomon, *New Directions in Emissions Trading: The Potential Contribution of New Insti-*

are based on targets established by the OTC and include emissions reductions by large stationary sources. The program is known as the Northeast Ozone Transport Region.[122]

EPA distributes NO_x allowances to each state, and states then allocate allowances to sources in their jurisdictions. Each source receives allowances equal to its restricted percentage of 1990 emissions, and sources must turn in one allowance for each ton of NO_x emitted over the ozone season. Sources may buy, sell, and bank allowances. Potential compliance cost savings of 40% to 47% have been estimated for the period 1999–2003, compared with a base case of continued command-and-control regulation without trading or banking.[123]

d. Maximum Available Control Technology for Air Toxics

The air toxics regulations necessitated by the 1990 CAA Amendments could be among the least cost-effective components of the CAA, depending on how they are implemented. The Amendments mandated that EPA issue standards for 188 toxic air pollutants, substances that are less common than the criteria pollutants for which NAAQS are promulgated, but may pose threats to human health.

Unlike in the case of the NAAQS, however, the Administrator of EPA is directed to require the maximum degree of emissions reduction achievable, taking costs into consideration. Despite the fact that EPA is allowed to take costs into account when determining standards for hazardous air pollutants, the type of regulation required by the CAA Amendments is a technology standard—Maximum Achievable Control Technology—not a market-based approach. From 1992 through August 2000, EPA issued technology standards for 45 of these substances, covering 82 categories of industrial sources.

While there are no estimates of the total monetized costs and benefits of this new set of technology standards for hazardous air pollutants, one analyst in 1990 estimated that when fully implemented, compliance costs would range from $7.9 to $13.2 billion per year, and benefits would range from $0 to $5.3 billion per year.[124] The lower bound of zero on potential benefits is indicative of the considerable uncertainty over risks posed by these pollutants to human health. Some analysts have been particularly

tutional Economics, 30 ECOLOGICAL ECON. 371 (1999).

[122] *See* Alex Farrell et al., *The NO_x Budget: Market-Based Control of Tropospheric Ozone in the Northeastern United States*, 21 RESOURCE & ENERGY ECON. 103 (1999).

[123] *Id.* at 117.

[124] *See* Portney, *supra* note 112, at 178–79. These figures were Portney's "educated guess" in 1990, based on the George H. W. Bush Administration estimates and those of a 1990 consulting firm study. We have converted them to 2000 dollars, assuming that they were originally stated in 1990 dollars.

critical of EPA's very conservative estimates of risks to human health from air toxics in its promulgation of standards.[125]

The mix of market-based and command-and-control regulations within the 1990 CAA Amendments demonstrates that while cost-effective-ness was increasingly accepted by the Administration and Congress, application to actual policies was inconsistent. In reality, market-based policy instruments are used to implement only a very small fraction of environmental regulation in the United States.

2. *Cost-Effectiveness of Selected EPA Regulations*

Most of the "stock" of regulations currently on the books were created without regard to choosing least-cost compliance alternatives, and the cost-effectiveness criterion influences only a small portion of the "flow" of regulations. To keep this fact firmly in mind, we provide the cost per statistical life saved of selected EPA rules from the 1980s and the 1990s in Table 4.

IV. Increasing Role of Distributional Equity

The increase in attention to efficiency and cost-effectiveness in environmental regulation is correlated with the substantial increase in the cost of such regulations to the U.S. economy from the 1970s through the 1990s.[126] There has also been an increase in the benefits of environmental regulation over the same period. The third theme in our analysis suggests that as both costs and benefits of environmental and natural resource regulation have increased, attention to the *distribution* of these costs and benefits has increased as well.

A. *Environmental Justice and the Distribution of Environmental Benefits*

In addition to requiring RIAs, Clinton's EO 12,866 instructed agencies to select regulatory approaches that would maximize net benefits, *including distributive impacts and equity*, unless a statute required otherwise.[127] This was the first time that distributional concerns had been included within the series of Presidential EOs dealing with regulatory analysis.

---

[125] *See* Richard L. Stroup, *Air Toxics Policy: Liabilities from Thin Air*, *in* Cutting Green Tape: Toxic Pollutants, Environmental Regulation and the Law 59 (Richard L. Stroup & Roger E. Meiners eds., 2000); George M. Gray & John D. Graham, *Risk Assessment and Clean Air Policy*, 10 J. Pol'y Anal. & Mgmt. 286 (1991).

[126] *See* Paul R. Portney, *Counting the Cost: The Growing Role of Economics in Environmental Decisionmaking*, Env't, Mar. 1998 at 14; Adam B. Jaffe et al., *Environmental Regulation and the Competitiveness of U.S. Manufacturing: What Does the Evidence Tell Us?*, 33 J. Econ. Literature 132 (1995).

[127] Exec. Order No. 12,866, 58 Fed. Reg. 51,735 (Sept. 30, 1993).

Increased attention to equity concerns during the 1990s was frequently characterized under the rubric of "environmental justice." In 1994, EO 12,898 instructed federal agencies to identify and address "disproportionately high and adverse human health or environmental effects of its programs, policies, and activities on minority populations and low-income populations."[128]

In practice, agencies have responded to the two EOs by including a separate distributional impact analysis within RIAs. Subsequent to EO 12,898, environmental justice was mentioned in RIAs for rules in which agencies were required to address the issue, but only infrequently was quantitative analysis included.[129] In no case did the Administration's explicit concern for equity clearly alter proposed policies.

B. Property Rights Movement and the Distribution of Regulatory Costs

Increased attention to the distribution of the costs of environmental and natural resource regulation in the 1990s was exemplified by the rise of the "property rights" movement, concerned with costs to private landowners, especially in Western states, of laws such as the Endangered Species Act ("ESA")[130] and wetlands regulations under Section 404 of the CWA.[131] In addition, concern about the distribution of costs may partly underlie continued inefficient subsidization of natural resource extraction during the 1990s.

1. Endangered Species Act

The distributional implications of the ESA were the focus of much debate during the 1990s. Private landowners objected to restrictions they claimed amounted to *de facto* seizures of private property ("takings") under the Fifth Amendment to the U.S. Constitution. Such interpretation of regulatory restrictions on private land use under the ESA as "takings" has generally not been upheld by the courts, but from an economic perspective, the concern of private property owners that they bear the costs of public goods provision is a distributional issue.

Attempts to reauthorize the ESA in the 1990s failed, but the Clinton administration made substantive administrative changes, aimed at rationalizing the incentives for private landowners under the Act.

The Administration implemented four provisions that had been included within many of the unsuccessful Congressional reauthorization attempts and had broad bipartisan support. First, the Administration em-

---

[128] Exec. Order No. 12,898, 59 Fed. Reg. 7,629 (Feb. 11, 1994).

[129] In some cases, RIAs mention that distributional impact analysis was conducted, but the analysis is not presented.

[130] 16 U.S.C. §§ 1531–1543 (2000).

[131] 33 U.S.C. § 1344 (2000).

phasized habitat conservation plans ("HCPs") as a tool to manage endangered and threatened species on non-federal lands. Under Section 10 of the ESA, private landowners applying for an "incidental take" permit must submit a HCP, in which they agree to restrict some uses in the interest of species and habitat protection in exchange for the permit.[132] More than 250 HCPs were completed between 1992 and 2000, compared to 14 between 1982 and 1992.[133] HCPs are considerably more flexible than direct enforcement of the Act. Second, voluntary "safe harbor" agreements guarantee that increases in species populations on private lands will not restrict future land use decisions.[134] Third, the "no surprises" rule guarantees that a landowner properly carrying out a habitat conservation plan will not experience further restrictions or costs without mutual consent. Fourth, "candidate conservation agreements" allow landowners to protect declining species that are not yet listed, in exchange for assurance that no additional measures will be required if species are listed.[135] The changes had broad bipartisan support in Congress.

2. *Wetlands Regulation*

The debate over land-use restrictions governed by wetlands regulation under Section 404 of the CWA in the 1990s was similar in nature to the ESA "takings" debate. Congress did not pass any major changes to federal wetlands regulation, although a series of actions by the Clinton administration during the decade exemplify conflicts over distributional concerns within the regulatory framework. In 1998, the Army Corps of Engineers greatly reduced the scope of nationwide permit 26, which authorizes discharges into non-tidal headwaters and isolated waters, a change that resulted in lawsuits by the development and commercial communities.[136] In addition, the Clinton administration endorsed the concept of wetlands mitigation banking in 1993. Mitigation banking would likely reduce the costs of wetlands regulation to private land owners and developers, but it has been opposed by environmental advocacy groups on the grounds that it does not adequately protect these ecologically valuable areas.

---

[132] 16 U.S.C. § 1539(a) (2000).

[133] Timothy Beatley, *Habitat Conservation Plans: A New Tool to Resolve Land Use Conflicts*, LAND LINES (Lincoln Inst. of Land Policy) Sept. 1995.

[134] *See* EUGENE H. BUCK ET AL., ENDANGERED SPECIES: DIFFICULT CHOICES 13 (CRS Issue Brief for Congress IB10072, 2003).

[135] *Id.*

[136] *See* COPELAND, *supra* note 88. The so-called "nationwide permits" authorize landowners to proceed with specified categories of activities without obtaining individual permits, reducing regulatory burdens.

3. Natural Resource Extraction Subsidies

Within its first budget proposal to Congress, the Clinton administration proposed reducing a variety of natural resource extraction subsidies, including those for logging, mining, and grazing livestock on public lands. These efforts were opposed vigorously by advocates of the "property rights" movement. Congress opposed all of the natural resource initiatives in the Clinton proposal, with one exception: the 104th Congress established a framework for user fee demonstration projects within the National Park Service.[137]

C. Efficiency and Equity as Issues of Political Convenience

The Clinton administration's focus on environmental justice in the 1990s could be seen as the desire of a Democratic administration to reach out to minority and low-income communities. The Administration's many attempts to introduce greater efficiency in natural resource management through subsidy reduction could be seen as an attempt to support efficiency where efficient policies were in close alignment with the preferences of the environmental community, a strong base of Democratic support.[138]

Similarly, Congressional opposition to natural resource subsidy reduction, when compared with its strong support for efficiency in environmental pollution control regulation, could be seen as the desire of a Republican legislature to forward the interests of supporters in the regulated community, typically conservative voters. Congressional support for extensive subsidies to grazing, timber extraction, mining, and other activities expanded the message of regulatory reform from the traditional industry association community to working-class, resource-based com-

---

[137] Omnibus Consolidated Rescissions and Appropriation Act of 1996, Pub. L. No. 104-134, 110 Stat. 1321 (1996). Congress also opposed, in one important case, the application of the cost-effectiveness criterion to natural resource management. The Sustainable Fisheries Act of 1996, 18 U.S.C. § 1853(d)(1) (2000), amended the Magnuson-Stevens Fishery Conservation and Management Act, 16 U.S.C. § 1881d(e) (2000), imposing a four-year moratorium on new individual transferrable quota programs among the nation's eight regional fishery management councils and repealing one such program that had been created in 1995. *See* Eugene H. Buck, *Magnuson Fishery Conservation and Management Act Reauthorization*, (Congressional Research Service Issue Brief for Congress IB95036, 1996), *available at* http://www.ncseonline.org/nle/crsreports/marine/mar-3.cfm (last visited Apr. 25, 2003). The Act did not, however, repeal the five other existing ITQ programs.

[138] The views of economists on natural resource extraction and pricing are closely aligned with those of strict conservationists, while economists' views on pollution control often contradict those of strict conservationists. That is, current rates of natural resource extraction in many countries are likely greater than the efficient rates, due to substantial subsidies and unregulated negative externalities. Thus, the economist's call for efficiency in resource management often supports higher prices and slower extraction. In contrast, the economist's call for efficiency in environmental regulation may often support a decrease in existing pollution control standards, as most industrialized countries have experienced a period of increasing stringency of environmental pollution control regulation over the past thirty years, and some of this regulation may have costs that exceed associated benefits.

munities, particularly in the Western United States. Congress in the 1990s appears to have supported efficiency when efficient policies were in close alignment with the preferences of its conservative base.

The notion of using benefit-cost analysis as a guide to regulation for environmental protection and natural resource management does not appeal to most interest groups or policy partisans, except where it is seen as a tool to achieve pre-determined goals. Politicians may thus endorse the use of the efficiency criterion only where its results are likely to be compatible with their own ideological agendas. The inconsistent application of efficiency analysis to environmental and natural resource regulation in the 1990s is part of a wider pattern of focus on the distribution of the costs and benefits of environmental and natural resource regulation in the United States.

D. Distribution Becomes More Salient as the Economic Impacts of Policies Increase

The tremendous increase in the aggregate costs and benefits of environmental and natural resource regulation over the past thirty years has focused substantial attention on the efficiency and cost-effectiveness of regulation. In addition, the presence of large costs and benefits from regulation has focused the attention of lawmakers and other participants in the policy process on the distribution of these costs and benefits.

Where pollution damages are highly localized, regulations that set aggregate standards for pollution emissions or concentrations can have differential distributional impacts that may be unappealing on equity grounds.[139] Policies that restrict natural resource management alternatives have inherently differential distributional impacts in the United States, where economic dependence upon resource extraction is highly localized. Even where it may be efficient to proscribe specific commercial activities or other resource uses from a national perspective, some local communities will experience substantial net losses from such policies.

An example may be the USFS Roadless Areas Initiative ("Roadless Rule"). The USFS regulatory impact analysis for the rule did not quantify benefits and costs. Hence, no definitive efficiency conclusions can be drawn. But inventoried roadless areas comprise about two percent of the U.S. landmass, and thirty-one percent of the USFS's property. These areas are characterized by rugged terrain and low-value timber, and they may be ecologically sensitive. These characteristics may suggest relatively low costs to leaving them in their current state, and relatively high

---

[139] Uneven distributional impacts can have implications for the efficiency of a regulation, as well, if damages are nonlinear. If marginal damages increase at an increasing rate, total damages (hence total benefits of regulation) may increase when damages are concentrated in certain areas.

environmental benefits of preservation.[140] Nonetheless, any reduction in commercial timber harvest associated with the Roadless Rule negatively affects some communities.[141]

Given that natural resource management regulations will necessarily have uneven distributional impacts, Congressional opposition to increasing efficiency and cost-effectiveness in natural resource management during the 1990s is not surprising. When the "winners" from a natural resource management policy are American citizens as a whole and the "losers" are identifiable members of particular Congressional districts, members of Congress are reluctant to impose those losses on their own district or a colleague's district. Similarly, as the substantial gains from thirty years of environmental pollution control regulation have been seen to accrue disproportionately to some communities over others, the debate has shifted somewhat from efficiency to distributional equity.

The implications of the increased focus on distribution in environmental and natural resource policy are twofold from the perspective of economics. First, while economists can analyze the distribution of costs and benefits from a regulation, they have little to contribute to the debate over how costs and benefits *should* be distributed. Second, in some cases, attempts to meet distributional goals (whether they succeed or not) may interfere with attempts to satisfy criteria of efficiency and cost-effectiveness.

V. Conclusions

Three conclusions emerge from our review of the role of economic analysis in environmental and natural resource policy during the 1990s. First, the use of efficiency as a criterion for assessing environmental and natural resource rules and regulations was controversial in the Clinton administration, while economic efficiency emerged as a central goal of the regulatory reform movement in Congress. Second, cost-effectiveness as a criterion for adopting specific policy instruments was embraced by

---

[140] Clinton Forest Service Chief Mike Dombeck pointed out that these areas were the 58.5 million acres of Forest Reserves created between 1891 and 2000, many of which had remained roadless through twenty presidencies. In addition, by USFS calculations, less than 0.3% of the U.S. timber harvest and less than 0.4% of U.S. oil and natural gas reserves will be affected by the Roadless Rule. Mike Dombeck, *Roadless Area Conservation: An Investment For Future Generations*, *at* http://roadless.fs.fed.us/documents/rule/dombeck_stmt.htm (last visited Apr. 25, 2003) (Jan. 5, 2001) (on file with the Harvard Environmental Law Review). Any benefit-cost calculation would also have to account for the costs of maintaining forest system roads. In 2000, USFS maintained a road system of more than 386,000 miles, with a maintenance backlog in excess of $8 billion. *Id.*

[141] The state of Idaho, the Kootenai Indian tribe, and logging groups challenged the Roadless Rule in federal court. In May 2001, a U.S. District Court judge in Idaho issued a preliminary injunction blocking the rule. Kootenai Tribe of Idaho v. Veneman, 142 F.Supp.2d 1231 (D. Idaho 2001). The Bush administration declined to appeal the ruling. In December 2002, the U.S. Court of Appeals for the Ninth Circuit overturned the District Court ruling, reinstating the Roadless Rule. Kootenai Tribe of Idaho v. Veneman, 313 F.3d 1094 (9th Cir. 2002).

both the Administration and Congress in the 1990s. Most interest groups in the environmental community and the regulated community could support cost-effectiveness because it reduced the burden of compliance on industry and made stringent environmental targets more affordable. But benefit-cost analysis raised the issue of goals or standards, as well as costs, and the process of setting goals was, and is, inherently more controversial than minimizing the costs of achieving them.

Third, during the 1990s, equity concerns played increasing roles in environmental and natural resource policy debates. Both the efficiency and the cost-effectiveness criteria may be hard to swallow when the distributional impacts of regulation are highly skewed. Examples continue to surface regularly in debates over the fairness of policies such as individual transferable quota systems for fisheries management, differential exposure to environmental hazards, and impacts on western farming communities of reduced availability of irrigation water to protect endangered species. The focus on equity in environmental policy debates is likely to intensify as the costs and benefits of regulation continue to rise.

APPENDIX

TABLE 1. U.S. EMISSIONS OF SEVEN MAJOR POLLUTANTS, 1970–1998

Year	SO2	NOx	VOCs	CO	Lead	PM10	PM2.5
1970	100	100	100	100	100	N/A	N/A
1980	83	117	85	91	34	N/A	N/A
1989	75	114	73	82	3	100	N/A
1990	76	115	68	76	2	54	100
1991	74	116	68	78	2	53	97
1992	73	118	67	75	2	53	96
1993	72	119	67	76	2	50	92
1994	70	121	70	79	2	56	100
1995	62	119	67	72	2	48	90
1996	61	118	60	74	2	61	103
1997	63	119	61	73	2	63	107
1998	63	117	58	69	2	64	105

Notes: Figures are indexed from EPA data, with 1970 aggregate U.S. emissions equal to 100 for all pollutants except PM10 (1989=100) and PM2.5 (1990=100). Data for 1970 and 1980 drawn from U.S. EPA, Pub. No. 454/R-00-002, NATIONAL AIR POLLUTANT EMISSION TRENDS 1900-1998 (2000). Data for 1989, 1991–1995, and 1997 drawn from U.S. EPA, Pub. No. 454/R-00-003, NATIONAL AIR QUALITY AND EMISSIONS TRENDS REPORT, 1998 (2000). Data for 1990, 1996, and 1998 appear in both reports. (Data for PM10 differ between the two reports—for this pollutant, the 1998 Report data were used exclusively.) Data for particulate matter ("PM") include only directly emitted PM. No figures are shown for PM10 and PM2.5 in 1970 or 1980; while estimates exist, they do not include natural sources, agriculture, forestry, fugitive dust and other sources which together comprise almost ninety percent of directly emitted PM10 and almost seventy percent of directly emitted PM2.5 in 1990.

TABLE 2. SHIFTS IN ORGANIZATIONAL LOCATION OF ECONOMIC ANALYSIS AT EPA

Years	Location of Core Economics Staff at EPA
1980–1983	Benefits Staff, Office of Policy Evaluation, Office of Policy and Resource Management
1983–1987	Benefits Branch, Office of Policy Analysis, Office of Policy, Plan-ning and Evaluation
1987–1990	Economic Analysis Branch, Office of Policy Analysis, Office of Policy, Planning and Evaluation
1990–1996	Economic Analysis and Research Branch, Office of Policy Analysis, Office of Policy, Planning and Evaluation
1996–1999	Economy and Environment Division, Office of Economy and Environment, Office of Policy, Planning and Evaluation
1999–2000	Economic and Policy Analysis Division and Economy and Environment Division, Office of Economy and Environment, Office of Policy and Reinvention
2000–2001	National Center for Environmental Economics, Office of Policy, Economics and Innovation

Source: U.S. EPA, National Center for Environmental Economics World Wide Web site, *available at* http://www.yosemite.epa.gov/ee/epa/eed.nsf/pages/aboutncee#OrganizationalStructureandHistory (last visited Mar. 16, 2003) (on file with the Harvard Environmental Law Review).

TABLE 3. BENEFITS AND COSTS, REVISED NAAQS FOR OZONE AND PARTICULATE MATTER

NAAQS (1997)	**Annual Monetized Benefits**	**Annual Monetized Costs**
Ozone	$2.0 to $11.2 billion	$12.7 billion
Particulate Matter	$26.4 to $145 billion	$48.8 billion

Source: U.S. OMB, REPORT TO CONGRESS ON THE COSTS AND BENEFITS OF FEDERAL REGULATIONS (1998). EPA estimates were in constant 1990 dollars; those reported here are 2000 dollars. Cost and benefit estimates assume full attainment.

TABLE 4. COST OF SELECTED EPA REGULATIONS PER STATISTICAL LIFE SAVED

Environmental Protection Agency Regulation	Year	Cost per Statistical Life Saved (millions of 2000 $)
Benzene fugitive emissions	1984	5
Radionuclides at uranium mines	1984	11
Asbestos prohibitions: manufacture, importation, processing and distribution in commerce (total)	1989	21
National primary and secondary water regulations—Phase II: MCLs for 38 contaminants	1991	28
Hazardous waste management system—wood preservatives	1990	57
Sewage sludge use and disposal regulations, 40 CFR Part 503	1993	215
Land disposal restrictions for third scheduled waste	1990	215
Hazardous waste management system: final solvents and dioxins land disposal restrictions rule	1986	226
Prohibition on land disposal of first third of scheduled wastes ("second sixth" proposal)	1988	452
Land disposal restrictions, Phase II: universal treatment standards and treatment standards for organic toxicity, characteristic wastes, and newly listed wastes	1994	1,030
Drinking water regulations, synthetic organic chemicals, phase V	1992	10,800
Solid waste disposal facility criteria, 40 CFR Parts 257 and 258	1991	40,700

Source: ROBERT W. HAHN ET AL., DO FEDERAL REGULATIONS REDUCE MORTALITY? 16–17 (AEI-Brookings Joint Center for Regulatory Studies, Washington, D.C., 2000). "Cost per statistical life saved" refers to net costs (costs minus cost savings, but not taking into account benefits in terms of reduced mortality risk) of discounted lives saved. The estimates for the first two rules in the table (both 1984) are from W. Kip Viscusi, *Regulating the Regulators*, 63 U. CHI. L. REV. 1423 (1996), noting that all values are millions of 2000 dollars annually. These final rules are ranked in order of decreasing cost-effectiveness.

[3]

Corporate Social Responsibility Through an Economic Lens

Forest L. Reinhardt*, Robert N. Stavins**, and Richard H. K. Vietor***

Introduction

Business leaders, government officials, and academics are focusing more and more attention on the concept of "corporate social responsibility" (CSR). The central issue is the appropriate role of business. Everyone agrees that firms should obey the law. But beyond the law—beyond full compliance with environmental regulations—do firms have additional moral or social responsibilities to (voluntarily) commit resources to environmental protection?

One of the challenges of examining the concept of CSR is simply identifying a consistent and sensible definition from among a bewildering range of concepts and definitions that have been proposed in the literature.[1] We adopt a simple definition originally offered by Elhauge (2005): sacrificing profits in the social interest. This definition has the merit of being consistent with some of the most useful prior perspectives (Graff Ziven and Small 2005; Portney 2005; Reinhardt 2005), while focusing the discussion on the most interesting normative and positive questions.

Of course, questions regarding sacrificing profits in the social interest apply beyond the environmental sphere. The academic debate over the legality of sacrificing profits in the public interest appears to have begun in 1932 with opposing articles (Dodd 1932; Berle 1932) in a *Harvard Law Review* symposium on "For Whom Are Corporate Managers Trustees?" The debate in economics began more recently, with Milton Friedman's 1970 article, "The Social Responsibility of Business Is to Increase Its Profits," in the *New York Times Magazine*. Since then, the debate has continued, and CSR has received considerable attention from both scholars and the public, especially in the environmental protection area.

*Harvard Business School
**John F. Kennedy School of Government, Harvard University, Resources for the Future, National Bureau of Economic Research; E-mail: robert_stavins@harvard.edu
***Harvard Business School

Exceptionally valuable research assistance was provided by Matthew Ranson, and our research benefited greatly from conversations with William Alford, Max Bazerman, Robert Clark, Joshua Margolis, and Mark Roe. The authors are grateful to Suzanne Leonard, Charlie Kolstad, and an anonymous referee for valuable comments on a previous version of the manuscript, but all remaining errors are our own.

[1]See reviews by Wood and Jones (1996) and Mohr, Webb, and Harris (2001).

Review of Environmental Economics and Policy, volume 2, issue 2, summer 2008, pp. 219–239
doi:10.1093/reep/ren008
Advance Access publication on July 11, 2008

The purpose of this article, which is part of a three-article symposium on Corporate Social Responsibility and the Environment,[2] is to introduce and provide an overview of the major issues related to CSR, synthesize what is known about CSR in the environmental arena, and thereby identify where the greatest uncertainties remain. To this end, we address four key questions about the issue of firms sacrificing profits in the social interest.[3] *May* they do so within the scope of their fiduciary responsibilities to their shareholders? *Can* they do so on a sustainable basis, or will the forces of a competitive marketplace render such efforts and their impacts transient at best? *Do* firms, in fact, frequently or at least sometimes behave this way, reducing their earnings by voluntarily engaging in environmental stewardship? And finally, *should* firms carry out such profit-sacrificing activities? In other words, is this an efficient use of social resources?

This article is organized as follows. We begin by examining legal thinking about whether firms *may* sacrifice profits to benefit individuals other than their shareholders, and then look at the legality of CSR in the United States and other countries. Next, we draw on theories of industrial organization and management to identify circumstances under which firms *can* sacrifice profits without being punished by market forces. We then turn to positive questions about whether firms actually *do* engage in CSR. Here we review and synthesize empirical evidence to assess whether some firms truly exceed full compliance with the law, and if so, whether their "socially responsible" actions actually sacrifice profits. To address our fourth question, *should* firms—from a societal perspective—be carrying out such activities, we examine CSR in a normative light and consider economic arguments on both sides of the issue. The final section summarizes our findings and offers some conclusions.

May Firms Sacrifice Profits in the Social Interest?

The prevailing view among most economists and business scholars is that corporate directors have a fiduciary duty to maximize profits for shareholders. While this view underlies many economic models of firm behavior, its legal basis is actually not very strong. The judicial record, although supportive of a duty to maximize profits for shareholders, also leaves room for the possibility that firms may sacrifice profits in the public interest. The courts' deference towards the judgment of businesspeople—the "business judgment rule"—prevents many public-minded managerial actions from being legally challenged.

The Legal Purpose of the Corporation

The most widely accepted position on the legal purpose of the corporation—known as shareholder primacy (Springer 1999; Fisch 2006; Ehrlich 2005)—was articulated by Milton Friedman in 1970:

> In a free-enterprise, private-property system, a corporate executive is an employee of the owners of the business. He has direct responsibility to his employers. That responsibility is to conduct the business in accordance with their desires, which

[2]The other two articles in the symposium, by Lyon and Maxwell (forthcoming) and Portney (forthcoming), discuss CSR from the theoretical and empirical perspectives, respectively.
[3]These four questions were originally identified by Hay, Stavins, and Vietor (2005).

> generally will be to make as much money as possible while conforming to the basic rules of the society, both those embodied in law and those embodied in ethical custom (Friedman 1970).

A more subtle version of the shareholder primacy argument is the "nexus of contracts" approach (Jensen and Meckling 1976; Easterbrook and Fischel 1991), which views the corporation as a nexus of legal contracts between the suppliers of various factors of production, who agree to cooperate in order to generate monetary returns. These agreements specify that in exchange for their contributions, the owners of most factors of production—labor, land, intellectual property rights, etc.—will receive set payments with little risk. Shareholders—the suppliers of capital—accept the residual financial risk of doing business, and in return receive the residual profits. Since shareholders have no contractual guarantee of a fixed payment from the firm's activities, any profits that are diverted towards other activities, such as pursuit of "the social good," come directly out of their pockets (Butler and McChesney 1999). Thus, from this perspective, CSR is close to theft.

A second view of the role of the corporation is found in the team-production model (Blair and Stout 1999), which views the corporation as the solution to the moral hazard problem that arises when the owners of factors of production must make firm-specific investments but fear they will not be rewarded *ex post.* To solve this problem, the board of directors of the corporation functions as a neutral "mediating hierarch" that allocates residual profits to all of the factors of production (team members) according to their relative contributions.[4] Under the team-production model, sacrificing profits in the social interest is legal, as long as the profits are allocated to a deserving factor of production.

A third view of the purpose of the corporation is the "operational discretion" model, which holds that the law grants corporate managers discretion to comply with social and moral norms, even if doing so reduces shareholder profits (Elhauge 2005). The judiciary's unwillingness to second-guess matters of business judgment has the practical effect of shielding managers who choose to sacrifice profits in the public interest.

A fourth and final position is the "progressive view" that the corporation is organized for the benefit of society at large, or at the very least, corporate directors have fiduciary responsibilities that extend to a wide variety of stakeholders (Sheehy 2005; Gabaldon 2006). Under this view, sacrificing profits in the public interest is entirely legal. The progressive view, however, is not well rooted in either statutes or case law (Clark 1986).

The Legality of CSR in the United States

In the United States, a variety of legal requirements define the responsibilities of the corporation (and its board of directors) to shareholders and other stakeholders. However, as discussed below, these requirements are limited in practice.

[4] For example, many US states have enacted statutes that permit corporate directors to consider the interests of stakeholders other than shareholders.

Corporate Responsibilities to Shareholders and Other Stakeholders

Although corporations in the United States are granted the "legal fiction of separate corporate personality," a corporation's decisions are made by its board of directors, or by executives who have been delegated decision-making authority (Clark 1986). To ensure that directors and managers do not act negligently or subvert corporate resources for their own benefit, the legal system imposes fiduciary duties of care and loyalty.

The duty of loyalty requires directors to act "in good faith and in the best interests of the corporation" (Scalise 2005), and places limitations on the motives, purposes, and goals that can legitimately influence directors' decisions (Cox and Hazen 2003). The duty of care complements the duty of loyalty by requiring managers to "exercise that degree of skill, diligence, and care that a reasonably prudent person would exercise in similar circumstances" (Clark 1986, p. 123). Violation of fiduciary duties can result in personal liability for directors (Scalise 2005). Legal formulations of fiduciary duties typically refer to the "best interests of the corporation," but whether the corporation's "best interests" include only its shareholders or a wider set of constituents is not immediately clear (Cox and Hazen 2003). The prevailing opinion is that fiduciary duties are owed to shareholders (Blomquist 2006), but a minority supports the view that corporations can be managed in part for the benefit of other stakeholders (Lee 2005).

Every US state recognizes the right of businesses to make charitable contributions. Seven states allow charitable donations regardless of corporate benefit, and nineteen other states allow donations that benefit the business or advance the public welfare (Choper, Coffee, and Gilson 2004). Statutes in the remaining 24 states (including Delaware) include similar language, but without legal clarification about whether donations are permitted when they do not benefit the firm (Donohue 2005).[5]

State corporate statutes grant corporations legal powers similar to those of people, and allow corporations to participate in lawful activities (Clark 1986). As a result, corporations presumably have the power (but not necessarily the *right*) to undertake CSR activities (Donohue 2005). Corporations can write their own corporate charters to explicitly authorize themselves to participate in CSR. For example, the *New York Times* is incorporated to pursue objectives other than profit maximization (Donohue 2005).

These statutory requirements and judicial precedents place limits on the actions of corporations and their boards. But an important judicial construct—the business judgment rule—creates substantial deference to firms' managerial decisions.

The Business Judgment Rule

The business judgment rule "acts as a presumption in favor of corporate managers' actions" (Branson 2002). It requires courts to defer to the judgment of corporate managers, as long as their decisions satisfy certain basic requirements related to negligence and conflict of interest. The basic premise is that since corporate managers are far more skilled at making business judgments than courts, allowing courts to second-guess managers' decisions would create potentially large transactions costs (Elhauge 2005).

[5]In addition, twenty-nine states have statutes that allow managers to consider the interests of non-shareholders such as employees, customers, suppliers, creditors, and society at large (Springer 1999).

The business judgment rule makes fiduciary duties difficult to enforce, and it effectively grants managers discretion to "temper business decision making with their perceptions of social values" (Clark 1986; Fisch 2006; Scalise 2005; Blair and Stout 1999).[6] As a practical matter, as long as managers can plausibly claim that their actions are in the long run interests of the firm, it is almost impossible for shareholders to challenge the actions of managers who act in the public interest.

The business judgment rule also offers managers protection from accusations of conflict of interest, primarily because it does not recognize most nonfinancial incentives as conflicts (Elhauge 2005; Branson 2002). Corporate managers' decisions can be regarded as irrational—and thus not protected by the business judgment rule—only if they "go so far beyond the bounds of reasonable business judgment that their only explanation is bad faith" (Blomquist 2006, p. 699). Donohue (2005) cites the extreme example of a Delaware court that ruled that the business judgment rule protected the 1989 decision by Occidental Petroleum to spend $120 million, slightly less than half of the company's yearly net profit, on an art museum named after its 91-year-old CEO, Armand Hammer.

So, are firms in the United States prohibited from sacrificing profits in the public interest? And if so, is the prohibition enforceable? The answers to these two questions appear to be "maybe" and "no," respectively. "While case law falls short of unequivocally mandating shareholder wealth maximization, it also falls short of unambiguously authorizing the pursuit of non-shareholder interests other than instrumentally for the benefit of the shareholders" (Lee 2006, p. 557). And as long as managers claim some plausible connection to future profitability, the business judgment rule grants them substantial leeway to commit corporate resources to projects that benefit the public.

The Legality of CSR in Other Countries

With their cultural traditions of social democracy or firm loyalty to employees, most European countries and Japan have legal systems that differ from the system in the United States. The legal systems in these other countries place a greater emphasis on stakeholder participation, and sometimes codify this by legalizing various forms of profit-sacrificing behavior. Europeans have sought to incorporate CSR into their investment climate, both at the institutional and individual level (Sutton 2004), and in strong social democracies, such as Germany and France, stakeholders (particularly employees) have much stronger legal positions than in the United States (Roe 2000). Corporations in Europe and Asia are also more likely to have a few large shareholders, who may take social responsibilities seriously, particularly those towards employees (Roe 2000). This contrasts with the pattern of highly dispersed share ownership in the United States.

Industrialized Countries

Common law industrialized countries, primarily former British possessions, share many legal features with the United States. Corporations in these countries have similar board

[6]For example, Clark cites the 1968 case of Shlensky v. Wrigley, in which the Illinois Court of Appeals allowed William Wrigley, Jr., the president and majority shareholder of the Chicago Cubs, to refuse to install lights at Wrigley Field because of his belief that night games would be bad for the surrounding neighborhood (1986).

structures, face similar legal requirements, and even share some legal precedents. In such countries, CSR is discouraged, but permitted. Under Australian corporate law, for example, corporate managers are required to make decisions in the best interest of the corporation, while a statutory business judgment rule grants managers considerable discretion (Corfield 1998). Likewise, Canadian law requires that directors and officers of corporations act in the best interests of the corporation, but the director is not permitted to ignore the collective interests of shareholders (Borok 2003). The United Kingdom's legal system permits corporate managers to engage in socially beneficial activities, as long as there is a plausible rationale that the activities are in shareholders' interests (Lynch-Fannon 2007).

In contrast with common law jurisdictions, countries with civil law systems tend to place a greater emphasis on stakeholder participation in corporate governance. Corporate boards often include employee representatives, and cultural traditions emphasize loyalty to employees. In these countries, more forms of CSR are permissible. In France, corporate directors have both a duty of care and a duty of loyalty (Fanto 1998). Although there has been a shift towards more investor friendly laws, the French legal code explicitly allows directors to make decisions based on the interests of all constituencies. German law does not even give management an explicit obligation to maximize shareholder value (Marinov and Heiman 1998), and large German corporations have a two-tiered board structure that encourages the board to consider the interests of parties other than shareholders (Corfield 1998).

Japanese corporate law is similar to corporate law in the United States, in that directors have duties of care and loyalty, which, if violated, can be grounds for shareholder lawsuit. But Japanese corporations have a strong tradition of CSR oriented towards their employees. In years with high profits, large corporations usually retain their earnings and reinvest them for the benefit of employees (Miwa 1999). The shares of many firms are owned by banks who handle firms' credit or by important business partners (Corfield 1998). These shareholder-creditors have financial goals that are similar to those of long-term employees, particularly in terms of corporate stability and minimizing risk (Roe 2000).

Developing Countries and Multinational Enterprises

Corporate law in developing countries has a number of special characteristics. First, the corporate legal system is often new. As a result, businesses have little experience complying with the law, and there are fewer judicial precedents mapping out the law's boundaries. Second, legal institutions in developing countries are often weak. Regulations can go unenforced; agency problems can be a serious issue; and members of the judiciary may be corrupt. Third, the operations of multinational corporations in these countries can lead to conflicts between the interests of home and host states.

Thus, both the laws governing CSR and the degree to which those laws are enforced may vary substantially across developing countries. Assuming that the laws in most developing countries allow some scope for managerial activity that may sacrifice profits, the question remains whether firms can do so in view of competitive pressures in the markets for their outputs and inputs. It is to this question that we now turn.

Can Firms Sacrifice Profits in the Social Interest?

Just because the legal system may allow firms to sacrifice profits in the social interest does not mean that firms *can* do so on a sustainable basis in the face of competitive pressures. Under what conditions is it economically feasible for firms to sacrifice profits in the social interest? Before turning to this question, we address a somewhat broader question: under what conditions might it be sustainable for firms to produce goods and services, such as public goods, that benefit individuals other than their customers (Lyon and Maxwell 2004; Vogel 2006)?

We identify six conditions that would facilitate the production of such goods and services. All six of these conditions involve government intervention, imperfect competition, or both. First is the imposition of regulatory constraints that require a firm as well as its competitors to carry out some socially beneficial actions. Second is the possibility that such production is not costly to the firm. For example, restaurants frequently donate leftover food to homeless shelters. The third condition is that the socially beneficial actions may reduce a firm's business expenses by an amount greater than the cost of the actions themselves. For example, installation of energy-saving (climate friendly) technologies may generate long-term cost savings that outweigh upfront costs. Fourth, in some cases socially beneficial actions may yield an increase in revenue. It is easy to think of goods and services that are differentiated along environmental lines, such as clothing made of organic cotton, or wood from forests managed in accordance with some principles of sustainability. Socially beneficial actions could also generate goodwill, improving a firm's reputation and sales. Fifth, firms may choose to go beyond full compliance with environment, health, or safety laws in order to improve their position in current or future regulatory negotiations. By doing so, they may be able to deflect or influence future regulation or deflect enforcement of existing regulation. Sixth, some firms may use overcompliance to spur future regulation, which would provide a competitive advantage over less adaptable firms.

We now turn to our more restrictive definition of CSR and address the question raised above: under what conditions is it economically feasible for firms to sacrifice profits in the social interest?

When Is It Feasible for Firms to Engage in Profit-Sacrificing CSR?

In some cases firms undertake CSR actions voluntarily, while in others they engage in CSR only under pressure from market participants or other social forces. In practice, it is difficult to discern voluntary from "reluctant" CSR. Whether CSR initiatives are voluntary or reluctant, their economic sustainability depends on the market pressures and social expectations confronted by the firm (Borck, Coglianese, and Nash 2006).

Voluntary CSR

The first possibility—that stakeholders voluntarily sacrifice profits—is what some observers would think of as the "purest form" of CSR. The primary economic agents who could fund such activities are shareholders and employees.

Some shareholders may be willing to subsidize firms' profit sacrificing behavior. Stock issued by socially responsible firms is a composite commodity, which combines a financial investment product with a charitable giving vehicle (Graff Zivin and Small 2005). When investors purchase the stock, they may be motivated by self-interest or by altruistic motives. As long as investors are willing to fund CSR activities, firms can participate in them. But whether investors are willing to accept lower returns may depend on whether the firm already enjoys an economic position that allows it to obtain rents, such as through natural monopolies, niche markets, imperfect information, regulatory distortions, anti-takeover laws, and other market imperfections.[7] In this case, investors sacrificing profit may still earn returns above the market norm.

Willingness to accept below market returns may depend on whether investors hold stakes in publicly—or privately—held companies. Investors with large private holdings are more likely to take an interest in their companies' activities and be able to influence the companies' actions. Whether this additional interest and influence would have a positive or negative influence on CSR is an open question.

Evidence suggests that some individuals are willing to pay more for socially responsible goods (Jensen et al. 2002). The existence of such "ethical investors" could—in principle—have consequences for firms that do not participate in CSR activities (Heinkel, Kraus, and Zechner 2001). For example, if ethical investors' choices increase the cost of capital for "irresponsible" firms, some of these firms might be forced to adopt more socially responsible practices. If the share price differential becomes sufficiently large, these firms may decide to participate in CSR activities to increase their own stock price (Heinkel, Kraus, and Zechner 2001; Graff Zivin, and Small 2005). But the effect of green investors on the cost of capital may be small. Because irresponsible firms will generate higher returns (relative to their stock price), investors in these firms will accumulate capital more quickly than socially responsible investors, and over time may dominate the capital market. This would lead to a decrease in the cost of capital for irresponsible businesses (Heinkel, Kraus, and Zechner 2001).

Employees may sacrifice part of the returns to labor to further the social good. This could occur explicitly if employees are given the opportunity to use their own salary and benefits to fund CSR projects. For example, some executives may be able to channel part of their compensation towards the cost of CSR activities, or lawyers may be able to donate their time to pro bono work. Employees may also fund CSR implicitly, such as when a firm works in a field that employees perceive as socially responsible (e.g., providing services to the elderly, remediating oil spills). Employees may be willing to accept less than the fair market value of their labor (as determined by the wage they would receive for working in a less socially responsible industry), because they are compensated in other ways through the knowledge that their work benefits society at large (Frank 1996).

Unfortunately, empirical evidence on CSR and wages is inconclusive. Most revealed preference studies show that wages are lower at non-profit firms than at for-profit firms, but this non-profit wage penalty disappears in econometric analyses that control for worker and firm-specific characteristics (Francois 2004). If non-profit status is a proxy for social

[7] Firms have strong economic incentives to take advantage of any market power available to them. If a firm maintains market power, it can—in principle—pass on the costs of CSR to its suppliers and/or customers. For example, regulated public utilities, which are granted geographic monopolies on specific conditions such as provision of universal service, may decide to engage in CSR activities and use the firm's monopoly power to pass resulting costs on to consumers.

responsibility, then socially responsible firms may not enjoy a significant discount on labor prices. This conclusion is supported by findings that CEO compensation at firms listed in the Domini Social Index (DS 400) is not significantly different from CEO compensation at other firms in similar industries (Frye, Nelling, and Webb 2006).

Reluctant CSR

Corporate decisions are actually made by individual managers and directors, not by the "firm" *per se*. Those decisions often further the interests of profit-minded shareholders, but not always. Investors may have little choice but to accept some degree of CSR profit-sacrificing activities. It may be less costly to accept a degree of principal-agent "slack" than to eliminate it completely, because managers who are excessively constrained may be ineffectual.

Investors may also be forced to accept profit-sacrificing activities that are the result of external constraints. This may be particularly relevant in the developing world, where environmental regulatory standards lag behind those of industrialized countries. For example, equipment purchased from industrialized countries may incorporate pollution control technology that meets standards considerably stricter than those in effect in a developing country (Jaffe et al. 1995).

The magnitude of the profits that managers can sacrifice against investors' wishes depends on the structure of managers' compensation and the strength of shareholder oversight. Principal/agent problems can be costly. Managers have been observed to "satisfice" profits, that is, they seek to achieve an adequate rate of return for shareholders and then divert the firms' resources to their personal ends (Choper, Coffee, and Gilson 2004; Clotfelter 1985).

Unsustainable CSR

Under many conditions, firms that participate in costly CSR activities will have to raise prices, reduce wages and other costs, accept smaller profits, or pay smaller dividends—and accept the economic consequences. For example, a firm's stock price may decline until it is proportional to returns, and attracting new capital may be difficult because returns are below market averages. Other short-term economic consequences may include loss of market share, increased insurance costs, increased borrowing costs, and loss of reputation. In the long term, the firm may face shareholder litigation, corporate takeover, or closure. Such consequences simply illustrate the general proposition and observation that (financially) inefficient firms tend to disappear (Alchian 1950; Altman 1999).

This process of economic survival of the fittest suggests that firms that engage in unsustainable CSR may find themselves being pushed out of business. The forces of globalization only increase this pressure. Given the seemingly inevitable outcome of this process, why would any firm choose to participate in unsustainable CSR activities? First, principal/agent problems may lead managers to make decisions that commit the firm to short-term CSR actions, even if those activities will not be continued in the long run. Second, managers may misjudge the potential profitability of certain actions, leading them to invest in actions that benefit society but harm the firm's bottom line. Neither the managers' probability assessments nor their motivations are transparent to outside observers, making it very difficult to distinguish between them (Baron 2006).

Economic, Structural, and Organizational Constraints

A variety of factors influence the economic actors who make decisions about engaging in CSR activities. These factors include managerial incentive and monitoring constraints, and organizational structure and culture.

Whether or not firms are able and likely to engage in CSR depends on managers' incentives and constraints, which in turn are determined by managers' preferences, ethical beliefs, contracts, and goals. The most direct incentives managers face are their employment agreements. For managers whose compensation is designed to align their incentives with those of shareholders, sacrificing profits means reducing their own compensation. In the United States, chief executive officers (CEOs) are usually paid in a mixture of stocks, stock options, and salary, with their compensation linked to explicit measures of the firm's performance (Prendergast 1999). But the relationship between compensation and firm performance may be close to flat at some levels of firm performance, which means CEOs may be able to trade off compensation against CSR activities at a rate they judge acceptable.

Organizational culture may also be significant (Howard-Grenville, Nash, and Coglianese 2006). Organizational identity influences how individuals within the firm view the purpose of the firm, what it stands for, and its future goals. Organizational self-monitoring affects how an organization interacts with outside stakeholders. Firms that are more self-conscious about their image may expend greater effort to communicate and interact in "socially appropriate" ways than other firms, even if their core values related to socially beneficial behavior are similar.

Other factors may also affect whether firms can sacrifice profits in the social interest. For example, firm size appears to matter, with evidence that larger firms can sacrifice proportionately more profits (Adams and Hardwick 1998). Further, public visibility may increase pressure on firms to participate in CSR activities: firms in notoriously "dirty" industries may find themselves under heavy pressure from public advocacy groups to reduce their emissions or to participate in offsetting activities (Brown, Helland, and Kiholm Smith 2006). Finally, firms cannot participate in CSR if their work provides no scope for it. CSR activities are simply much more plausible for firms in certain industries (Porter and Kramer 2006).

Do Firms Sacrifice Profits in the Social Interest?

As described above, there are specific circumstances in which firms can sacrifice profits in the social interest without suffering serious adverse economic consequences. Whether they actually do so is another matter. This section discusses empirical evidence about the existence of such profit-sacrificing behavior.

Before interpreting the evidence, it is important to be aware of several challenges to making inferences about CSR. First, it is difficult to test whether firms' actions actually go beyond ordinary compliance with environmental regulations. Data on environmental performance are typically very limited, and because of the difficulty of observing appropriate counterfactuals, it is difficult to demonstrate that firms sacrifice profits. Whole industries often engage in CSR together, leaving behind no comparison group. Even when firms act individually, it is difficult to know whether unobservable characteristics explain differences in both socially responsible activity and profitability. Studies that link profitability to CSR practices are particularly vulnerable to this problem. For example, because many high-technology companies have low pollutant emissions (in contrast with firms engaged in electricity generation, heavy manufacturing, or resource extraction), the high-tech boom in the 1990s created a perceived but spurious correlation between market measures of "socially responsible business practices" and stock returns. Furthermore, as discussed above, there are a variety of ways in which firms can profit from investments in socially beneficial projects. Finally, the effects of many actions differ in the short versus the long term, with a short-term decrease in profits followed by a more-than-compensatory increase in the long-term profits. Thus, demonstrating that an action has truly sacrificed profits in the social interest is exceptionally difficult.

Of course, distinguishing between motivations and outcomes is even more difficult. Although most firms are likely motivated by a combination of social and financial concerns, managers may cite social responsibility as the motive for actions that were actually driven by profitability. Or managers may use profitability to justify socially responsible business choices, even when those choices result in smaller profits (Baron 2006).

Do Firms Overcomply?

A first step in evaluating whether firms participate in CSR is to determine whether they overcomply with regulations or participate in other costly activities that benefit society. We consider five sources of evidence: voluntary government programs, voluntary industry initiatives, voluntary action by individual firms, corporate charitable donations, and shareholder resolutions.

Voluntary Government Programs

In principle, the willingness of a firm to participate in a voluntary government program could be evidence of CSR activity. A variety of studies have evaluated the determinants of participation in voluntary government programs (e.g., Borck, Coglianese, and Nash 2006). Several patterns emerge. First, larger firms are more likely to participate in voluntary programs. Second, participation is more likely for firms that either produce final goods or experience more pressure from NGOs and consumers. Third, firms with higher emissions or poor compliance records are more likely to participate in voluntary programs. And fourth, participation may be positively influenced by factors such as industry association membership, R&D expenditures, organizational culture, and managerial discretion. However, there is no consensus that voluntary government programs have generated environmental benefits net of the opportunity cost of the resources required to implement them.

Voluntary Industry Initiatives

In addition to voluntary programs administered by governments, industry associations have created voluntary initiatives. For example, the Responsible Care program, established in 1989 by the US Chemical Manufacturers Association, requires participating facilities to adopt ten guiding principles and six codes of management practices related to the environmental and social dimensions of community interactions, facility management, and customer and supplier interactions. By and large, the program was ineffective because it did not provide strong incentives for compliance (King and Lenox 2000). Similarly, the Institute of Nuclear Power Operations (INPO) was created in the wake of the 1979 reactor meltdown at Three Mile Island, a nuclear power plant in Pennsylvania. A third example is Sustainable Slopes, a voluntary program for reporting and encouraging improved environmental performance at ski resorts. The evidence indicates that firms took advantage of positive publicity, although the actual environmental benefits are debatable (Rivera and de Leon 2004).

In general, industry-sponsored programs exhibit the same kinds of participation patterns as government-administered voluntary programs. That is, larger firms, more prominent firms, and firms with poorer environmental records are more likely to participate. Again, there is no systematic evidence of positive environmental impacts net of social costs.

Voluntary Action by Individual Firms

An indicator of firm participation in independently developed CSR activities is whether firms adopt CSR plans, environmental management systems, or other plans that seek to encourage socially beneficial decision making within the firm. These plans often have the nominal goal of taking a holistic management approach towards compliance with environmental and safety laws, contractual and voluntary environmental obligations, management of environmental and social impacts and risk, and other issues (Clark 2005). These systems may benefit firms by allowing them to manage the business aspects of environmental and social issues, but they may also serve as a mechanism for firms to improve environmental quality or otherwise benefit society.

One such mechanism is ISO 14001, an international standard that provides guidelines for monitoring environmental outputs, controlling environmental processes, and improving environmental performance (US Environmental Protection Agency 2006). To demonstrate that its environmental management system complies with the standard, a business (or any other organization) must receive a third-party audit. Capital intensity, intensity of competition, and dependence on overseas markets are all positively associated with voluntary compliance with the standard (Chapple et al. 2001).

The best source of evidence about whether firms participate in CSR activities on their own initiative is independent studies of socially responsible actions. Perhaps surprisingly, many studies of individual beyond-compliance behavior analyze firms in developing countries (e.g., Hartman, Huq, and Wheeler 1995; Hettige et al. 1996; Pargal and Wheeler 1996; Blackman and Bannister 1998; Dasgupta, Hettige, and Wheeler 2000). One possible reason for this focus is that firms in industrialized countries are subject to a wide range of environmental regulations that make it difficult to judge whether their actions are legally required, risk-averting, or voluntarily beyond compliance. In contrast, in the developing world, pollution regulations may be poorly enforced or even nonexistent, making it easier to identify individual beyond-compliance behavior.

Corporate Charitable Contributions

Evidence of corporations making financial contributions to charity supports the general hypothesis that corporations can and do commit corporate resources to CSR. Average contributions as a percent of net income before taxes increased, from less than 0.5 percent in the

1930s to 1.1 percent in the 1960s and 1970s (Harris and Klepper 1976). In general, CEOs and other high-level corporate officers have a high degree of control over the amount and destination of corporate charitable contributions, even if their company has established a separate charitable foundation (Kahn 1997). But charitable giving can be curtailed by debtholders (Brown, Helland, and Kiholm Smith 2006; Adams and Hardwick 1998). Overall, the evidence shows that charitable giving is more likely when financial and monitoring constraints are weak. Corporate charitable giving is also sensitive to firm income and marginal tax rates (Clotfelter 1985).

Shareholder Resolutions

Shareholders sometimes request that corporations comply with ethical or other requirements. In 2005, the shareholders of public US corporations proposed 348 resolutions on social and environmental issues, of which 177 reached a proxy vote (Social Investment Forum 2006). On average, these resolutions have received support from 10–12 percent of all votes cast. Of the 25 social policy resolutions in the United States that gained the highest percentage of votes during the years 2003–2005, only six gained a majority of all votes cast. But winning even a modest share of votes in a shareholder resolution can influence management policies.

Is There Evidence of Profit-Sacrificing Behavior?

According to our strict definition of CSR, beyond-compliance behavior is a necessary but not sufficient condition for CSR because, under some conditions, such behavior can be profitable. One way to measure the profit sacrificed by socially responsible companies would be to calculate the difference in profitability between firms that do and do not participate in socially responsible activities. In fact, a large literature, consisting of at least seventeen review articles, has explored this relationship.[8]

The most recent and comprehensive review is by Margolis, Elfenbein, and Walsh (2007). In a meta-analysis of the results from 167 studies of the relationship between financial performance and socially responsible business practices (ignoring the mechanism and direction of causality), they find that 27 percent of the analyses show a positive relationship, 58 percent show a non-significant relationship, and 2 percent show a negative relationship.[9] Margolis, Elfenbein, and Walsh argue that the evidence indicates that CSR, in general, has little effect on profitability. However, they note that there is stronger evidence to suggest some causality in the opposite direction: companies that are profitable are more likely to engage in more CSR activities.

The finding that there is little relationship between CSR and profitability is consistent with a market equilibrium in which firms invest in socially responsible projects until the marginal returns decline to the overall market rate of return. In this situation, investing in CSR is not profitable (in the sense that it does not generate economic rents), but neither is it a losing proposition. Instead it means that for most firms, CSR "pays for itself."

[8]See, for example, Aupperle, Carroll, and Hatfield 1985; Wood and Jones 1996; Griffin and Mahon 1997; Orlitzky, Schmidt, and Rynes 2003.

[9]Thirteen percent did not report a sample size that could be used to test significance.

These conclusions require a number of caveats. First, when evaluating studies of the relationship between social responsibility and profitability, it is important to keep in mind that not all companies that are classified as socially responsible actually sacrifice profits. Many operate in industries, such as software development, that by their very nature have little environmental or social impact. Second, many of the measures of CSR used in such studies are not consistent with CSR as we define it in this article. Thus, measured effects on profitability may have more to do with advertising, charitable contributions, or other tangentially relevant factors than with CSR.

In summary, evidence on sacrificing profits in the social interest is lacking. The bulk of the available evidence suggests that most firms view socially responsible actions in the same way that they view more traditional business activities, such as advertising and R&D. Instead of altruistically sacrificing profits, they engage in a more limited—but more profitable—set of socially beneficial activities that contributes to their financial goals. Hence, although proponents of sustainable business practices may argue that being environmentally responsible will inevitably lead to higher profits in the long term, the relationship between socially responsible activities and profitability may be best characterized as *some* firms will generate long-term profits from *some* socially responsible activities *some* of the time (Reinhardt 2000).

Should Firms Sacrifice Profits in the Social Interest?

Even if firms may, can, and do sacrifice profits in the social interest, an important normative question remains, namely, *should* they? In other words, is it really in the broadly defined social interest for firms to carry out such activity? There are two main approaches to answering this question. First, we can compare firms' actual CSR choices with the CSR alternatives available to them. For any firm, such alternatives include a broad range of projects addressing various private and public issues, costing different amounts, and resulting in varying degrees of environmental protection and profitability. For example, a power plant could reduce its emissions of carbon dioxide, sulfur dioxide, or particulate matter; switch to a renewable source of fuel; implement a job training program to benefit local community members; make a donation to a charitable organization; or take any number of other "socially responsible" actions. The question of interest here is whether firms' actual CSR choices are likely to be optimal relative to available alternatives.

A second approach takes a public policy perspective, where a comparison is made between allowing CSR (i.e., permitting firms to sacrifice profits in the social interest) and prohibiting CSR (i.e., requiring firms exclusively to maximize profits for shareholders). To evaluate these two approaches, we employ a variety of criteria, including social welfare and legal, political, and social considerations.

Social Welfare

In the context of CSR, the social welfare criterion suggests that: (1) firms should invest in projects that produce the highest level of social welfare; and (2) it is preferable to allow CSR if aggregate welfare is likely to be higher when CSR is allowed than when it is prohibited.

The benefits of CSR include direct welfare gains to individuals, such as asthmatics living near a power plant that voluntarily reduces its emissions. More broadly, if firms voluntarily internalize externalities, a more efficient allocation of resources may result. Of course, there is no reason, *ex ante*, to anticipate that firms will reduce externality-producing activities to efficient levels.

The direct costs of CSR are the loss of consumer surplus resulting from firms producing less output at higher cost and hence at higher prices. In addition, shareholders receive reduced financial returns. On the other hand, some shareholders may gain utility from the knowledge that their profits have been invested in socially responsible projects.

There are a number of reasons to believe that firms do not make socially optimal CSR investments, in the sense of choosing activities that generate the greatest net social benefits, subject to budgetary constraints. This is because firms' CSR decisions are influenced by a number of factors that are unrelated to social benefits and costs.

First, firms' CSR investment choices are influenced by managers' personal preferences and firm characteristics. For example, some managers may favor building art museums, while others favor the provision of affordable housing. This idiosyncratic element of personal preference is particularly likely if principal/agent issues drive CSR (Butler and McChesney 1999). Similarly, firms' choices about CSR activities are affected by the nature of their industry, firm size, technical capabilities, and relevant expertise, geographic location, and existing regulatory limits. To the extent that these factors are unrelated to the social benefits and costs of CSR, their influence on firm decisions about CSR may result in social inefficiency.

Second, although firms may be well informed about the private costs of CSR, they may have little experience evaluating its social benefits, leading them to choose inefficient levels of environmental protection effort. Third, firms may fail to consider alternative mechanisms to achieve their social goals. For example, firms may be able to achieve higher social returns by donating profits to charities, which are dedicated exclusively to the task of improving social welfare and thus presumably are well suited to the task. If this is the case, then firms that fund CSR activities effectively "crowd out" their own donations to more efficient charities (Graff Zivin and Small 2005). Finally, choice of CSR activity is affected by the firm's ability to sacrifice profits. Firms that are the most profitable are also the most able to sacrifice profits in the public interest. However, the opportunity cost of sacrificing profits may also be greatest for these firms, assuming they could otherwise invest the resources in their businesses and earn similarly high returns.

Although there are reasons to doubt the optimality of firms' decisions about CSR, there are also reasons to believe that firms' CSR investment decisions may increase welfare. First, firms have access to private information about their current and future pollution activities, including control costs. Such information can lead firms to identify better policies than less well-informed government agencies. Second, firms have relevant expertise and operational capacity. Third, government policies are driven by a variety of objectives, only one of which may be maximizing social welfare. Hence, compared with the counterfactual of prohibiting CSR but leaving government policy otherwise unchanged, allowing CSR may generate higher net social benefits.

Many types of potential CSR activities—from reducing particulate emissions to preserving open space—are mandated to some degree by federal, state, or local laws and regulations. To the extent that such regulations require a level of environmental protection that is below the socially optimal level, additional corporate investment in these activities can increase social welfare (if incremental social benefits exceed incremental social costs). In addition, there may be socially responsible activities that address environmental issues that are unregulated but of significant scientific or political concern (e.g., global climate change). In such cases (i.e., in the absence of government policies), CSR activities may lead to positive net social benefits. However, given that it appears to be relatively rare for firms to actually sacrifice profits in the social interest, the overall net welfare flow from CSR, whether positive or negative, is unlikely to be large.

Legal, Political, and Social Considerations

Although legality is not synonymous with social desirability (as evidenced by the legality of many socially undesirable activities), some observers would surely identify legality as a normative criterion by which to judge many actions. In the second section, we argued that in the United States and other common law countries, sacrificing profits in the social interest is not strictly legal, although in practice CSR is not prohibited because of the business judgment rule and problems of enforcement.

One argument that can be made against CSR is that it is not a democratic process. There is no particular reason to believe that society should prefer firms' choices and priorities to the choices and priorities of a democratic government. Some observers might also argue that corporations already dominate too many aspects of modern life, and that it would be undesirable for them to control the supply of public goods as well.

Under a broader interpretation of the idea of social responsibility, however, it can be argued that businesses have a moral commitment to hold themselves to higher ethical standards and to engage in activities that benefit society. In fact, in a poll of citizens' attitudes towards the responsibility of businesses in 23 developed and developing countries, public opinion seems to support the notion that corporations in the West should "set higher ethical standards and help build a better society." In countries such as China and Kazakhstan, however, the notion that corporations should "make profits, pay taxes, create jobs, and obey all laws" dominates (Environics International Ltd. 1999).

The Special Case of Developing Countries

Given that economic, social, and environmental conditions in developing countries are so different from those in industrialized countries, one would expect the answers to normative questions about CSR to also be different. For example, environmental regulations in the developing world are often not well enforced. Hence, many relatively cost-effective interventions that have already been implemented in industrialized countries may still be available to businesses that operate in the developing world. This suggests that CSR could lead to significant gains in net social welfare.

Other concerns about CSR arise in the developing country context. Precisely because legal and contractual systems often operate poorly in developing countries, it is important to prevent activities that could erode the basis for future economic growth. Thus, strong investor protections may be particularly desirable in developing economies (Marinov and Heiman 1998) to help buttress the political viability of privatization and market-based systems. Allowing managers to divert profits to socially responsible projects means giving managers substantial discretion. However, there is also the risk that this discretion could tempt managers to use corporate resources for personal gain.

Summary and Conclusions

This article has examined the concept of firms sacrificing profits in the social interest in the environmental realm. In this section, we summarize our answers to the four questions posed at the outset: *May* they do so within the scope of their fiduciary responsibilities to their shareholders? *Can* they do so on a sustainable basis? *Do* firms behave this way? And, finally, *should* firms carry out such profit-sacrificing activities?

Our starting point for examining the first question—may they—was the prevailing view among economists and business scholars that corporate directors have a fiduciary duty to maximize profits for shareholders. Surprisingly, the legal basis for this view is not very strong. Although the judicial record is supportive of a duty to maximize profits for shareholders, it leaves room for firms to sacrifice profits in the public interest. Moreover, the "business judgment rule" effectively protects many public-minded managerial actions from successful legal challenge.

Are firms in the United States prohibited from sacrificing profits in the public interest? And if so, is the prohibition enforceable? The answers to these two sub-questions appear to be "maybe" and "no," respectively. US corporate law is consistent with the shareholder primacy model, but as long as managers claim some plausible connection to future profitability, the business judgment rule grants them leeway to commit corporate resources to projects that benefit the public.

Just because the legal system may allow firms to sacrifice profits in the social interest does not mean that firms *can* do so on a sustainable basis in the face of competitive pressures. Under many conditions, firms that participate in costly CSR activities will have to raise prices, reduce wages and other costs, accept smaller profits, or pay smaller dividends—and accept the economic consequences. After taking such measures, a firm's stock price may decline until proportional to returns, and attracting new capital may be difficult because returns are below market averages. Other short-term economic consequences may include loss of market share, increased insurance costs, increased borrowing costs, and loss of reputation. In the long term, the firm may face shareholder litigation, corporate takeover, or closure.

This process of economic survival of the fittest suggests that firms that engage in unsustainable CSR may find themselves being pushed out of business. Given the seemingly inevitable outcome of this process, why would any firms choose to participate in unsustainable CSR activities? First, the firms that engage (or say they engage) in CSR are often active in markets that are imperfect or distorted by government intervention, so that they are protected from Friedman's evolutionary imperatives. Second, principal/agent problems may lead managers to make decisions that commit the firm to short-term CSR actions, even if those activities will not be continued in the long run.

Despite a large and growing literature on CSR, evidence of firms actually sacrificing profits in the social interest is lacking. The bulk of the available evidence suggests that most firms view socially responsible actions in the same way that they view more traditional business activities. Instead of altruistically sacrificing profits, they engage in a more limited—but more profitable—set of socially beneficial activities that contributes to their financial goals.

Although proponents of sustainable business practices may argue that being environmentally responsible will inevitably lead to higher profits in the long-term, the relationship between socially responsible activities and profitability may be best characterized as *some* firms will generate long-term profits from *some* socially responsible activities *some* of the time.

Is it in the social interest for firms to engage in CSR? More to the point, should governments allow such activity? To the extent that existing regulations require a level of environmental protection that is below the socially optimal level, additional corporate investment in CSR activities may increase social welfare. In this context, CSR should be viewed as a complement to, rather than a substitute for, increasingly effective government regulation.

References

Adams, Mike, and Phillip Hardwick. 1998. An analysis of corporate donations: United Kingdom evidence. *Journal of Management Studies* 35(5): 641–54.

Alchian, A. A. 1950. Uncertainty, evolution, and economic theory. *Journal of Political Economy* LVIII: 211–21.

Altman, Morris. 1999. The methodology of economics and the survivor principle revisited and revised: Some welfare and public policy implications of modeling the economic agent. *Review of Social Economy* 57(4): 427–49.

Aupperle, Kenneth, Archie Carroll, and John Hatfield. 1985. An empirical examination of the relationship between corporate social responsibility and profitability. *Academy of Management Journal* 28(2): 446–63.

Baron, David P. 2006. *Business and Its Environment*, Fifth Edition. New York: Pearson Prentice-Hall.

Berle, A. A. Jr. 1932. For whom corporate managers are trustees: A note. *Harvard Law Review* 45(8): 1365–72.

Blackman, Allen, and Geoffrey Bannister. 1998. (A community pressure and clean technology in the informal sector: An econometric analysis of the adoption of propane by traditional Mexican brickmakers. *Journal of Environmental Economics and Management* 35(1): 1–21.

Blair, Margaret M., and Lynn A. Stout. 1999. A team production theory of corporate law. *Virginia Law Review* 85(March): 247.

Blomquist, Robert. 2006. Six thinking hats for the lorax: Corporate responsibility and the environment. *Georgetown International Environmental Law Review* 18(4): 691–705.

Borck, Jonathan, Cary Coglianese, and Jennifer Nash. 2006. Why do they join? An exploration of business participation in voluntary environmental programs. In *Beyond Compliance: Business Decision Making and the US EPA's Performance Track Program." Regulatory Policy Program Report RPP-10.* Cambridge, MA: John F. Kennedy School of Government, Harvard University.

Borok, Tuvia. 2003. A modern approach to redefining in the best interests of the corporation. *Windsor Review of Legal and Social Issues* 15(March): 113.

Branson, Douglas M. 2002. The rule that isn't a rule—The business judgment rule. *Valparaiso University Law Review* 36(Summer): 631.

Brown, William, Eric Helland, and Janet Kiholm Smith. 2006. Corporate philanthropic practices. *Journal of Corporate Finance* 12(5): 855–77.

Butler, Henry N., and Fred S. McChesney. 1999. Why they give at the office: Shareholder welfare and corporate philanthropy in the contractual theory of the corporation. *Cornell Law Review* 84(July): 1195.

Chapple, Wendy, Andrew Cooke, Vaughn Galt, and David Paton. 2001. The determinants of voluntary investment decisions. *Managerial and Decision Economics* 22(8): 453–63.

Choper, Coffee, and Gilson. 2004. Cases and materials on corporations, 6th ed. London: Little, Brown & Co.

Clark, Matthew. 2005. Corporate environmental behavior research: Informing environmental policy. *Structural Change and Economic Dynamics* 16(3): 422–31.

Clark, Robert. 1986. *Corporate Law.* Boston, MA: Little, Brown & Co.

Clotfelter, Charles. 1985. *Federal Tax Policy and Charitable Giving.* Chicago, IL: University of Chicago Press.

Corfield, Andrea. 1998. The stakeholder theory and its future in Australian corporate governance: A preliminary analysis. *Bond Law Review* 10(2): 213.

Cox, James, and Thomas Hazen. 2003. Corporations. New York: Aspen Publishers.

Dasgupta, Susmita, Hemamala Hettige, and David Wheeler. 2000. What improves environmental compliance? Evidence from Mexican industry. *(A Journal of Environment Economics and Management* 39(1): 39–66.

Dodd, E. Merrick Jr. 1932. For whom are corporate managers trustees? *Harvard Law Review* 45(7):1145–63.

Donohue, John. 2005. Does greater managerial freedom to sacrifice profits lead to higher social welfare? In *Environmental Protection and the Social Responsibility of Firms*, eds. Bruce Hay, Robert Stavins, and Richard Vietor. Washington, DC: Resources for the Future.

Easterbrook, Frank, and Fischel Daniel. 1991. The economic structure of corporate law.

Ehrlich, Craig. 2005. Is Business ethics necessary? *DePaul Business & Commercial Law Journal* 4(Fall): 55.

Elhauge, Einer. 2005. Corporate managers' operational discretion to sacrifice corporate profits in the public interest. In *Environmental Protection and the Social Responsibility of Firms*, eds. Hay Bruce, Robert Stavins, and Richard Vietor. Washington, DC: Resources for the Future.

Environics International Ltd. 1999. The millennium poll on corporate social responsibility: Executive briefing. http://www.globescan.com/news_archives/MPExec Brief.pdf (accessed September 16, 2007).

Fanto, James A. 1998. The role of corporate law in French corporate governance. *Cornell International Law Journal* 31: 31.

Fisch, Jill E. 2006. Measuring efficiency in corporate law: The role of shareholder primacy. *Iowa Journal of Corporation Law* 31(Spring): 637.

Francois, Patrick. 2004. Making a difference: Labor donations in the production of public goods. CMPO, University of Bristol, Working Paper Series No. 04/093.

Frank, Robert. 1996. Can socially responsible firms survive in a competitive market? In *Codes of Conduct: Behavioral Research into Business Ethics*, eds. David Messick and Ann Tenbrunsel, 214–27. New York, NY: Russell Sage Foundation.

Friedman, Milton. 1970. The social responsibility of business is to increase its profits. *New York Times Magazine*. September 13.

Frye, Melissa B., Edward Nelling, and Elizabeth Webb. 2006. Executive compensation in socially responsible firms. *Corporate Governance: An International Review* 14(5): 446–55.

Gabaldon, Theresa A. 2006. Like a fish needs a bicycle: Public corporations and their shareholders. *Maryland Law Review* 65: 538.

Graff Zivin, Joshua, and Arthur Small. 2005. A Modigliani–Miller theory of altruistic corporate social responsibility. *B. E. Journals in Economic Analysis and Policy: Topics in Economic Analysis and Policy* 5(1): 1–19.

Griffin, Jennifer, and John Mahon. 1997. The corporate social performance and corporate financial performance debate: Twenty-five years of incomparable research. *Business and Society* 36(1): 5–31.

Harris, James, and Anne Klepper. 1976. Corporate philanthropic public service activities. New York: The Conference Board.

Hartman, Raymond, Mainul Huq, and David Wheeler. 1995. Why paper mills clean up: Determinants of pollution abatement in four Asian countries. World Bank Policy Research Department Working Paper 1710.

Hay, Bruce L., Robert N. Stavins, and Richard H. K. Vietor. 2005. The four questions of corporate social responsibility: May they, can they, do they, should they? In *Environmental Protection and the Social Responsibility of Firms*, eds. Bruce Hay, Robert Stavins, and Richard Vietor. Washington, DC: Resources for the Future.

Heinkel, Robert, Alan Kraus, and Josef Zechner. 2001. The effect of green investment on corporate behavior. *Journal of Financial and Quantitative Analysis* 36(4): 431–49.

Hettige, Hemamala, Mainul Huq, Sheoli Pargal, and David Wheeler. 1996. Determinants of pollution abatement in developing countries: Evidence from South and Southeast Asia. *World Development* 24(12): 1891–1904.

Howard-Grenville, Jennifer, Jennifer Nash, and Cary Coglianese. 2006. Constructing the license to operate: Internal factors and their influence on corporate environmental decisions. In *Beyond Compliance: Business Decision Making and the US EPA's Performance Track Program*. Regulatory

Policy Program Report RPP-10. Cambridge, MA: John F. Kennedy School of Government, Harvard University.

Jaffe, Adam B., Steven R. Peterson, Paul R. Portney, and Robert N. Stavins. 1995. Environmental regulation and the competitiveness of US manufacturing: What does the evidence tells us? *Journal of Economic Literature* 33(1995):132–63.

Jensen, Kim, Paul Jakus, Burt English, and Jamey Menard. 2002. Willingness to pay for environmentally certified hardwood products by Tennessee consumers. University of Tennessee, Department of Agricultural Economics, Study Series No. 01–02.

Jensen, Michael, and William Meckling. 1976. Theory of the firm: Managerial behavior, agency costs and ownership structure. *Journal of Financial Economics* 3(4): 305–60.

Kahn, Faith Stevelman. 1997. Pandora's box: Managerial discretion and the problem of corporate philanthropy. *UCLA Law Review* 44(February): 579.

King, Andrew, and Michael Lenox. 2000. Industry self-Regulation without sanctions: The chemical industry's responsible care program. *Academy of Management Journal* 43(4): 698–716.

Lee, Ian B. 2006. Efficiency and ethics in the debate about shareholder primacy. *Delaware Journal of Corporate Law* 31: 533.

Lee, Ian B. 2005. Corporate law, profit maximization, and the responsible shareholder. *Stanford Journal of Law, Business, and Finance* 10(Spring): 31.

Lynch-Fannon, Irene. 2007. The corporate social responsibility movement and law's empire: Is there a conflict? *Northern Ireland Legal Quarterly* 58(1).

Lyon, Thomas, and John Maxwell. 2004. *Corporate Environmentalism and Public Policy*. Cambridge, UK: Cambridge University Press.

Lyon, Thomas, and John Maxwell. Corporate Social Responsibility and the Environment: A Theoretical Perspective. *Review of Environmental Economics and Policy* doi:10.1093/reep/ren004.

Margolis, Joshua, Hillary Elfenbein, and James Walsh. 2007. Does it pay to be good? A meta-analysis and redirection of research on the relationship between corporate social and financial performance. Working Paper, Harvard Business School.

Marinov, Boris, and Bruce Heiman. 1998. Company law and corporate governance renewal in transition economies: The Bulgarian dilemma. *European Journal of Law and Economics* 6: 231–61.

Miwa, Yoshiro. 1999. CSR: Dangerous and harmful, though maybe not irrelevant. *Cornell Law Review* 84(July): 1227–54.

Mohr, Lois, Deborah Webb, and Katherine Harris. 2001. Do consumers expect companies to be socially responsible? The impact of corporate social responsibility on buying behavior. *Journal of Consumer Affairs* 35(1): 45–72.

Orlitzky, Marc, Frank Schmidt, and Sara Rynes. 2003. Corporate social and financial performance: A meta-analysis. *Organization Studies* 24(3): 403–41.

Pargal, Sheoli, and David Wheeler. 1996. Informal regulation of industrial pollution in developing countries: Evidence from Indonesia. *Journal of Political Economy* 104(6): 1314–27.

Porter, Michael, and Mark Kramer. 2006. The link between competitive advantage and corporate social responsibility. *Harvard Business Review* 84(12): 78–92.

Portney, Paul. 2005. Corporate social responsibility: An economic and public policy perspective. In *Environmental Protection and the Social Responsibility of Firms*, eds. Bruce Hay, Robert Stavins, and Richard Vietor. Washington, DC: Resources for the Future.

Portney, Paul R. The (Not So) New Corporate Social Responsibility: An Empirical Perspective. *Review of Environmental Economics and Policy* doi:10.1093/reep/ren003.

Prendergast, Candice. 1999. The provision of incentives in firms. *Journal of Economic Literature* 37(March): 7.

Reinhardt, Forest. 2005. Environmental protection and the social responsibility of firms: Perspectives from the business literature. In *Environmental Protection and the Social Responsibility of Firms*, eds. Bruce Hay, Robert Stavins, and Richard Vietor. Washington, DC: Resources for the Future.

Reinhardt, Forest. 2000. Down to earth. Boston: Harvard Business School Press.

Rivera, Jorge, and Peter deLeon. 2004. Is Greener Whiter? Voluntary environmental performance of western ski areas. *Policy Studies Journal* 32(3): 417–37.

Roe, Mark. 2000. Political preconditions to separating ownership from corporate control. *Stanford Law Review* 53(December): 539.

Scalise, Elisa. 2005. The code for corporate citizenship: States should amend statutes governing corporations and enable corporations to be good citizens. I 29(Fall): 275.

Sheehy, Benedict. 2005. Scrooge the reluctant stakeholder: Theoretical problems in the shareholder–stakeholder debate. *University of Miami Business Law Review* 14(Fall/Winter):193.

Social Investment Forum. 2006. 2005 Report on socially responsible investing Trends in the United States: Ten year review. Washington, DC: Social Investment Forum.

Springer, Jonathan. 1999. Corporate law, corporate constituency statues: Hollow hopes and false fears. In *New York University School of Law Annual Survey of American Law* 1999: 85.

Sutton, Michele. 2004. Between a rock and a judicial hard place: Corporate social responsibility reporting and potential legal liability under kasky v. nike. *University of Missouri-Kansas City School of Law Review* 72(Summer): 1159.

US Environmental Protection Agency. 2006. Environmental management systems/ISO 14001—Frequently Asked Questions. http://www.epa.gov/owm/iso14001/isofaq.htm (accessed August 14, 2007).

Vogel, David. 2006. *The Market for Virtue.* Washington, DC: Brookings Institution Press.

Wood, Donna, and Raymond Jones. 1996. Research in corporate social performance: What have we learned. In *Corporate Philanthropy at the Crossroads*, eds. Dwight Burlingame and Dennis Young. Bloomington: Indiana University Press.

PART II

METHODS OF ENVIRONMENTAL POLICY ANALYSIS

[4]

Available online at www.sciencedirect.com

SCIENCE DIRECT®

Economics Letters 79 (2003) 339–343

economics letters

www.elsevier.com/locate/econbase

Interpreting sustainability in economic terms: dynamic efficiency plus intergenerational equity

Robert N. Stavins[a,*], Alexander F. Wagner[b], Gernot Wagner[c]

[a] *John F. Kennedy School of Government, Harvard University, 79 John F. Kennedy Street, Cambridge, MA 02138, USA*

[b] *Department of Economics, Harvard University, Cambridge, MA 02138, USA*

[c] *Department of Economics, Stanford University, Stanford, CA 94305, USA*

Received 20 September 2002; accepted 20 November 2002

Abstract

Economists have confined the concept of 'sustainability' to intertemporal distributional equity. We propose a broader definition, combining dynamic efficiency and intergenerational equity, and relate it to two concepts from neoclassical economics: potential Pareto-improvements and inter-personal compensation.

Keywords: Sustainability; Sustainable development; Intergenerational equity; Kaldor–Hicks criterion; Environmental quality

JEL classification: Q01; D90

1. Introduction

There has been much debate among economists, and between economists and nearly everyone else regarding the meaning of the frequently employed concept of 'sustainability'. As one popular periodical wrote in regard to the World Summit on Sustainable Development, held in Johannesburg, South Africa, in 2002, sustainability 'risks being about everything and therefore, in the end, about nothing' (The Economist, 2002). Ecologists and many others outside the economics profession have taken sustainability to be the unique and comprehensive criterion that can and should guide global

*Corresponding author. Tel.: +1-617-495-1820; fax: +1-617-496-3783.

E-mail address: robert_stavins@harvard.edu (R.N. Stavins).

0165-1765/03/$ – see front matter
doi:10.1016/S0165-1765(03)00036-3

development. In contrast, economists have tended to define sustainability as being purely about intertemporal distribution, that is, intergenerational equity.[1] As such, economists have viewed sustainability as no more than one element of a desirable development path.

In this article, we suggest that a more normatively useful notion of sustainability, which would have broad appeal outside of the economics community, would combine two components: dynamic efficiency and intergenerational equity. This broader definition can facilitate discussions about sustainability, both within the economics community and as part of an interdisciplinary discourse.

2. Dynamic efficiency

In 1987, the Brundtland Commission placed sustainability on international political and scientific agendas with its report, 'Our Common Future' (World Commission on Environment and Development, 1987). The Commission proposed the widely embraced definition that 'development is sustainable when it meets the needs of the present without compromising the ability of future generations to meet theirs' (World Commission on Environment and Development, 1987).

We take the World Commission's definition as our point of departure.[2] In the absence of efficiency, however, constant consumption at no more than a subsistence level could satisfy this requirement, yet it surely would not be accepted as meeting reasonable social goals for public policy. Rather, any appealing normative criterion for public policy in this domain ought to include the notion of 'non-wastefulness'. Translated into economic terms, a meaningful definition of sustainability which has normative standing ought to include dynamic efficiency, that is, the choice of a feasible consumption path such that the economy is on the Pareto frontier. Using this path, total welfare can be evaluated as:

$$W(t) \equiv \int_{t}^{\infty} U(c(\tau))\, \mathrm{e}^{-r(\tau - t)}\, \mathrm{d}\tau \qquad (1)$$

where $U(c(\tau))$ denotes the most general, idealized utility function comprising both direct consumption as well as the enjoyment of non-market goods and services, and ρ is the social rate of time

[1] For example, Arrow et al. (2002) make a clear distinction between 'optimality', defined as the maximized discounted present value of future well being, and sustainability, defined as 'the maintenance or improvement of well being over time'. One exception in the current literature is provided by Asheim et al. (2001), who impose so-called efficiency and equity axioms and show that if social preferences fulfil these two axioms, any optimal path will lead to an efficient and non-decreasing path, thus implicitly including dynamic efficiency in the definition of sustainability. For an earlier discussion of sustainability and optimality, see Pezzey (1992). For a timely review of the major issues involved, see Pezzey and Toman (2001, 2002).

[2] Farrow (1998) begins from this same notion, and concludes that benefit–cost tests should be based not only upon potential Pareto-improvements but on *actual* compensation to identified sub-groups of the population. Such actual compensation is very difficult to envision in the realm of intergenerational equity (with respect to utility). We therefore take Farrow's conclusion as our starting point, and propose an alternative operational criterion for policy.

preference.[3] So, if it is desirable to avoid unnecessarily degrading resources and if sustainability is to have normative standing as a policy goal, then dynamic efficiency is a necessary condition for (such a normatively meaningful interpretation of) this concept.[4]

3. Intergenerational equity

Although dynamic efficiency is necessary for what we consider to be the most normatively useful definition of sustainability, dynamic efficiency is most surely not a sufficient condition. It is also necessary—for consistency with widely embraced definitions of this concept—that the total welfare function not decrease over time. Formally, a consumption path fulfils the condition of intergenerational equity if

$$\frac{dW(t)}{dt} \geq 0 \tag{2}$$

where $W(t)$ represents the total welfare function from Eq. (1).[5]

So, in this broader economic definition of sustainability, an economy is sustainable if and only if it is dynamically efficient and the resulting stream of total welfare functions is non-declining over time.

4. Sustainability

We acknowledge that the above definition provides a demanding pair of decision criteria, and some might argue that this is too demanding to be useful as a guide for public policy. The same is true, however, of the benchmark of a Pareto-improving policy—one which makes some members of society better off, but makes *no one* worse off (Pareto, 1896). Actual Pareto-improvements are exceptionally rare, of course, perhaps even non-existent. Hence, the strict Pareto criterion is virtually never taken as a guide for public policy, despite its considerable normative appeal. Economists resort instead to seeking 'potential Pareto-improvements' in the Kaldor–Hicks sense—the world is viewed as being made better off if the magnitude of gains and the magnitude of losses are such that the gainers can fully compensate the losers for their losses and still be better off themselves.[6]

---

[3]This formulation and the notation used in Eq. (2) are consistent with Arrow et al. (2002), who refer to the solution of this problem as the 'present value of felicities'. Weitzman (2002) refers to it as a measure of 'welfare-equivalent sustainability'. Of course, $W(t)$ must capture total welfare. Omitting contributions to welfare of any kind of capital will lead to an underestimate, and omitting any form of capital depreciation will lead to an overestimate. The theoretical implications of technological and population change have been examined in this context, and the theory regarding ideal measures of $W(t)$ has been explored extensively. Heal (1998, 2001), Solow (1991), and Weitzman (2002) also give reasons why narrow definitions of 'economic' capital should be expanded to include, for example, human and natural capital.

[4]Despite being the ultimate goal, *maximization* of $W(t)$, that is, intertemporal *optimality*, is too much to require of a normatively useful criterion.

[5]Note that this condition does not assure that felicity U in each future period will be as high as its current level. It only refers to a given aggregation (discounted value) of felicities.

[6]The notion that a welfare improving change ought to be associated with a 'potential Pareto improvement' was introduced by Kaldor (1939) and Hicks (1940).

What is key is that the Kaldor–Hicks criterion is a necessary condition for satisfying the strict Pareto criterion. If a policy proposal fails the Kaldor–Hicks test, it cannot pass the Pareto test. If a proposed change is not a potential Pareto-improvement, it cannot be a Pareto-improvement. This is the fundamental theoretical foundation—the normative justification—for employing benefit–cost analysis, that is, for searching for policies that maximize the positive difference between benefits and costs.

Similarly, we can think of an economy as having the *potential* to become sustainable if it fulfils the criterion of dynamic efficiency. It can then, in principle, be made sustainable by appropriate intergenerational transfers to achieve a non-declining total welfare path. One such economy that *can* be made sustainable has been formalized by Hartwick (1977), in which there exists the *possibility* of turning exhaustible resources into capital stock, a particular type of intergenerational transfer. If the Hartwick rule of investing all rents from exhaustible resources in reproducible capital is followed, the economy can be made sustainable.[7]

Much as economic analysis has long focused on potential rather than actual Pareto-improvements, leaving the allocation of net gains among individuals (and, hence, the resolution of debates regarding distributional equity) to the political process, similar reasoning may lead to an analogous approach to sustainability debates. In theory, it may be argued that sustainability is ultimately the most desirable policy goal, but in practice it may be more reasonable to aim for potential sustainability in the form of dynamic efficiency (of an all-encompassing societal welfare function).[8]

This may open us to criticism for being excessively focused on efficiency rather than equity.[9] Our view, however, is *not* that economists should ignore distributional issues, whether cross-sectional or intertemporal, but rather that potential sustainability—that is, dynamic efficiency—provides a reasonable focus for much (but not all) economic analysis of long-term public policies, because it is a necessary—though not sufficient—condition for sustainability.

5. Conclusion

Sustainability is a broad concept, but it does not need to be 'vague', as Solow (1991) has argued. Interpretations that are acceptable both to natural scientists and economists should be possible. We find that sustainability can be conceptualized in a way that has intuitive appeal both to economists and non-economists by viewing a sustainable growth path as one which is both dynamically efficient and non-decreasing over time. Much as a potential Pareto-improvement in the Kaldor–Hicks sense can

[7]The conditions under which the Hartwick rule holds, however, are restrictive, as discussed by Asheim and Buchholz (2000).

[8]Except for the elusive case of the Hartwick economy, utility transfers between generations are difficult to operationalize. Their abstractness provides an additional reason why more useful policy statements can be made by being satisfied with potential transfers.

[9]One of the most prominent critiques of the traditional focus of economics on efficiency has been offered by Sen (1970). He points out that a society may be efficient 'even when some people are rolling in luxury and others are near starvation, as long as the starvers cannot be made better off without cutting into the pleasures of the rich. In short, a society can be Pareto optimal and still be perfectly disgusting'. Our definition of sustainability involves the notion of distributional equity by including both dynamic efficiency and intergenerational equity. We argue only that the comparative advantage of economics may lie in its focus on the first element, without trying to diminish the overall importance of the latter.

yield Pareto optimality when combined with appropriate compensation of losers by winners, so too can dynamic efficiency lead to the more ambitious goal of sustainability when it is combined with appropriate intergenerational transfers. And much as practical economic analyses often resort to seeking potential Pareto-improvements, leaving the final allocation to the political process, so too may intertemporal analyses focus on dynamic efficiency, leading to the possibility, at least, of actual sustainability.

Acknowledgements

We thank Geir Asheim, Partha Dasgupta, Lawrence Goulder, John Hartwick, John Pezzey, and Martin Weitzman for helpful comments on a previous draft. The authors are responsible for any and all remaining errors.

References

Arrow, K., Daily, G., Dasgupta, P., Ehrlich, P., Goulder, L., Heal, G., Levin, S., Mäler, K., Schneider, S., Starrett, D., Walker, B., 2002. Are We Consuming Too Much?. Beijer International Institute of Ecological Economics, Stockholm, Discussion paper.

Asheim, G., Buchholz, W., 2000. The Hartwick Rule: Myths and Facts. University of Regensburg, Discussion paper.

Asheim, G., Buchholz, W., Tungodden, B., 2001. Justifying sustainability. Journal of Environmental Economics and Management 41, 252–268.

2002. Sustaining the poor's development. The Economist, 11, August 31st.

Farrow, S., 1998. Environmental equity and sustainability: rejecting the Kaldor–Hicks criteria. Ecological Economics 27, 183–188.

Hartwick, J.M., 1977. Investment of rents from exhaustible resources and intergenerational equity. American Economic Review 67, 972–974.

Heal, G., 1998. Valuing the Future: Economic Theory and Sustainability. Columbia University Press, New York.

Heal, G., 2001. Optimality or sustainability. In: Plenary address to the Annual Conference of the European Association of Environmental and Resource Economists, Southampton.

Hicks, J.R., 1940. The valuation of the social income. Economica 7, 105–124.

Kaldor, N., 1939. Welfare propositions of economics and interpersonal comparisons of utility. The Economic Journal 49, 549–552.

Pareto, V., 1896. In: Vol. 2. Cours d'Economie Politique, Lausanne.

Pezzey, J.C.V., 1992. Sustainable Development Concepts: An Economic Analysis. World Bank, Washington, DC, World Bank environment paper no. 2.

Pezzey, J.C.V., Toman, M.A., 2001. Progress and problems in the economics of sustainability. In: Tietenberg T.H., Folmer, H. (Eds.), International Yearbook of Environmental and Resource Economics 2002/2003, Edward Elgar, Cheltenham.

Pezzey, J.C.V., Toman, M.A., 2002. Making Sense of 'Sustainability', Issue Brief 02-25. Resources for the Future, Washington, DC.

Sen, A.K., 1970. Collective Choice and Social Welfare. Holden-Day, San Francisco.

Solow, R.M., 1991. Sustainability: an economist's perspective, the eighteenth J. Seward Johnson lecture to the Marine Policy Center, Woods Hole Oceanographic Institution. In: Dorfman, R., Dorfman, N.S. (Eds.), Economics of the Environment: Selected Readings. Norton, New York, pp. 179–187.

Weitzman, M.L., 2002. Income, Capital, and the Maximum Principle. Harvard University Press, Cambridge, in press.

World Commission on Environment and Development, 1987. Our Common Future. Oxford University Press, Oxford.

[5]

An eye on the future

Lawrence H. Goulder and
Robert N. Stavins

Decisions made today usually have impacts both now and in the future. But in carrying out policy evaluations to help decision-makers, economic analysts typically discount future impacts. In the environmental realm, many of the future impacts are benefits from policy-induced improvements. Thus, in the environmental-policy context, future benefits (as well as costs) are often discounted.

This is controversial, partly because discounting can seem to give insufficient weight to future benefits and thus to the well-being of future generations. But does it actually shortchange the future? As economists, we have often encountered scepticism about discounting, particularly from non-economists. Some of this scepticism seems valid, yet some reflects misconceptions about the nature and purpose of discounting. By examining here how discounting affects the evaluation of environmental policies, we hope to clarify this concept.

H. ARMSTRONG ROBERTS/CORBIS

Laying a golden egg? Discounting allows potential investors to weigh up their potential returns.

It helps to begin by considering the use of discounting in private investments. Here, the rationale stems from the fact that capital is productive — money earns interest. Consider a company deciding whether to invest \$1 million in the purchase of a copper mine, and suppose that the most profitable strategy involves extracting the available copper three years from now, yielding revenues (net of extraction costs) of \$1,150,000. Would investing in this mine make sense? Assume that the company has the alternative of putting the \$1 million in the bank at 5% annual interest. Then, on a purely financial basis, the company would do better with the bank, as after three years it will have \$1,157,625 ($\$1,000,000 \times (1.05)^3$), compared with only \$1,150,000 if it invests in the mine.

Future returns

We compared these alternatives by compounding to the future the up-front cost of the project. It is mathematically equivalent to compare the options by discounting to the present the future revenues or benefits from the mine. Discounting offers a quick way to check whether the return on a project is greater or less than the interest rate by taking future revenues and translating them into present units, using the 'alternative rate of return' (the bank's rate of interest in our example) as the discount rate. So the discounted revenue in this case is \$1,150,000 divided by $(1.05)^3$, or \$993,413 — less than the cost of the investment. Thus, the project would not earn as much as the alternative of putting money in the bank. If the discounted revenue exceeded the cost of the project, then the project would yield a higher return than the bank, and the company would be better off investing in the mine.

This simple example suggests a general formula to determine whether an investment offers a return that is greater or less than the alternative of putting money in the bank. Suppose a project involves benefits (revenues) and costs over a time span from the present (time 0) to T years from now. Let B_t and C_t refer, respectively, to the benefit and cost t years from now, and let r represent the annual rate of return on a standard investment. The present value of the net benefit (PVNB) is given by

$$\text{PVNB} = \sum_{t=0}^{T} (B_t - C_t)/(1+r)^t$$

If this value is positive, the project will yield a return that is higher than the market interest rate.

Discounting

How economists' controversial practice of discounting really affects the evaluation of environmental policies.

Discounting translates future sums of money into equivalent current sums; it undoes the effects of compound interest. It is not aimed at accounting for inflation, as even if there were no inflation it would still be necessary to discount future revenues to account for the fact that a dollar today translates (through interest) into more dollars in the future.

Sums for society

Can the same kind of thinking be applied to investments made by the public sector for the benefit of society? Consider the following hypothetical public-sector investment: a potential climate policy. Our purpose is to convey key issues in the starkest terms, so we will intentionally oversimplify some aspects of what follows. Suppose that a policy, if introduced today and maintained, would avoid significant damage to the environment and human welfare 100 years from now. The 'return on investment' is the avoidance of future damage to the environment and to people's well-being. Suppose that this policy costs \$4 billion to implement, and that this cost is borne in its entirety today. Suppose also that the beneficial impacts — avoided damages to the environment — will be worth \$800 billion to people alive 100 years from now. Should the policy be implemented?

The answer will depend, of course, on the evaluation criteria used. Consider first the criterion of whether the winners have the potential to compensate the losers and still be no worse off. For this condition to be met, the benefit to the winners, after being translated to equivalent dollars, needs to be larger than the losses of the losers. After compensation from winners to losers, the policy would yield what economists call a 'Pareto improvement': some individuals would be better off, and no individual would be worse off.

Are the benefits great enough that the winners could potentially compensate the losers and still be no worse off? Here, discounting is helpful. If, over the next 100 years, the average rate of interest on ordinary investments is 5%, a gain of \$800 billion to people 100 years from now is equivalent to \$6.08 billion today. (Equivalently, \$6.08 billion today, compounded at an annual interest rate of 5%, will become \$800 billion in

concepts

100 years.) The project satisfies the principle of potential compensation if it costs the current generation less than \$6.08 billion.

Because the up-front cost of \$4 billion is indeed less than this figure, the benefit to future generations is more than enough to offset the cost to the current one. More generally, a positive PVNB means that the policy has the potential to yield a Pareto improvement. More realistic policies involve costs and benefits that occur at all points in time. For these policies, discounting serves the same purpose, as we convert costs and benefits from various periods into their equivalents at a given time (such as the present, for example).

Applying a discount rate does not mean giving less weight to the welfare of future generations. Rather, the process simply converts the (full) values of the impacts that occur at different points of time into common units. In our example, the full benefit to future generations is translated into a current monetary sum, which then allows us to compare this benefit with the full cost to the present generation.

Winners and losers

Even if one accepts the idea of discounting as a mechanism to translate impacts into equivalent monetary units, one might be uneasy about PVNB analysis, which is based on the potential Pareto improvement (PPI) criterion — whether the winners from a given policy could compensate the losers and still be better off. If a policy's benefits exceed its costs, and compensation is introduced so that no one is worse off, then the attractiveness of the policy seems clear. But if compensation is not actually made, the appeal seems considerably weaker.

In our climate-policy example, discounted benefits to future generations will exceed the loss to the current generation, so the potential exists for a Pareto improvement — the PPI criterion is met. But if future generations do not actually compensate the present one, is it still appropriate to enact the policy? Perhaps so, but many would argue that if actual compensation cannot be made, a positive PVNB has less merit as an evaluation criterion.

Suppose a proposed climate policy fails the PPI test — the future benefits are not large enough to offset current costs. Do current generations nevertheless have an obligation to undertake the policy? They might. The PPI criterion deserves to be given weight, but in almost all policy evaluations — especially when compensation is not actually carried out — it is important to consider other evaluation criteria, as there are bound to be some cases in which there are compelling reasons for adopting a policy even when the PPI criterion is not satisfied, or for rejecting a policy even when it is.

Should a lower discount rate be used to incorporate considerations of intergenerational equity more fully in the PVNB calculation? Suppose that, when the market interest rate is used for discounting, a policy that would benefit future generations fails to generate a positive PVNB. Using a lower discount rate would give greater weight to future benefits (and costs), possibly making the PVNB positive. Such adjustments are problematic, however: they blur the distinction between the PPI (efficiency) criterion and other legitimate policy-evaluation criteria, such as distributional (in this case, intergenerational) equity. In evaluating policies, it seems better to use the market interest rate so that the PVNB calculation provides a meaningful indication of whether the PPI criterion is satisfied, while at the same time judging intergenerational fairness by direct examination.

Even if one accepts the use of the PPI criterion and discounting in principle, estimates of PVNB are necessarily imprecise. There is uncertainty about the denominator — the discount rate. Theoretically, this should reflect the market interest rate but, of course, future market rates are impossible to predict. There is also considerable uncertainty about the elements in the numerator — the benefits and costs that current and future generations will experience from a policy that is introduced today. This uncertainty is derived both from scientific uncertainty about the biophysical impacts of policies and from uncertainty about future generations' tastes and preferences — how much they will value the biophysical impacts.

Much scepticism about discounting and, more broadly, the use of benefit–cost analysis, is connected to these uncertainties. Consider the difficulties of ascertaining, for example, the benefits that future generations would enjoy from a regulation that protects certain endangered species. Some of the gain to future generations might come in the form of medical products (such as serums or vaccines) derived from the protected species, but such future impacts are impossible to predict. Moreover, benefits reflect the value that future generations will attach to the protected species — the enjoyment of observing them in the wild or just knowing of their existence. But how can we predict future generations' values? Economists and other social scientists try to infer them through surveys (such as the contingent valuation method) and by inferring preferences from individuals' behaviour. But these approaches are far from perfect, and at best they indicate only the values or tastes of people alive today.

The uncertainties are substantial and unavoidable. They do not invalidate the use of discounting or benefit–cost analysis, but they do oblige analysts to acknowledge them in their policy evaluations. It is crucial to evaluate policies using a range of values for discount rates and for future benefits and costs. We should have less confidence in a project for which the sign of the PVNB is highly sensitive to the discount rate or to small changes in projected future benefits and costs, compared with a project with a PVNB that is not very sensitive to these elements.

The discounting debate

The application of discounting to environmental-policy evaluation is controversial, partly because of misunderstanding outside the economics community of what discounting actually does, which is to translate the values of future impacts into equivalent values in today's monetary units. The PPI criterion, which provides the rationale for discounting and calculation of the PVNB, deserves weight in evaluating environmental policies, although it is also important to consider other criteria (such as distributional equity), especially when the potential harm to 'losers' is substantial. Moreover, it is crucial to acknowledge any uncertainties about benefits, costs and interest rates. Some may argue that these complications invalidate PVNB calculations, but in our view such calculations — when carefully executed and thoughtfully interpreted — can provide useful information for making environmental-policy decisions. ■

Lawrence H. Goulder is in the Department of Economics and the Institute for International Studies, Stanford University, Stanford, California 94305, USA. Robert N. Stavins is at the John F. Kennedy School of Government, Harvard University, 79 John F. Kennedy Street, Cambridge, Massachusetts 02138, USA.

Uncertainty is unavoidable. This does not invalidate the use of discounting or benefit–cost analysis, but it does oblige analysts to acknowledge it in their policy evaluations.

FURTHER READING

Arrow, K. J. *et al.* in *Climate Change 1995: Economic and Social Dimensions of Climate Change* (eds Bruce, J. P. *et al.*) 125–144 (Cambridge Univ. Press, 1996).
Arrow, K. J. *et al. Science* **272**, 221–222 (1996).
Fullerton, D. & Stavins, R. *Nature* **395**, 433–434 (1998).
Portney, P. R. & Weyant, J. P. *Discounting and Intergenerational Equity* (Resources for the Future, Washington DC, 1999).
Weitzman, M. L. *J. Environ. Econ. Management* **36**, 201–208 (1998).

[6]

Journal of Regulatory Economics; 28:2 157–179, 2005

Using Revealed Preferences to Infer Environmental Benefits: Evidence from Recreational Fishing Licenses*

LORI S. BENNEAR
Duke University
Nicholas School of the Environment and Earth Sciences,
Box 90328 Durham, NC 27708 USA
E-mail: lori.bennear@duke.edu

ROBERT N. STAVINS
Harvard University and Resources for the Future
John F. Kennedy School of Government
79 John F. Kennedy Street
Cambridge, Massachusetts 02138, USA
E-Mail: robert_stavins@harvard.edu

ALEXANDER F. WAGNER
Swiss Banking Institute, University of Zurich
Plattenstrasse 14, CH-8032 Zurich, Switzerland
E-mail: alex.wagner@post.harvard.edu

Abstract
We develop and apply a new method for estimating the economic benefits of an environmental amenity. The method is based upon the notion of estimating the derived demand for a privately traded option to utilize an open access good. In particular, the demand for state fishing licenses is used to infer the benefits of recreational fishing. Using panel data on state fishing license sales and prices for the continental United States over a 15-year period, combined with data on substitute prices and demographic variables, a license demand function is estimated with instrumental variable procedures to allow for the potential endogeneity of administered prices. The econometric results lead to estimates of the benefits of a fishing

* Financial support of this work was provided by Resources for the Future, the U.S. Environmental Protection Agency, and the Dean's Research Fund, John F. Kennedy School of Government. The analysis benefitted from discussions with Andrew Metrick, Richard Newell, and Kerry Smith, and comments from participants at seminars at the National Bureau of Economic Research and Harvard University, and very helpful comments from two anonymous referees. The data set, which builds upon previous work by Wolf (1988), was assembled with the help of a series of research assistants, including: Rod Bender, Nancy Bilica, Alisha Bloom, Diane Cherry, Peter Condon, Andrew Hoffman, and Michael Susanto. The authors alone are responsible for any errors.

license, and subsequently to the expected benefits of a recreational fishing day. In contrast with previous studies, which have utilized travel cost or hypothetical market methods, our approach provides estimates that are directly comparable across geographic areas. Our findings show substantial variation in the value of a recreational fishing day across geographic areas in the United States. This suggests that current practice of using benefits estimates from one part of the country in national or regional analyses may lead to substantial bias in benefits estimates.

Key words: revealed-preference valuation, environmental benefits, recreational fishing day

JEL Classification: Q26, Q21, Q22, H41

1. Introduction

When considering regulatory actions for a number of disparate environmental and natural resource problems, policy makers may wish to have estimates of the economic value of a day of recreational fishing. Such values can be used to measure the recreational benefits of proposed regulatory or other policy actions. In the past, such estimates have been used in economic impact analyses of: new dams and reservoirs, improvements in water quality, cleanups of abandoned hazardous waste sites, and reductions in the magnitude of global climate change. Virtually all of these estimates have drawn on one of two methods: contingent valuation, a direct survey approach employing hypothetical constructed markets; or travel-cost, an indirect market-based method. The use of the first of these approaches has generated considerable controversy within economics;[1] and both approaches require large quantities of geographically specific data. The majority of analyses by government agencies—including benefits estimation for Regulatory Impact Analyses—do not employ new site-specific or policy-specific studies of the value of a recreational fishing day. Rather, these analyses typically employ "benefit transfer methods," whereby estimates from a previous study are applied—sometimes with modifications—to a new and different policy scenario (Desvousges et al. 1998).

In this context, it may be of interest to have an additional set of estimates—based upon a conceptually distinct, revealed-preference approach—of the economic benefits of a recreational fishing-day (Stavins 1992). Our methodology differs from previous studies in two important respects. First, we estimate willingness-to-pay for a recreational fishing day from observed behavior regarding the purchase of fishing licenses, rather than observed behavior regarding travel to sites or stated preferences regarding those sites. Second, the two existing approaches[2] use detailed

1 For an overview of the controversy (see Diamond and Hausman 1994; Hanemann 1994; Portney 1994).

2 Other direct, revealed-preference methods that have been used for examining other environmental amenities and that require detailed micro-data—hedonic property and wage models—have not

micro-data (of observations of opportunity costs of travel or respondents' explicit estimates of willingness-to-pay) to develop benefit estimates specific to particular bodies of water and particular changes in water quality. This is both their advantage and disadvantage. In contrast, the approach developed in this paper uses aggregate data at the state level to derive estimates—in a national modeling framework—of state averages of recreational benefits.[3] As a result, our state estimates are directly comparable among one another, allowing inferences to be made about relative recreational benefits across geographic areas with more confidence than is possible based on previously available methods.

In Part 2 of the paper, we outline the theoretical framework underlying our estimation strategy. In Part 3, we describe our data, and in Part 4, we describe the econometric analysis, including the results from generalized least squares (GLS) and instrumental variables (IV) regressions. In Part 5, we use the econometric results from the IV equations to derive estimates of average expected recreational fishing day values, and we compare these estimates with results from previous studies. In Part 6, we conclude.

2. Overview of Method

Throughout the United States, a state fishing license is required for recreational fishing on any and all bodies of water, with the exception of privately own ponds. Thus, apart from the possibility of illegal fishing activity, to which we return below, a license is a necessary condition for deriving benefits from a day of recreational fishing. Likewise, apart from the relatively rare urbanite who may enjoy displaying to others an (unused) fishing license, experiencing some fishing days is a necessary prerequisite for deriving benefits from owning a fishing license.[4]

Building upon the household production approach to consumer behavior (Becker 1965), Bockstael and McConnell (1983) identified the conditions under which empirical knowledge of the demand function for a private, market good could be used to infer the benefits derived from a related public good. In a simple model,[5] let X be the number of fishing days experienced, L a fishing license, and Z a composite of other goods and services. If utility is defined by the function,

---

been applied to estimating the value of recreational fishing days (Freeman 2003). In principle, a sport fishing demand function could also be estimated in some cases by drawing on data from pay-for-use facilities, such as private, managed trout ponds, where users are charged for access or use (Vaughan and Russell 1982).

3 Seneca and Davis (1976), in an analysis of the factors affecting participation in recreational activities, carried out a county-level, cross-sectional econometric analysis of the factors affecting fishing license sales in West Virginia in 1970. Because there was no variation in license prices in the cross section, price could not be included as an explanatory variable.

4 We return later to the possibility that the license provides its owner with the *option* to go fishing, and thereby that simply *expecting* to go fishing is all that is required for a person to derive benefits from owning a license.

5 For a detailed theory of the utilization of recreational fisheries (see Anderson 1993).

$$U = U(X, L, Z) \tag{1}$$

then the above situation can be represented by the following pair of marginal utility relationships:

$$\frac{\partial U(X, 0, Z)}{\partial X} = 0 \tag{2}$$

$$\frac{\partial U(0, L, Z)}{\partial L} = 0 \tag{3}$$

defining what McConnell (1992) termed "joint weak complementarity." Thus, we can employ information about peoples' revealed valuation of fishing licenses, measured by the appropriate area under the respective demand functions, to draw inferences regarding their revealed valiation of (expected) recreational fishing days (Smith 1991).

The first step is to estimate econometrically a set of demand functions for state annual recreational fishing licenses. By measuring the appropriate area under the (state-specific) inverse demand function, we can estimate the average benefits per capita of fishing licenses. Further manipulation leads to an estimate of the average benefits of fishing licenses *per license* (again, specific to states and years). From this we derive a revealed-preference estimate of the expected value of a recreational fishing day. This value can be compared with estimates derived in previous studies by contingent valuation or travel cost methods.

3. Data

Recreational fishing licenses are sold by all 50 states. In all cases, prices are administratively set by state governments, and licenses are sold without limit.[6] This study focuses on a panel of licenses sold in 48 states[7] over a 15-year period (1975–1989). We aggregated the numerous types of fishing licenses that exist into 10 categories. All states offered both resident and nonresident licenses, the former at lower prices. Resident annual licenses were by far the most popular type, with sales of more than 257 million over the 15-year sample period, representing about two-thirds of all licenses sold (Table 1).[8] Second in numerical importance were resident "combination licenses" that allow for both hunting and fishing during a given year.

---

6 Because fishing licenses are sold without limit, they are not used to limit fishing activity or to correct environmental externalities. Rather they are used primarily as a revenue generating source for states.

7 Hawaii is excluded from the analysis because the structure of license demand does not match the rest of the United States, and Tennessee is excluded because it does not sell a fishing license *per se*, but a "Sportsman License," which can be used for both hunting and fishing.

8 Two states—Montana and Wyoming—required the purchase of a "conservation stamp" in addition to a fishing license. The cost of the stamp was added to the price of a license.

Table 1. Major Categories of Recreational Fishing Licenses Continental United States, 1975–1989

	Total License Sales (T = 15 Years, N = 48 States)		
License Type	Number[a]	Share (%)	Total Number of Observations (NT)
Resident Annual	257,054,000	67.6	720
Resident Combination[b]	52,690,000	13.9	481
Resident Short-Term 1[c]	9,661,000	2.5	203
Resident Short-Term 2[d]	349,000	0.1	71
Resident Short-Term 3[e]	986,000	0.3	29
Nonresident Annual	20,059,000	5.3	709
Nonresident Combination	795,000	0.2	118
Nonresident Short-Term 1	17,022,000	4.5	378
Nonresident Short-Term 2	15,034,000	4.0	422
Nonresident Short-Term 3	6,374,000	1.7	203
All License Categories	380,024,000	100	3,305

[a] Rounded to the nearest 1,000.
[b] "Combination" licenses cover both fishing and hunting on an annual basis.
[c] "Short-Term Type 1" licenses are temporary fishing licenses, ranging in length from 1 to 3 days.
[d] "Short-Term Type 2" licenses are temporary fishing licenses, ranging in length from 4 to 9 days.
[e] "Short-Term Type 3" licenses are temporary fishing licenses, ranging in length from 10 to 15 days.

Various duration short-term fishing licenses were also available to residents in many states; these allowed for as little as a single day or as much as two weeks of fishing, but—in total—made up only 3 percent of all resident fishing license sales. In contrast, nonresident license sales were much more heavily weighted toward short-term permits. In fact, about 65 percent of all nonresident fishing license sales during the sample period were of short-term licenses (Table 1).

There was substantial variation in aggregate and per capita sales of the various types of licenses. The variation was greatest across states, but was also significant within states over time. For example, in the final year of the sample, resident annual license sales ranged from about one percent to over 25 percent of state population. There was also considerable variation in (real) license prices across states and over time. In the case of resident annual licenses, the range in the sample was from a minimum of \$3.79 to a maximum of \$32.80[9] (Table 2). In the final year of the sample, the range was from \$7.63 (Minnesota) to \$26.73 (Colorado). The variation was even greater for some of the other license categories. For example, in 1989, the minimum price of a nonresident annual license was \$16.66 (South Dakota) and the maximum was \$69.44 (California).

9 These and all other monetary amounts in this paper are expressed in year 2000 dollars.

In addition to the license price and quantity information, data were assembled on various demographic, political, and environmental variables that were thought to be relevant to license demand or necessary for establishing the links between license demand and participants' expected benefits of a recreational fishing day. These data are summarized in Table 2 and discussed in the next section.[10]

4. Econometric Analysis of License Demand

Since license prices are set administratively by states, and licenses are sold without limit, the quantity sold at various prices traces out a demand function if the administratively set prices are exogenously determined and if any relevant variables that are omitted are uncorrelated with the license price.[11] The first set of variables—in addition to the license price itself—that would seem to be relevant are the prices of major substitutes. In this analysis we focus on the demand for resident annual fishing licenses, and therefore the relevant substitute prices include the price of resident short-term fishing licenses, the price of resident combination fishing and hunting licenses, and the price of nonresident licenses in adjacent states.[12]

Presumably, the characteristics of demanders are also relevant, and we therefore included the following demographic variables in the resident license demand estimation: median family income; mean years of education; and the share of the population living in urban areas. Finally, the nature of available recreational fishing resources in states should affect demand for state fishing licenses. We dealt with this in two ways. First, we included a variable that measures acres of "fishable waters" per state,[13] but this treats all fishing resources as being homogeneous in terms of the experiences they offer. Clearly, this is not correct. An acre of pristine, high-quality Colorado mountain stream is not equivalent to an acre of Ohio reservoir. The omitted variable—quality of fishing waters—is likely to be important and may be correlated with license price, hence causing biased estimates of demand

10 A detailed table of fishing license prices by year and state is available from the authors upon request.

11 Under these conditions, the observed price-quantity combinations are the intersections of infinitely elastic license supply functions (one for each price) and an assumed downward-sloping license demand function. A graph of the change in quantity of fishing licenses demanded against the change in license prices shows a negative correlation resembling a downward sloping demand curve. This is the pattern that would be expected if prices were administratively set. This data pattern is not surprising because revenue generation, rather than regulation of environmental externalities, is the primary goal of selling fishing licenses.

12 A problem arises in specifying which state licenses are relevant as substitutes. We constructed a variable that is a weighted average of prices of specified types of nonresident licenses in adjacent states and Canadian provinces.

13 Note that this variable varies not only across states, but also over time, reflecting both development of new reservoirs and changes in water quality.

elasticity. The problem presented by such unobserved quality is not insurmountable, because although quality variation is dramatic across states, quality variation within states over time is trivial by comparison. Hence, we can model this unobserved factor as a fixed effect.[14]

This leaves one concern regarding the possibility of inferring a true demand relationship from econometric estimates: is it supply, demand, or some combination of the two that is being observed? In other words, are price and quantity simultaneously determined, or are prices *exogenously* set by states? A reasonable first approximation is that the administered prices are set exogenously by state officials, and so we begin with simple GLS estimates. But it is not difficult to posit theories of administered prices that support the notion that these prices are endogenous. Hence, we follow the GLS estimates with a set of specifications in which we treat the license price as endogenous, and estimate the relationships with IV methods.

4.1. Generalized Least Squares (GLS) Estimation

Resident annual licenses comprise approximately two-thirds of all fishing licenses sold in the United States.[15] For resident license demand, the dependent variable was expressed as sales per capita.[16] The variance in license sales per capita is likely to be smaller in states with larger populations, thus violating the homoskedasticity assumption of an ordinary least squares (OLS) estimator. To correct for this, we employ a GLS procedure that weights each observation by the square root of the state's population (Bowes and Loomis 1980). In addition, fixed effects were employed to control for constant differences among states in the quantity and quality of their recreational fishing resources. Thus, the demand for resident annual licenses is estimated as:

$$\frac{Q_{it}}{N_{it}}=f\left(P_{it}, P_{it}^{S1}, D_{it}^{S1}, P_{it}^{S2}, D_{it}^{S2}, P_{it}^{S3}, D_{it}^{S3}, P_{it}^{NR}, F_{it}, U_{it}, E_{it}, Y_{it}, D_i, D_t, \varepsilon_{it}, \beta\right) \quad (4)$$

where Q_{it} = quantity of sales of resident annual license in state i in year t;
N_{it} = population of state i in year t;
P_{it} = price of resident annual license in state i in year t;

---

14 If preferences for fishing opportunities differ significantly across states, the parameters of the demand functions should likewise differ. The data do not allow us to estimate separate demand functions for each state, so we allow differences in preferences to be captured by fixed effects and by demographic control variables. Our approach assumes that each individual within a state has the same willingness to pay. However, in the absence of individual data, we cannot directly address within-state heterogeneity.

15 To whatever degree the holders of annual licenses have greater or lesser fishing-day valuations than holders of other categories of fishing licenses, the eventual results will tend to over or under-estimate average state valuations.

16 These equations were also estimated with sales as the dependent variable and state population as an independent variable; the estimated parameters on population were not significantly different from 1.0 (in the primitive equations).

P_{it}^{S1} = price of short-term, type 1 (1–3 day) resident license in state i in year t;
D_{it}^{S1} = dummy variable which equals unity if a short-term, type 1 resident license is not offered in state i in year t, and otherwise equals zero;
P_{it}^{S2} = price of short-term, type 2 (4–9 day) resident license in state i in year t;
D_{it}^{S2} = dummy variable which equals unity if a short-term, type 2 resident license is not offered in state i in year t, and otherwise equals zero;
P_{it}^{S3} = price of short-term, type 3 (10–15 days) resident license in state i in year t;
D_{it}^{S3} = dummy variable which equals unity if a short-term, type 3 resident license is not offered in state i in year t, and otherwise equals zero;
P_{it}^{NR} = average price of adjacent state nonresident annual licenses for state i in year t;
F_{it} = area of fishable waters (acres) in state i in year t;
U_{it} = share of population living in urban areas in state i in year t;
E_{it} = mean years of education of population in state i in year t;
Y_{it} = median family income in state i in year t;
D_i = state fixed effects;
D_t = annual fixed effects;
ε_{it} = an independent, but not necessarily homoskedastic error term;
β = parameters to be estimated.

The results of estimating the fixed effects model of demand for resident annual fishing license are reported in Table 3.[17] We report two specifications: one includes the prices of all relevant substitutes as explanatory variables;[18] and the other includes only the price of short-term Type 1 licenses plus dummy variables for each year.

In general, estimated own price effects were negative and statistically significant, and substitute price effects were positive, as expected.[19] The parsimonious

17 The reported results are for a semi-log functional form. The demand function was also estimated using linear and log-log functional forms. We report the semi-log results because we believe the demand function is unlikely to be linear, as that would imply that at a zero price there is a finite demand, and because the semi-log results consistently had a better fit than the log-log results. Of course, a more fully specified, structural model of demand would—in principle—be preferred, but it would be necessary to make other assumptions of questionable plausibility. We choose instead to estimate several demand specifications, and trace out the range of likely results with robustness tests. A full set of results for all functional forms is available from the authors upon request.

18 Not all states offer all categories of substitute (short-term) licenses during all time periods. In a sense, the "prices" of these non-existent licenses are infinite. The various specifications allow for the effect of some type of license not being available through the inclusion of dummy variables, D_{it}, where $(1\text{-}D_{it})$ is interacted with the respective license price, so that for each observation either a substitute price effect or a lack-of-substitute effect is estimated. In theory, both should be positive, which they consistently were.

19 But when the full menu of substitute prices was included, some of the respective parameters were insignificant and negative. The prices of nonresident licenses in neighboring states performed particularly poorly. This could be because the simple arithmetic average of neighboring state prices does not correctly capture the role that neighboring state fishing opportunities play in the demand for resident annual licenses. If sufficient data were available, it would be preferable to allow the econometrics to determine the appropriate weighting of the neighboring state prices.

specification that included only short-term Type 1 resident licenses as substitutes consistently yielded positive and statistically significant coefficients (for all functional forms).[20] The attempt to capture (partially) resource-quality effects with the fishable waters variable met with limited success. Presumably, much of the variation in the quality of fishing resources across states was picked up by the fixed effects. Finally, the demographic variables seem to have had some effect on fishing license demand. In several specifications, income was positive and statistically significant. In other specifications, years of education was negative and significant. Goodness-of-fit statistics were reasonably good for these fixed-effects models, with R^2 on the order of 0.15–0.23; not surprisingly, the complete models—including the fixed effects—explain a greater share of the observed variance.[21]

4.2. Potential Problems

These results raise two major concerns regarding the effect of illegal fishing activity and the possibility of endogeneity of license prices. Since the purpose of econometrically estimating fishing license demand is to derive implied valuations of expected recreational fishing days, it is necessary to consider the implications of illegal fishing, that is, fishing without a license.[22] This, by itself, need not be a problem given the approach that is taken below to derive valuation from license demand, but it can lead to bias in the econometric estimates.

A theoretically desirable way of treating illegal fishing in the license demand equation would be to allow for this particular substitute activity. Hence, we would want to include as an explanatory variable the "price" of illegal fishing, which may be thought of as the magnitude of fines multiplied by the probability of being fined. Unfortunately, these fines are typically set by courts, not by statute or regulation, and data even for proxies of the probability of being fined (enforcement levels) are rarely available.[23] Although it is reasonable to assume that the true "price of illegal fishing" is positively correlated with the demand for

---

20 The latter specification includes dummy variables for each year in the sample. On average, nominal license prices changed only 2.5 times per state over the 15-year sample period. Therefore, much (but not all) of the intertemporal variation in prices consists of gradually declining *real* prices. If, at the same time, sales were drifting upward, this would yield a negative correlation, but not one due to price-quantity demand effects. The yearly dummy variables were included to examine this potential problem. However, their inclusion did not materially affect the results, and so it appears unlikely that the observed negative price elasticities were due to such spurious correlation.

21 A Hausman specification test consistently rejected the hypothesis that state-level variation could be adequately modeled as a random effect.

22 If we were concerned with the demand for fishing licenses per se, then illegal fishing would not be a problem for the econometrics; indeed, in that case it is important to exclude illegal fishing, as the data implicitly do. Why might one be interested in license demand in and of itself? One reason is that such an analysis can provide the relevant elasticities for examining revenue and other effects of fishing license taxes of various forms.

23 To whatever degree these factors and their effects vary across states but are constant over time, they are picked up by the state fixed effects.

Table 2. Recreational Fishing Licenses, 48 States, 1975–1989 Descriptive Statistics

Variable	Mean[a]	Standard Deviation	Minimum	Maximum	Number of Observations
Quantity of Resident Annual Licenses per State	357,019	362,301	10,925	2,293,671	720
Quantity Per Capita of Resident Annual Licenses	0.091	0.050	0.010	0.249	720
Price of Resident Annual Licenses	$14.89[b]	3.48	$3.79	$32.80	720
Quantity of Resident Combination Licenses per State	109,543	104,386	5,059	724,990	481
Quantity Per Capita of Resident Combination Licenses	0.044	0.038	0.001	0.207	481
Price of Resident Combination Licenses	$29.02	7.84	$10.44	$72.02	481
Quantity of Resident Type 1 Short-Term Licenses per State	47,590	64,300	194	307,893	203
Quantity Per Capita of Resident Type 1 Short-Term Licenses	0.015	0.015	0.0004	0.077	203
Price of Resident Type 1 Short-Term Licenses	$9.01	3.47	$1.74	$31.77	203
Quantity of Resident Type 2 Short-Term Licenses per State	4,909	5,217	151	17,947	71
Quantity Per Capita of Resident Type 2 Short-Term Licenses	0.005	0.007	0.0001	0.026	71
Price of Resident Type 2 Short-Term Licenses	$13.43	3.36	$5.30	$22.69	71
Quantity of Resident Type 3 Short-Term Licenses per State	34,000	20,657	93	74,141	29
Quantity Per Capita of Resident Type 3 Short-Term Licenses	0.007	0.005	0.0001	0.018	29
Price of Resident Type 3 Short-Term Licenses	$6.69	1.20	$4.14	$10.44	29
Quantity of Nonresident Annual Licenses per State	28,292	36,570	1,232	224,850	709
Price of Nonresident Annual Licenses	$35.65	11.09	$6.40	$85.96	709
Quantity of Nonresident Combination Licenses per State	6,741	20,344	10	103,921	118
Price of Nonresident Combination Licenses	$97.52	40.10	$7.86	$250.08	118
Quantity of Nonresident Type 1 Short-Term Licenses per State	45,031	42,310	562	216,568	378
Price of Nonresident Type 1 Short-Term Licenses	$11.07	3.58	$2.64	$31.77	378
Quantity of Nonresident Type 2 Short-Term Licence's per State	35,625	38,964	661	180,712	422
Price of Nonresident Type 2 Short-Term Licenses	$16.93	5.37	$3.79	$43.33	422
Quantity of Nonresident Type 3 Short-Term Licences per State	31,398	30,568	1	185,187	203
Price of Nonresident Type 3 Short-Term Licenses	$22.19	7.26	$6.15	$47.13	203

(continued overleaf)

Table 2. (Continued)

Variable	Mean[a]	Standard Deviation	Minimum	Maximum	Number of Observations
Median Income (for four-person families)	$51,585	5,072	$36,956	$91,269	720
Share of State Population Residing in Metro. Areas Variable	0.614	0.230	0.126	1.000	720
Mean Years of Schooling	12.67	0.40	11.54	13.54	720
State Population	4,673,888	4,911,251	377,000	29,063,000	720
State Area (in square miles)	74,377	88,748	1,212	591,004	720
Total Area of Fishable Waters (in acres)	1,845,732	4,054,958	5,983	25,416,000	720

[a]For states and time periods where particular types of licenses did not exist, "zero observations" have been eliminated before calculation of descriptive statistics. The unit of observation for the table is a state in a given year; there is no weighting to account for the underlying samples of unequal size.
[b]All monetary amounts throughout the paper are expressed in 2000$.

Table 3. Demand For Resident Annual Fishing Licenses

	Semi-Log (fixed effects)		Semi-Log (instrumental variables)
	1	**2**	**3**
Price of Residential Annual License	−0.0175746*** (0.00199)	−0.0177431*** (0.00168)	−0.0224241*** (0.00844)
Price of Short-Term Type 1 License	0.0049107*** (0.00181)	0.0059081*** (0.00205)	0.0063156*** (0.002277)
Dummy/No Short-Term Type 1 License	0.0153356*** (0.00354)	0.01447*** (0.00387)	0.0146133*** (0.00402)
Price of Short-Term Type 2 License	−0.004802 (0.00573)		
Dummy/No Short-Term Type 2 License	−0.0060926 (0.01198)		
Price of Short-Term Type 3 License	0.004835 (0.00435)		
Dummy/No Short-Term Type 3 License	0.017926** (0.00779)		
Price of Adjacent N-R Annual Licenses	−0.0158293*** (0.00207)		
Quantity (Acres) of Fishable Waters	0.000558 (0.00333)	0.004332*** (0.00312)	0.0048068 (0.00324)
Share of Population Living in Urban Areas	−0.036721* (0.00642)	−0.007402 (0.00654)	−0.008021 (0.00667)
Median Family Income	0.0158995** (0.00509)	0.025432* (0.00711)	0.0245089*** (0.00761)
Mean Years of Education	−0.077769*** (0.01912)	0.027214 (0.04313)	−0.010737 (0.05656)
Annual Fixed Effects	No	Yes	Yes
Number of Observations	720	720	720
R^2 (Within)	0.274	0.233	n.a.

Note: Robust Standard Errors are in parentheses next to respective parameter estimates
*** Significant at the 1% level
** Significant at the 5% level
* Significant at the 10% level

fishing licenses, it is much less clear how it is correlated with license prices, if at all. Hence, omitting this variable may not seriously bias the elasticity estimates.[24]

There is also the possibility that license prices, administratively set by governments, are endogenous, that is, that causality runs not only from price to quantity, but vice-versa. How could this be the case? One potential source of such a causal linkage would be state budgets. States might seek to set license prices at levels that cover annual budgets of fish and game services, assumed to be more or less fixed over time (Whitehead 1983; Walsh 1986). In the time series for a single state, this could yield a negative correlation between quantity of license sales and administered price (recognizing that this assumes that states can predict sales).

Another potential explanation for a spurious, negative price-quantity correlation is associated with persons sorting themselves for residence among states. People with strong preferences for fishing may be expected to exert political pressure to keep license prices low. If people with strong preferences for fishing move to relatively good fishing states (or develop preferences for fishing as a result of having been born and raised in such a state), then states with large quantities of license sales could tend to have relatively low prices, suggesting a source of spurious, negative cross-sectional correlation. We allow for price endogeneity by identifying a set of instruments for license price, and by using instrumental variables in what is essentially a reduced form approach.[25]

4.3. Instrumental Variable (IV) Estimation

To address possible price endogeneity, we wanted an instrument or set of instruments that would be correlated with resident annual license prices, but uncorrelated with unobservable determinants of license sales (for example, preferences for fishable waters). That is, we wanted instruments that are exogenous to the demand for fishing licenses. Preferences for fishable waters can affect overall expenditures on water quality improvements, but such expenditures can be funded through user fees (fishing license sales) or through general tax revenues. The relative degree to which these two sources of revenue are utilized is determined largely by bureaucratic and political proclivities. Thus, it would be desirable to have instruments that represent states' (possibly changing) proclivities to employ user fees as opposed to taxes and regulations. These proclivities might be correlated with administrative prices of fishing licenses, but are less likely to be correlated with the quantity of fishing licenses sold. The set of instruments used in this analysis were: cigarette taxes (cents per package); motor fuels taxes (cents per gallon); general sales taxes (percent); and state expenditures (millions of dollars).

---

24 A more serious problem arises, however, if illegal fishing increases when license prices increase, perhaps as a form of protest. If this is the case, then demand responsiveness will be overestimated and the benefits of licenses will be underestimated.

25 A third potential problem with the fixed effects estimation results presented in the previous section is that the fixed effects model allows the intercept to vary, but other demand parameters are constrained to be equal across states. The data do not permit estimation of separate demand functions for each state.

The IV regression results are reported for the parsimonious specification in Table 3.[26] The results are robust to different specifications, including changes in the list of substitute prices and changes in the demographic variables. The parameters on the price and fishable waters variables are all of the logical sign and statistically significant, although several are quite small in magnitude. The demographic variable parameters are all of the expected sign, although some are not statistically different from zero. The own-price elasticity of demand is consistently greater (in absolute value) in the IV estimates than in the GLS estimates.[27]

5. Estimating The Value of a Recreational Fishing Day

Three steps were required to derive the (state and year specific) average expected value of a recreational fishing day from the econometrically estimated demand functions for fishing licenses: derive average benefits of fishing licenses per capita from an estimated demand function; calculate average benefits per licensee; and estimate average expected value of a recreational fishing day.

5.1. Estimating Average Benefits of Owning Fishing Licenses from the Demand Function

To derive the average benefits of a fishing license (in per capita terms), we begin with an equation for which the parameters have been estimated econometrically. From our econometric estimates, we obtain both point estimates and uncertainty estimates (in the form of the variance-covariance matrix) on the key parameters of the demand for fishing licenses. We begin by using the point estimates to derive our measure of benefits and then use the uncertainty estimates to calculate 90% confidence intervals for those benefit measures.

To estimate the value of a recreational fishing day, we begin by estimating the total benefits derived from fishing licences as measured by the area under the inverse demand curve. To calculate this area, we first set all variables—with the exception of the (annual resident) license price and the dependent variable (sales per capita)—equal to their actual values for a given state and year. Also, we set all parameters at

---

26 The instruments are good predictors of fishing license prices. The F-statistic for the first stage regression is 10.98, and the instruments are jointly significant at the one-percent level.

27 This suggests that whatever endogeneity underlies these results, it is not due to either of the linkages posited above, both of which suggested a negative bias for the GLS results. One test of the potential endogeneity of the price variable regresses the license price on the instrumental variables and then includes the predicted price from this first-stage regression as an explanatory variable (in addition to the licence price itself) in the quantity regression. If the coefficient on the predicted price variable is statistically significant, then it is drawing variance from the error term that would otherwise be attributed to the price variable. This would be what we would expect if the price variable was endogenous. We found that the predicted price variable was marginally significant (at the 10% level) when included in the quantity regression. This provides some evidence that price is endogenous. As a result, we use the IV regressions in our benefits estimation.

their econometrically estimated values. The inverted form of the resulting equation (that is, the inverse demand function) is then integrated between the actual per capita demand (sales per capita) and zero, yielding an estimate of the (revealed) benefits per capita of fishing licenses for each state and year. Multiplying by the state's population produces an estimate of the total benefits of licenses; and dividing this by license sales yields an estimate of the average revealed benefits of owning a fishing license *per licensee*:

$$B_L = \left[\int_c^{q_{it}} f(\hat{\alpha}_{it}, \hat{\beta}_0, q) dq \right] \bullet \frac{N}{Q} \tag{5}$$

where $f(\cdot)$ = inverse demand function;

q_{it} = per capita sales of resident annual licenses in state i in year t;

c = appropriate cutoff (zero for linear and semi-log specifications of the demand curve);

α_{it} = the fitted value from setting all variables—other than (annual resident) license price and the dependent variable—equal to their actual values for state i in year t, and all parameters, including the relevant fixed, state effect, at their econometrically estimated values;

β_0 = the estimated own-price elasticity of demand; and

B_L = average benefits of owning a fishing license (per licensee).

These average benefits of a recreational fishing license can be combined with information on the expected number of fishing days to derive a benefit estimate for a recreational fishing day. This is the focus of the next section.

5.2. Estimating the Expected Value of a Recreational Fishing Day

Since an annual fishing license is essentially an option to purchase (through direct expenditures plus opportunity costs of time) some number of days of recreational fishing experience (up to the total number of days in the season), it would seem that the stochastic relationship between a fishing license and the experiences it can facilitate would bring forth an important component of "option value." After all, no one knows with certainty how many days they will be able to go fishing during a season. Two conditions, however, essentially undue the option value, and make it possible to infer user value directly. First, annual license sales continue throughout the season, up until the very last day. Hence, there is no necessity to purchase a license *before* the time of one's first expedition. Second, numerous short-term licenses (substitutes) are always available.[28]

28 It is also true that an annual fishing license could offer option value for someone who decided he wanted to go fishing (for the first time in a season) in the middle of the night, when local outlets for purchasing a short-term license are closed. But this does not represent an important class of exceptions.

Thus, if short-term licenses did not exist *and* if there were an early deadline for purchasing annual licenses, then it could be argued that any valuation linked with licenses would include both user and option value. In the absence of those conditions, however, it seems more reasonable to assume that the benefits of a fishing license are linked with expected use value. If we assume that consumers are risk neutral in regard to their fishing license purchases and that the duration of a fishing season is short enough that discounting is not a significant issue, then we can derive the approximate conditional value of an expected fishing day in a very direct manner from the license demand evidence. First, we can say that:

$$B_L = E[B_{FD}|B_{FD} > 0] \bullet pr[B_{FD} > 0] \bullet S \tag{6}$$

where B_{FD} is the benefit (value) of a recreational fishing day; and S is the number of days in the season. The probability that the benefits are positive will be equivalent to the expected number of days of recreational fishing experienced, $E[d]$, divided by the length in days of the season, S. Therefore, we have the following:

$$E[B_{FD}|B_{FD} > 0] = \left[\frac{B_L}{E[d]}\right] \tag{7}$$

Thus, we can approximate the conditional value of an expected recreational fishing day by dividing the revealed valuation of an annual license by the expected number of fishing days. This is subject to four caveats. First, these relationships assume risk neutrality. If license purchasers are risk averse, then we will over-estimate the daily valuation. This seems, however, to be a second-order problem. Second, we have ignored discounting, but it is unlikely to amount to a significant error (relative to econometric and other sources of error), considering the length of the fishing season. Third, this assumes independence of the valuation of each day, but if there is declining marginal valuation of fishing days by license holders and serial correlation among days of participation, then such independence does not hold. Fourth, depending upon the nature of unobserved heterogeneity among licensees within states, the aggregation may not produce the correct weighted average.

This takes us to the point of estimating the average expected number of fishing days per licensee per state per year, $E[d_{it}]$. Various approaches to this problem exist, but a reasonable approximation is simply to use the actual numbers, which are periodically compiled at the state level by the U.S. Fish and Wildlife Service (FWS).[29] This may impart a slight upward bias to the final results, since the FWS numbers refer to all (resident) participants, not only annual license holders. Some of these are short-term license holders, who likely go fishing less frequently.

29 As reported in U.S. Fish and Wildlife Service (1975, 1980, 1985, and 1991), the surveys provide annual estimates of state-level total days of recreational fishing (separately by residents and nonresidents) and the number of participants. The estimates from those five years were interpolated to provide a set of annual estimates for the period, 1975–1989. The U.S. average over this time period is about 20 days per year for residents and 10 days per year for nonresidents.

But the overwhelming majority of licenses sold are annual licenses, implying that this problem will not be important. A final point concerns uncertainty estimates. We use Monte Carlo procedures to simulate confidence intervals for the benefits of a recreational fishing day.

The results from this analysis are summarized in Table 4, which provides benefit estimates derived with the semi-log IV regression, averaged over the sample period. Our results provide evidence of substantial heterogeneity among states in the expected value of a recreational fishing day. The mountain states plus Alaska, Arkansas, and Minnesota exhibit valuations that are 10 or even 20 times the magnitude of the estimates for the lowest value states, such as Delaware, Massachusetts, and Rhode Island. This should not be surprising, and any absence of such dramatic contrasts in previous studies might even be a source of concern. Our results reflect considerable uncertainty, however. The 90% confidence intervals are typically large, both in absolute terms and relative to the estimated benefits of recreational fishing.

5.3. Comparisons of National Revealed Preference Estimates with Site-Specific Estimates

The primary innovation in this study is to use revealed preference methods to derive the value of a recreational fishing day from market data on a national scale. Other methods, both revealed and stated preference, rely on detailed site-specific data. While these studies are often well-constructed and can provide good estimates of the value of a recreational fishing day for a specific site, the applicability of these benefit estimates to changes at other sites is questionable. To illustrate the potential problems, we contrast our estimates with previous estimates of the value of a recreational fishing day in Table 5, drawing upon a number of earlier studies that used either contingent valuation or travel-cost methods (but typically do not report confidence bounds).[30]

Our results appear to be generally lower than previous estimates.[31] It should be noted, however, that in 10 out of 13 cases, the 90% confidence interval of our valuations include the point estimates of respective previous valuations. Likewise, in four out of the seven instances in which previous studies provided confidence bounds, our 90% confidence limits on valuations fall within the respective 90% confidence limits from the previous analyses.

Perhaps more important than the absolute numbers, however, are the relative estimates of the benefits of recreational fishing across states. No matter which specification we use, the findings imply that fishing in states such as Colorado,

30 We only consider states in Table 5 for which previous studies exist of the time period employed in our analysis and for which our results yield estimates that are statistically different from zero.

31 An exception is that our estimate is higher than the only contingent valuation (CV) study in the comparison group. This is consistent with findings from Loomis et al. (2000), who use a similar method with hunting licenses.

Table 4. Estimates of the Value of a Freshwater Recreational Fishing Day (1975–1989 averages, 2000$, Based on Weighted Semi-Log Instrumental Variables Regressions)

	Estimated Value	90% Confidence Interval from Monte Carlo Analysis	
Alabama	28.86	8.10	53.39
Alaska	1,078.53	247.74	2,422.94
Arizona	7.52	1.91	15.37
Arkansas	469.48	104.19	1,096.76
California	11.46	2.52	27.28
Colorado	50.21	10.66	123.35
Connecticut	1.91	0.47	4.07
Delaware	1.13	0.31	2.13
Florida	1.54	0.40	3.11
Georgia	10.43	2.97	18.93
Idaho	37.91	8.27	90.75
Illinois	2.99	0.86	5.40
Indiana	4.84	1.31	9.29
Iowa	14.71	3.74	30.03
Kansas	14.03	3.14	32.69
Kentucky	15.92	3.89	33.97
Louisiana	3.53	1.32	4.87
Maine	14.92	2.92	39.74
Maryland	1.97	0.49	4.08
Massachusetts	1.61	0.31	4.31
Michigan	12.13	2.94	25.82
Minnesota	243.89	51.97	594.17
Mississippi	1.81	0.54	3.14
Missouri	15.65	4.04	31.47
Montana	1,418.78	293.44	3,535.67
Nebraska	20.83	4.55	49.59
Nevada	17.34	3.21	48.69
New Hampshire	5.81	1.11	15.93
New Jersey	1.69	0.34	4.38
New Mexico	1.69	0.38	3.89
New York	1.83	0.42	4.12
North Carolina	1.64	0.36	3.95
North Dakota	66.45	15.62	147.61
Ohio	10.11	2.42	21.90
Oklahoma	19.94	4.92	42.06
Oregon	62.26	11.79	171.50
Pennsylvania	8.66	1.71	22.92
Rhode Island	0.94	0.25	1.82
South Carolina	3.34	0.81	7.18
South Dakota	34.58	8.71	71.75
Texas	4.33	1.13	8.60
Utah	31.70	6.06	86.80
Vermont	11.73	3.07	23.33
Virginia	4.84	1.18	10.30
Washington	23.27	4.38	64.72
West Virginia	4.78	1.09	10.88
Wisconsin	50.96	12.01	112.74
Wyoming	3,793.74	854.97	8,731.08

Minnesota, Oregon, and Wisconsin, for example, conveys much greater recreational value than fishing in many other areas.

These differences in relative benefit estimates reinforce concerns about the validity of so-called "benefit transfer" methods, which extrapolate findings from site-specific studies to other specific sites or to a national scale. While concern about these methods is not new (U.S. Environmental Protection Agency 2000), the lack of viable national alternatives on which to base benefit estimates has increased reliance on benefit transfer methods. This study provides a supplemental approach, and the results indicate that a policy maker who were to extrapolate findings from site-specific studies of fishing in Oregon, for example, to other parts of the country would be employing a highly biased estimate.

Throughout the above discussion, we emphasized factors that could alter our findings. First, if there is significant option value of licenses, the actual (use value) benefits of recreational fishing would be even less than estimated here. Second, if illegal fishing increases when license prices increase, demand responsiveness will be overestimated, implying that the benefits of licenses are underestimated. Third, to the extent that other license holders may have systematically different valuations than annual license holders, our estimates will be affected. Fourth, our calculation of the expected number of fishing days relies upon FWS statistics, which may contain errors of their own.

6. Conclusions

Economic analyses of many proposed environmental and resource regulations rely critically on accurate estimates of recreational benefits. To date, the source of most information on the value of a recreational fishing day has come from site-specific methods. The data requirements for site-specific contingent valuation and travel cost methods are severe, and hence the expense of carrying out such analyses is a major impediment to their use. Furthermore, there is ongoing controversy surrounding the use of the CV and other hypothetical market methods for environmental benefits estimation. For these reasons, government agencies such as the U.S. Environmental Protection Agency rarely carry out original analyses, typically relying instead on "benefit transfers" from other studies. But the validity of transferring values derived for one specific site to other sites is often equally suspect. Given these realities, it is of considerable value to have access to a conceptually distinct method of estimating environmental values that is based upon a revealed-preference, econometric framework. As such, the method developed here holds promise.

Our numerical estimates of recreational fishing-day values suggest great variation across geographic areas. Since previous studies have been of single sites or single states, there was inevitably some question as to whether and to what degree any observed variations were due to real differences in values, as opposed to differences among respective models. Although our approach may suffer from being a macro-oriented approach—in contrast with survey methods and travel-cost

Table 5. Comparing Previous Estimates of the Value of a Freshwater Recreational Fishing day with those Derived in this Study

State	Previous Estimates				Valuation from this Study (2000 $, based on semi-log specification) [90% conf.]
	Estimation Method[a]	Types of Fishing	Study	Valuation (2000 $)	
Alabama	TC	Trout	King and Hof (1985)	24.57	22.40 [8.81 56.61]
Arizona	TC	All	Miller and Hay (1980)	73.14	4.24 [1.56 11.60]
Colorado	CV	Cold water	Walsh et. al. (1980)	22.01	28.84 [8.64 96.19]
Georgia	TC	Warm water	Ziemer et. al. (1980)	27.65	6.29 [2.54 15.53]
Idaho	TC	All Cold/warm water	Miller and Hay (1980) Loomis and Sorg (1986)	56.42 45.59/47.04	26.07 [7.61 89.69]
Maine	TC	All	Miller and Hay (1980)	48.06	6.96 [1.98 24.87]
Minnesota	TC	All	Miller and Hay (1980)	60.60	157.22 [46.81 522.78]
Missouri	TC	Trout	Haas and Weithman (1982)	27.97	10.09 [3.66 28.17]
Ohio	TC	Cold water Perch/Walleye	Dutta (1984) Hushak et. al. (1988)	8.73 4.58/5.60	6.03 [2.04 18.29] 8.32 [2.69 25.89]
Oregon	TC	Salmon Steelhead	Brown and Shalloof (1984)	36.47 49.72 7	33.59 [8.92 126.36]
Wisconsin	TC	All	Kealy and Bishop (1986)	51.60	39.09 [12.47 123.11]

(continued overleaf)

Table 5. Continued.

US FWS Regions				Valuation [90% conf.]	Average 1975-1989 [90% conf.]
CA, ID, NV, OR, WA	CV	Trout	US FWS (1996)	138.29 [−38.41 314.99]	30.45 [6.03 80.59]
AZ, NM, OK, TX	CV	Bass and Trout	US FWS (1996)	1266.53 [−909.45 3438.50]	8.37 [2.09 17.48]
IA, IL, IN, MO	CV	Bass	US FWS (1996)	243.65 [141.58 344.62]	9.55 [2.09 17.48]
AL, AR, FL, GA, KY, LA, MS, NC, SC, TN[1]	CV	Bass	US FWS (1996)	57.07 [−301.82 414.86]	59.62 [13.62 136.15]
CT, DE, MA, MD, ME, NH, NJ, NY, PA, RI, VA, VT, WV	CV	Bass and Trout	US FWS (1996)	164.62 [91.09 238.16]	4.76 [1.05 11.39]
CO, KS, MT, NE, UT, WY	CV	Bass and Trout	US FWS (1996)	317.18 [280.96 354.50]	888.21 [195.47 2093.20]
AK	CV	Trout	US FWS (1996)	411.57 [391.81 432.42]	1078.53 [247.74 2422.94]

[a]CV refers to contingent valuation (survey) methods; TC refers to travel-cost (Hotelling-Clawson-Knetch) methods.
[1]Tennessee is not included in this study.

models that focus on single sites—this is also an advantage, because it facilitates the development of a set of mutually consistent estimates that can be effectively compared with one another over time and space.

References

Anderson, L. G. 1993. "Toward a Complete Economic Theory of the Utilization and Management of Recreational Fisheries." *Journal of Environmental Economics and Management* 24: 272–295.

Becker, G. S. 1965. "A Theory of the Allocation of Time." *Economic Journal* 75: 493–517.

Bockstael, N. E., and K. E. McConnell. 1983. "Welfare Measurement in the Household Production Framework." *American Economic Review* 73: 806–814.

Bowes, M. D. and J. B. Loomis. 1980. "A Note on the Use of Travel Cost Models with Unequal Zonal Populations," *Land Economics* 56(4): 465–470.

Brown, W. G., and F. M. Shalloof. 1984. "Recommended Values for Columbia River Salmon and Steelhead for Current Fishery Management Decisions." Oregon State University.

Desvousges, W. H., F. R. Johnson, and H. S. Banshaf. 1998. *Environmental Policy Analysis with Limited Information: Principles and Applications to the Transfer Method.* Cheltenham, United Kingdom: Edward Elgar.

Diamond, P. A., and J. A. Hausman. 1994. "Contingent Valuation: Is Some Number Better than No Number?" *Journal of Economic Perspectives* 8(4): 45–64.

Dutta, N. 1984. "The Value of Recreational Boating and Fishing in the Central Basin of Ohio's Portion of Lake Erie." Ohio State University, Sea Grant Program.

Freeman, A. M. III. 2003. *The Measurement of Environmental and Resource Values: Theory and Methods.* Second Edition. Washington, D.C.: Resources for the Future.

Haas, M. A., and S. A. Weithman. 1982. "Socioeconomic Value of the Trout Fishery in Lake Teneycomo, Missouri." *Transactions of the American Fisheries Society* 111: 223–230.

Hanemann, W. M. 1994. "Valuing the Environment Through Contingent Valuation." *Journal of Economic Perspectives* 8(4): 19–43.

Hushak, L. J., J. M. Winslow, and N. Dutta. 1988. "Economic Value of Great Lakes Sportfishing: The Case of Private-Boat Fishing in Ohio's Lake Erie." *Transactions of the American Fisheries Society* 117:363–373.

Kealy, M. J., and R. C. Bishop. 1986. "Theoretical and Empirical Specifications Issues in Travel Cost Demand Studies." *American Journal of Agricultural Economics* 68: 660–667.

King, D. A. and J. G. Hof. 1985. "Experimental Commodity Definition in Recreation Travel Cost Methods." *Forest Science* 31: 519–529.

Loomis, J., C. Pearce, and M. Manfredo. 2000 "Using the Demand for Hunting Licenses to Evaluate Contingent Valuation Estimates of Willingness-to-Pay," *Applied Economic Letters* , 7: 435–438.

Loomis, J. B., and C. F. Sorg. 1986. "Economic Value of Idaho Sport Fisheries With an Update on Valuation Techniques." *North American Journal of Fisheries Management* 6: 494–503.

McConnell, K. E. 1992. "The Problem of Onsite Time in the Demand for Recreation." *American Journal of Agricultural Economics* 74: 918–925.

Miller, J. R. and M. J. Hay. 1980. "Estimating Substate Values of Fishing and Hunting." *Transactions of the North American Wilderness and Natural Resources Conference* 49: 345–355.

Portney, P. R. 1994. "The Contingent Valuation Debate: Why Economists Should Care." *Journal of Economic Perspectives* 8(4): 3–17.

Seneca, J. J., and R. K. Davis. 1976. "A Cross Section Analysis of State Recreation Activity." *Journal of Leisure Research* 8: 88–97.

Smith, V. K. 1991. "Household Production Functions and Environmental Benefit Estimation." In *Measuring the Demand for Environmental Quality*, edited by J. B. Braden and C. D. Kolstad, New York: North-Holland, pp. 41–76.

Stavins, R. N. 1992. "Private Options to Use Public Goods: The Demand for Fishing Licenses and the Benefits of Recreational Fishing." Paper presented at Public Economics and Environment Summer Institute, National Bureau of Economic Research, Cambridge, Massachusetts, August 11–13.

U.S. Environmental Protection Agency. 2000. *Guidelines for Preparing Economic Analyses.* Washington, D.C.: U.S. Environmental Protection Agency.

U.S. Fish and Wildlife Service. 1975, 1980, 1985, 1991. *National Survey of Fishing, Hunting, and Wildlife-Associated Recreation, State Overview*. Washington, D.C.: U.S. Department of the Interior.

Vaughan, W. J., and C. S. Russell. 1982. *Freshwater Recreational Fishing: The National Benefits of Water Pollution Control*. Washington, D.C.: Resources for the Future.

Walsh, R. 1986. *Recreation Economic Decisions*. State College, PA: Venture Publishing Inc.

Walsh, R., R. Ericson, D. Arosteguy, and M. Hansen. 1980. "An Empirical Application of a Model for Estimating the Recreation Value of Instream Flow." Colorado Water Resources Research Institute, Colorado State University: Colorado.

Whitehead, C. 1983. "State Fish and Wildlife Agency Responses to Funding Challenges." *Wildlife Management Institute, Transactions of 48th North American Wildlife and Natural Resources Conference*, pp. 139–148.

Wolf, S. A. 1988. "The Sport Fishing Conservation Stamp: A Proposal for Wetland Protection." A.B. Thesis, Harvard College.

Ziemer, R., W. Musser, and C. Hill. 1980. "Recreation Demand Equations: Functional Form and Consumer Surplus." *American Journal of Agricultural Economics* 62: 136–141.

[7]

Regulation & Governance (2007) 1, 154–171 doi:10.1111/j.1748-5991.2007.00008.x

On the value of formal assessment of uncertainty in regulatory analysis

Judson Jaffe* and Robert N. Stavins**

*Analysis Group, Inc., Boston, MA, USA, and **John F. Kennedy School of Government, Harvard University, Cambridge, MA, USA

Abstract

The US Office of Management and Budget introduced in 2003 a new requirement for the treatment of uncertainty in Regulatory Impact Analyses (RIAs) of proposed regulations, requiring agencies to carry out a formal quantitative uncertainty assessment regarding a regulation's benefits and costs if either is expected to reach $1 billion annually. Despite previous use in other contexts, such formal assessments of uncertainty have rarely been employed in RIAs or other regulatory analyses. We describe how formal quantitative assessments of uncertainty – in particular, Monte Carlo analyses – can be conducted, we examine the challenges and limitations of such analyses in the context of RIAs, and we assess how the resulting information can affect the evaluation of regulations. For illustrative purposes, we compare Monte Carlo analysis with methods typically used in RIAs to evaluate uncertainty in the context of economic analyses carried out for the US Environmental Protection Agency's Nonroad Diesel Rule, which became effective in 2004.

Keywords: benefit–cost analysis, Monte Carlo analysis, regulation, uncertainty.

1. Introduction

Benefit–cost analysis is a core component of a Regulatory Impact Analysis (RIA), which Presidential Executive Order 12866 requires for all "economically significant" Federal regulatory actions.[1] Similar requirements for RIAs have been in place since the Reagan administration[2] and prior procedures for regulatory review date back to the administration of Richard Nixon.[3] In carrying out analyses to estimate proposed regulations' benefits and costs, analysts must frequently rely on inputs to those analyses that are uncertain – sometimes substantially so.[4] Such uncertainties in underlying inputs are propagated through analyses, leading to uncertainty in ultimate benefit and cost estimates. But despite such uncertainty, the most prominently displayed results in RIAs are typically single, apparently precise estimates of benefits, costs, and net benefits,[5] masking uncertainties inherent in their calculation and possibly obscuring tradeoffs among competing policy options (Arrow *et al.* 1996).[6]

Historically, efforts to address uncertainty in RIAs have been very limited, but new guidance set forth in the US Office of Management and Budget's Circular A-4 on Regulatory Analysis has the potential to enhance dramatically the information provided

Correspondence: Professor Robert N. Stavins, John F. Kennedy School of Government, Harvard University, 79 John F. Kennedy Street, Cambridge, MA 02138, USA. Email: robert_stavins@harvard.edu

Accepted for publication 8 February 2007.

in RIAs regarding uncertainty in benefit and cost estimates (OMB 2003). Circular A-4 requires the development of a *formal quantitative analysis of uncertainty* regarding a regulation's economic impact if either annual benefits or costs are expected to reach $1 billion.

Over the years, the use of formal quantitative uncertainty assessments – in particular, Monte Carlo analysis[7] – has become common in a variety of fields, including engineering, finance, and several scientific disciplines.[8] It has also been found to be useful in certain regulatory contexts. For example, the US Environmental Protection Agency has recognized that it can be an important element of risk assessments (EPA 1997). But efforts to quantify uncertainties formally have rarely been made in the context of RIAs. Instead, uncertainty typically has been addressed qualitatively or through sensitivity analysis.

In this paper, we examine the role that formal quantitative assessment of uncertainty can play in regulatory analysis. In Section 2, we provide a brief overview of how such assessments of uncertainty can be developed, focusing on the execution of Monte Carlo analyses. In Section 3, we assess the potential benefits of uncertainty assessments for policy-making and in Section 4 we compare formal quantitative assessments with typical means of addressing uncertainty in RIAs, drawing on the results of three Monte Carlo analyses of uncertainty in the economic analyses of EPA's Nonroad Diesel Rule, which became effective in 2004. We carried out one of these Monte Carlo analyses (Stavins *et al.* 2004) and EPA, two others as a part of the RIA that accompanied its final rulemaking. In Section 5, we consider the challenges and limitations of public agencies carrying out Monte Carlo analyses and in Section 6 we examine the potential value of making assessments of uncertainty more prominent in RIAs. We conclude in Section 7.

2. Implementation of Monte Carlo analysis in an RIA

The approach most commonly used by public agencies to characterize uncertainty in RIAs and other benefit–cost analyses is sensitivity analysis, which typically involves examining the effects on estimated net benefits of changing the values chosen for particular inputs to their highest and lowest possible values within some plausible range. For example, in a "base case," the net benefits of a regulation might be computed with the expected (or the best-guess of the) number of lives saved per year by the regulation, say 15; and the sensitivity analysis might then carry out the same calculation of net benefits, sequentially using the analyst's estimates of the lowest and highest possible number of lives that the regulation might save each year, say 5 and 25.

Such sensitivity analysis, particularly this sort of common, extreme-case sensitivity analysis, shows three critical problems.[9] First, it fails to take account of important available information regarding assumed values of parameters. Typically, it is believed that values near the base-case assumptions are much more likely to occur than are values near the extremes of the range of plausible values. Therefore, the worst case and the best case are highly unlikely to occur (because they actually require the joint occurrence of a number of different low-probability events). In the example above, life-saving outcomes of 5 and 25 lives per year may both be exceptionally unlikely, whereas, outcomes between 10 and 20 may be much more likely to occur.

A second major problem with conventional sensitivity analysis is that it does not provide information about the dispersion (variance) of the distribution of net benefits.[10] If, for example, there are two policies that cannot be distinguished in terms of the expected values of their net benefits, then a policy-maker might wish to choose the policy with the smaller variance (because that policy would have a greater likelihood of producing net benefits similar to the predicted expected value).[11] With conventional sensitivity analysis, the policy-maker will not know which alternative has the lesser or greater variance, and thus which policy is more likely to achieve the desired result.

A third limitation of conventional sensitivity analysis is that such analysis typically involves the perturbation of one or perhaps two parameters at a time in isolation. But in the real world, uncertainty in net benefits is driven by the interaction of many variables and corresponding parameters. Only detailed simulations – in which many, if not all, parameters are allowed to vary simultaneously – can provide analysts with a reliable sense of potential outcomes, that is, resulting uncertainty in net benefits.[12]

Monte Carlo analysis provides a means of overcoming all three of these problems through a formal quantitative uncertainty assessment. The first step involves the development of probability distributions of uncertain inputs to an analysis.[13] These probability distributions reflect the implications of uncertainty regarding an input for the range of its possible values and the likelihood that each value is the true value.

There are multiple sources of uncertainty in any regulatory impact analysis. Several taxonomies of such uncertainty have been developed (Finkel 1990; Morgan & Henrion 1990; Linkov & Burmistrov 2003). A useful composite taxonomy is offered by Krupnick *et al.* (2006), which recognizes four categories of uncertainty: variability, parameter uncertainty, model uncertainty, and decision uncertainty. *Variability* refers to the inherent diversity of an empirical quantity across space, time, or individuals. Such uncertainty cannot be reduced by further research (although there can be uncertainty about the distribution of variability). *Parameter uncertainty* refers to lack of knowledge about some empirical quantity, due to measurement error (including both random error and statistical variation and systematic bias), unpredictability, lack of data, or extrapolation errors. *Model uncertainty*, like parameter uncertainty, is due to lack of knowledge, but about fundamental, systematic behavior. Whereas parameter uncertainty is due to the practical limits of available data, model uncertainty is due to limitations in the ability of modelers to represent real-world systems in mathematical models. Finally, *decision uncertainty* refers to uncertainty regarding aspects of an analysis that reflect social valuation of regulatory outcomes, such as the choice of discount rate, for example.

These diverse sources of uncertainty can be accounted for by using a variety of methods that rely on existing data, expert judgment, or a combination of the two.[14] But it should be recognized that only those sources of uncertainty that the analyst can characterize quantitatively can find their way into a formal quantitative analysis.[15]

Once probability distributions of inputs to a benefit–cost analysis are established, a Monte Carlo analysis is used to estimate the resulting probability distribution of the regulation's net benefits by carrying out the calculation of benefits and costs thousands of times. With each iteration of the calculations, new values are randomly drawn from each input's probability distribution and used in those calculations. Over the course of these repeated calculations, the frequency with which any given value is drawn for a particular input is governed by that input's probability distribution. For example, if an input has a 30% chance of being 10, then, on average, 10 will be used as the value for that input in 30 out of every 100 calculations. If a sufficient number of calculations are carried out, the range of resulting net benefit estimates and the frequency of particular estimates within that range characterizes the probability distribution of net benefits.

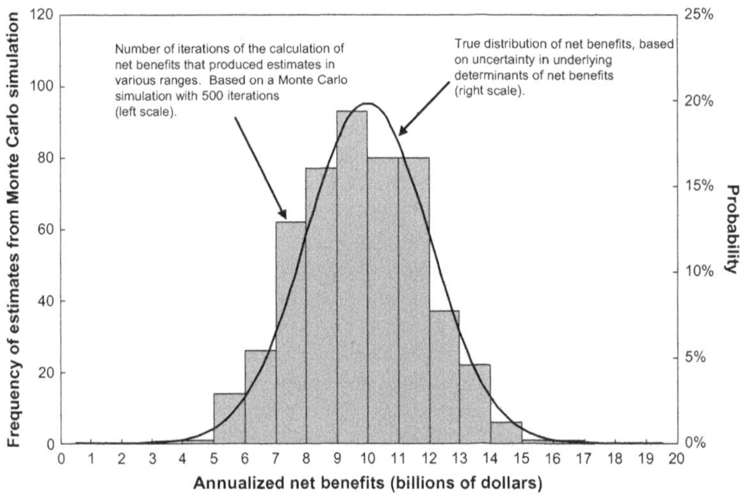

Figure 1 A histogram of estimated net benefits from a Monte Carlo simulation and the associated true probability distribution of net benefits, based on uncertainty in inputs to the net benefits calculation.

Note: The probability distribution show *n* here is symmetric, but relevant empirical distributions are frequently highly skewed (asymmetric).

The probability distribution of net benefits can be effectively approximated by dividing the range of net benefit estimates into a number of equal increments and then counting the frequency with which the repeated calculations produce net benefits falling into each increment. These frequencies can be represented pictorially by a histogram, providing a graphic image of the estimated distribution of net benefits (Fig. 1). The more random the draws that make up the Monte Carlo simulation, the more likely is it that the resulting histogram is a good approximation of the "true" distribution of net benefits.[16] Such a histogram can provide an effective visual display of the distribution, so that a decision-maker can readily perceive not only the expected value, but also the spread and skewness of net benefits.[17] Equally important, the numerical results of the Monte Carlo analysis can be used directly to calculate the expected value, the variance, and other important summary statistics characterizing uncertain net benefits.

3. Value of uncertainty assessments for policy-making

Assessments of uncertainty in regulatory analysis can lead to both short-term and long-term gains for policy-making. In the short term, such assessments allow more informed evaluations of proposed regulations and comparisons among regulatory alternatives. In the long term, such assessments can help establish research priorities.

Assessments of uncertainty can provide valuable information for policy-makers evaluating proposed regulations in at least two respects. First, findings regarding uncertainty in a net benefits estimate can provide a context for interpreting an estimate of the expected value of a regulation's net benefits. Second, consideration of uncertainties in

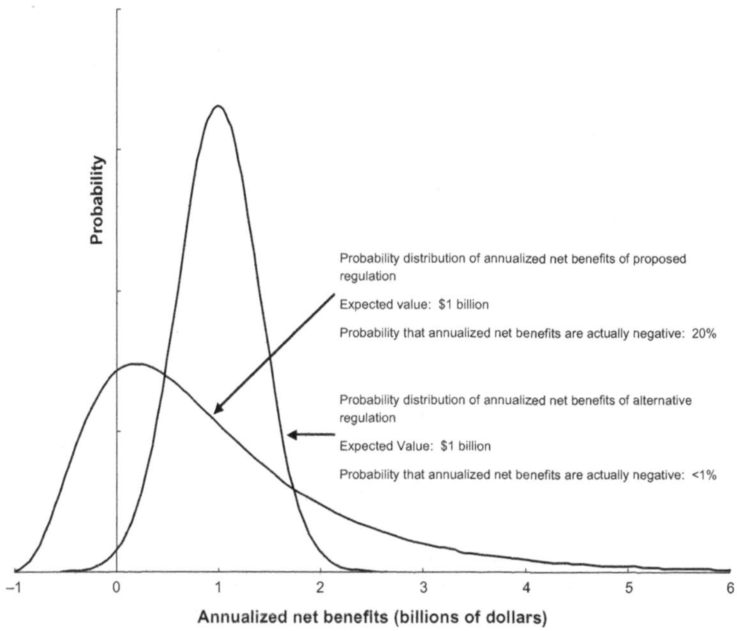

Figure 2 Probability distributions of annualized net benefits of a hypothetical proposed regulation and of an alternative regulation with the same expected value of net benefits.

underlying inputs, and how those uncertainties interact, can lead to an estimate of the expected value of a regulation's net benefits that differs from the estimate that would be produced if uncertainty in underlying inputs was not accounted for and single values were used for each input in calculating net benefits.

Uncertainty assessments provide insights regarding the distribution of possible net benefits associated with a regulation. Similar point estimates can be associated with vastly different distributions of possible outcomes around those estimates and the distribution of possible outcomes can significantly affect perceptions of an estimate of a regulation's net benefits. For example, Figure 2 depicts the probability distributions of annualized net benefits for a hypothetical proposed regulation and for an alternative to that proposed regulation. The expected value of the annualized net benefits for both the proposed regulation and the alternative regulation are $1 billion. But there is a 20% probability that the proposed regulation will have negative net benefits, whereas there is less than a 1% probability that the alternative regulation will have negative net benefits.

There may even be occasions where the distribution of possible outcomes associated with a regulation should be given equal or greater consideration than the most likely or average outcome. Such occasions could arise when there is a low probability of an outcome involving either extraordinarily high costs or benefits. An obvious example would have been an analysis of investment in levee maintenance in New Orleans before Hurricane Katrina, as a thorough analysis of uncertainty would have shown that there was a low probability of a catastrophic outcome (due to the confluence of a number of factors). This potentially important characteristic of a regulation could not be discerned without going beyond an estimate of the expected value of net benefits to consider the full distribution of possible outcomes.

Assessments of uncertainty in benefit and cost estimates also can aid the evaluation of alternative approaches to proposed regulatory actions. As differences in the benefits or costs of alternatives may be one of many factors that influence which alternative is chosen, it is important that policy-makers understand the degree of confidence that can be placed in estimates of those differences. If two alternatives cost the same amount, and the first alternative is estimated to yield greater benefits than the second, the weight that policy-makers give to this difference, relative to other factors, should be influenced by the probability that the second alternative's benefits could, in fact, be no less than those of the first.[18]

In addition to providing important context for interpreting point estimates from a benefit–cost analysis, there are circumstances in which consideration of uncertainty can lead to different point estimates than would be developed in a deterministic analysis (that does not account for uncertainty). First, careful consideration of uncertainty in inputs to benefit–cost analyses, which is necessary in the context of Monte Carlo analysis, can lead to different point estimates of the expected values for those inputs than may be used in a deterministic analysis. For example, the point estimate of an input used in a deterministic analysis may represent that input's expected value using one particular estimation model. But, consideration of model uncertainty may show that other plausible models would yield significantly different point estimates. Therefore, a point estimate of the input's expected value that accounts for the distribution of possible values resulting from model uncertainty would differ from an estimate based on just one model, which might be used in a deterministic analysis.

Second, an input may factor into an analysis in a manner in which the ultimate net benefit estimate will differ depending on whether only a point estimate is used for the input or the distribution of that input's possible values is explicitly incorporated in the analysis. This is because for some mathematical functions applied to an uncertain input, an estimate of the function's expected value that accounts for the distribution of the input's possible values will differ from an estimate that results from applying the function to the expected value of the input.[19] For example, if an uncertain input has an equal probability of being 2 or 6, the square of that input would have an equal probability of being 4 and 36, leading to an expected value of 20 (the average of the two possible outcomes). However, if one simply developed an estimate of the square of that input by determining the variable's expected value, which is 4, and then squaring it, the result would be 16.

Third, correlations among uncertain inputs can cause the expected value of net benefits resulting from those uncertain inputs to differ from a net benefits estimate developed by using point estimates for the inputs.[20] For example, if the values of two uncertain inputs are positively correlated, then the expected value of their product – accounting for uncertainties and this correlation – would be greater than the product of point estimates of each input's expected value. Thus, if that product were a component of benefits, an analysis that does not account for uncertainties in the inputs and correlations between them would underestimate the expected value of benefits.

Uncertainty assessments also can help identify the most significant determinants of uncertainty regarding a regulation's net benefits and thereby the potential for future research to reduce that uncertainty. In the short term, this information can be used to determine whether there may be benefits to delaying a rulemaking, in anticipation of future improvements in knowledge that will allow for more informed decision-making. In the long term, the cumulative information gained from uncertainty assessments in RIAs can be used to identify sources of uncertainty that have the greatest effect on estimates of the economic influence of regulations. This knowledge can be used to help establish research priorities.

4. Comparison of formal quantitative uncertainty assessments with conventional means of addressing uncertainty in RIAs

Sensitivity analysis is one of the most common methods used to characterize uncertainty in regulatory analysis. To compare sensitivity analysis with formal quantitative assessments of uncertainty, we examine the application of these two approaches in a recent regulatory analysis.

During the same period that OMB was finalizing its new guidance for regulatory analysis contained in Circular A-4, EPA was finalizing its Nonroad Diesel Rule. This Rule establishes new emissions standards for land-based, non-road diesel engines, such as those used in construction and in agricultural equipment.[21] To facilitate the use of technologies needed to meet these standards, which are phased in beginning in 2008, the Rule requires significant reductions in the sulfur content of diesel fuel used in non-road engines. The Rule also requires reductions in the sulfur content of locomotive and marine diesel fuels.

With other investigators, we carried out a Monte Carlo analysis of uncertainty in the net benefits of the proposed Nonroad Diesel Rule (Stavins *et al.* 2004).[22] In the RIA that accompanied EPA's final rulemaking, EPA itself presented the results of two Monte Carlo analyses that address uncertainty in particular determinants of that Rule's benefits (EPA 2004, appendix 9B). In the RIAs for the proposed and final rulemaking, EPA also used methods of assessing uncertainty that are more typical of RIAs. Thus, the Nonroad Diesel Rule provides a useful basis for assessing the merits of Monte Carlo analysis, relative to typical means of addressing uncertainty in RIAs.[23]

Our Monte Carlo analysis of EPA's proposed Rule incorporated probability distributions for more than 60 inputs to EPA's analysis, including inputs to both cost and benefit estimates.[24] These probability distributions were largely limited to characterizations of statistical variation, using readily available data that EPA relied on in developing the RIA, and so we did not address all sources of uncertainty. Therefore, the resulting probability distributions of the Rule's costs and benefits likely understate the extent of uncertainty in those estimates. The EPA's Monte Carlo analyses only address uncertainty in the Rule's benefits resulting from reductions in ambient concentrations of particulate matter (PM). In particular, EPA analyzes the implications for its benefits estimate of uncertainty in concentration–response (C–R) functions for certain PM-related health effects – most importantly premature mortality – and uncertainty in economic values of reductions in those effects.[25]

There is one important difference between EPA's two Monte Carlo analyses. In one of EPA's analyses (and in our analysis), only statistical variation is considered in characterizing uncertainty in the C–R function for PM-related premature mortality. These characterizations of uncertainty are based on the results of a single, observational epidemiological study, and thereby do not account for uncertainty as to whether the model used in the study is the correct model for estimating the C–R function. Model uncertainty, however, can be a major source of overall uncertainty.[26] In EPA's other Monte Carlo analysis, its characterization of uncertainty in the C–R function for premature mortality draws on expert judgments that account for some aspects of model uncertainty (EPA 2004, pp. 9–218).

4.1. Estimation of the range of possible economic impacts associated with a regulation

Because of the number of sources of uncertainty in a regulatory analysis and the complexity of their interactions, assessments of the extent of uncertainty in a regulation's net benefits – or components thereof – that are conducted with ordinary sensitivity analysis are unlikely to represent the true extent of uncertainty accurately. For example, in the draft RIA for the proposed Nonroad Diesel Rule, EPA's quantitative consideration of uncertainty regarding the Rule's cost is limited to an assessment of the effect on net benefits of actual costs being 20% lower and 20% higher than its primary estimate (EPA 2003, pp. 9–49). But our Monte Carlo analysis indicates that there is more than a 5% probability that costs could differ from EPA's primary estimate by more than 20%.[27]

The EPA also presents an assessment of the implications for its benefits estimate of scenarios in which reductions in fine $PM_{2.5}$ emissions resulting from the Rule are 5% below and 5% above its primary estimate. Our Monte Carlo analysis finds, however, that there is more than a 75% probability that emission reductions will be outside EPA's defined range of plus or minus 5%.

It is unlikely that a Monte Carlo analysis will comprehensively address all sources of uncertainty in the estimation of a regulation's economic impact. Therefore, the results of such analyses will likely understate the range of possible outcomes. Nevertheless, ranges produced by such Monte Carlo analyses provide substantially more reliable information than do less rigorous means of addressing uncertainty.

4.2. Evaluation of the likelihood of particular outcomes within a range

By providing a full characterization of the distribution of possible outcomes, Monte Carlo analysis provides information on the probability of particular outcomes within a range. Such understanding of the likelihood of values within a range is essential for meaningful interpretation of the range. For example, when a range of possible net benefits is provided, some may assume that all values within that range are equally likely to be the ultimate outcome. But this is rarely the case. Others may assume that the distribution of possible net benefits is symmetric.[28] This, too, is typically not the case. EPA's second Monte Carlo analysis shows that, given the sources of uncertainty that it assesses, there is a 90% probability that actual benefits in 2030 will fall within the first half of its estimated range of possible benefits.[29]

Conventional methods of addressing uncertainty in RIAs, such as sensitivity analysis, are less likely to provide meaningful guidance regarding the probability that a regulation's net benefits will fall above or below certain values.[30] In investigating uncertainty in its estimate of the proposed Nonroad Diesel Rule's benefits, EPA examines the implications of uncertainty in three income elasticities for the economic values of avoiding particular health effects.[31] To do this, EPA calculates the benefit estimate that results when all three elasticities are set to their lowest value within the range that EPA believes to be plausible for each. Likewise, EPA examines the consequences of setting all elasticities at the high end of their respective ranges.

In conducting this exercise, the analysts are examining extremely unlikely scenarios. Even if, for each income elasticity, there were as much as a 20% probability that its true value is no more than the low estimate that EPA uses, there would be less than a 1% probability that the true values of all three elasticities are at or below these low values.[32] But the fact that the specific scenarios examined are highly improbable does not imply that there is a low probability that benefits could be outside the established range. Although the probability of the particular scenarios may be small, uncertainties in many other inputs not considered in the exercise can lead to a high probability of benefits being outside the range. Indeed, the range that EPA establishes in this exercise is from 82 to 141% of its primary point estimate for 2030 benefits, but our Monte Carlo analysis indicates that there is at least a 73% probability that 2030 benefits will be *outside* that range.[33]

These are striking examples of how ordinary sensitivity analysis can offer misleading information regarding the range of net benefits that could result from a regulation, and the likelihood of various outcomes within that range. However, sensitivity analysis can be useful for indicating the extent to which uncertainty *in particular inputs* contributes to overall uncertainty in net benefits. But the implications of uncertainty in one input cannot be understood fully without the use of a formal quantitative assessment to characterize overall uncertainty.

4.3. Effect of uncertainty assessments on point estimates

In some circumstances, uncertainty assessments can change point estimates. EPA's second Monte Carlo analysis provides an example. In that analysis, EPA develops a characterization of uncertainty in the C–R function for premature mortality that goes beyond uncertainty arising from statistical variation and incorporates other sources of uncertainty, such as model uncertainty. This leads to a probability distribution for anticipated reductions in premature mortality whose expected value is nearly 30% less than the point estimate EPA calculates in an analysis that does not account for these broader sources of uncertainty (Fig. 3).[34]

As this shows, Monte Carlo analysis can indicate when uncertainties in inputs to a benefit–cost analysis cause the expected value of a regulation's net benefits to differ from what would be suggested by a deterministic analysis. Sensitivity analysis may examine the implications of uncertainties that can bring about such results, but it cannot address all possible outcomes resulting from those uncertainties or indicate the probability associated with any one outcome. Therefore, unlike Monte Carlo analysis, sensitivity analysis cannot provide policy-makers with insight regarding whether uncertainties in inputs cause the expected value of a regulation's net benefits to differ substantially from the primary net benefits estimate that would result from a deterministic analysis.

5. Challenges and limitations of Monte Carlo analysis

Several concerns have been raised regarding the potential use of Monte Carlo analysis in the context of Regulatory Impact Analysis, as well as within the broader context of risk assessment (Ferson 1996; Poulter 1998).

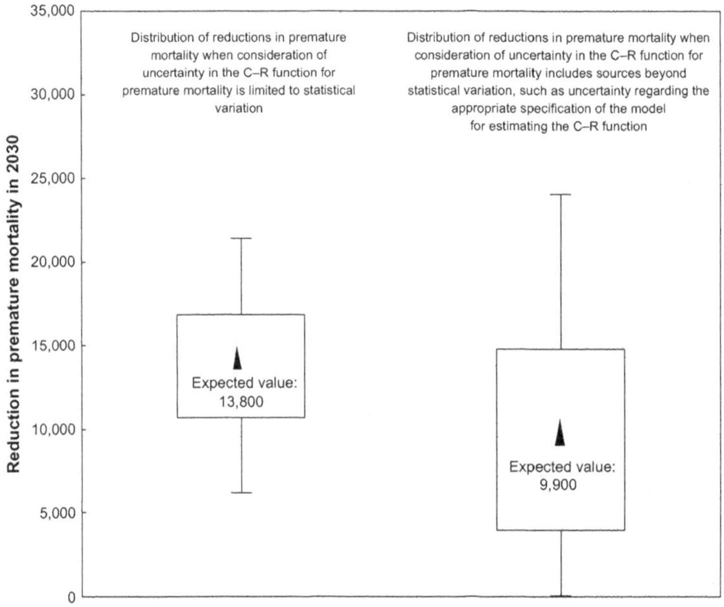

Figure 3 Effects on the probability distribution of reductions in premature mortality in 2030 from EPA's non-road rule when sources of uncertainty in the concentration–response function for premature mortality beyond statistical variation are considered.

Note: For each distribution, the box defines the range within which 50% of the Monte Carlo simulation's estimates of premature mortality reduction fell; 25% of the estimates fell at or below the bottom of the box, and the remaining 25% of the estimates fell at or above the top of the box. The lines define the range within which 90% of the Monte Carlo simulation's estimates of premature mortality reduction fell; 5% of the estimates fell at or below the bottom of the line and 5% fell at or above the top of the line.

Source: Adapted from US EPA (2004, pp. 9–240, figure 9B-4). These estimates are based on a preliminary EPA analysis of the Nonroad rule and are therefore not directly comparable to EPA's final estimate of the Rule's benefits.

5.1. Is additional computational effort required?

Implementation of a Monte Carlo analysis imposes two requirements that are not strictly necessary to develop point estimates of benefits and costs. First, instead of requiring a single point estimate for each input to a benefit–cost analysis, Monte Carlo analysis requires the development of probability distributions for key uncertain inputs. Second, numerous repetitions of the calculations in a benefit–cost analysis must be carried out. These requirements may appear burdensome (Poulter 1998), but a reasonable balance can be struck that accounts for the additional information that can be gained from increasingly comprehensive uncertainty assessments, and the additional effort that is required to develop such assessments, relative to what is already expended on RIAs.

First, as with benefit–cost analysis, a Monte Carlo analysis does not need to be exhaustive to offer valuable insights. For example, EPA's Monte Carlo analyses for its

Nonroad Diesel Rule addressed the implications of limited types of uncertainty in just a few of the numerous uncertain inputs in its regulatory analysis. Nonetheless, those Monte Carlo analyses offer a much more informative assessment of uncertainty than is offered by sensitivity analysis alone.

Second, in developing probability distributions for uncertain inputs in a regulatory analysis, uncertainty from statistical variation can often be characterized with little additional effort relative to that needed to develop point estimates. Much of the data necessary for such characterizations already will have been collected for the development of point estimates.

Characterizing other sources of uncertainty in inputs, however, can require more effort. For example, EPA's second Monte Carlo analysis required an expert elicitation to characterize uncertainty in the C–R function for premature mortality beyond that attributable to statistical variation.[35] The amount of additional effort necessary to develop a Monte Carlo analysis can be minimized through careful consideration of which input uncertainties are worthwhile addressing in the analysis. The importance of more fully characterizing uncertainty in the C–R function for premature mortality in EPA's analysis of the Nonroad Diesel Rule is made clear by the fact that reductions in premature mortality account for ninety percent of the Rule's quantified benefits (EPA 2004, pp. 9–28).

Although a Monte Carlo analysis can require additional effort to characterize uncertainty in inputs to a benefit–cost analysis, that effort often may be warranted even in the absence of the needs of a Monte Carlo analysis. Such characterizations of uncertainty may be necessary just to develop an accurate point estimate for an input.

Developments in computer performance and software over the years have dramatically reduced the amount of effort required to conduct calculations for a Monte Carlo analysis, once input uncertainties have been characterized. Widely available software allows the execution of Monte Carlo analysis in common spreadsheet programs on a desktop computer.[36] Also, modern programming techniques allow the writing of Monte Carlo computer programs with minimal additional effort, relative to that needed to produce point estimates. Nevertheless, Monte Carlo analysis may be resource intensive in some contexts, for example, in the case of EPA's environmental systems models. Such resource demands call for striking a balance between thoroughness of analysis, on the one hand, and implementation costs, on the other.

5.2. Will greater use of Monte Carlo analysis bring about "paralysis by analysis"?

Will OMB Circular A-4's requirement that formal quantitative uncertainty analyses be conducted for proposed regulations with annual benefits or costs expected to reach $1 billion tie up the regulatory system, leading to the so-called "paralysis by analysis?" In contrast, is the number of rules that exceed the $1 billion threshold so small as to render the requirement meaningless? Here are the facts. In a recent 1-year period, between October 2004 and September 2005, 3,980 final rules were published in the *Federal Register.* Of these, 292 were reviewed by OMB, 45 of which were deemed "major" final rules, given their anticipated economic impacts.[37] Of these 45 major rules, only six had estimates of annual benefits or costs that exceeded $1 billion (OMB 2006).

Given that analysts can strike a balance between how comprehensive and resource intensive a Monte Carlo analysis is, it is unlikely that a requirement affecting so few regulations per year need significantly delay, let alone paralyze the regulatory process. However, because these analyses are targeted at the most significant rules – those exceeding a threshold 10 times greater than the threshold for an RIA – the insights they bring to the regulatory process can be of considerable value.

5.3. Is Monte Carlo analysis strictly appropriate only for certain types of uncertainty?
It is generally recognized, even by its critics, that Monte Carlo analysis is "relatively straightforward" for assessing the effects of parameter uncertainty and natural variability, characterizations of which are frequently available in useful statistical form (Ferson 1996). But model uncertainty, from this perspective, can be a greater challenge. This should not be denied. But model uncertainty can be addressed quantitatively, such as through expert elicitation methods, as shown by EPA's own work in this regard. More importantly, it is crucial to keep in mind that failure to be comprehensive in including all sources of uncertainty by no means renders invalid a Monte Carlo analysis, which includes some important sources of uncertainty.

5.4. Does Monte Carlo analysis favor less stringent regulation?
A share of the general distrust that some advocates have for benefit–cost analysis appears to have spilled over to uncertainty quantification in this realm. It seems that some fear that simply acknowledging explicitly the degree of uncertainty that underlies the findings in RIAs will cause policy-makers and others to be less willing to impose regulations (Poulter 1998). If this is the case, it hardly seems that the appropriate answer would be to hide such uncertainty from regulators and the public. More to the point, however, is the reality that there is no reason a priori to expect an uncertainty analysis to favor benefits or costs. Indeed, formal quantification of uncertainty can be and has been supportive of greater regulatory stringency.[38]

Too much should not be claimed for Monte Carlo analysis, however. It should go without saying that "Monte Carlo methods cannot do everything, and they cannot solve all problems" (Ferson 1996). As with any analytical tool, formal quantification of uncertainty can yield results that are incorrect or unjustifiable if based on assumptions that are problematic or if the data employed are not empirically justified. Furthermore, if a Monte Carlo analysis is not appropriately explained, it can reduce the transparency of regulatory analysis. However, if properly executed and described, such analysis can be fully supportive of appropriate stakeholder participation (Farrow *et al.* 2001).

6. Prominence of uncertainty assessments in RIAs

Point estimates of regulations' net benefits have been given far greater prominence in RIAs than discussions of uncertainty associated with them. Uncertainty assessments are frequently relegated to appendices and discussed in a manner that makes it difficult for readers to discern their significance. This may be inevitable, given that single point estimates can be communicated more easily than lengthy qualitative assessments of uncertainty or a series of sensitivity analyses. A Monte Carlo analysis produces a quantitative probability distribution of a regulation's net benefits, a concise and effective summary of uncertainty that can be communicated almost as briefly as a point estimate.[39] If a summary of uncertainty in an estimate is not given equal prominence with the estimate itself, essential context for interpreting the estimate is lost.

Some resistance to the use of Monte Carlo analysis and prominent presentation of its results may be due to the perception that such analysis requires more judgment and therefore makes results more speculative (National Research Council 2002). Also, some have expressed concern that, given the inevitably incomplete nature of any Monte Carlo analysis, prominently presenting its results would incorrectly lead readers to conclude that results are more certain than they are. Both concerns seem unfounded. First, developing characterizations of uncertainty in inputs for a Monte Carlo analysis often involves making explicit and transparent judgments that must be made in any event. Moreover, to the extent that a Monte Carlo analysis is incomplete in its characterization of uncertainty regarding a regulation's net benefits, that fact can surely be communicated qualitatively.

7. Conclusions

Uncertainty is inevitable in estimates of regulations' economic impacts and assessments of the extent and nature of such uncertainty can provide important information for policy-makers evaluating proposed regulations. Such information offers a context for interpreting benefit and cost estimates and can lead to point estimates of regulations' benefits and costs that differ from what would be produced by purely deterministic analyses. In addition, these assessments can help establish priorities for research.

Due to the complexity of interactions among uncertainties in inputs to RIAs, an accurate assessment of uncertainty can be gained only through the use of formal quantitative methods. Such analysis involves relatively straightforward extensions to benefit–cost analysis. Its use can offer significant insights, while requiring only limited additional effort relative to that already expended on RIAs. Many of these insights cannot be gained using methods of addressing uncertainty that have typically been employed in RIAs, such as extreme-case sensitivity analysis. The analysis of EPA's Nonroad Diesel Rule provides an example of the differences between the information that can be offered by Monte Carlo analysis and that offered by conventional means of addressing uncertainty in RIAs. We compared findings from the two methods and described how formal quantitative methods offer policy-makers a richer and more reliable characterization of uncertainty regarding a regulation's economic impact even if such formal methods are not completely exhaustive in their treatment of uncertainty. The differences in information provided by the two methods result, in part, from the capacity of Monte Carlo analysis to account appropriately for interactions among the numerous uncertain inputs to a benefit–cost analysis.

The new requirement for formal quantitative assessments of uncertainty that is incorporated in OMB's Circular A-4 can mark a significant step forward in enhancing regulatory analysis under Executive Order 12866. It has the potential to improve substantially our understanding of uncertainty regarding the economic influence of regulations and thereby to lead to more informed policy-making.

Acknowledgments

Helpful comments were provided by Cary Coglianese, Jody Freeman, Jason Johnston, Sheila Olmstead, the editors, and several anonymous referees. Financial support for our previous work on this subject was provided by ExxonMobil and the American Chemistry Council. The authors alone are responsible for any errors.

Notes

1 Executive Order 12866 requires RIAs for all rules that are likely to "have an annual effect on the economy of $100 million or more or adversely affect in a material way the economy, a sector of the economy, productivity, competition, jobs, the environment, public health or safety, or State, local, or tribal governments or communities." See Executive Order 12866, 58 Fed. Reg. 51,735 (30 September 1993). If a regulation meets this criterion, or any of three other criteria, it is deemed a "significant regulatory action" under Executive Order 12866. However, Executive Order 12866 only requires an RIA for those significant regulatory actions that meet the first criterion, described above. In a 20 September 2001 memorandum, Dr. John Graham, the Administrator of the US Office of Management and Budget's Office of Information and Regulatory Affairs, defined an "economically significant" rule as one that meets that first criterion, leading to the requirement for an RIA (Graham 2001). The $100 million threshold is not indexed for inflation and has not been modified over time. Although Executive Order 12866 was originally issued by the Clinton administration (Hahn *et al.* 2003), its RIA requirements have continued to be enforced by the administration of President George W. Bush, which chose not to issue a new Executive Order (Graham 2001).

2 Throughout the Reagan and George H. W. Bush administrations, RIAs were required under Reagan Executive Orders 12291 and 12498. Under Executive Order 12291, 46 Fed. Reg. 13,193 (17 February 1981), agencies were required to conduct RIAs for all proposed and final rules that were anticipated to have an effect on the national economy in excess of $100 million. This Executive Order has been called the "foremost development in administrative law of the 1980s" (Morgenstern 1997).

3 President Richard Nixon required a "Quality of Life" review of selected regulations, beginning in 1971 (Hahn 2000) and President Gerald Ford formalized this process in 1974 with Executive Order 11821, 39 Fed. Reg. 41,501 (29 November 1974). Subsequently, President Jimmy Carter's Executive Order 12044 required analysis of proposed rules and centralized review by his Regulatory Review Group (Hahn 2000).

4 In cases where benefits cannot be monetized adequately, cost-effectiveness analysis can provide an economic framework for evaluating a regulation (OMB 2003, pp. 9–10).

5 Net benefits are benefits minus costs, all of which are expressed either as the present discounted value of future streams or as annualized amounts. For a general discussion of discounting in this context, see Goulder and Stavins (2002). Normative issues regarding the use of benefit–cost analysis as part of the policy process are beyond the scope of this essay. For a broad, recent review, see Adler and Posner (2001).

6 It should also be noted that some RIAs do not provide even single quantitative estimates of net benefits, because they fail to quantify or monetize impacts (Hahn *et al.* 2000).

7 Monte Carlo analysis is a means of statistical evaluation of mathematical functions using random samples. The probability of certain outcomes is approximated by running multiple trial runs, called iterations, using random variables. The "Monte Carlo" designation is a reference to the casino in Monaco. The method's use of randomness and the repetitive nature of the process are analogous to the activities conducted in a casino. The phrase "Monte Carlo analysis" is often used (as it is here) to refer generally to probabilistic modeling that is conducted using sampling-based simulation approaches. However, among specialists it is sometimes used to refer to a specific method of drawing random values in the context of a sampling-based simulation.

8 Monte Carlo analysis may be best known for its use during World War II in the design of the atomic bomb. It has since been used in a wide range of applications, including the analysis of traffic flow on highways, the development of models for the evolution of stars, and attempts to predict fluctuations in the stock market (Hammersley & Handscomb 1964). For more recent texts on the use of Monte Carlo methods in engineering and finance see Ayyub (1997, 2003), Glasserman (2003), and McLeish (2005).

9 Despite the severe problems documented here that are inherent in sensitivity analysis (in contrast with formal quantitative assessments of uncertainty through Monte Carlo analysis),

public agencies ought not be condemned for having relied in the past on this approach. The vast majority of textbooks and practical guides to carrying out benefit–cost analysis make no mention of Monte Carlo analysis, instead recommending conventional sensitivity analysis. See, for example, Brent (1996), Gramlich (1998), and Loomis and Helfand (2001). A key exception is the text by Boardman *et al.* (2001). EPA's own revised guidelines for carrying out economic analysis mention the possibility of using Monte Carlo analysis, but in the most recent version (EPA 2000) do not provide practical guidance. Whereas Monte Carlo analysis is superior to sensitivity analysis when both are viable options, sensitivity analysis still has value in addressing the implications of uncertainty in inputs for which a reasonable probability distribution cannot be developed. As Circular A-4 indicates "Sensitivity analysis is especially valuable when the information is lacking to carry out a formal probabilistic simulation" (p. 41).

10 The variance of a random variable is a measure of its dispersion, indicating how far the possible values of that variable typically are from its expected value.

11 When the true value of a variable such as net benefits is uncertain, the "mean" or "expected value" is the weighted average of all possible values for that variable, where the weight assigned to each possible value reflects the probability that that value is the true value. The "most likely" value of that variable is the value that has the highest probability of being the true value.

12 For example, the severe loss of life in New Orleans due to Hurricane Katrina in 2005 was a function of interactions among several disparate factors: (i) a low-probability storm intensity; (ii) a low-probability storm trajectory; (iii) low-probability failures of the levee system; and (iv) difficulties in achieving high evacuation rates. Simulations focused on a scenario that incorporated just one or even a pair of these factors would have masked the full range of possible outcomes associated with a hurricane making landfall on the Gulf Coast.

13 A probability distribution is the mathematical expression (function) that gives the probabilities that a particular variable equals each of a sequence of possible values.

14 For a discussion of methods of characterizing these uncertainties, see Morgan and Henrion (1990).

15 Also, the types of uncertainty that should be accounted for in a Monte Carlo analysis depend, to some extent, on the objectives of the analysis. For example, although there may be uncertainty regarding the influence of a regulation on individual households and the variability across households, analysts may wish to exclude consideration of variability across households if the goal of the analysis is to understand uncertainty regarding general impacts, such as the median household.

16 By the law of large numbers, the frequencies in the histogram will converge to the "true" underlying probabilities as the number of random draws (iterations) approaches infinity.

17 Skewness refers to the degree to which a distribution departs from symmetry around its mean value. Referring to Figure 2, the probability distribution of annualized net benefits of the proposed regulation is highly skewed to the right (with some relatively small probability of very high values), whereas the probability distribution of annualized net benefits of the alternative regulation is symmetric.

18 In some areas of regulation, such as the environmental realm, there is frequently greater uncertainty surrounding estimates of benefits than estimates of costs. Whereas a Monte Carlo analysis may show this difference, it does not favor either side of the ledger. In other words, it does not bias the resulting analysis in either direction.

19 See Borodovsky and Lore (2000).

20 Two uncertain inputs are correlated if the probability that one input's true value falls within a particular range of its possible values is related to, rather than independent of, another uncertain input's true value. Correlations can result from either the direct effect of one input's value on that of the other input, or from indirect relationships. Two inputs are positively (negatively) correlated if "high" values of one tend to be associated with "high" ("low") values of the other.

21 For further details on the final rule, see 69 Fed. Reg. 38,958 (29 June 2004).

22 See Jaffe and Stavins (2004).

23 The EPA's use of Monte Carlo analyses in evaluating this rule was intended to be illustrative only. Thus, the analyses did not lead to different regulatory decisions than EPA would have reached absent the Monte Carlo illustration. EPA states that the Monte Carlo analyses were designed to "*illustrate* how some aspects of ... uncertainty ... can be handled in a benefits analysis" (EPA 2004, pp. 9–29), and subsequently states that these analyses were not used to inform any regulatory decisions in the rulemaking (pp. 9–203).

24 In our analysis, no changes were made to the underlying benefit–cost analysis conducted by EPA. Therefore, the results from our Monte Carlo analysis can be compared with the results from other methods used by EPA to address uncertainty.

25 The C–R functions describe the relations between changes in ambient PM levels and changes in the incidence of particular health effects, such as premature mortality. They are used to estimate reductions in adverse health effects that can be expected from emissions reductions following the Rule's implementation.

26 See, for example, Koop and Tole (2004).

27 Given that our analysis omits several significant sources of uncertainty in the cost estimate, it is likely that our analysis understates the probability that costs will be outside the range defined by EPA.

28 That is, some may assume that while various outcomes may have different likelihoods of occurring, actual net benefits would be just as likely to fall somewhere in the first half of the range of possible values as in the second half.

29 US Environmental Protection Agency (2004, pp. 9–245, figure 9B-9). EPA's analysis of the Nonroad Diesel Rule's benefits focused on benefits in 2 years, 2020 and 2030. Because it will take several decades for the fleet of existing non-road engines to be completely replaced with new engines incorporating the required emissions controls, benefits in 2030 are more representative of the Rule's long-run annual benefits.

30 There are circumstances in which sensitivity analysis may provide insights of this type, particularly when there are very few uncertain inputs and the sensitivity analysis examines the implications of uncertainties in all inputs simultaneously. In practice, however, benefit–cost analyses in RIAs are rarely this simple.

31 US Environmental Protection Agency (2003, pp. 9-196–9-197). These income elasticities are used to estimate how the value that individuals place on avoiding PM-related health effects will change with changes in per-capita income.

32 This assumes that these income elasticities are uncorrelated.

33 US Environmental Protection Agency (2003, pp. 9–197, table 9B-6).

34 US Environmental Protection Agency (2004, pp. 9-213–9-245). In particular, see figure 9B-4.

35 Expert elicitation involves a "formal, highly structured and well documented process whereby expert judgments ... are obtained" (EPA 2004, pp. 9–214).

36 Examples of such software include Crystal Ball and @Risk, which function as companions to spreadsheet programs, such as Excel.

37 To be deemed a major final rule by OMB, a rule typically must be expected to have annual costs or benefits of at least $100 million.

38 Poulter (1998) provides an example of how probabilistic analysis was used by the Sierra Club to oppose licensing of the Diablo nuclear power plant.

39 See Figure 1 for an illustrative example. The means of communicating uncertainty in such a fashion have been addressed in published work. See, for example, Morgan and Henrion (1990, Ch. 9); US EPA (1997); and National Research Council (2002, Ch. 5).

References

Adler MD, Posner EA (eds) (2001) *Cost-Benefit Analysis: Legal, Economic, and Philosophic Perspectives*. The University of Chicago Press, Chicago.

Arrow KJ, Cropper ML, Eads GC, Hahn RW, Lave LB, Noll RG *et al.* (1996) Is There a Role for Benefit-Cost Analysis in Environmental, Health, and Safety Regulation? *Science* 272, 221–222.

Ayyub B (1997) *Uncertainty Modeling and Analysis in Civil Engineering*. CRC Press, London, UK.

Ayyub B (2003) *Risk Analysis in Engineering and Economics*. CRC Press, London, UK.

Boardman AE, Greenberg DH, Vinig AR, Weimer DL (2001) *Cost-Benefit Analysis: Concepts and Practice*. Prentice Hall, Upper Sadle River, NJ.

Borodovsky L, Lore M (eds) (2000) *The Professional's Handbook of Financial Risk Management*. Elsevier, Oxford, UK.

Brent RJ (1996) *Applied Cost-Benefit Analysis*. Edward Elgar, Cheltenham, UK.

Farrow RS, Wong E, Ponce RA, Faustman EM, Zerbe RO (2001) Facilitating Regulatory Design and Stakeholder Participation. In: Fishbeck PS, Farrow RS (eds) *Improving Regulation: Cases in Environment, Health, and Safety*, pp. 429–442. Resources for the Future Press, Washington, DC.

Ferson S (1996) What Monte Carlo Methods Cannot Do. *Human and Ecological Risk Assessment* 2, 990–1007.

Finkel AM (1990) *Confronting Uncertainty in Risk Management: A Guide for Decision-Makers*. Resources for the Future, Center for Risk Management, Washington, DC.

Glasserman P (2003) *Monte Carlo Methods in Financial Economics*. Springer, New York.

Goulder LH, Stavins RN (2002) An Eye on the Future: How Economists' Controversial Practice of Discounting Really Affects the Evaluation of Environmental Policies. *Nature* 419, 673–674.

Graham JD (2001) *Presidential Review of Agency Rulemaking by OIRA*. Memorandum for the President's Management Council, Regarding Presidential Review of Agency Rulemaking by OIRA. 20 Sept. [Cited 8 Sept 2005.] Available from URL: www.whitehouse.gov/omb/inforeg/oira_review-process.html/

Gramlich EM (1998) *A Guide to Benefit-Cost Analysis*, 2nd edn. Waveland Press, Prospect Heights, IL.

Hahn RW (2000) The Impact of Economics on Environmental Policy. *Journal of Environmental Economics and Management* 39, 375–399.

Hahn RW, Burnett JK, Chan Y-HI, Mader EA, Moyle PR (2000) Assessing Regulatory Impact Analyses: The Failure of Agencies to Comply With Executive Order 12,866. *Harvard Journal of Law & Public Policy* 23, 859–885.

Hahn RW, Olmstead SM, Stavins RN (2003) Environmental Regulation in the 1990s: A Retrospective Analysis. *Harvard Environmental Law Review* 27, 377–415.

Hammersley JM, Handscomb DC (1964) *Monte Carlo Methods*. John Wiley & Sons, New York.

Jaffe J, Stavins RN (2004) *The Value of Formal Quantitative Assessment of Uncertainty in Regulatory Analysis*. Related Publication No.: 04-22. AEI-Brookings Joint Center for Regulatory Studies, Washington, DC.

Koop G, Tole L (2004) Measuring the Health Effects of Air Pollution: To What Extent Can We Really Say That People Are Dying From Bad Air? *Journal of Environmental Economics and Management* 47, 30–54.

Krupnick A, Morgenstern R, Batz M, Nelson P, Burtraw D, Shih J-S, McWilliams M (2006) *Not a Sure Thing: Making Regulatory Choices Under Uncertainty*. Resources for the Future, Washington, DC.

Linkov I, Burmistrov D (2003) Model Uncertainty and Choices Made by Modelers: Lessons Learned from the International Atomic Energy Agency Model Intercomparisons. *Risk Analysis* 23, 1297–1308.

Loomis J, Helfand G (2001) *Environmental Policy Analysis for Decision Making*. Kluwer Academic Publishers, Boston.

McLeish DL (2005) *Monte Carlo Simulation and Finance*. John Wiley & Sons, New York.

Morgan MG, Henrion M (1990) *Uncertainty: A Guide to Dealing with Uncertainty in Quantitative Risk and Policy Analysis*. Cambridge University Press, Cambridge.

Morgenstern RD (1997) The Legal and Institutional Setting for Economic Analysis at EPA. In: Morgenstern RD (ed.) *Economic Analysis at EPA: Assessing Regulatory Impact*, pp. 5–23. Resources for the Future Press, Washington, DC.

National Research Council (2002) *Estimating the Public Health Benefits of Proposed Air Pollution Regulations*. The National Academies Press, Washington, DC.

Poulter SR (1998) Monte Carlo Simulation in Environmental Risk Assessment – Science, Policy and Legal Issues. *Risk: Health, Safety & Environment* 7 (Winter), pp. 7–26.

Stavins R, Jaffe J, Hess P, Crouch E, Green L, Sahay S (2004) *Monte Carlo Analysis: An Introduction to and Application of Formal Quantitative Assessment of Uncertainty in Regulatory Impact Analysis.*

US Environmental Protection Agency (EPA) (1997) *Guiding Principles for Monte Carlo Analysis.* Mar EPA/630/R-97/001. EPA, Washington, DC.

US Environmental Protection Agency (EPA) (2000) *Guidelines for Preparing Economic Analyses.* Sept EPA 240-R-00-003. Office of the Administrator, EPA, Washington, DC.

US Environmental Protection Agency (EPA) (2003) *Draft Regulatory Impact Analysis: Control of Emissions from Nonroad Diesel Engines.* Apr EPA420-R-03-008. EPA, Washington, DC.

US Environmental Protection Agency (EPA) (2004) *Final Regulatory Impact Analysis: Control of Emissions from Nonroad Diesel Engines.* May EPA420-R-04-007. EPA, Washington, DC.

US Office of Management and Budget (OMB) (2003) Washington, DC. 17 Sept Circular A-4.

US Office of Management and Budget (OMB) (2006) *Draft 2006 Report to Congress on the Costs and Benefits of Federal Regulations.* OMB, Washington, DC. Apr

PART III

ECONOMIC ANALYSIS OF ENVIRONMENTAL POLICY INSTRUMENTS

[8]

Vintage-Differentiated Environmental Regulation

Robert N. Stavins*

---

*Albert Pratt Professor of Business and Government, John F. Kennedy School of Government, and Director, Environmental Economics Program at Harvard University; and University Fellow, Resources for the Future. Research assistance was provided by Barry Newstead. Lori Bennear, Dallas Burtraw, Robert Hahn, Louis Kaplow, Nathaniel Keohane, Peter Nagelhout, Richard Newell, Steven Shavell, Matthew Stephenson, participants at the Law and Economics Seminar at Harvard Law School, and the editors of this journal provided helpful comments on a previous version. Financial support was provided by the National Center for Environmental Economics of the United States Environmental Protection Agency, but the author is solely responsible for any errors.

I. INTRODUCTION

Vintage-differentiated regulation (VDR) is a common feature of many environmental and other regulatory policies in the United States. Under VDR, standards for regulated units are fixed in terms of the units' respective dates of entry, or "vintage," with later entrants facing more stringent regulation. In the most common application, often referred to as "grandfathering," units produced prior to a specific date are exempted from new regulation or face less stringent requirements.

The vintage-differentiated approach has long appealed to many participants in the policy community, for reasons associated with efficiency, equity, and simple politics. First, it is frequently more cost-effective—in the short term—to introduce new pollution-abatement technologies at the time that new plants are constructed than to retrofit older facilities with such technologies. Second, it seems more fair to avoid changing the rules of the game in midstream, and hence to apply new standards only to new plants. Third, political pressures tend to favor easily identified existing facilities rather than undefined potential facilities.

On the other hand, VDRs can be expected—on the basis of standard investment theory—to retard turnover in the capital stock (of durable plants and equipment), and thereby to reduce the cost-effectiveness of regulation in the long term, compared with equivalent undifferentiated regulations.[1] A further irony is that, when this slower turnover results in delayed adoption of new, cleaner technology, VDR can result in higher levels of pollutant emissions than would occur in the absence of regulation.

In this Article, I survey previous applications and synthesize current thinking regarding VDRs in the environmental realm, and develop lessons for public policy and for future research. In Part II, I describe the ubiquitous nature of VDRs in American regulatory policy, and examine the reasons why VDRs are so common. In Part III, I establish a theoretical framework for analysis of the cost-effectiveness of alternative types of environmental policy instruments to provide a context for the analysis of VDRs. In Part IV, I focus on the effects of VDRs, and

1. *See generally* RICHARD A. BREALEY & STEWART C. MYERS, PRINCIPLES OF CORPORATE FINANCE 15-29 (5th ed. 1996) (providing background on investment theory).

describe a general theory of the impacts of these instruments in terms of their effects on technology adoption, capital turnover, pollution abatement costs, and environmental performance. In Parts V and VI, I examine empirical analyses of the impacts of VDRs in two significant sectors: Part V focuses on the effects of VDRs in the American auto industry, and Part VI on the effects of new source review, which is a form of VDR, in power generation and other sectors. In Part VII, I examine implications for policy and research, and conclude in Part VIII by recommending avenues for improvements in both.

II. THE PREVALENCE OF VINTAGE-DIFFERENTIATED REGULATION

VDRs are found throughout American regulatory policy. They are prominent features of a diverse set of federal environmental statutes and regulations, state and local environmental laws, as well as of a host of non-environmental regulations (Table 1). In this section of the Article, I summarize the use of VDRs in the environmental realm, and then examine explanations for the prevalence of these approaches.

A number of important federal environmental laws make use of VDRs. For example, VDRs appear within the Clean Air Act in its standards for emissions from new versus existing stationary sources, motor vehicle and motor vehicle engines, non-road engines and vehicles, and commercial vehicles; within the Clean Water Act in a wide variety of aspects, including in effluent limits for public treatment plants; within the Safe Drinking Water Act; and within laws affecting the generation and disposal of hazardous and solid waste.[2] State and local environmental laws also make frequent use of VDRs, for stationary and mobile source emissions limits and energy efficiency standards in new construction, among other instances.[3] In the non-environmental realm, VDRs appear in a variety of occupational health and safety laws, automotive safety regulations, consumer product safety laws, and building codes.[4]

---

2. *See infra* Table 1.

3. For example, California has established differentiated emissions standards for new motor vehicles. *See infra* Table 1.

4. For example, the Occupational Safety and Health Act of 1970 seeks to assure men and women of safe working conditions, but some regulations under the statute exempt facilities built before specific dates. *See* Occupational Safety and Health Act of 1970, 29 U.S.C. §§ 651-678 (2005). Likewise, many cities have adopted regulations from fire codes written by national technical organizations, such as the National Fire Protection Association (NFPA); some older facilities—such as dry cleaning shops—are not required

Why have VDRs been such a common feature of American regulatory policy? As suggested above, proponents frequently claim that VDRs are efficient and equitable. Indeed, it is typically cheaper to build a new plant with required technology in the first place than to retrofit an older, existing plant. In terms of equity, it seems fair to avoid changing the rules for facilities that have already been built or for products that have already been manufactured. On the other hand, the claim of efficiency is based on a static view that ignores the perverse dynamic incentives that such a time differentiated standard can put in place.[5] Such standards give firms an incentive to continue using (exempt) older polluting plants, rather than investing in new, more environmentally friendly plants that would fall under more stringent regulations. And, in terms of equity, VDRs by definition lack a "level playing field," which hardly appears fair from the perspective of those entering the market and facing the more stringent component of an age differentiated regulation.

A broader explanation for the prevalence of VDRs is fundamentally political. Both regulated constituents and legislators have very strong incentives to favor these policies.[6] First, on the regulatory demand side, some types of regulations can augment firms' profits through the generation of rents[7] and erection of entry barriers. Consider a simple model of an industry made up of

---

to comply with codes that were passed after the shops began operation. Consumer product safety laws, such as regulations promulgated under the Consumer Product Safety Act of 1972, 15 U.S.C. §§ 2051-2084 (2005), typically control (either by design standard or outright banning) new products manufactured for sale, offered for sale, distributed in commerce, or imported into the United States, but do not control products already purchased and in use. Similarly, automotive safety regulations—such as passive passenger restraint requirements—affect new cars, not the entire existing stock of automobiles. And, obviously, most building codes—including those regulating lot size and setback requirements—affect new construction, not previously built structures. *See generally* REDWOOD KARDON, MICHAEL CASEY, AND DOUGLAS HANSEN, CODE CHECK BUILDING: A FIELD GUIDE TO BUILDING CODES. (2005) (providing a review of national building codes).

5. It is by no means impossible, however, to design a VDR that accelerates investment. The nitrous oxide (NO_x) requirements in Titles I and IV of the Clean Air Act Amendments of 1990, for example, provided an inducement in the form of a less ambitious standard for early adopters. *See* Clean Air Act Amendments, Pub. L. No. 101-549, 104 Stat. 2399 (1990).

6. *See generally* Nathaniel Keohane, Richard Revesz & Robert Stavins, *The Choice of Regulatory Instruments in Environmental Policy*, 22 HARV. ENVTL. L. REV. 313 (1998) (developing a political economy model of the supply and demand for alternative environmental policy instruments).

7. Economic rents are payments made to a factor of production—labor, land, or capital—that are in excess of what is required to elicit the supply of that factor.

identical firms in long run competitive equilibrium. If the government imposes a command-and-control standard that sets an allowable level of pollution for each firm, and firms can meet the standard only by reducing their output, the standard will lead to reduced total production and therefore an increase in price along the aggregate demand curve.[8] If the environmental restriction is not exceptionally severe, the new price will be above average cost for all firms, and firms will earn rents. Firms, however, are not limited to the single response of cutting output. They can also reduce emissions by adopting a new technology or changing their input mix. Yet, in this more realistic scenario, command-and-control standards can still have the effect of conveying rents to regulated firms, depending on the stringency of the standards and other factors.[9]

These rents, however, are only sustainable to the extent that other participants cannot freely enter the market—enhanced industry profitability resulting from rents is sustainable *only* in the presence of entry restrictions.[10] In theory, such a mechanism might prohibit new entry outright, but a more politically feasible approach simply imposes higher costs on new entrants.[11] This is precisely what VDRs do. Accordingly, private firms and their trade associations demand command-and-control standards that are more stringent for new sources. This is one important explanation for the prevalence of VDRs in American environmental laws.

Another demand-side explanation for the prevalence of VDRs arises from the differential costs of environmental compliance across firms resulting from varying pollution abatement costs from one firm to another. Because of this heterogeneity, a firm may

---

8. James M. Buchanan & Gordon Tullock, *Polluters' Profits and Political Response: Direct Controls Versus Taxes*, 65 AM. ECON. REV. 139, 139-41 (1975) (setting out the basic theoretical model of how conventional regulations can convey rents to firms).

9. Michael T. Maloney & Robert E. McCormick, *A Positive Theory of Environmental Quality Regulation*, 25 J.L. & ECON. 99, 105, 122 (1982).

10. Higher than normal profits in an industry, due to rents, will encourage other firms to enter the market until competition drives down prices, eliminating the rents. Only at this point will other firms cease to enter the market. Hence, barriers to entry are necessary for the sustainability of such rents.

11. George J. Stigler, *The Theory of Economic Regulation*, 2 BELL J. ECON. 3, 5 (1971) (examining how the powers of the state can be used to assist firms through the control over entry of potential rivals); *see generally* Eric Rasmusen & Mark Zupan, *Extending the Economic Theory of Regulation to the Form of Policy*, 72 PUB. CHOICE 167 (1991). (examining how firms choose desired policies from the set including entry barriers, price floors, subsidies, and demand stimulation).

support policy instruments that impose costs on it, as long as those costs affect it less than the industry average and thus give it a competitive advantage.[12] One form of cost differential arises as a result of the erection of barriers to entry. It is important to distinguish here between the entry of new firms and the expansion of existing firms, but the entry barriers of environmental regulation generally apply to both situations. Thus, within an industry, firms with no plans to expand would derive a greater benefit from entry barriers, which could discourage further growth by their competitors.

Turning to the supply side of regulation, legislators are typically more concerned with the distribution of costs and benefits—in particular their geographic distribution— than with a comparison of total benefits and costs.[13] Politicians are likely to oppose instruments that may induce firms to close business in their districts and relocate elsewhere, leading to localized unemployment.[14] Although there will be winners as well as losers from such relocation, potential losers are likely to be more certain of their status, and thus exert greater political pressure, than potential gainers.[15] This asymmetry creates a bias in favor of the status quo.

For the same reason, grandfathering is likely to be a politically expedient option for legislators, since it allows leeway in rewarding firms and in distributing the costs and benefits of regulation among jurisdictions. By limiting the scope of regulation to new capital assets, the burden of regulatory compliance is concentrated on a small subset of the electorate and the cost is transferred to unspecified, future "new sources." In addition, the costs of such regulation are less obvious when combined with major investments in new capital.[16] In contrast, if regulations were to require retrofits to existing capital assets, the costs would be vastly more

12. Robert A. Leone & John E. Jackson, *The Political Economy of Federal Regulatory Activity: The Case of Water-Pollution Controls*, *in* STUDIES IN PUBLIC REGULATION 231, 247, 254, 260 (Gary Fromm ed., 1981); Sharon Oster, *The Strategic Use of Regulatory Investment by Industry Sub-Groups*, 20 ECON. INQUIRY 604, 610-611 (1982).

13. Robert W. Hahn & Robert N. Stavins, *Incentive-based Environmental Regulation: A New Era from an Old Idea?*, 18 ECOLOGY L.Q. 1, 40-41 (1991).

14. Robert W. Hahn & Roger G. Noll, *Environmental Markets in the Year 2000*, 3 J. RISK & UNCERTAINTY 351, 358 (1990).

15. When a firm leaves an area, the "losers"—such as those losing their jobs—are easily identified, whereas there is considerable uncertainty regarding the "gainers," i.e. the identification of those individuals who will actually succeed in getting the new jobs.

16. *See* Keohane, Revesz, & Stavins, *supra* note 6 at 359, 364.

transparent.[17]

In summary, the prevalence of VDRs can be understood not simply through claims made about the efficiency and fairness of these approaches, but also through fundamental political forces. Demand for VDRs comes from existing firms, which seek to erect entry barriers to restrict competition and to protect the rents created by command-and-control standards. Environmentalists often support strict standards for new sources because they represent environmental progress, at least symbolically.[18] On the supply side, more stringent standards for new sources allow legislators to protect existing constituents and interests by placing the bulk of the pollution control burden on future facilities.

Recognizing the ubiquity of VDRs in environmental regulation in the United States, I next develop an analytic framework within which these mechanisms can be normatively assessed.

17. In the context of automobile emissions regulation, Robert W. Crandall and his co-authors noted that "owners of existing equipment can organize to fight regulation that adversely affects their interests. For instance, California automobile owners successfully fought retrofit requirements in the 1960s. In contrast, future purchasers of new equipment do not constitute as cohesive an interest group; therefore regulation tends to be focused upon new vehicles." ROBERT W. CRANDALL ET AL., REGULATING THE AUTOMOBILE 89 (1986).

18. *See* Keohane, Revesz, & Stavins, *supra* note 6 at 354-56.

TABLE 1: EXAMPLES OF VINTAGE-DIFFERENTIATED REGULATIONS IN AMERICAN ENVIRONMENTAL LAW

Regulatory Area	**Application**	**Statute**
Air Quality	Performance Standards for New and Existing Stationary Sources	Clean Air Act, 42 U.S.C. § 7411 (2005) (establishing differentiated program for stationary sources); 42 U.S.C. §§ 7521-7554 (2005) (establishing differentiated standards for mobile sources).
	Motor Vehicle and Motor Vehicle Engine Emissions Standards	Clean Air Act, 42 U.S.C. § 7521 (2005) (establishing emissions standards for motor vehicles based on model year).
	Non-Road Engines and Vehicles	Clean Air Act, 42 U.S.C. § 7547 (2005) (applying standards only to new sources).
	Commercial Vehicle Standards	Clean Air Act, 42 U.S.C. § 7554 (2005) (setting standards that apply only to buses manufactured subsequent to the 1993 model year).
Water Quality	Effluent Limits for Public Treatment Plants	Clean Water Act, 33 U.S.C. § 1311(a)(1)(B)-(C) (2005) (setting differential standards for plants in operation prior to July 1, 1977).
	Drinking Water Treatment	Safe Drinking Water Act, 42 U.S.C. § 300g-9(d)(4) (2005) (requiring EPA to set standards for new plants).
Waste Management	Generation and Disposal of Hazardous and Solid Waste	42 U.S.C. § 6925 (2005) (limiting new hazardous waste management to new facilities).
State & Local Environmental Laws	Stationary and Mobile Source Emission Limits; Energy-Efficiency Standards for New Construction; etc.	*See, e.g.*, CAL. HEALTH & SAFETY CODE § 39665-39669 (2005) (California Air Resources Board regulations setting differentiated emissions standards for new vehicles).

III. THE COST-EFFECTIVENESS OF ALTERNATIVE ENVIRONMENTAL POLICY INSTRUMENTS

In order to assess the role of VDRs in environmental policy, it is necessary to identify normative criteria by which alternative policy instruments can be judged. A variety of normative criteria have been posited as relevant for choosing environmental policy instruments, including: (1) whether the policy instrument will achieve the stated goal or standard; (2) whether it will do so at the lowest possible cost, including both private sector compliance and public sector monitoring and enforcement; (3) whether it provides government with the information it needs to implement the policy; (4) whether the instrument will be flexible in the face of changes in tastes and technology; (5) whether the instrument provides dynamic incentives for research, development, and adoption of better pollution abatement technologies; (6) whether the implementation of the policy instrument will result in an equitable distribution of the benefits and costs of environmental protection; and (7) whether the policy will be politically feasible in terms of enactment and implementation.[19] Items (1) through (5) together refer to a comprehensive description of the criterion of *cost-effectiveness*, while item (6) refers to *distributional equity*, and item (7) refers to *political feasibility*.[20] This section focuses on the cost-effectiveness criterion.

Cost-effectiveness refers to the allocation of pollution control requirements among sources which results in an aggregate pollution target being achieved at the lowest possible cost. It is well known that the necessary condition for cost-effectiveness is that all sources (that exercise some degree of control) experience the same marginal abatement costs.[21] Recall that the marginal costs of pollution control are the additional or incremental costs of achieving an additional unit of pollution reduction. If these marginal costs of control are not equal across sources, then the

19. *See* Peter Bohm & Clifford F. Russell, *Comparative Analysis of Alternative Policy Instruments*, *in* 1 HANDBOOK OF NATURAL RESOURCES AND ENERGY ECONOMICS 395, 399-402 (Allen V. Kneese & James L. Sweeney eds., 1985) (proposing a list of criteria for environmental regulations). For further details, see Richard L. Revesz & Robert N. Stavins, *Environmental Law and Policy*, *in* HANDBOOK OF LAW AND ECONOMICS (A. Mitchell Polinsky & Steven Shavell eds.) (forthcoming 2005).

20. *See* Bohm & Russell, *supra* note 19.

21. WILLIAM J. BAUMOL & WALLACE E. OATES, THE THEORY OF ENVIRONMENTAL POLICY 165-67 (Cambridge University Press 1988) (1975).

same aggregate level of pollution control could be achieved at lower overall cost simply by reallocating the pollution control burden among sources so that the low-cost sources controlled proportionately more pollution and the high-cost sources controlled proportionately less. Additional savings could theoretically be achieved through such reallocations until marginal costs were identical at all sources. Thus, when examining alternative types of environmental policy instruments, a key question is whether particular instruments are likely to result in marginal abatement costs being equated across sources—the point at which they are cost-effective.

The most frequently employed division of environmental policy instruments is between command-and-control and market-based approaches. Conventional approaches to regulating the environment—frequently characterized as command-and-control[22]—allow relatively little flexibility in the means of achieving goals. Such policy instruments tend to force firms to take on similar shares of the pollution control burden, regardless of the cost, sometimes by setting uniform standards for firms, the most prevalent of which are technology- and performance-based standards.[23]

Market-based instruments, including pollution charges, tradeable permits, market-friction reductions, and government subsidy reductions, encourage behavior through market signals, rather than through explicit directives regarding pollution control levels or methods.[24] These policy instruments can be described as

22. The phrase "command-and-control" is by far the most commonly employed characterization of conventional environmental policy instruments, including uniform performance and technology standards. Admittedly, the phrase has an inescapable negative stigma associated with it, and so a better, more neutral description of this category of policy approaches might be "prescriptive instruments." Nonetheless, "command-and-control" is the generally accepted name for this category.

23. Note that uniform standards can specify the amount of pollution that can be released into the environment (an emission standard) or the permissible concentration of pollution in the air, water, or soil (an ambient standard). The cost-effective allocation consistent with ambient standards requires equalization of the marginal costs to reduce a unit of ambient concentration, rather than emission.

24. *See* Hahn & Stavins, *supra* note 13, at 7-15. *See also* Robert N. Stavins, *Experience with Market-Based Environmental Policy Instruments*, *in* 1 THE HANDBOOK OF ENVIRONMENTAL ECONOMICS 355, 360-61 (K.G. Mäler & J. Vincent eds., Elsevier 2003). Liability rules can also be thought of as a market-based instrument, because they provide incentives for firms to take into account the potential environmental damages of their decisions, allowing full flexibility in technology and control practices. *See generally* RICHARD L. REVESZ, FOUNDATIONS IN ENVIRONMENTAL LAW AND POLICY (1997) (considering liability rules

"harnessing market forces,"[25] because, when well designed and properly implemented, they encourage firms or individuals to undertake pollution control efforts through generation of an economic interest in compliance.

Where there is significant heterogeneity of costs, holding all firms to the same target will be unduly expensive, and command-and-control methods will not be cost-effective. Real world control costs can vary enormously due to production design, physical configuration, age of assets, and other factors. Indeed, abatement cost heterogeneity is a ubiquitous and important characteristic of many environmental problems.[26] The cost of controlling a unit of a given pollutant may vary by a factor of 100 or more among sources.[27] This heterogeneity argues for the use of market-based instruments in many cases.

If properly designed and implemented, market-based instruments can allow any desired level of pollution cleanup to be realized at the lowest overall cost to society by providing incentives for the greatest reductions in pollution to those firms that can achieve the reductions most cheaply. Rather than equalizing pollution levels among firms, market-based instruments have the effect of encouraging firms to equalize their marginal abatement

among other market-based instruments). Other taxonomies of regulatory instruments are possible, and some take a more inclusive view, including, for example, contractual approaches. *See generally* ENVIRONMENTAL LAW (Peter S. Menell, ed. 2002) (including contractual approaches as a category of environmental policy instruments).

25. *See generally* J.B. OPSCHOOR & HANS B. VOS, ECONOMIC INSTRUMENTS FOR ENVIRONMENTAL PROTECTION (Org. for Econ. Coop. and Dev. 1989) (surveying economics-based environmental policy instruments); ECON. INCENTIVES TASK FORCE, U.S. ENVTL. PROT. AGENCY, PUB. NO. 21P-2001, ECONOMIC INCENTIVES: OPTIONS FOR ENVIRONMENTAL PROTECTION (1991) (discussing economic incentives in solid waste, climate change, and water resources contexts); ALAN CARLIN, U.S. ENVTL. PROT. AGENCY, PUB. NO. EPA-230-R-92-001, THE UNITED STATES EXPERIENCE WITH ECONOMIC INCENTIVES TO CONTROL ENVIRONMENTAL POLLUTION (1992) (for early assessments of the use of market-based instruments in environmental policy).

26. TOM H. TIETENBERG, EMISSIONS TRADING: AN EXERCISE IN REFORMING POLLUTION POLICY 16, 42-43 (1985) (describing heterogeneity in abatement costs).

27. For numerical examples of the heterogeneity of incremental costs of pollution control, see Robert W. Crandall, *The Political Economy of Clean Air: Practical Constraints on White House Review*, *in* ENVIRONMENTAL POLICY UNDER REAGAN'S EXECUTIVE ORDER: THE ROLE OF BENEFIT-COST ANALYSIS 205, 210-15 (V. Smith ed. 1984). But where costs are similar among sources, command-and-control instruments may perform equivalently to (or better than) market-based instruments, depending on transactions costs, administrative costs, possibilities for strategic behavior, political costs, and the nature of the pollutants. *See generally* Richard G. Newell & Robert N. Stavins, *Cost Heterogeneity and the Potential Savings from Market-Based Policies*, 23 J. REG. ECON. 43 (2003) (analyzing the effects of abatement cost heterogeneity on the magnitude of aggregate abatement costs).

costs.[28] Command-and-control approaches could, in theory, achieve this cost-effective solution, but this would require that different standards be set for each pollution source, and, consequently, that policy makers obtain detailed information about the compliance costs each firm faces. Such information, which is privately generated by firms when it is known at all, is simply not available to government. Thus, potentially cost-effective, non-uniform performance standards are generally considered to be infeasible, both because of the asymmetric information problem identified above and because unequal standards would raise questions of fairness among policy makers.

If the government has reason to believe that abatement costs are highly correlated with the (more easily observed) vintage of sources, despite the fact that it cannot observe source specific abatement costs, then an appropriately designed vintage-differentiated regulation could theoretically serve as an approximation of a cost-effective, non-uniform standard. The quality of that approximation would depend, of course, on the degree of correlation of abatement costs with the VDR's quantitative pollution standards. Given that abatement cost heterogeneity is typically distributed continuously and as a function of many factors, it is unlikely that a simple, one-dimensional and dichotomous VDR would serve as a good approximation.[29] Nevertheless, this is an empirical question.[30]

Even if a VDR were perfectly correlated with underlying abatement cost heterogeneity, and therefore cost-effective in the short term, such an instrument, like any non-uniform performance standard, would still fail to provide dynamic incentives for sources

---

28. David W. Montgomery, *Markets in Licenses and Efficient Pollution Control Programs*, 5 J. ECON. THEORY 395, 400-01 (1972); Baumol & Oates, *supra* note 21, at 168, 182.

29. This assessment of the potential static cost-effectiveness of VDRs is specific to the realm of pollution abatement, for the reasons given above. For an analysis of the economic implications of a much broader class of policies, of which VDRs are only one category, and environmental VDRs a still smaller subset, see generally Steven Shavell, *Optimal Legal Change When Adjustment to Legal Change is Costly* (Harvard Law School, Working Paper, 2005) (on file with author). The key previous work along those lines is by Louis Kaplow. *See* Louis Kaplow, *An Economic Analysis of Legal Transitions*, 99 HARV. L. REV. 509, 582-92 (1986); Louis Kaplow, *Government Relief for Risk Associated with Government Action*, 94 SCANDINAVIAN J. ECON. 525, 534-35 (1992).

30. The existing theoretical and empirical literature on environmental VDRs (surveyed in subsequent Parts of this paper) gives very little attention to the potential static cost-effectiveness of VDRs as dichotomous approximations to unobservable continuous distributions of abatement-cost heterogeneity. Rather, this possibility seems to be taken for granted, a reality that was brought home to me by Shavell, *supra* note 29.

to adopt cheaper and better pollution control technologies over time. With market-based instruments, most clearly with emission taxes, firms benefit by cleaning up a bit more than is required if a sufficiently low-cost method (technology or process) of doing so can be identified and adopted.[31]

Furthermore, VDRs, unlike market-based instruments, and indeed unlike ordinary non-uniform standards, present another problem: an environmentally negative dynamic incentive for regulated firms to extend the useful lives of their plant and equipment, thereby reducing cost-effectiveness and possibly retarding environmental progress. This issue—the effect of VDRs on investment and maintenance decisions and the implications for abatement costs and environmental performance—is the focus of the remainder of this article.

IV. IMPACTS THEORY OF VINTAGE-DIFFERENTIATED REGULATION

The core theory of the effects of vintage-differentiated regulation on the lives of durable plant and equipment comes from the general theory of the determinants of the age of capital.[32] Consider a simple model in which a firm's production decision can be characterized by the choice of two variables: the number of pieces of durable equipment to be operated and the lives of each of these devices. Inevitably, these pieces of durable equipment (for example, machines) degrade over time. In other words, maintenance costs rise as the machines age. If competitive firms seek to maximize their profits, they will place new machines in service if and only if the present discounted value of the expected future net revenue per machine (revenue from its use minus maintenance costs) exceeds a new machine's initial cost. This

---

31. Paul B. Downing & Lawrence J. White, *Innovation in Pollution Control*, 13 J. ENVTL. ECON. & MGMT. 18, 21-22 (1986); Scott R. Milliman & Ronald Prince, *Firm Incentives to Promote Technological Change in Pollution Control*, 17 J. ENVTL. ECON. & MGMT. 247, 257-58 (1989); Adam B. Jaffe & Robert N. Stavins, *Dynamic Incentives of Environmental Regulations: The Effects of Alternative Policy Instruments on Technology Diffusion*, 29 J. ENVTL. ECON. & MGMT. S43, S45-46 (1995).

32. Robert N. Stavins, *The Effects of Vintage-Differentiated Environmental Regulation*, 2005 AEI-BROOKINGS JOINT CTR. FOR REGULATORY STUDIES, Related Pub. 05-03, 4 (explicating a mathematical model of differentiated regulation); *see also* Michael T. Maloney & Gordon L. Brady, *Capital Turnover and Marketable Pollution Rights*, 31 J.L. & ECON 203, 207-09 (1988) (describing a simpler model that this article employs); *see generally* Howard K. Gruenspecht, Differentiated Social Regulation in Theory and Practice (1981) (unpublished Ph.D. dissertation, Yale University) (providing a detailed theoretical model of differentiated regulation).

theory leads to the intuitive profit-maximizing rule that machines should be run until maintenance costs equal net revenues.

The impacts of vintage-differentiated environmental regulation can be understood with this simple model of capital replacement. VDR endows old machines with a value that cannot be transferred, since new machines are rendered systematically more costly due to more stringent regulation. The effective price of new capital is increased, and old capital is accordingly retired later. Hence, the average age of capital increases and the rate of investment decreases.[33]

A more specialized model, with explicit representations of product and pollutant output reinforces the finding that the systematic bias against new sources in VDR has the effect of retarding the rate of turnover of the capital stock.[34] Pollution abatement costs are therefore higher under a VDR scheme than under an equivalent undifferentiated regulation scheme. Strikingly, the relationship between aggregate pollutant emissions and regulatory stringency can be perverse, with more stringent regulation leading to increased aggregate emissions. This can occur when VDR is applied to technologies in which newer vintages have lower pollutant emissions (even in the absence of the vintage-differentiated regulation).[35] Requiring pollution control technologies only for new sources forces a choice between continuing to operate older, uncontrolled equipment and adopting new, cleaner equipment with the additional expense of a pollution control technology. Such conditions are typical in the environmental regulatory sphere.[36] Thus, increasing the stringency

---

33. A more complex model would also allow for endogenous maintenance. As is discussed in Parts V and VI below, VDRs create incentives to extend the useful lives of equipment in two ways: first, by keeping it longer (as in the model above), and second by investing more in maintenance throughout the life of the equipment than would otherwise be the case. The latter incentive reinforces the capital investment effect examined above. For a general discussion of endogenous maintenance, *see* Nathaniel Keohane et al., *Controlling Stocks and Flows to Promote Quality: The Environment, with Applications to Physical and Human Capital* (NAT'L BUREAU OF ECON. RESEARCH 31-36, Working Paper No. 7727, 2000) (discussing optimal maintenance of physical capital).

34. *See generally* Gruenspecht, *supra* note 32. (describing and using such a model).

35. These effects are heightened, of course, if the older technology in question exhibits increasing emissions (degraded abatement) as it ages.

36. As market forces push technological developments to reduce economic waste, the result can be more efficient use of energy, fewer discharges, and so on. Hence, in the environmental realm, it is frequently the case that new generations of technologies exhibit lower pollutant emissions (even in the absence of vintage differentiated regulations intended to bring down emissions over time).

of a VDR can drive up costs *and* simultaneously drive down benefits (by increasing emissions due to the continued use of old equipment without emission control).

In summary, VDRs can have significant impacts on decisions regarding capital investment. A regulatory bias against new sources will reduce investment in new facilities and lengthen the economic lives of existing facilities. Most remarkable, more stringent VDRs can, at least in the short term, be environmentally counterproductive, increasing the aggregate emissions level, rather than reducing it, if they discourage the use of cleaner new technology strongly enough.[37] Granted, in theory, a more stringent VDR would ultimately lead to lower aggregate emissions as the older, retained vintage of equipment is retired. Nevertheless, the perverse effects on investment can lead to short-term emissions increases.

A key question is, therefore, how long the "short term" will be. If the short term is truly short, perverse effects of VDRs will be minimized. However, the trade-off between short term and long-term effects becomes problematic when the sources subject to regulation have especially low rates of deterioration and technical obsolescence—conditions that are met by power plants, for example. In this case, the "short term" is very long, on the scale of decades, and vintage-differentiated regulations are likely to be particularly problematic. It is an empirical question, however, as to whether the extreme, counterproductive outcome actually occurs in any given case. I examine empirical evidence of the effects of VDRs in the auto industry (where turnover might be expected to be relatively high) and in the power generation industry (where turnover is very low) in parts V and VI, below.

V. VINTAGE-DIFFERENTIATED REGULATION IN THE AMERICAN AUTO INDUSTRY

Vintage differentiation has been a prominent feature of American motor vehicle regulation, including automobile emissions standards under the Clean Air Act, automotive safety standards under the National Traffic and Motor Vehicle Safety Act (1966), and the Corporate Average Fuel Economy (CAFE) standards under the Energy Policy and Conservation Act

37. Howard K. Gruenspecht, *Differentiated Regulation: The Case of Auto Emissions Standards*, 72 AM. ECON. REV. 328, 328 (1982).

amendments (1975).[38] In all three cases, more stringent standards apply to vehicles of later vintages. Research in this area suggests the following conclusions. First, VDRs have extended the useful lives of cars on the road, and therefore theoretically may have increased aggregate emissions.[39] But empirical analyses addressing whether motor vehicles—with their relatively rapid rate of turnover—have actually exhibited counterproductive emissions increases due to VDRs have been less clear-cut. Whereas cost-effectiveness was certainly compromised, impacts on aggregate emissions appear to have been small (although statistically significant) and temporary.[40] Furthermore, the costs of retrofitting motor vehicles for pollution control or fuel efficiency are typically prohibitive (less so for some safety standards),[41] and so in this realm the simple replacement of VDRs with undifferentiated regulation (that is, elimination of grandfathering) is not an attractive or even feasible alternative. Rather, a number of alternative regulatory strategies may hold some promise.[42]

A. *Theoretical Analysis of the Effects of Automobile VDR*

In general terms, the anticipated effects of automobile VDR are linked with a retirement decision regarding a durable good. With well-functioning markets for both new and used cars, it is reasonable to anticipate that, on average, cars will be scrapped if and only if the net market value of a car in operable condition is less than that car's repair cost.[43]

Since new and used cars are substitutes for each other, any increase in the price or operating costs of new cars causes substitution of used cars, increasing demand in that relatively inelastic[44] market and thereby driving up the price of used cars.

38. Clean Air Act, Pub. L. No. 95-95, 91 Stat. 865, (codified as amended at 43 U.S.C. §§ 7401-7661) (1977); National Traffic and Motor Vehicle Safety Act, Pub. L. No. 89-563, 80 Stat. 718 (codified as amended at 15 U.S.C. §§ 1381-1431); Energy Policy and Conservation Act Amendments, Pub. L. No. 94-163, 89 Stat. 871 (codified as amended at42 U.S.C. §§ 6231-6247) (1975).

39. See notes 52-55 *infra*, and accompanying text.

40. See notes 52-54 *infra*, and accompanying text.

41. *See* Crandall et al., *supra* note 17, at 89.

42. *See infra* Part VII.

43. *See* Gruenspecht, *supra* note 37 at 328. For further details on the underlying mathematical model, see Stavins, *supra* note 32.

44. Elasticity of demand refers to the responsiveness of demand to changes in price. An inelastic market is one in which a one percent change in price leads to less than a one

The result is that the economic lives of used cars are extended, scrapping rates fall, and there are more cars of the older vintage on the road.[45] Since the emissions of older vintages of cars are generally greater than those of newer vintages of cars, the delay in retiring vehicles could result in counterproductive increases in aggregate emissions. Again, whether such emission increases actually occur is an empirical question; two of the key variables required to answer it are the price elasticities of scrappage and of new vehicle sales.[46]

In a comprehensive assessment of automotive regulation, Crandall et al. raised three primary concerns.[47] First, because VDR increases the price of new cars, it affects the used car market (as outlined above) and provides incentives to extend the useful lives of automobiles. Such extensions can "partially or fully offset the direct emissions-reducing impact of tighter standards for a considerable period."[48] Second, since VDR establishes a direct link between the rate of fleet turnover and environmental performance, when new car sales fall below expectations environmental progress can be compromised.[49] Third, VDRs for automobiles provide no incentives for individual car owners to maintain their emissions control systems (in the absence of inspection programs).[50]

B. *Empirical Analysis of the Effects of Automobile VDR*

Analyzing the effects of motor vehicle VDRs provides an

---

percent change in quantity demanded.

45. *See* Gruenspecht, *supra* note 37 at 328.

46. Researchers have conducted theoretical and empirical investigations of "accelerated vehicle retirement programs" as a means of reducing aggregate pollutant emissions by increasing the rate of turnover of the automobile fleet . *See, e.g.*, Robert W. Hahn, *An Economic Analysis of Scrappage*, 26 RAND J. ECON. 222, 238-40 (1995) (examining the impact of policies aimed at encouraging the retirement of older, high-polluting vehicles); *see generally* Anna Alberini et al., *Determinants of Participation in Accelerated Vehicle-Retirement Programs*, 26 RAND J. ECON. 93 (1995) (presenting an empirical analysis of the factors affecting participation in accelerated vehicle-retirement programs).

47. *See* CRANDALL ET AL., *supra* note 17 at 89-90. In an earlier theoretical study, Kwoka demonstrated that CAFE alters the mix and total number of vehicles in the fleet in ways that could "partially or even fully offset the fuel savings envisioned by a rising CAFE." John E. Kwoka Jr., *The Limits of Market-Oriented Regulatory Techniques: The Case of Automotive Fuel Economy*. 98 Q. J. ECON 695, 696 (1983).

48. CRANDALL ET AL., *supra* note 17, at 89.

49. CRANDALL ET AL., *supra* note 17, at 89-90.

50. CRANDALL ET AL., *supra* note 17, at 90.

opportunity to test a centrally important empirical question: do VDRs have counterproductive effects on aggregate emissions even when applied to a durable good with a relatively rapid rate of capital turnover? To investigate this, Gruenspecht estimated the key demand elasticities of scrappage and new vehicle sales from empirical data and combined these with an independent estimate of the cost penalty associated with the 1981 new car emissions standards for carbon monoxide (CO) and nitrogen oxides (NO_x).[51] This enabled him to calculate the effect of the adoption of the more stringent new car standards on auto scrappage rates and then estimate the net effects of the VDR on aggregate emissions.

Gruenspecht found that the VDR had depressed new car sales by between 2% and 4% over the first five years after the regulation came into force.[52] As a result, CO emissions were actually increased by the regulation by approximately 1% per year over the first four years before declining in year five. Hydrocarbon emissions were increased up to 2% per year over the first five years before declining in year six, while NO_x emissions were uniformly decreased by regulation. Of course, the CO and hydrocarbon effects were short term. Six years after the regulation had come into effect, emissions of all three pollutants had declined.[53] By the end of 1990, the level of emissions had fallen 5.3% below baseline for CO and 16% below baseline for HCs.[54] Thus, counterproductive environmental performance was a temporary phenomenon.[55] Nevertheless, by delaying turnover in the vehicle fleet, the VDR did not reduce emissions as quickly as the legislative standards would suggest[56] and the cost-effectiveness of the policy was reduced.

A few authors have contrasted the vintage-differentiated regulations in motor vehicle emissions laws and the VDR-based Corporate Average Fuel Economy (CAFE) standards[57] with

51. *See generally* Gruenspecht, *supra* note 37 (for details of his analysis).

52. Gruenspecht, *supra* note 37, at 330.

53. *Id.*

54. *Id.*

55. *Id* at 330-31. The long term favorable effects on emissions of the retirement of older vehicles when new car emissions standards become more stringent are documented by Matthew E. Kahn, *New Evidence on Trends in Vehicle Emissions*, 27 RAND J. OR ECON. 183, 194-95 (1996).

56. CRANDALL ET AL., *supra* note 17, at 91-97.

57. The Energy Policy and Conservation Act of 1975 established fuel efficiency standards for automobiles and light trucks, requiring manufacturers to meet minimum

alternative approaches. For example, Gruenspecht considered a $250 "bounty" for scrapping 15-year old cars,[58] for which he predicted an environmental performance better than either the actual VDR or the absence of it. Similarly, Crandall contrasted the CAFE program with commensurate gasoline taxes, finding CAFE to be the more costly regulatory option for improving fuel economy.[59] This comparison goes well beyond VDRs, of course: there are other reasons why CAFE standards are a relatively costly means of achieving fuel conservation in addition to the component of CAFE that features vintage differentiation.[60] Nevertheless, it should be noted that CAFE adds to the cost of new vehicles, thereby reducing new car sales and extending the life of old vehicles while having no fuel-use reducing effects on old vehicles. This contrasts with a gasoline tax, which increases the marginal cost of driving for all vehicles while simultaneously inducing car owners to trade in old fuel-inefficient cars for newer, more fuel-efficient ones by increasing the cost of operation of older cars.[61]

---

sales-weighted average fuel efficiency for their fleets sold in the United States. A penalty is charged per car sold per unit of average fuel efficiency below the standard. The program operates like an intra-firm tradeable permit system, since manufacturers can undertake efficiency improvements wherever they are cheapest within their fleets. Light trucks, which are defined by the Federal government to include "sport utility vehicles," face weaker CAFE standards.

58. Gruenspecht, *supra* note 37, at 329-30.

59. Robert W. Crandall, *Policy Watch: Corporate Average Fuel Economy Standards*, 6 J. ECON. PERSPECTIVES 171, 175-79 (1992).

60. For varying assessments of the CAFE program's costs relative to equivalent gasoline taxes, see generally CRANDALL ET AL. (1986), *supra* note 17; Pinelopi K. Goldberg, *The Effects of the Corporate Average Fuel Efficiency Standards* (Nat'l Bureau of Econ. Research, Working Paper No. 5673, 1997) (finding a 780% increase in gas tax would be required to make the emissions reductions gains reached through CAFE) National Research Council, *Effectiveness and Impact of Corporate Average Fuel Economy (CAFE) Standards* (The National Academies Press 2002) (finding fuel tax to be a superior alternative to CAFE); Ian W. Parry et al., *Should Corporate Average Fuel Economy (CAFE) Standards be Tightened?* (Resources for the Future, Discussion Paper 04-53, 2004) (discussing factors affecting efficacy of higher CAFE standards).

61. How strong have these effects, which result both from CAFE's VDR component and its other aspects, been? Estimates of the welfare costs of CAFE range from $0.41 to $0.63 per gallon of gasoline conserved, compared with welfare costs of $0.08 per gallon conserved for a 2.4 cents per gallon tax, indicating a five to eight-fold cost differential. *See* Robert W. Crandall & John D. Graham, *The Effect of Fuel Economy Standards on Automobile Safety*, 32 J. L. & ECON. 97, 115-18 (1989); David L. Greene, *CAFE or Price?: An Analysis of the Effects of Federal Fuel Economy Regulations and Gasoline Price on New Car MPG, 1978-89*, 11 ENERGY J. 37 (1990) Andrew N. Kleit, *The Effect of Annual Changes in Automobile Fuel Efficiency Standards*, 2 J. REG. ECON. 151, 164 (1990). Not too much should be made of this comparison, however, since the CAFE program is much more stringent than a 2.4 cent per

The perverse incentives of VDRs also arose in the context of the California Air Resources Board's 1990 requirement that automobile manufacturers include zero-emissions vehicles[62] (ZEVs) in their 1998 California sales mix.[63] Although the requirement would have slightly reduced emissions from the average new car sold in the state, the program would also have had the effect of increasing prices of both electric (ZEV) and non-electric vehicles, as manufacturers spread the costs of developing the ZEVs and the subsidies needed to induce consumers to buy them. Gruenspecht found that new car purchases would consequently fall by 2% to 3%, commensurate with an increase in the retention of less efficient older cars in the fleet.[64] Gruenspecht's analysis indicates that the extra emissions thereby generated would more than offset the emissions reductions from the ZEVs.

Have VDRs had counterproductive effects on aggregate emissions in the automotive sector, despite the relatively rapid rate of capital turnover that characterizes this realm? It appears clear from the existing body of research—both theoretical and empirical—that VDRs have extended the useful lives of cars on the road, and therefore *can* increase aggregate emissions. Empirical analysis addressing whether motor vehicles have exhibited counterproductive emissions increases due to VDRs indicates that, as would be expected, these impacts have been temporary and relatively short term. Nevertheless, cost-effectiveness was compromised. But because the costs of retrofitting motor vehicles

---

gallon gasoline tax. If the marginal costs of fuel efficiency increase over the relevant domain, which is the case, then on this basis alone, one would anticipate some cost differential.

62. Zero Emissions Vehicles (ZEVs) are those certified to meet the most stringent emission standards established by the California Air Resources Board, which require zero regulated emissions of nonmethane organic gases, carbon monoxide, and nitrogen oxides. For a summary of California's ZEV regulations, including links to regulations for each model year, *SEE* California Air Resources Board, Zero Emissions Vehicle Legal and Regulatory Activities (2005), http://www.arb.ca.gov/msprog/zevprog/zevregs/zevregs.htm (on file with author). A ZEV is most likely powered by electricity, fuel cells, or hydrogen.

63. For historical and current requirements and methods of compliance, see: California Air Resources Board. Memorandum from Alan Lloyd, Chairman, regarding "Zero-Emission Vehicle (ZEV) Credit Reporting and Tracking System." Manufacturers Advisory Correspondence (MAC) 2004-01, www.arb.ca.gov/msprog/macs/mac0401/mac0401.pdf, March 24, 2004 (on file with author).

64. Howard K. Gruenspecht, *Zero Emission Vehicles: A Dirty Little Secret*, 142 RESOURCES 7, 8 (2001).

for pollution control or fuel efficiency would be prohibitive, the simple replacement of VDRs with undifferentiated regulation is not an attractive or even feasible alternative in the automotive realm. Rather, alternative regulatory approaches merit ongoing exploration, a topic to which I return in Part VII of this article.

VI. NEW SOURCE REVIEW

The Clean Air Act's New Source Review (NSR) program is a widely studied example of vintage-differentiated regulation, particularly as it relates to the coal-fired electricity generation sector.[65] This is an important and controversial application of VDRs in American environmental policy, and it contrasts with the automotive sector because of the longer life of the relevant capital. NSR sets technology-based emissions control requirements for new sources and for sources that are being expanded or modified significantly. The regulations thereby exclude existing facilities from emissions control requirements, essentially requiring new or upgraded power plants and some other types of facilities to be cleaner than old ones.[66] Because of the incentives this creates to extend the life of older plants rather than build new, more stringently regulated, facilities, concern exists that NSR wastes resources and can retard environmental progress.[67]

The NSR program applies to any new source whose potential emissions at full utilization are high enough to qualify it as a major source.[68] The program also applies to any modification at an existing major source resulting in an emissions increase. In

65. In 2003, the Bush administration initiated changes in the New Source Review program. The research considered in this paper deals with the program prior to those changes. The basic requirements of the NSR program were established in Parts C and D of Title I of the Clean Air Act. A series of regulatory changes in NSR, not analyzed in this article, are detailed at: EPA New Source Review Regulatory Actions, http://www.epa.gov/nsr/actions.html (on file with author).

66. 42 U.S.C. § 7411 (B)(1)(b) (2005); *see also* 42 U.S.C. § 7411 (D)(1) (2005) (establishing standards for existing sources); *see generally* U.S. ENVTL. PROT. AGENCY, Document No. II-A-01, NSR 90-DAY REVIEW BACKGROUND PAPER (2001) *available at* http://www.epa.gov/ nsr/documents/nsr-review.pdf (on file with author).

67. *See, e.g.*, Howard K. Gruenspecht & Robert N. Stavins, Op-Ed., *A Level Field on Pollution at Power Plants*, BOSTON GLOBE, Jan. 26, 2002 at A15; Howard K. Gruenspecht & Robert N. Stavins, *New Source Review Under the Clean Air Act: Ripe for Reform*, 147 RESOURCES 19, 19 (2002) [hereinafter Gruenspecht and Stavins, *New Source Review*].

68. The cut-off value used to determine whether a source is major generally varies between 10 and 100 tons of emissions per year, depending on the source category and the severity of any air quality problem where the source is located. *See* U.S. ENVTL. PROT. AGENCY, note 66 *supra*, at 3.

addition to securing a permit prior to commencing construction or modification, sources subject to NSR must achieve emission rates that reflect the performance of the best-available emissions-control technology.[69]

The NSR requirements under the Clean Air Act date back to the 1970s. Those who wrote the Act presumably thought they could secure greater environmental progress by imposing tougher emissions standards on new power plants (and certain other emission sources) than on old ones. The frequently voiced theory was that emissions would fall as old plants were retired and replaced by new, cleaner plants.[70] Critics claim that experience over the past 25 years has shown this approach to be both excessively costly and environmentally counterproductive, because firms have been motivated to keep old (and dirty) plants operating and to hold back on investments in new (and cleaner) power generation technologies.

A. *NSR and New Power Plants*

In theory, NSR can create perverse environmental incentives, especially when major technology advances make new plants much cleaner than old ones. Several studies have empirically examined the hypothesis that NSR in the electricity generation sector has extended the useful lives of existing plants and delayed the construction of replacement capacity, thereby having a perverse effect on emissions abatement.

One study examined the relationships between regulation and capital turnover, and capital age and emissions.[71] In research carried out across time and across firms, the researchers found a statistically significant inverse relationship between the rate of new plant investment and the presence of air quality regulation, controlling for other relevant variables. In other words, increased air quality regulation was correlated with decreased new plant investment even when allowing for the effects of other factors. The

---

69. NSR sources located in areas that do not meet national ambient air quality standards are also required to secure offsets for their emissions. *See* U.S. Envtl. Prot. Agency, note 66 *supra*, at 5.

70. For a historical perspective on the development of the 1971 and 1979 new source standards, *see generally* BRUCE A. ACKERMAN & WILLIAM T. HASSLER, CLEAN COAL/DIRTY AIR (1981). For an early assessment of the effects of the regulations, *see generally* Frank M. Gollop & Mark J. Roberts, *Environmental Regulations and Productivity Growth: The Case of Fossil-fueled Electric Power Generation*, 91 J. POL. ECON. 654 (1983).

71. *See generally* Maloney & Brady, *supra* note 32.

authors conclude that "these results convince us that regulation has induced delay in the retirement of capital and that this delay has been detrimental to the improvement of the environment."[72]

A second study, working with different data, provided partial validation of the presence of a regulatory-induced lengthening of the life of capital, finding that "increases in regulatory intensity lead firms to delay the construction of new steam-generating plants."[73] In this case, the researchers found that over the period from 1969 to 1983 environmental regulations increased plant age by more than twenty-two percent.[74] However, for the most part, the study did not find a significant effect of plant age on emissions.[75] The reason for this may be found in an earlier study by Joskow and Schmalensee, which noted that fuel efficiency in coal-fired plants improved throughout the 1960s, but stabilized in the 1970s (and may even have declined for some new units).[76] If fuel efficiency was related to emissions, then older plants may not have been more polluting than newer ones during that period.

A much more recent analysis by Swift illustrates how NSR requirements can impede the adoption of cleaner and more efficient energy technologies, such as combined heat and power (CHP) systems.[77] Swift found that NSR rules pose a deterrent to the

---

72. *Id.* at 222. In particular, Maloney & Brady found that in the ten states with the greatest regulatory expenditures, the regulation-induced component of capital longevity was approximately eight years (of an average longevity of thirty-three years), and emissions rates were elevated twenty-seven percent because of the impacts of regulation on the age of capital. *See also* Randy A. Nelson, *Regulation, Capital Vintage and Technical Change in the Electric Utility Industry*, 66 REV. ECON. & STAT. 59, 66 (1984).

73. Randy A. Nelson, Tom Tietenberg, & Michael R. Donihue, *Differential Environmental Regulation: Effects on Electric Utility Capital Turnover and Emissions*, 75 REV. ECON. & STAT. 368, 371 (1993).

74. *Id.* at 373.

75. *Id.*

76. Paul L. Joskow & Richard Schmalensee, *The Performance of Coal-Burning Electric Generating Units in the United States: 1960-1980*, 2 J. APPLIED ECONOMETRICS 85, 107 (1987).

77. In a modern CHP system, fuel is burned in a turbine to generate electricity, and the waste heat from combustion, which in conventional stand-alone generation systems is vented to the atmosphere, is used in commercial or industrial processes at the site. A new CHP installation using a gas-fired turbine with low-nitrogen oxide burners and no end–of–pipe emissions controls substantially reduces nitrogen oxide emissions from levels that would result from the continued operation of an existing onsite boiler to provide process heat and an offsite power plant to provide power. CHP also allows for a substantial reduction in the total primary energy input required to meet heat and power needs, yielding economic benefits and lower carbon dioxide emissions. *See generally* Byron Swift, *Grandfathering, New Source Review and NOx - Making Sense of a Flawed System*, 31 BUREAU NAT'L AFF. ENV'T REP. 1538 (2000) (describing CHP systems and assessing the impacts of NSR).

spread of CHP technology, although it is difficult to disentangle the effects of the regulation's technology-standard approach from the effects of the regulation's NSR component *per se*.[78] In any event, the rules require the application of end-of-pipe control technology to an already clean turbine with very low emissions.[79] This requirement can significantly increase the cost of a CHP project and removes only a small amount of pollution, resulting in a very high cost-per-ton of removal, more than $25,000 by Swift's estimate for the most modern plants, or twenty-five to seventy-five times the cost of emissions reductions available from existing sources.[80]

In addition to delaying capital stock turnover, NSR can promote environmentally perverse decisions regarding the use of new capacity. NSR can change the operating costs of plants of various vintages in ways that theoretically provide incentives for companies to favor the utilization of older plants over newer ones fitted with more effective emissions control technologies. For example, new coal-fired power plants built following passage of the 1977 Clean Air Act Amendments were essentially required to install scrubbers to remove sulfur dioxide (SO_2) emissions. But the "costs of running scrubbers were high enough that new coal-fired plants were more expensive to operate than many existing ones, which were not regulated under a new source standard."[81] "Under these conditions, utilities might be expected to reduce output from their new, scrubbed units while operating older plants at full capacity during off-peak seasons and time periods."[82]

These effects were validated empirically in an econometric analysis by Stanton, which found that plants with laxer environmental requirements were utilized more intensely.[83] Thus, by reversing the usual preference for maximizing use of modern plants, vintage-differentiated regulation of SO_2 emissions under the Clean Air Act reduced the environmental benefits resulting from the mandated investment in expensive scrubbing

78. *See* Swift, *supra* note 77, at 1539-42.

79. *Id.* at 1541.

80. *Id.* at 1539, 1542.

81. *Id.*

82. *Id.* at 21.

83. Timothy J. Stanton, *Capacity Utilization and New Source Bias: Evidence from the U.S. Electric Power Industry*, 15 ENERGY ECON. 57, 58-60 (1993). (examining statistically the relationship between operating costs at American electric power plants and utilization rates).

equipment.[84]

B. *NSR and Existing Power Plants*

NSR applies to existing plants only if they carry out a major modification which results in a net increase in emissions.[85] This approach—which at first may sound perfectly reasonable—has several problems.[86] First, old plants typically emit the vast majority of total pollution in any sector,[87] but NSR does not provide a continuous and effective incentive for emissions reductions at these plants. As a result, many of the most cost-effective emissions reduction opportunities are simply left untouched.[88] Second, because the lengthy and costly NSR process is triggered by "modifications," the NSR program provides a powerful disincentive for improvements and efficiency upgrades at old plants.[89] Since adjusting existing equipment to perform more efficiently can be a source of pollution reductions as well as cost savings, this effect of NSR can be both economically and environmentally harmful.

Finally, NSR creates an uncertain regulatory environment with potentially high transaction costs.[90] NSR may discourage

84. The SO_2 allowance trading program, implemented in 1995 under the Clean Air Act Amendments of 1990, overcame these inefficiencies in SO_2 regulation associated with NSR-type programs. *See* Byron Swift, *How Environmental Laws Work: An Analysis of the Utility Sector's Response to Regulation of Nitrogen Oxides and Sulfur Dioxide Under the Clean Air Act*, 14 TUL. ENVTL. L.J. 309, 390-91 (2001).

85. *See* U.S. ENVTL. PROT. AGENCY, *supra* note 66 , at 3-5.

86. *See* Gruenspecht & Stavins, *New Source Review*, *supra* note 67, at 21-22.

87. *See id. supra* note 67, at 21.

88. *Id.*

89. The finding of a major modification inevitably raises tricky issues in situations where changes are made to an existing plant. The combination of delay costs, control technology costs, and the costs of emissions offsets creates a powerful incentive for existing sources to avoid triggering NSR. Existing plants can avoid triggering NSR requirements by demonstrating that a modification, even if major, does not increase emissions. *See* U.S. ENVTL. PROT. AGENCY, *supra* note 66 , at 3-5. For steam-electric generating units, actual emissions before the change are compared with projected emissions after the change to determine whether a modification increases emissions. *See* 57 Fed. Reg. 32314 (July 21, 1992). Modifications that allow a plant to produce more electricity per unit of fuel burned can lead to an increase in its projected future emissions, since better efficiency will often result in higher projected utilization. Such projects can trigger NSR, even if they reduce emissions in the region, because they induce changes in the utilization of other facilities.

90. Routine maintenance, repair, and replacement activities are recognized as falling outside the scope of the NSR program, but the line separating these activities from a change that would be covered by the NSR process is subject to uncertain and changing interpretation. *See* Gruenspecht & Stavins, *New Source Review*, *supra* note 67, at 21. Several enforcement actions against electric generators were initiated in 1999, alleging that the

companies from maintaining their existing facilities. Plant owners contemplating maintenance activities must weigh the possible loss of considerable regulatory advantage if the work crosses the murky line between upkeep and improvement.[91] Protracted, costly, and time-consuming legal wrangling over whether maintenance activities have crossed a threshold sufficient to justify forcing an old plant to meet new plant standards is inevitable.[92] In the electricity sector, the deferral of maintenance compromises plant reliability and thereby increases the risk of outages.[93]

C. *NSR in Other Sectors*

Although most empirical analyses regarding NSR have focused on the electricity generation sector, a number of studies have been carried out in other areas as well. As early as 1979, Hartman, Bozdogan, and Nadkarni analyzed the economic impacts of the Clean Air Act, including NSR, on the American copper industry and detected what may have been the first evidence of negative effects on capacity growth.[94] Similarly, in the early 1980s, an analysis by Crandall of eight industrial sectors offered quantitative estimates of what was then characterized as "new source bias."[95]

---

utilities had evaded NSR requirements by improperly classifying major upgrade and life-extension projects as activities that do not count as major modifications under EPA's regulations. *See, e.g.*, Patrick D. Traylor, *New Source Review Inoculation: Environmental Risk Management in the Face of Regulatory Uncertainty*, HOGAN & HARTSON LLP FOCUS ON ENVIRONMENT, Mar. 2005.

91. In 2003, the Bush administration initiated changes in the NSR program, and some of those changes were intended to sharpen the threshold between routine maintenance and upgrades that would trigger NSR. As indicated previously, the research surveyed in this paper predates those changes, and hence this paper takes no position with regard to the impacts of the 2003 changes. For a description of these changes and preliminary analyses of their effects, *see generally* NATIONAL RESEARCH COUNCIL, INTERIM REPORT OF THE COMMITTEE ON CHANGES IN THE NEW SOURCE REVIEW PROGRAMS FOR STATIONARY SOURCES OF AIR POLLUTANTS (2005) (describing initial results of congressionally mandated investigation), *available at* http://www.nap.edu/books/0309095786/html/ (on file with author).

92. *See* Gruenspecht & Stavins, New Source Review, *supra* note 67, at 21-22.

93. For further discussion of the ways in which NSR may impede reinvestment in existing capital stock due to the uncertainty implicit in the NSR process, see UNITED STATES ENVIRONMENTAL PROTECTION AGENCY, NEW SOURCE REVIEW: REPORT TO THE PRESIDENT 1 (2002), *available at* http://www.epa.gov/nsr/documents/nsr_report_to_president.pdf (on file with author).

94. Raymond Hartman et al., *The Economic Impact of Environmental Regulations on the U.S. Copper Industry*, 10 BELL J. ECON. 589, 593 (1979) (using an econometric model to simulate the impacts of environmental regulations on the American copper industry).

95. ROBERT W. CRANDALL, CONTROLLING INDUSTRIAL POLLUTION: THE ECONOMICS

A more recent study examined whether the timing of plant investments was affected by the nature of regulation.[96] In a study of several industries over the period between 1963 and 1992, the researchers found that NSR significantly depressed the "birth" of new plants, keeping old plants in use.[97] For example, in an empirical analysis of the organic industrial chemicals industry, Becker and Henderson found that grandfathering of plants contributed to environmental degradation by raising survival rates, reducing plant turnover rates, and keeping otherwise unprofitable operations in business.[98] This also slowed improvements in air quality by prolonging the lives of older, dirtier plants.[99] They concluded "a more uniform policy with respect to age would have encouraged retrofitting and other antipollution activities of existing VOC[100] and NO_x emitters much earlier in the regulatory process."[101]

What can be concluded from the empirical analyses that have been carried out on the effects of New Source Review, both in the electricity-generating sector and in other sectors? First, there is considerable evidence that this form of VDR has retarded turnover of the capital stock, increased the age of plants, driven up the costs of environmental improvement, and had counterproductive effects on emissions in the electricity-generating sector. Second, there is also evidence that NSR has promoted environmentally perverse decisions regarding the use of new (high-cost) capacity. Third, because NSR does not provide a continuous and effective incentive for emissions reductions at the old plants that typically are the most polluting in any sector, it has reduced cost-effectiveness of pollution control among existing plants. Fourth, although there has been no statistical analysis to date of the impacts of NSR on maintenance and improvements at old plants, theoretical arguments—if not common sense—provide support for these concerns. Fifth and finally, in a series of studies going back forty

AND POLITICS OF CLEAN AIR 39-44 (The Brookings Institution 1983) (examining the effects of "new source bias" in regulations on compliance costs per unit of economic activity).

96. *See generally* Randy Becker & Vernon Henderson, *Effects of Air Quality Regulation on Polluting Industries*, 108 J. POL. ECON. 379 (2000) (examining unintended impacts of air quality regulation).

97. *Id.*

98. *Id.*

99. *Id.*

100. "VOC" refers to volatile organic compound.

101. Becker and Henderson, *supra* note 96, at 415.

years or more, researchers have found statistical evidence in a number of other sectors of similar negative impacts of NSR regulations.

VII. IMPLICATIONS FOR POLICY AND RESEARCH

The theoretical and empirical findings regarding VDRs reported above have implications for public policy in the environmental realm. The results from previous studies also point the way to promising areas for future research.

Previous research has shown that VDRs drive up costs and can result in worse environmental quality than would have occurred if firms faced cost-effective regulation without a disincentive to invest in new, cleaner technologies. In some cases, where capital turnover rates are low and rates of technologically driven environmental improvements are high, VDRs may actually lead to worse environmental quality than would result from an absence of regulation. Environmental regulation is often necessary to achieve emissions reductions, however. If in some cases it would make sense to abandon VDRs, what should be put in their place?

In the context of point sources, such as in the electricity-generation sector, one approach would be to move toward a level playing field through even-handed regulation that motivates both old and new plants to cut emissions in order to achieve environmental objectives. This might be accomplished through a cap on total pollutant emissions together with the use of an allowance trading system to ensure that any emissions increases at one plant are balanced by offsetting reductions at another. The SO_2 allowance trading program in the 1990 Clean Air Act is one among many possible models.[102] No matter how emissions are initially allocated across plants, the owners of existing facilities and those who wish to build new ones would face appropriate incentives with respect to retirement decisions, investment decisions, and decisions regarding the use of alternative fuels and technologies to reduce pollution.[103] Thus, in principle, it should be possible to strengthen environmental laws by selectively and

102. *See generally* Robert N. Stavins, *What Can We Learn from the Grand Policy Experiment? Lessons from SO_2 Allowance Trading*, 12 J. ECON. PERSP. 69 (1998) (reviewing performance and implications of the SO_2 allowance trading program).

103. Presumably requirements for localities to meet ambient air quality standards would remain in effect, preventing a concentration of emissions in any geographic region that would conflict with needs to protect public health.

gradually phasing out reliance on VDRs, replacing some of these age-discriminatory mechanisms with binding environmental constraints, sometimes implemented through cost-effective, market-based policy instruments.

How might such selective and gradual phasing out proceed? Research has identified important trade-offs in VDR use, including short-term emissions increases versus long-term emissions reductions, and increases in unit abatement costs versus reductions in pollution levels.[104] Hence, it is necessary to ask *which* VDRs should actually be targeted for phase out. Much remains to be learned if such priorities are to be established. Research should focus on investigating the circumstances in which VDRs are most likely to be welfare-reducing compared with viable alternatives. Some of the relevant factors affecting the conditions under which VDRs are particularly troublesome can be identified, but future research will need to establish the relationships with greater rigor and quantify the magnitude of their respective effects. The list of candidates for further analysis includes the following topics.

First, as the natural economic life of the relevant durable good increases, so too does the potential for VDRs to reduce the rate of turnover of the capital stock. This is one of the reasons why the investment-retarding effects of the NSR program have been so pronounced in the electricity generation sector, particularly for large-scale, coal-fired plants. It is also the reason why the impacts of VDRs in the automotive sector may be becoming more severe over time. As the quality of motor vehicles has greatly improved, the average life of vehicles on the road has nearly doubled over the past twenty-five years. What, empirically, is the relationship between the durability of a capital good and the effectiveness of VDRs?

Second, and conversely, a combination of circumstances may argue in favor of the retention of VDRs in spite of the impact they have on the lives of durable goods and thereby on abatement costs. If abatement cost heterogeneity is great, cost heterogeneity is highly correlated with vintage, and a cost-effective instrument is not feasible, then the static cost-effectiveness associated with an appropriate VDR may overwhelm the perverse dynamic effects on costs (and environmental performance). In these circumstances, VDRs may be the better choice. For example, the cost of

104. *See supra* Parts III and IV.

retrofitting new pollution-abatement devices to existing fleets of cars is typically prohibitive (a situation that contrasts with the situation of retrofitting power plants). What degree of vintage-correlated abatement cost heterogeneity would make VDRs the most cost-effective policy instrument?

Third, if the rate of positive relevant technological change is great, that is, if newer vintages of given machines are also much cleaner, then VDRs can lead to an unintended inverse relationship between the stringency of regulation and aggregate emissions. For example, if the overall energy efficiency of some type of fuel-burning equipment is increasing with newer vintages, then newer vintages of the machine in question will likely have lower levels of air pollutant emissions per unit of services provided. By providing an incentive to keep the older equipment in place, a VDR can theoretically lead to an inverse relationship between the stringency of regulation and aggregate emissions. What rate of environmentally positive technological change in a given capital good renders VDR application problematic?

Fourth, the negative impacts of VDRs on pollution levels may be increased if the older technology in question exhibits increasing emissions (degraded abatement) as it ages. For example, if the energy efficiency of the equipment degrades as it ages (and hence the emissions increase as the equipment ages), then the negative consequences of VDRs when they extend the lives of such durable goods will be heightened. What is the empirical relationship between the rate of age-correlated emissions increases and the potential perverse effects of VDRs?

Fifth, market structure matters. For example, selective deregulation of electricity markets has presumably rendered affected generators more sensitive to environmental compliance costs than they were in a world of rate-of-return regulation (when such compliance costs could typically be passed on to rate payers). This suggests that the negative consequences of VDRs may have increased, all else being equal, in this sector during the period of deregulation—a correlation that could be tested empirically. Which market structures are likely to amplify the negative affects of VDRs?

Sixth and finally, the underlying regulatory structure itself seems significant. In theory, VDRs can be employed within the context of command-and-control instruments, such as technology-based standards, or within market-based systems, such as tradable permits. For example, a command-and-control instrument such as a uniform performance standard can involve VDRs, whereby old facilities are permitted one emissions rate and new facilities are limited to a more stringent rate. Likewise, in a tradable permit system, older units can be allocated larger numbers of permits, while newer units are allocated smaller numbers (or even none at all). How does the underlying regulatory structure in which a VDR is placed affect its effectiveness?

In sum, while theoretical analysis regarding the use of VDRs suggests clear hypotheses, there is need for more empirical analysis examining the actual effects of these age-discriminatory mechanisms in practice. Of the research possibilities discussed above, three areas stand out: the role of market structure in determining the effects of VDRs, the impact of VDRs on motor vehicle turnover rates, and the performance of VDRs in combination with alternative policy instruments.

A. *The Role of Market Structure*

One area of research that has not been explored despite strong theoretical evidence to suggest its relevance is the role of market structure in determining the effects of VDRs. Theoretically, more competitive markets should exhibit more severe perverse consequences from VDRs: the competitive nature of these industries forces firms to seek cost savings wherever possible, including through prolonging the life of capital equipment in order to avoid incurring higher pollution abatement costs. Less competitive industries should be less vulnerable to such cost-saving pressures. A natural experiment on the effects of market structure on VDR outcomes occurred in the United States when several regions deregulated electricity markets while other regions maintained rate-of-return regulations.[105] It would be illuminating to compare the rates of capital turnover in states with deregulated systems to states with regulated systems.

A variety of challenging methodological issues will need to be addressed in such research. First, there is the possibility that the policy variable—deregulation—is itself a function of various economic variables. The concern would be that unobservable

105. Dallas Burtraw et al., *Electricity Restructuring: Consequences and Opportunities for the Environment*, *in* 5 INTERNATIONAL YEARBOOK OF ENVIRONMENTAL AND RESOURCE ECONOMICS, pp. 40-89 (Henry Folmer &Tom Tietenberg, eds. 2001).

factors (such as general economic conditions in a jurisdiction) that affect whether or not a plant is subject to deregulation also affect the frequency of pollution control equipment installation. For example, states that had the greatest costs from rate-of-return regulation[106] may have been those that opted to deregulate. If this were true, it would be essential to test whether the degree of inefficiency in the regulated system is correlated with the rate of turnover. If relevant deregulation decisions were largely political, in contrast, then they were exogenous to the rate of capital turnover. In either case, a careful consideration of the endogeneity issue is required.[107] The gradual and partial deregulation of the electricity generation industry is a promising natural experiment for analyzing the effect of market structure on capital turnover rates under VDRs.

B. *Impacts on Turnover of Mobile Capital*

A critical issue with VDRs is their effect on the rate of retirement of durable goods. Research is needed in general on the rate of turnover of mobile capital, automobiles in particular. The longevity of American cars has increased, in part due to decreasing replacement part prices but new car sales continue to be significant.[108] One can define the gross turnover rate for a given policy jurisdiction to be car retirements less new car sales and the net turnover rate to be the gross turnover rate plus transfers out of the area. The increased longevity of automobiles combined with an increased sales rate then implies that the gross turnover rate has been declining. But it is possible that the *net* turnover rate of automobiles has actually increased over the last decade. This would be the case if more people are buying newer cars *and* if the older cars, which have a longer life, are leaving the country (or

106. Under conventional rate-of-return regulation, electricity generators were granted a local monopoly and thus protected from competition. Hence, regulators had to make determinations about what price could be charged for electricity; in general, they allowed firms to charge a price that would give them a reasonable rate of return on their capital.

107. In addition, there might be a concern about the validity of the natural experiment if states that were not deregulated anticipated being deregulated in the near future. If there was uncertainty over future deregulation, then regulated states might not be a good control for what would have happened to the rate of capital turnover in the absence of deregulation.

108. *See generally* Bruce W. Hamilton & Molly K. Macauley, *Competition and Car Longevity* (Resources for the Future Discussion Paper 98-20, 1998) (documenting increase in commercial longevity of American automobiles).

state, in the case of California regulations).[109]

A second natural experiment that might illuminate the effect of VDRs on the rate of automobile turnover comes from selective state implementation of vintage-differentiated safety and emissions tests. In the northeastern states, for example, emissions test requirements are less stringent for older automobiles. Other states do not have emissions tests at all. This variation across states provides for the possibility of examining the effects of VDRs on automobile turnover rates. As with the possible study of the effect of deregulation on capital turnover in the electricity sector, this study would have to address the potential endogeneity of the policy variable, that is, the degree to which the policy variable is itself affected by economic forces. In particular, the density of urban areas in the northeast and the existence of reliable public transportation services may lead to lower turnover rates for automobiles in these states, even in the absence of VDRs.

C. *Performance of VDRs Combined with Alternative Policy Instruments*

A third area of promising research concerns the interaction of VDRs with alternative policy instruments. As indicated above, Swift demonstrated that NSR requirements slowed the adoption of combined heat and power (CHP) systems.[110] In that case, combining vintage-differentiation with a technology standard in NSR created perverse incentives. Despite the fact that CHP systems were already highly efficient with low emissions, the technology standard required the addition of costly abatement equipment.[111] Hence, plants reasonably decided to forgo adopting CHP systems altogether. This negative effect of vintage-differentiation might have been reduced if the underlying regulation were performance-based rather than technology-based; that is, if it had emissions constraints rather than technology requirements.

One can imagine VDRs combined with alternative policy instruments in which the negative effects of VDRs are reduced, if

---

109. In some policy situations, the relevant turnover rate would be the net rate. Older cars leaving the state or being sold in Mexico would not affect the ultimate policy outcome within a given jurisdiction. This would be the case for local air pollution in Los Angeles, for example. But if California is employing CO_2 standards for climate change policy then old cars going to Mexico constitute a problem. In this case, the relevant turnover rate would be the gross rate.

110. *See* Swift, *supra* note 77 at 1539-42. Swift refers to CHP plants as "combined cycle" plants, but the term refers to the same technology.

111. *Id.* at 1539.

not eliminated. The use of the SO_2 allowance trading system in conjunction with NSR under the Clean Air Act Amendments of 1990 may provide the basis for empirical analysis of the effects of VDRs under alternative policy regimes. Prior to the Clean Air Act Amendments of 1990, affected facilities were regulated under a command-and-control system combined with VDR. After the 1990 Amendments came into force, some plants were regulated under the allowance trading (cap and trade) system.[112] This combination of temporal and spatial variation may provide useful data for research on how effectively combination VDR and non-VDR regulatory schemes operate.

VIII. Conclusion

Vintage-differentiated regulations, in which later vintages face more stringent standards, play prominent roles in major Federal environmental statutes, state and local environmental laws, and a variety of non-environmental regulations. This is true despite the fact that economists and others have long argued that such age-discriminatory regulations can retard turnover of the capital stock, drive up the cost of environmental protection, and may retard pollution abatement. American environmental policy frequently relies on VDRs because of powerful political incentives for key constituencies to demand VDRs and for legislators to supply them.

With roots in standard investment theory, models of the impacts of vintage-differentiated regulation have been developed over the past twenty-five years. Empirical analysis in the environmental realm has accelerated over that time. Much of that empirical analysis has focused on VDRs in the automotive sector and on New Source Review in the electric power generation sector. Overall, these empirical studies have validated theoretical findings that, all else being equal, vintage-differentiated regulations have retarded the turnover of durable goods and thereby increased aggregate pollution abatement costs. In some limited cases, empirical studies have also validated the particularly perverse potential consequence of VDRs—that environmental progress has itself been slowed. These theoretical and empirical findings point the way to what could be significant improvements in public policy

112. *See* Byron Swift, *How Environmental Laws Work: An Analysis of the Utility Sector's Response to Regulation of Nitrogen Oxides and Sulfur Dioxide Under the Clean Air Act*, 14 TUL. ENVTL. L.J. 309, 319-22 (2001).

in the environmental realm. At the same time, much more remains to be learned, and additional research efforts in this area are likely to produce unusually large dividends.

Given the ubiquitous political forces that have made vintage-differentiated regulations in the environmental realm so attractive to interest groups and legislators, it is reasonable to question whether the phasing out of VDRs in select environmental policy applications advocated here really stands any chance of being implemented. In response, we should acknowledge that the battle will be difficult. However, the same political forces that have brought about VDRs can also help bring about their reform. Many in the regulated community (particularly new entrants and existing firms that are growing rapidly) are disadvantaged by VDRs. As they come to recognize this, they can become a force for change. Equally important, when and where the environmental advocacy community comes to recognize that VDRs are systematically not cost-effective, they may begin to favor alternative, lower-cost regulatory approaches, knowing that lower compliance costs can translate through the political process into the achievement of more ambitious goals. And finally, when environmentalists accept that VDRs can in some circumstances be counterproductive, retarding rather than advancing environmental progress, then environmental advocates may themselves become enthusiastic agents for regulatory reform.

[9]

Journal of Regulatory Economics; 23:1 43–59, 2003

Cost Heterogeneity and the Potential Savings from Market-Based Policies*

RICHARD G. NEWELL
Resources for the Future
1616 P Street, NW, Washington, DC 20036
E-mail: newell@rff.org

ROBERT N. STAVINS
Harvard University
John F. Kennedy School of Government,
79 John F. Kennedy Street, Cambridge, MA 02138
E-mail: robert_stavins@harvard.edu

Abstract

Policy makers and analysts are often faced with situations where it is unclear whether market-based instruments hold real promise of reducing costs, relative to conventional uniform standards. We develop analytic expressions that can be employed with modest amounts of information to estimate the potential cost savings associated with market-based policies, with an application to the environmental policy realm. These simple formulae can identify instruments that merit more detailed investigation. We illustrate the use of these results with an application to nitrogen oxides control by electric utilities in the United States.

1. Introduction

Over the past decade, policy makers in many parts of the world have given increasing attention to the use of markets to address a variety of social problems. This has been

* Richard G. Newell is a Fellow at Resources for the Future; and Robert N. Stavins is the Albert Pratt Professor of Business and Government at the John F. Kennedy School of Government, Harvard University, and a University Fellow at Resources for the Future. We thank Michael Batz and Nathaniel Keohane for excellent research assistance; Alan Krupnick for information on nitrogen oxides control; and anonymous referees, Timothy Besley, James Hammitt, Cathy Kling, Ian Parry, William Pizer, Jim Sanchirico, David Simpson, Martin Weitzman, and seminar participants at the Allied Social Science Associations meetings and Harvard University for very helpful comments on previous versions of the manuscript. We also acknowledge financial support from U.S. Environmental Protection Agency grant CR825095, but none of these institutions or individuals are thereby implicated for any errors that remain.

particularly true in the case of environmental protection efforts, where various types of tradeable permit and charge systems have begun to be employed (Stavins 2001). Indeed, market-based environmental policy instruments, which were controversial just ten years ago, have now evolved in political circles to the point of becoming conventional wisdom, at least in the United States (Keohane et al. 1998). This change may please most economists, but it also highlights the importance of identifying the appropriate policy instrument for each environmental problem that is faced in its particular socio-economic context. In some cases, market-based instruments may be highly desirable, but in other cases their advantages may be small. We provide some relatively simple formulae for exploring the costs of alternative policy instruments, with the intention of increasing intuition and understanding of the sources of cost savings, and to help identify instruments that merit more detailed investigation.

While we focus here on policymaking for environmental problems, the basic issue we explore is quite general. Namely, what is the relationship between potential gains from trade and heterogeneity in technology and/or preferences? There are many areas where this generic issue arises, including: international trade and cross-country heterogeneity in technology, tastes, and endowments; gains from localized regulatory control and heterogeneity across regional jurisdictions; welfare effects of tax and public spending harmonization across a set of countries, such as the European Union; the question of whether potential renewable energy performance standards for electric utilities should be tradable; and, our chosen example, potential cost savings due to market-based policy instruments and heterogeneity of pollution abatement costs.

As with any policy goal, a variety of criteria can and have been brought to bear upon the choice of policy instruments to achieve environmental protection. As the stringency of environmental targets has increased, cost-effectiveness has become a more important criterion for instrument choice.[1] A key factor affecting relative aggregate costs under alternative policy instruments is the heterogeneity of pollution control costs across sources. There are many reasons why the costs of complying with environmental regulations tend to be heterogeneous, including differences in plant location, size, age, and production technology. Location, for example, can affect costs due to differences in the quality and price of inputs (for example, proximity to clean inputs), physical characteristics (for example, urban or rural), and political jurisdiction (for example, pre-existing regulations).

While it is widely recognized that abatement-cost heterogeneity is a fundamental determinant of the potential cost-savings associated with market-based policy instruments, there is surprisingly little analysis of the general relationship between the nature and magnitude of such heterogeneity and the prospective cost savings. Despite the notable absence of such analysis, there is a relevant, albeit small, theoretical literature on the relationship between potential gains from trade and the underlying heterogeneity of

1 For a list of additional candidate criteria for policy instrument choice, see Bohm and Russell (1985). The aggregate level and distribution of the benefits of pollution control may also be affected by the choice of instrument, depending on the degree to which the pollutant is uniformly mixed within the relevant region.

consumer preferences and production technology. Most prominent in this literature are studies by Weitzman (1977) and Suen (1990) on the effects of diversity in consumer preferences on the relative efficiency of the price system, and Krueger and Sonnenschein (1967) on the relationship between price divergence across countries and the gains from international trade. In the environmental area, Nichols (1984) explores optimal policies with heterogeneous costs and Mendlesohn (1986) assesses pollution regulation in the presence of heterogeneous benefits and costs. None of these, however, provides a simple framework for directly estimating the cost-savings associated with using market-based policy instruments relative to more commonly used uniform performance standards.

There is also an empirical literature that explores the costs of using alternative policy instruments to address particular environmental problems.[2] These studies are concerned with specific environmental pollutants in specific contexts, ranging from dissolved oxygen in the Fox River of Wisconsin to particulates in Santiago, Chile. Each study uses different types of data at different levels of aggregation, each makes different simplifying assumptions, and each employs different methods of analysis. Although this differentiation in data and methods may be appropriate for specific cases, little intuition thereby emerges regarding the general relationship between the nature and magnitude of cost heterogeneity and the potential cost savings associated with market-based instruments.[3] Relatively simple formulae that can be employed with a small set of summary statistics can provide a useful lens for conducting initial screenings of environmental problems, so that analysts and decision makers can focus their attention on cases where potential cost savings are greatest. In addition, by clearly demonstrating the relationship between cost savings and different sources of underlying heterogeneity—such as the slope of marginal costs, baseline emissions, and firm size—our analytical results can provide guidance for the design of relatively cost-effective policies in the event ''pure'' market-based policies are not an option.

Such analysis is important because market-based instruments are by no means a panacea; in some cases, they hold tremendous promise of providing environmental protection cost effectively, but in other cases they are not well suited, for a variety of reasons. It might seem that if costs are homogeneous, a market-based instrument will perform identically to a conventional uniform standard, and that therefore policy makers ought always to choose market-based instruments (because they can presumably perform no worse in terms of aggregate costs and may perform better than command-and-control approaches). But market-based instruments may not perform better than conventional standards along a number of dimensions we do not consider here, including: relative administrative costs of the various instruments; possibilities of strategic behavior (Hahn

2 See, for example, Atkinson and Lewis (1974); Atkinson and Tietenberg (1982); Carlson et al. (2000); Coggins and Swinton (1996); Gollop and Roberts (1983, 1985); Hahn and Noll (1982); Kolstad (1986); Krupnick (1986); Maloney and Yandle (1984); McConnell and Schwartz (1992); O'Neil et al. (1983); O'Ryan (1996); Perl and Dunbar (1982); and Seskin et al. (1983).

3 Note that the question we are addressing—the relative *cost-effectiveness* of market-based versus conventional policy instruments—is distinct from the question of the overall *efficiency* of price versus quantity instruments (under conditions of uncertainty), a question addressed in a literature which originated with Weitzman (1974).

1984; Misolek and Elder 1989; Malueg 1990) and transaction costs (Stavins 1995) with specific market-based instruments; political costs attendant to moving toward innovative instruments with different distributional consequences (Keohane et al. 1998); systematic "over-control" that results from particular applications of command-and-control measures (Oates et al. 1989); and smaller aggregate benefits of market-based instruments where "hot spots" of emissions or concentrations combine with non-linear damage functions. Therefore, it is valuable to predict the potential cost savings that may be associated with using a market-based instrument for a particular environmental problem.[4]

Our approach is to develop three stylized models of alternative means of achieving an aggregate environmental target. Two are performance-based, command-and-control instruments: a uniform emission rate standard and a uniform percentage reduction standard. We model the uniform performance standards as being in terms of an allowable emission rate per unit of product output or percentage emissions reduction, because these are typical of actual command-and-control regulation (Russell et al. 1986; Helfand 1991).[5] The third instrument is market-based, such as an emissions fee or tradable permit system.

A fundamental feature of the models is that a subset of their parameters represents the nature and degree of cost heterogeneity. We solve each model for the aggregate cost it would imply, and then by comparing results, we derive expressions for the absolute and percentage cost-savings attributable to adopting a market-based instrument. Subsequently, we demonstrate the potential use of these expressions with a specific application using information on nitrogen oxides control in the eastern United States.

Economists frequently develop analytical models of the costs of environmental and other regulation. These range from small-scale, sector-specific, econometric cost-side models to large-scale, integrated assessment models of global climate change. Despite the frequently acknowledged importance of cost heterogeneity, it is a characteristic of the real world that is sometimes (understandably) ignored in such modeling, because of the complexity and the data requirements that it inevitably would introduce.[6] In this context,

4 Even among those environmental problems for which market-based instruments would appear preferable strictly on compliance-cost grounds, there is tremendous variation in the cost-savings that could be anticipated (essentially because of variation in the degree of cost heterogeneity). Tietenberg (1985) assimilated the results from ten analyzes of the costs of air pollution control, and in a frequently-cited table, indicated the ratio of actual command-and-control programs to least-cost benchmarks. The resulting ratios ranged from 22.0 to 1.1.

5 Note in addition that our analysis of a uniform emission rate standard will also hold when the standard is in terms of emissions per unit of some input—assuming that the production function uses the input in fixed proportions to output. Another possibility is a uniform emission or abatement quantity standard, which has been employed in theoretical models, but we know of no examples of its use in actual policy. A uniform quantity standard may be useful as a point of comparison in a model with homogeneous firms, but—as evidenced by its absence from real policy—it becomes unreasonable when firms are heterogeneous, especially in terms of size.

6 In the simplest market model, for example, one would anticipate that firms with relatively high abatement costs would be driven out of the market by firms with lower abatement costs. In reality, of course, this may not happen because those same firms may enjoy cost advantages along other dimensions, and because there are a variety of frictions in the relevant markets, including those due to the regulatory environment.

our work also provides modelers with a parsimonious structure for incorporating key elements of heterogeneity into their analytical models of costs for a variety of environmental and other problems.

The cost-savings potentials that are identified with the approach we offer may not reflect precisely the magnitude of realized savings, because we abstract from several other dimensions along which the costs of market-based instruments may differ from those of command-and-control approaches. Many of these additional effects would tend to further increase the cost-effectiveness of market-based policies due to the flexibility and incentives they provide.[7] On the other hand, real-world command-and-control approaches may contain exemptions or other types of ''fine-tuning'' that increase their cost-effectiveness relative to our stylized representation.[8] The basic message is the same: increased regulatory flexibility can yield increased cost-savings, and the degree of such savings depends on the degree of heterogeneity.

In section 2, we develop our stylized models of three basic types of policy instruments and identify key general findings. In section 3, we apply the approach to the policy problem of reducing nitrogen oxide emissions in the eastern United States, and in section 4, we conclude.

2. A Model of Cost Heterogeneity and Policy Choice

2.1. A Model of Heterogeneous Costs

We posit the common situation in which the government seeks to limit the aggregate emissions, Q, of a set of sources. The output, x_i, of each source is given, as is aggregate

7 First, firms can potentially reduce emissions through three types of activities: product output reduction; input substitution; and end-of-pipe abatement (Spulber 1985; Goulder et al. 1999). We abstract from the first type of activity by treating output as exogenous, which seems reasonable, because pollution-control costs are typically a very small fraction of total production costs (Jaffe et al. 1995). Further, we do not differentiate between input substitution and end-of-pipe treatments, considering both as emission abatement. Hence, our stylized command-and-control policy instruments are (uniform) performance standards, not technology mandates. This is important to recognize, because even if firms were perfectly homogeneous, a true technology mandate would not be cost-effective, because it provides no latitude for firms to substitute ''cleaner'' inputs. Second, our focus is exclusively on static cost-effectiveness, but it is well known that the incentive-structure of market-based instruments can lead to dynamic efficiency gains, both in terms of decreased abatement costs through technological change (Jaffe et al. 2001) and firm entry and exit (Spulber 1985; Helfand 1991). Third and finally, our analysis is partial-equilibrium in nature. In a general-equilibrium context, particular types of market-based instruments, such as those that combine revenue generation with cuts in pre-existing distortionary taxes, can enjoy additional cost advantages over commensurate command-and-control regulations (Goulder et al. 1999).

8 For such regulatory tailoring to increase cost-effectiveness, the characteristics along which regulations are differentiated must coincide with the characteristics along which costs are differentiated, and the relationship must be in the correct direction—that is, the regulation faced by sources with low-cost characteristics must be relatively stringent. Although regulatory differentiation can be motivated by a variety of political and bureaucratic motivations other than costs, one could argue that political-economy incentives may lead high-cost sources to fight regulation more forcefully, leading to relatively ''efficient'' regulatory differentiation.

output, X, for all n sources.[9] Emission quantities for each source are given by q_i and are chosen by sources to minimize costs subject to policy constraints. Each source has baseline emissions of q_{0i} in the absence of emission regulation and faces increasing costs of reducing emissions from this baseline. Because we are striving for transparent results, we use a second-order approximation of these costs around baseline emissions:

$$C(q_i) = c_{0i} + c_{1i}(q_{0i} - q_i) + \frac{c_{2i}}{2}(q_{0i} - q_i)^2,$$

where c_{0i}, c_{1i}, and c_{2i} are source-specific parameters of the emission cost function $C(q_i)$. The marginal cost of emission reductions is therefore given by

$$-C'(q_i) = c_{1i} + c_{2i}(q_{0i} - q_i).$$

Note that $c_{0i} = 0$ because emission costs are zero in the baseline. Furthermore, because the cost-minimizing firm will equate the marginal cost of emission reduction with the price of emissions p, we have $c_{1i} = 0$ because $p = 0$ in the baseline. Finally, we multiply the cost function by x_i^2/x_i^2 in order to express the variables in per unit of output terms. This translation is useful because: (i) environmental policy discussions are often couched in terms of pollution intensity rather than absolute emissions levels; and (ii) it reduces the degree of correlation that would otherwise exist between the variables due to scale effects.

We are therefore left with the following simplified total and marginal cost functions:

$$C(q_i) = \frac{x_i}{2b_i}\left(a_i - \frac{q_i}{x_i}\right)^2, \tag{1}$$

$$-C'(q_i) = \frac{1}{b_i}\left(a_i - \frac{q_i}{x_i}\right), \tag{2}$$

where $a_i = q_{0i}/x_i$ is baseline emissions intensity per unit of output, and we have made the substitution $1/b_i = c_{2i}x_i$ in order to simplify the taking of expectations in subsequent parts of the analysis.[10] Solving for q_i,

$$q_i = (a_i - b_i p)x_i,$$

assuming firms are price-takers, we find that each source has a factor demand for emissions that is linear in the price, p, of emissions (relative to other factors). Therefore, b_i is both the slope of the emissions demand function and the reciprocal of the slope of the

9 Note that this framework requires a consistent measure of output for different sources, which is necessary given that we explore the relative cost of uniform performance standards measured in terms of emissions per unit of product output. Such a framework is relevant for exploring many cases where market-based policies have been or may be applied, including control of emissions of sulfur dioxide, nitrogen oxide, and carbon dioxide from electric utilities and large industrial boilers. The sources need not be from the same industry, however, as illustrated by the case of emissions from electric power plants and industrial boilers, both of which do or could face standards in terms of emissions per unit of heat throughput.

10 See table 1 for examples of units of the variables and parameters.

marginal cost function, expressed per unit of output. The parameter b_i represents the rate at which the source alters its pollution intensity per unit change in the price of pollution; different values of b alter the slope of the marginal cost function. Thus, a_i and b_i are source-specific demand parameters that allow for heterogeneity across sources in the intercept and slope of emission demand and marginal cost.[11] For the sake of clarity, we assume the heterogeneous variables are independently distributed, with mean values $\bar{a}, \bar{b}$, and $\bar{x}$. Allowing for correlation among the demand parameters does not change the aggregate compliance cost ordering of market-based compared with command-and-control instruments, but it could alter the magnitude of the cost difference between these types of policies depending on the magnitude and sign of the correlations between a, b, and x. We comment further on this below.

2.2. Alternative Policies for Emission Allocation

We consider three stylized policies for allocating aggregate emissions of Q among the sources. Two are performance-based instruments: a uniform emission rate standard and a uniform percentage reduction standard. The third is a market-based instrument, such as an emissions fee or tradable permit system.

2.2.1. Cost of a Uniform Emission Rate Standard

A uniform emission rate standard results in emissions from each source of $\tilde{q}_i$ equal to aggregate emissions per unit of output, weighted by source output:

$$\tilde{q}_i = \left(\frac{Q}{X}\right) x_i, \tag{3}$$

where X is aggregate output. This assumes perfect compliance, an assumption we also make for the other stylized instruments we consider. We also assume that the standard is binding for the relevant set of sources; that is, each source undertakes non-negative reductions. In addition, note that the same analysis and results will hold when the standard is in terms of emissions per unit of some input—assuming that the production function uses the input in fixed proportions to output.

Substituting (3) into (1) we find the cost to each source of the uniform emission rate standard:

$$C(\tilde{q}_i) = \frac{x_i}{2b_i}\left(a_i - \frac{Q}{X}\right)^2. \tag{4}$$

To estimate the aggregate cost of the uniform emission rate standard, we first find the

11 While we have incorporated heterogeneity directly into the parameters of the cost function, an alternative would be to characterize heterogeneity in underlying variables described earlier, such as location, age, size, and production technology. Because that approach would still involve a further mapping between these underlying variables and costs, it would considerably increase the complexity of attaining our goal, which is to provide a parsimonious, transparent representation of the relationship between cost heterogeneity and cost savings.

expected value of equation (4) using a second-order approximation around mean characteristics $\bar{a}, \bar{b}$, and $\bar{x}$:

$$E[C(\tilde{q}_i)] = \frac{\bar{x}}{2\bar{b}}\left(\bar{a} - \frac{Q}{X}\right)^2 + \frac{1}{2}\left(\frac{\bar{x}}{\bar{b}}V[a] + \frac{\bar{x}}{\bar{b}^3}\left(\bar{a} - \frac{Q}{X}\right)^2 V[b]\right), \quad (5)$$

where $V[\cdot]$ represents the variance of the bracketed parameter.[12]

Multiplying equation (5) by n, we find the aggregate expected cost of the uniform emission rate standard, denoted as $E[C(\tilde{Q})]$, which we further simplify by multiplying by $\bar{a}^2/\bar{a}^2$, yielding

$$E\left[C(\tilde{Q})\right] = \frac{X\bar{a}^2}{2\bar{b}}\left(R^2 + R^2\beta + \alpha\right), \quad (6)$$

where $R = (\bar{a} - Q/X)/\bar{a}$ is the fractional aggregate reduction in emissions from the baseline to the aggregate emission constraint $Q(0 \leq R \leq 1)$, and $\alpha = V[a]/\bar{a}^2$ and $\beta = V[b]/\bar{b}^2$ are dimensionless measures of heterogeneity or spread in a and b relative to their means; the square roots of α and β are known as coefficients of variation.

2.2.2. *Cost of a Uniform Percentage Reduction Standard*

A uniform percentage reduction standard results in emissions from each source of $\hat{q}_i$, where

$$\hat{q}_i = (1 - R)a_i x_i = \frac{a_i}{\bar{a}}\left(\frac{Q}{X}\right)x_i. \quad (7)$$

Substituting (7) into (1) we find the cost to each source of the uniform percentage reduction standard,

$$C(\tilde{q}_i) = \frac{x_i}{2b_i}\left(a_i - \frac{a_i}{\bar{a}}\frac{Q}{X}\right)^2,$$

which differs from the uniform emission rate standard (Q/X) by adjusting the standard for each source's baseline emissions relative to the mean $(a_i/\bar{a})$. Proceeding as above, we find the aggregate expected cost of the uniform percentage reduction standard, denoted as $E[C(\hat{Q})]$:

$$E\left[C\left(\hat{Q}\right)\right] = \frac{X\bar{a}^2}{2\bar{b}}R^2(1 + \beta + \alpha). \quad (8)$$

12 Higher-order approximations would involve terms for skewness, kurtosis, and higher-order moments of the distribution of b. We note that the skewness term for b would have a negative sign, indicating that positive skewness (i.e., skewed to the right, or a long upper tail) would tend to yield lower costs of a uniform performance standard. This makes intuitive sense because high values for b represent firms with low costs of emission control.

2.2.3. Cost of a Market-Based Policy Instrument

Under a market-based instrument, each source has an incentive to choose an emissions level q_i^* that minimizes its own costs, so that

$$q_i^* = (a_i - b_i p^*) x_i, \tag{9}$$

where an emissions market of size Q clears at price p^*. In principle, the cost-effective allocation could be implemented either by establishing a tradable permit market of size Q or by setting an emissions fee of p^*, where p^* will equal each source's marginal cost of emissions control (equation (2)) in equilibrium.

Taking expectations of (9), we find that average emissions will be demanded from a source that has average levels of each characteristic:

$$\mathrm{E}[q^*(p^*)] = \bar{q} = (\bar{a} - \bar{b}p^*)\bar{x}, \tag{10}$$

where $\bar{q}$ represents average emissions. Solving for p^*, and substituting $Q/X = \bar{q}/\bar{x}$, we find the emissions tax that will deliver aggregate emissions of Q, which also equals the market-clearing permit price for a permit market of size Q:

$$p^* = \frac{1}{\bar{b}}\left(\bar{a} - \frac{Q}{X}\right). \tag{11}$$

Substituting equation (11) back into equation (9), we find that each source's cost-effective emissions level is given by:

$$q_i^* = \left(a_i - \frac{b_i}{\bar{b}}\left(\bar{a} - \frac{Q}{X}\right)\right)x_i.$$

Following the same approach as before, we find the aggregate expected cost of the market-based instrument, denoted as $\mathrm{E}[C(Q^*)]$:

$$\mathrm{E}[C(Q^*)] = \frac{X\bar{a}^2}{2\bar{b}}R^2. \tag{12}$$

2.3. The Potential Cost Savings of Market-based Policy Instruments

Employing our model of the cost of alternative policy instruments, the aggregate cost savings, Δ, from using a cost-effective policy relative to a uniform emission rate standard is found simply by subtracting equation (12) from (6):[13]

---

13 Note that the cost savings approximation is likely to be acceptable for the policy-relevant range of emission reductions, but may not perform as well when aggregate reduction requirements are stringent enough to lead many sources to eliminate all of their emissions under a market-based policy. This can occur in our current model because we do not impose a non-negativity constraint on emissions under a

$$\tilde{\Delta} = E[C(\tilde{Q})] - E[C(Q^*)] = \frac{X\bar{a}^2}{2\bar{b}}(R^2\beta + \alpha). \tag{13}$$

We can also express the cost savings in percentage terms by dividing equation (13) by (6):

$$\%\tilde{\Delta} = 1 - \frac{E[C(Q^*)]}{E[C(\tilde{Q})]} = 1 - \frac{R^2}{R^2 + R^2\beta + \alpha}. \tag{14}$$

Likewise, the aggregate cost savings, $\hat{\Delta}$, from using a cost-effective policy relative to a uniform percentage reduction standard is found by subtracting equation (12) from (8):

$$\hat{\Delta} = E\left[C\left(\hat{Q}\right)\right] - E[C(Q^*)] = \frac{X\bar{a}^2}{2\bar{b}}R^2(\beta + \alpha), \tag{15}$$

or in percentage terms:

$$\%\hat{\Delta} = 1 - \frac{E[C(Q^*)]}{E[C(\hat{Q})]} = 1 - \frac{1}{1 + \beta + \alpha}. \tag{16}$$

Equations (13)–(16) provide the basis for evaluating the potential cost savings from adopting a market-based policy instrument. Minimal information is required about the relevant set of sources facing environmental regulation: aggregate production of the regulated industry, X; the emissions constraint, Q, which determines R; the mean and variance of the slope of the emissions demand function, $\bar{b}$ and $V[b]$, which determine β; and the mean and variance of baseline emissions intensity, $\bar{a}$ and $V[a]$, which determine α.

2.4. What Do the Expressions for Cost Savings Tell Us?

The first message from equations (13) to (16) is that the cost savings of market-based policies relative to uniform performance standards increase in a straightforward proportional manner as a function of greater abatement–cost heterogeneity. This cost heterogeneity comes from two sources. The first is heterogeneity in baseline emissions intensities, α, which indicates how much abatement each source will have to do, depending on the policy in place. The second is heterogeneity in the slope of the cost function, β, which describes how fast each source's costs rise as additional reductions are sought.

Each source of cost heterogeneity can have an effect independent of the other's presence. This makes it possible to identify a lower-bound of anticipated cost savings if information is available on only one source of heterogeneity, but not the other (assuming

market-based allocation. It is easily seen by evaluating the cost savings at the extreme of 100% reductions (i.e., $R = 1$). At 100% reductions the expressions suggest positive cost savings, although exact cost savings should be zero because every source would be required to emit zero emissions, facing the same costs under any policy. One could restrict the cost structure in order to avoid this limitation, for example by constraining the model parameters so that a_i/b_i is a constant equal to the marginal abatement cost at zero emissions.

distributional independence). In fact, information on baseline emission rates is likely to be more readily available than information on marginal costs. For example, from the United States Environmental Protection Agency's National Allowance Data Base, we can easily calculate that the baseline (1985) emission rate of sulfur dioxide for electric utility boilers that must reduce emissions under the sulfur dioxide allowance trading program was about 3.1 lbs/mmBtu on average ($\bar{a} = 3.1$), with a standard deviation of about 1.5 ($V[a] = 2.25$), resulting in $\alpha = 0.23$. Equation (14) therefore indicates an estimate of 38% cost savings from using a market-based policy relative to a uniform emission rate standard of 1.2 lbs/mmBtu ($R = 0.61$), which is the rate that was used to determine initial allocations of sulfur dioxide emission allowances. Heterogeneity in marginal cost functions ($\beta > 0$) would increase these savings. Using an econometric approach, with much greater data requirements, Carlson et al. (2000) estimated the potential cost savings for this program at about 43%.

It is interesting to note that heterogeneity in source size (x) does not enter the expressions because the market-based policy does not offer any more flexibility along the size dimension than performance standards, which themselves include an adjustment for size differences (on either a per unit of output or percentage basis). Of course, these results depend, to some degree, on the assumption that source size and other heterogeneous variables (a and b) are independent. We would expect any correlation with x to be small, because a and b are already expressed in rate terms.[14] The effect of correlation between a and b is straightforward: positive (negative) correlation in a and b will lower (raise) the cost savings of market-based instruments relative to either of the uniform standards. This makes intuitive sense, because if sources with relatively high baseline emissions tend also to have steeply sloped marginal costs, uniform standards will be even less cost-effective (and vice versa) than if these variables were independent. There is no a priori reason to expect the correlation to be positive or negative.

Next, we turn specifically to equations (13) and (15), which express the cost savings in dollar terms, and focus on the initial term multiplying the entirety of each of the expressions, $X\bar{a}^2/(2\bar{b})$. This term equals the aggregate cost of 100% emission reduction for a homogeneous set of sources. It serves to scale the expression to a degree appropriate for any particular environmental problem, depending on the size of the industry, baseline emission rates, and the slope of marginal control costs. The remaining variables in the expressions, α, β, and R, are dimensionless. When the cost savings are expressed in percentage terms, as in (14) and (16), only the dimensionless variables remain. Although expressing the cost savings in percentage terms has obvious appeal, one can lose an overall sense of the importance of the choice at hand. Hence, both forms are useful. Finally, we note that the effect of more stringent reduction rates, R, on absolute cost savings is amplified by higher degrees of cost heterogeneity.

---

14 Allowing for correlation among a, b, and x, the expressions for cost savings become:

$\tilde{\Delta} = X\bar{a}^2/2\bar{b}\left(R^2(\beta - \delta) + \alpha + 2R\gamma - 2R\lambda + R^2 - (R+\gamma)^2/1 + \delta\right)$, and

$\hat{\Delta} = X\bar{a}^2/2\bar{b}\left(R^2(\alpha + \beta - \delta + 2\gamma - 2\lambda) + R^2 - (R+\gamma)^2/1 + \delta\right)$ where

$\delta = \mathrm{COV}[b,x]/(\bar{b}\bar{x}), \gamma = \mathrm{COV}[a,x]/(\bar{a}\bar{x})$, and $\lambda = \mathrm{COV}[a,b]/(\bar{a}\bar{b})$.

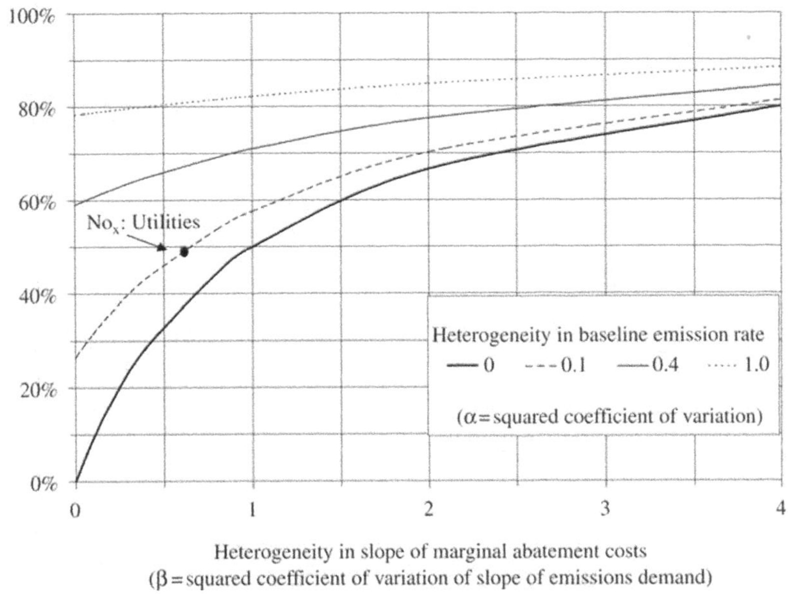

Figure 1. Cost savings of market-based policy relative to uniform emission rate standard (53% reduction in aggregate emissions).

These relationships between the cost savings of market-based policies and the various types and degrees of cost heterogeneity are illustrated in figures 1 and 2. Because general results can be illustrated for cost savings measured in percentage terms, we focus on such measures. Figure 1 portrays the anticipated cost savings (in percentage terms, on the vertical axis) of employing a fully cost-effective market-based policy instrument instead of a uniform emission rate standard in the case of an aggregate emission reduction target of 50%. Increasing heterogeneity in the slope of marginal costs is measured on the horizontal axis (in terms of our measure of spread of the slope of emissions demand, β). The general relationship between heterogeneity in the slope of the marginal cost function and cost savings is provided in the figure for four different degrees of heterogeneity in baseline emission rate (in terms of the respective measure of spread, α). Figure 2 provides the analogous cost savings results relative to a uniform percentage reduction standard.

As can be seen in figure 1, even in the absence of "slope heterogeneity," increasing heterogeneity in the baseline emission rate brings with it greater cost savings due to employing a market-based instrument (see the four vertical intercepts in the figure). Likewise, even when sources are identical in terms of their baseline emission rates ($\alpha = 0$), percentage cost savings increase with greater degrees of heterogeneity in marginal cost function slopes, β. In both cases, there are "decreasing returns" of cost-effectiveness from greater degrees of cost heterogeneity. That is, relatively small degrees of both types of abatement-cost heterogeneity result in market-based instruments enjoying

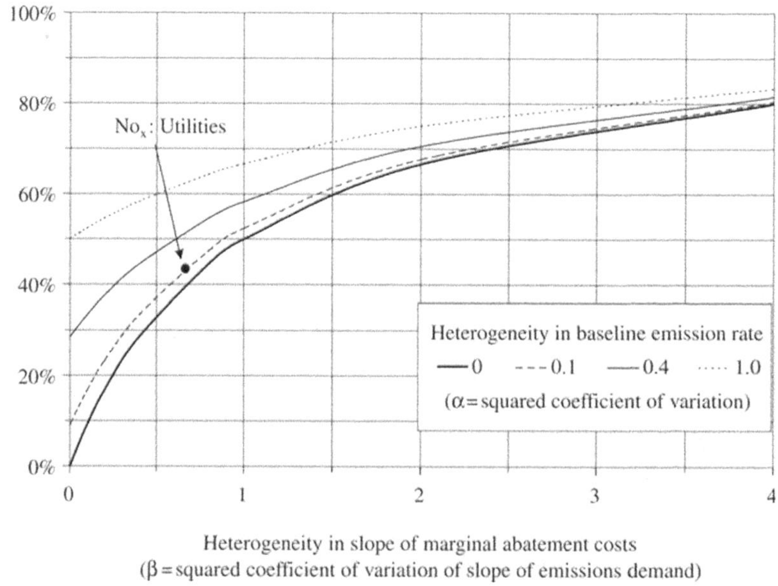

Figure 2. Cost savings of market-based policy relative to uniform percent reduction standard.
Note. The figure applies to any rate of emission reduction because *R* does not enter into the equation describing percent cost savings relative to a uniform percent reduction standard.

significant advantages over their command-and-control counterparts. But there are limits to these savings.[15]

The clarity of our results comes at the expense of some simplifying assumptions. Depending on the degree to which these assumptions are violated, our results will tend to over- or underestimate the extent of cost savings. In particular, the approximations may be less accurate where: there is a high degree of correlation among the heterogeneous variables; the distributions of these variables are highly skewed; the marginal cost function is highly nonlinear; or the policy goal is very stringent.

3. An Application to Nitrogen Oxides Control

As an example of how these relationships can be employed, we apply the framework to the case of nitrogen oxides (NO_x) reduction by electric utilities in a group of eastern states.

15 The cost-savings expressions and the related figures provide for convenient sensitivity analysis. In figure 1, it is apparent that when heterogeneity in baseline emission rates (as measured by α) is significant (for example, on the order of 0.4 or greater), potential cost savings are quite insensitive to changes in heterogeneity in the slope of marginal abatement costs (as measured by β).

When emitted, NO_x emissions and volatile organic compounds react in the presence of sunlight to form compounds that contribute to the formation of ground-level ozone. Ozone in the lower atmosphere can cause a variety of health problems by damaging lung tissue, reducing lung function, and adversely sensitizing the lungs to other irritants. Ozone tends to be a problem over broad regional areas, particularly in the eastern United States, where it can be transported by wind over hundreds of miles and across state boundaries. Through a multi-year effort known as the Ozone Transport Assessment Group (OTAG), the United States Environmental Protection Agency (EPA) worked with eastern states and the District of Columbia, private industry, and environmental advocacy groups to address ozone transport.[16]

We model several different emission control policies for large electric utility boilers in the relevant region, with each policy set to achieve a 53% reduction in aggregate emissions. The policies are (1) a 0.15 lb/mmBtu uniform emission rate standard; (2) a uniform 53% emission reduction standard; and (3) a cost-effective allocation among utilities. The virtue of the approach we have developed is its analytic simplicity and its use of a small set of summary statistics. The necessary summary statistics for these data are given in table 1. These summary statistics should be based on a representative sample of data describing the relevant regulated population. The difficulty of obtaining such information will depend on individual circumstances, but can be eased, for example, through the consideration of representative source types, weighted by their approximate shares in the overall population. Thus, while the data underlying the summary statistics we use in the present application are in fact from a relatively detailed and comprehensive database, the basis of the statistics generally need not be so detailed.[17]

The results are provided in table 1 and illustrated by the points labeled "NO_x: Utilities" in figures 1 and 2. Thus, figure 1 indicates that given the estimated degrees of heterogeneity of the two relevant cost function parameters for electric utility NO_x emissions, our model predicts 51% cost savings from employing a market-based policy instrument relative to a uniform emission rate standard. Our results are in the same range as those of Krupnick and McConnell (2000), who employ a source-by-source programming approach to evaluate cost-effective NO_x control, and find cost savings for

16 In 1998, EPA issued regulations requiring 22 eastern states and the District of Columbia to submit State Implementation Plans (SIP's) that address the regional transport of ground-level ozone through NO_x reductions (U.S. Environmental Protection Agency 1998). Building on OTAG recommendations, EPA established NO_x budgets for each state, but gave states the flexibility to decide which electric utility boilers, large industrial boilers, and other sources should be required to reduce NO_x emissions, by how much, and the specific policies they must follow to meet the projected budgets (for example, uniform emission rates, percentage reduction standards, or emissions trading). To determine the budgets, EPA chose a control level for large electric utility boilers based on a uniform emission rate standard of 0.15 lb/mmBtu (pounds of NO_x per million Btu of boiler heat input).

17 All of the necessary data on control costs, emissions, and output are available as projections for 2007 (Pechan Associates 1997). We used equation (1) to compute the slope of each source's emission demand function (*b*), taking an average in cases where there was more than one emission control scenario available. Costs for utilities are relative to an emission baseline consistent with meeting restrictions imposed by the 1990 Clean Air Act Amendments. We exclude from the analysis sources that already met the relevant performance standard in the baseline.

Table 1. Application to Nitrogen Oxides Control by Electric Utilities in the Eastern United States

Variable	Value
$\bar{a}$ (lb/mmBtu)	0.32
$V[a]$ (lb/mmBtu)^2	0.010
α	0.098
$\bar{b}$ (lb/mmBtu)/($/lb)	0.15
$V[b]$ (lb/mmBtu)^2/($/lb)^2	0.015
β	0.68
X (mmBtu)	9.85×10^9
Q/X (lb/mmBtu)	0.15
R	53%
$C(\tilde{Q})$ ($M)	1,990
$C(\hat{Q})$ ($M)	1,750
$C(Q^*)$ ($M)	980
p^* ($/lb)	1.16
$\tilde{\Delta}$ ($M)	1,010
$\%\tilde{\Delta}$	51%
$\hat{\Delta}$ ($M)	770
$\%\hat{\Delta}$	44%

utilities of about 46%. For comparison, we also show that the estimated cost savings relative to uniform percent reductions for utilities is 44% (figure 2). This result illustrates a point made earlier: that percent reduction standards tend to be less costly than uniform emission rate standards, unless, of course, there is no heterogeneity in baseline emission rates, in which case the two policies are identical. Finally, in this case, we found the degree of correlation between the variables to be small. Accounting for correlation in the estimates of cost savings lowered each of the estimates by about 5%.[18]

4. Summary and Conclusions

Policy makers and analysts in the environmental and other policy realms are often faced with situations where it is unclear whether market-based instruments hold significant promise of reducing costs, relative to conventional command-and-control approaches. We have developed some simple expressions that can be employed with modest amounts of information to estimate the potential cost savings associated with designing and implementing market-based policy instruments. Because our analytical models are simple, yet capture key properties of relevant cost functions, they can be used to predict potential cost savings through simple formulae and can improve understanding of the importance of cost heterogeneity and its policy implications in real-world environmental, resource, and other policy contexts.

We identified several key relationships between the forms and magnitude of cost

18 For this case, $\delta = 0.21, \gamma = 0.036$, and $\lambda = 0.07$, yielding $\%\tilde{\Delta} = 45\%$ and $\%\hat{\Delta} = 39\%$.

heterogeneity and the anticipated performance of alternative policy instruments. Two primary sources of heterogeneity—baseline emissions and the slope of marginal costs—can be expressed as having independent proportional effects on anticipated cost savings. Through an application to NO_x control in the eastern United States, we demonstrated the use of these results by estimating the anticipated cost savings from use of a market-based policy instrument, relative to both a uniform emission rate standard and a uniform percent reduction standard.

Our approach also provides modelers with a parsimonious structure for incorporating key elements of heterogeneity into their analytical models of costs for a variety of environmental problems. In addition, the framework we have developed may be usefully applied to address a number of other public policy questions. For example, what are the costs of maintaining one-size-fits-all environmental (and other) regulations across heterogeneous regions, such as the countries of Europe, the provinces of Canada, or the states of the United States? And what are the costs of uniform acreage control programs, such as in the European Union, where farms and farmers are heterogeneous across relevant dimensions?

References

Atkinson, S. E., and D. H. Lewis. 1974. ''A Cost-Effectiveness Analysis of Alternative Air Quality Control Strategies.'' *Journal of Environmental Economics and Management* 1: 237–250.

Atkinson, S. E., and T. H. Tietenberg. 1982. ''The Empirical Properties of Two Classes of Designs for Transferable Discharge Permit Markets.'' *Journal of Environmental Economics and Management* 9: 101–121.

Bohm, P., and C. S. Russell. 1985. ''Comparative Analysis of Alternative Policy Instruments.'' In *Handbook of Natural Resource and Energy Economics, Volume I*, edited by A. V. Knease, and J. L. Sweeney. Amsterdam: North-Holland, pp. 395–460.

Carlson, C., D. Burtraw, M. Cropper, and K. Palmer. 2000. ''SO2 Control by Electric Utilities: What are the Gains from Trade?'' *Journal of Political Economy* 108(6): 1292–1326.

Coggins, J. S., and J. R. Swinton. 1996. ''The Price of Pollution: A Dual Approach to Valuing SO_2 Allowances.'' *Journal of Environmental Economics and Management* 30: 58–72.

Gollop, F. M., and M. J. Roberts. 1983. ''Environmental Regulations and Productivity Growth: The Case of Fossil-fueled Electric Power Generation.'' *Journal of Political Economy* 91(4): 654–674.

Gollop, F. M., and M. J. Roberts. 1985. ''Cost-Minimizing Regulation of Sulfur Emissions: Regional Gains in Electric Power.'' *Review of Economics and Statistics* 67(1): 81–90.

Goulder, L. H., I. W. H. Parry, R. C. Williams, and D. Burtraw. 1999. ''The Cost-Effectiveness of Alternative Instruments for Environmental Protection in a Second-Best Setting.'' *Journal of Public Economics* 72: 329–360.

Hahn, R. W. 1984. ''Market Power and Transferable Property Rights.'' *Quarterly Journal of Economics* 99: 753–765.

Hahn, R. W., and R. G. Noll. 1982. ''Designing a Market for Tradable Emissions Permits.'' In *Reform of Environmental Regulation*, edited by W. A. Mayat. Cambridge, MA: Ballinger.

Helfand, G. E. 1991. ''Standards versus Standards: The Effects of Different Pollution Restrictions.'' *American Economic Review* 81(3): 622–634.

Jaffe, A. B., R. G. Newell, and R. N. Stavins. 2001. ''Technological Change and the Environment.'' NBER Working Paper 7970, prepared as a chapter for the forthcoming *Handbook of Environmental Economics*, edited by K.-G. Mäler and J. Vincent. Amsterdam: North-Holland/Elsevier Science.

Jaffe, A. B., S. R. Peterson, P. R. Portney, and R. N. Stavins. 1995. ''Environmental Regulation and the Competitiveness of U.S. Manufacturing: What Does the Evidence Tells Us?'' *Journal of Economic Literature* 33: 132–163.

Keohane, N. O., R. L. Revesz, and R. N. Stavins. 1998. ''The Choice of Regulatory Instruments in Environmental Policy.'' *Harvard Environmental Law Review* 22(2): 313–367.

Kolstad, C. D. 1986. ''Empirical Properties of Economic Incentives and Command-and-Control Regulations for Air Pollution Control.'' *Land Economics* 62(3): 250–268.

Krueger, A. O., and H. Sonnenschein. 1967. ''The Terms of Trade, the Gains from Trade, and Price Divergence.'' *International Economic Review* 8(1): 121–127.

Krupnick, A. J. 1986. ''Costs of Alternative Policies for the Control of Nitrogen Dioxide in Baltimore.'' *Journal of Environmental Economics and Management* 13: 189–197.

Krupnick, A. J., and V. McConnell. 2000. *Cost-Effective NO_x Control in the Eastern U.S.* Discussion Paper 00–18. Washington, DC: Resources for the Future.

Maloney, M. T., and B. Yandle. 1984. ''Estimation of the Cost of Air Pollution Control Regulation.'' *Journal of Environmental Economics and Management* 11: 244–263.

Malueg, D. A. 1990. ''Emission Credit Trading and the Incentive to Adopt New Pollution Abatement Technology.'' *Journal of Environmental Economics and Management* 16: 52–57.

McConnell, V. D., and G. E. Schwartz. 1992. ''The Supply and Demand for Pollution Control: Evidence from Wastewater Treatment.'' *Journal of Environmental Economics and Management* 23: 54–77.

Mendlesohn, R. 1986. ''Regulating Heterogeneous Emissions.'' *Journal of Environmental Economics and Management* 13: 301–312.

Misolek, W. S., and H. W. Elder. 1989. ''Exclusionary Manipulation of Markets for Pollution Rights.'' *Journal of Environmental Economics and Management* 16: 156–166.

Nichols, A. L. 1984. *Targeting Economic Incentives for Environmental Protection.* Cambridge: MIT Press.

Oates, W. E., P. R. Portney, and A. M. McGartland. 1989. ''The Net Benefits of Incentive-Based Regulation: A Case Study of Environmental Standard Setting.'' *American Economic Review* 79(5): 1233–1242.

O'Neil, W., M. David, C. Moore, and E. Joeres. 1983. ''Transferable Discharge Permits and Economic Efficiency: The Fox River.'' *Journal of Environmental Economics and Management* 10: 346–355.

O'Ryan, R. E. 1996. ''Cost-Effective Policies to Improve Urban Air Quality in Santiago, Chile.'' *Journal of Environmental Economics and Management* 31: 302–313.

Pechan, E. H. Associates, Inc. 1997. ''The Emission Reduction and Cost Analysis Model for NO_x (ERCAM-NO_x).'' EPA-68-D3-0035. Prepared for Ozone Policy and Strategies Group, U.S. Environmental Protection Agency, Research Triangle Park, NC.

Perl, L. J., and F. C. Dunbar. 1982. ''Cost Effectiveness and Cost-Benefit Analysis of Air Quality Regulation.'' *American Economic Review* 72(2): 208–213.

Russell, C. S., W. Harrington, and W. J. Vaughan. 1986. *Enforcing Pollution Control Laws.* Washington, DC: Resources for the Future.

Seskin, E. P., R. J. Anderson, Jr., and R. O. Reid. 1983. ''An Empirical Analysis of Economic Strategies for Controlling Air Pollution.'' *Journal of Environmental Economics and Management* 10: 112–124.

Spulber, D. F. 1985. ''Effluent Regulation and Long-Run Optimality.'' *Journal of Environmental Economics and Management* 12: 103–116.

Stavins, R. N. 1995. ''Transaction Costs and Tradeable Permits.'' *Journal of Environmental Economics and Management* 29: 133–147.

Stavins, R. N. 2001. ''Experience with Market-Based Environmental Policy Instruments.'' Prepared as a chapter for the forthcoming *The Handbook of Environmental Economics*, edited by K.-G. Mäler and J. Vincent. Amsterdam: North-Holland/Elsevier Science.

Suen, W. 1990. ''Statistical Models of Consumer Behavior With Heterogeneous Values and Constraints.'' *Economic Inquiry* 28(1): 79–98.

Tietenberg, Th. H. 1985. *Emissions Trading: An Exercise in Reforming Pollution Policy.* Washington, D.C.: Resources for the Future.

U.S. Environmental Protection Agency. 1998. *Final Rule for Reducing Regional Transport of Ground-level Ozone (Smog) and Two Related Proposals.* Washington, DC: Fact Sheet.

Weitzman, M. L. 1974. ''Prices vs. Quantities.'' *Review of Economic Studies* 41: 477–491.

Weitzman, M. L. 1977. ''Is the Price System or Rationing More Effective in Getting a Commodity to Those Who Need it Most?'' *Bell Journal of Economics* (Autumn): 517–524.

[10]

The Effect of Allowance Allocations on Cap-and-Trade System Performance

Robert W. Hahn *University of Oxford*

Robert N. Stavins *Harvard University*

Abstract

An implication of the Coase theorem is that under certain conditions, the market equilibrium in a cap-and-trade system will be cost-effective and independent of the initial allocation of tradable rights. That is, the overall cost of achieving a given aggregate emission reduction will be minimized, and the final allocation of permits will be independent of the initial allocation. We call this the independence property. This property is important because it means that the government can establish the overall pollution reduction goal for a cap-and-trade system by setting the cap and leaving it up to the legislature to construct a constituency in support of the program by allocating the allowances to various interests without affecting either the environmental performance of the system or its aggregate social costs. We examine the conditions under which the independence property is likely to hold—both in theory and in practice.

1. Introduction

The fiftieth anniversary of the publication of "The Problem of Social Cost" (Coase 1960) is an appropriate time to examine a key implication of that landmark study, which is of great importance not only for economics but for public policy as well. Our starting point is a well-known result from that article, namely, that bilateral negotiation between the generator and the recipient of an externality

Hahn is Director of Economics, Smith School of Enterprise and the Environment, University of Oxford, and Senior Fellow at the Georgetown Center for Business and Public Policy. Stavins is the Albert Pratt Professor of Business and Government at the John F. Kennedy School of Government, Harvard University; University Fellow at Resources for the Future; and Research Associate at the National Bureau of Economic Research. We benefited from comments by participants at the conference Markets, Firms, and Property Rights: A Celebration of the Research of Ronald Coase (University of Chicago Law School, December 4–5, 2009). We are grateful for helpful comments provided by Dallas Burtraw, Carolyn Fischer, Denny Ellerman, Suzi Kerr, Matti Liski, Juan-Pablo Montero, Sheila Olmstead, Tom Tietenberg, Alistair Ulph, and especially Thomas Hazlett. We thank James Sigel for superb research assistance. All remaining errors are our own.

[*Journal of Law and Economics*, vol. 54 (November 2011)]

will lead to the same efficient outcome regardless of the initial assignment of property rights, in the absence of transaction costs, income effects, and third-party impacts. This result, or a variation of it, has come to be known as the Coase theorem.[1]

We focus on an idea that is closely related to the Coase theorem, namely, that the market equilibrium in a cap-and-trade system will be cost-effective[2] and independent of the initial allocation of tradable rights (typically referred to as permits or allowances). That is, the overall cost of achieving a given emission reduction will be minimized, and the final allocation of permits will be independent of the initial allocation, under certain conditions. We call this the independence property. It is closely related to a core principle of general equilibrium theory (Arrow and Debreu 1954), namely, that when markets are complete, outcomes remain efficient even after lump-sum transfers among agents.

We are interested in the independence property because of its great political importance. It is a key reason that cap-and-trade systems have been employed and have evolved as the preferred instrument in a variety of environmental policy settings.

Our primary objective is to examine the conditions under which the independence property is likely to hold. Our analysis is limited to environmental cap-and-trade systems, in which the government specifies an overall level of pollution, distributes allowances corresponding to that level, and permits those allowances to be traded.[3] Furthermore, we examine only the consequences of alternative free allocations of allowances, not the impacts of free allocations versus auctioning of allowances, a topic that has received much attention from economists (Goulder and Parry 2008).

In Section 2, we examine the theory behind the independence property, starting with the key findings from partial and general equilibrium analyses. Then, in Section 3, we identify the conditions under which independence may be violated, including when there are transaction costs, when firms have market power, and when firms are subject to differential regulatory treatment.

Section 4 provides an assessment of the independence property in past and current cap-and-trade systems: lead trading; chlorofluorocarbons (CFCs) under the Montreal Protocol, the sulfur dioxide (SO_2) allowance trading program, the Regional Clean Air Incentives Market (RECLAIM) in Southern California, eastern nitrogen oxides (NO_X) markets, the European Union Emission Trading Scheme (EU ETS), and Article 17 of the Kyoto Protocol. Section 5 concludes.

[1] As Hazlett (2009) noted, Coase was primarily concerned with the importance of transaction costs, not a world of zero transaction costs. It was Stigler (1966) who dubbed the above result the Coase theorem.

[2] Cost-effectiveness refers to achieving an overall emissions objective at the lowest cost.

[3] We do not consider systems that primarily involve offsets, in which a firm can generate emission credits from a baseline that is not always well defined. While many of the cap-and-trade systems we examine allow for offsetting of pollution from sources outside of the cap-and-trade system, we do not focus on this aspect of these regimes.

2. The Fundamentals of Cap and Trade

One textbook environmental economics problem is to design policy instruments that are efficient in the sense that they equate marginal social benefits with marginal social costs. As is well known, both taxes and cap and trade can accomplish this in principle, provided the regulator has adequate information on damages. In the absence of such information or the political will to use it, another important problem remains for environmental economics—to design policy instruments that are likely to achieve given aggregate environmental targets at the lowest cost. This is the cost-effectiveness challenge, and—again—either taxes or cap and trade can, in principle, meet it.

For some 40 years prior to Coase (1960), the sole economic response to the problem of externalities was that the externality in question should be taxed. In principle, a regulator could ensure that emitters would internalize the damages they fostered by charging a tax on each unit of pollution equal to the marginal social damages at the efficient level of pollution (Pigou 1920). Any quantity of aggregate emissions could be achieved cost-effectively through an appropriate emissions tax, because such a tax would make it in the interest of each emitter to equate the tax to its own (increasing) marginal abatement cost, with the result that marginal abatement costs would thereby be equated among emitters and achieve the necessary condition for cost-effectiveness. Whenever abatement costs differ across emitters (a ubiquitous fact of the real world), conventional command-and-control pollution policies—such as uniform performance standards and technology standards—would not be cost-effective, but a uniform Pigouvian tax would be.

Despite the theoretical merits of Pigouvian taxes, they have rarely been used, for a number of reasons. First, it is difficult to identify the appropriate tax rate, because decision makers typically do not have good information on benefits and costs. Second, political problems are introduced by the distributional consequences of Pigouvian taxes for regulated sources. Despite the fact that such systems minimize the aggregate social costs, these systems may be more costly than are comparable command-and-control instruments for regulated firms. This additional expense occurs because with the tax approach, firms pay both their abatement costs and taxes on their residual emissions (Buchanan and Tullock 1975). In practice, some of these costs will be passed on to consumers, but many firms may still be worse off under a tax.

Following Coase (1960), it became possible to think about solving the problem of pollution as one of clarifying poorly defined property rights. If resources such as clean air and water could be recognized as a form of property, whose corresponding rights could be traded in a market, private actors could allocate the use of this property in a cost-effective way. Crocker (1966) and Dales (1968) each proposed a system of transferable discharge permits that could provide such a market solution: the regulator need only designate the total quantity of emissions allowed (the cap), distribute rights corresponding to this total, and allow individual sources of emissions to trade the permits until an optimal (cost-effective) allocation had been reached.

In theory, the final allocation of the permits (and therefore the location of the emissions) would not depend on the initial allocation. This result was implicit in the work of both Dales and Crocker, because such independence follows directly from the notion that by properly defining property rights, participating firms have incentives to exploit all potential gains from trade. In such a well-functioning market, the permit price itself will not be affected by the initial distribution of permits, because the initial allocation does not affect firms' marginal abatement cost functions. Therefore, the final allocation of permits, in which no firm can be made better off by buying or selling a permit at the market price, is unaffected. Montgomery (1972) rigorously proved that transferable permits would in theory achieve a cost-effective allocation of control responsibility and would do so independently of the initial allocation of permits.[4] Following Montgomery, this result was frequently cited and extended to other contexts (Tietenberg 1985).

This independence property is of central political importance and is an essential reason that cap-and-trade systems have been employed in representative democracies, where distributional issues may be of paramount importance in mustering support for a policy.[5] In principle, the government can set the overall emissions cap—whether on the basis of economic efficiency or, more likely, some other grounds—and then leave it up to the legislature to allocate the available number of allowances among sources (locations) to build a constituency of support for the initiative without reducing the system's environmental performance or driving up its cost.[6]

Experience has validated the political importance of this property. For example, in the Senate debate over the Clean Air Act Amendments of 1990, bonus allowances were awarded to electricity generators in Ohio, which were going to incur particularly high costs because of their reliance on high-sulfur coal (Joskow and Schmalensee 1998; Stavins 1998); the result was the key support of Senator John Glenn (D-Ohio) for the legislation. This pattern should be contrasted with

[4] In particular, Montgomery (1972) showed that a cap-and-trade system defined in terms of ambient air quality at many receptor points would exhibit the independence property. Subsequent research demonstrated the independence property for a system of emissions permits (Krupnick, Oates, and Van De Verg 1983).

[5] The importance of efficient redistribution is highlighted in Becker (1983). In general, the most efficient way to buy or sell influence is by using money; yet politicians frequently do not choose that approach, because such transfers are highly visible. Allocated permits, which are a close substitute for money, appear to be a politically acceptable form of redistribution in applications ranging from environmental regulation to airport takeoff and landing slots.

[6] The choice of an environmental goal and the choice of a particular policy instrument for achieving that goal may be connected, and similarly it is possible that the choice of the cap of a cap-and-trade system may be connected with the choice of a specific allocation. In principle, taxes could also be designed in ways to address distributional concerns. For example, the legislature could exempt emissions up to some level for specific firms, thereby matching some of the distributional properties of a cap-and-trade system with freely allocated permits.

most public policy proposals—environmental or otherwise—for which the normal course of events is that the political machinations that are necessary to develop sufficient legislative support reduce the effectiveness of the policy and/or drive up its costs.

3. Conditions Affecting the Independence Property

We consider six conditions under which the independence property may—in principle—break down in a cap-and-trade system: transaction costs, market power, uncertainty, conditional allowance allocations, non-cost-minimizing behavior by firms, and differential regulatory treatment of firms.

3.1. Transaction Costs

Transaction costs arise in the exchange of goods and services and are particularly important in thin markets. In bond markets, for example, transaction costs may give rise to expanded bid-ask spreads. In cap-and-trade markets, there are three major sources of transaction costs: search and information, bargaining and reaching a decision, and monitoring and enforcement.[7]

The first source, search and information, may be the most obvious. Because of the public good nature of some information, it can be underprovided by markets. Brokers may step in to provide information about firms' pollution control options and potential trading partners and thus reduce the transaction costs. Although less obvious, the second source of transaction costs, bargaining and reaching a decision, is potentially as important. There are real resource costs to a firm involved in entering into negotiations, including time and/or fees for brokerage, legal, and insurance services. The third source of transaction costs—monitoring and enforcement—can also be significant, but these costs are typically borne by the responsible governmental authority and not by trading partners and hence do not fall within the notion of transaction costs incurred by firms.

Transaction costs can lead to a violation of the independence property. Coase's original formulation noted that the outcome of bilateral negotiation in the presence of transaction costs is affected by the identity of the exclusive property right recipient. But is the outcome independent of the initial allocation when the quantity of allocation to a source changes? This is the key question for cap-and-trade systems.

The answer is that the effect of transaction costs on the independence of the final allocation from the initial permit allocation depends on the nature of the

[7] All three sources can be thought of as representing costs resulting from lack of information (Dahlman 1979).

transaction costs present. If marginal transaction costs are constant, the final allocation of allowances is independent of the initial distribution (Stavins 1995).[8]

In the case of increasing marginal transaction costs, independence does not hold. Under such conditions, as the initial allocation of allowances to a source is increased, its equilibrium control level will be reduced, thereby increasing the departure of the posttrading equilibrium outcome from the cost-effective outcome. However, increasing marginal transaction costs are unlikely to be an important case in practice, because such costs are unsustainable: parties can simply split their transactions into smaller trades to economize.[9]

Finally, under decreasing marginal transaction costs, a theoretically important case, independence does not hold. Such transaction costs might result when brokers offer quantity discounts on their services or when there are positive information externalities at the market level. In this case, a shift in the allowance allocation away from the cost-effective equilibrium leads to a posttrading outcome that is closer than otherwise to the cost-effective equilibrium (Stavins 1995). This counterintuitive result stems from the fact that decreasing marginal transaction costs mean that there are scale economies of trading of which firms can take advantage. These theoretical results have been supported with experimental evidence from a laboratory study (Cason and Gangadharan 2003).[10]

3.2. Market Power and Market Structure

The presence of market power of a firm in the allowance market of a cap-and-trade regime can prevent the final allocation of permits from being independent of the initial distribution. If the firm with market power is a likely allowance seller, it has an incentive to act as a monopolist and hold back allowances from the market to drive up allowances prices,[11] and if it is a likely allowance buyer, it has an incentive to act as a monopsonist and buy fewer allowances to keep the price low (Hahn 1984).[12] Because trading behavior depends critically

[8] This is true apart from the limited sense discussed above, that is, the identification of an exclusive property right recipient. Also, Montero (1998) observes that this result does not hold if firms' marginal cost functions are discontinuous, as they might be if a firm is choosing among discrete technologies. In such cases, the existence of constant marginal transaction costs may preclude independence.

[9] Liski (2001) notes that the notion that increasing marginal transaction costs are unsustainable is based on the assumption that contacting new trading partners is costless. Increasing marginal transaction costs are sustainable if they are combined with sufficient fixed transaction costs between parties.

[10] If transaction costs are a function of market size—which is plausible if a thicker market results in lower costs of gathering information and greater opportunity for brokers to reduce friction in the market—then independence holds only if the pretrade allocation of permits is significantly different from the cost-effective allocation (Liski 2001).

[11] Malik (2002) shows that this dominant firm may actually choose to hold more permits than it needs, retiring the excess permits without their being used.

[12] According to Liski and Montero (2011), the independence property is not at risk in a dynamic permit market in which firms can bank allowances if the large players in the market are on the buyer side. These large buyers cannot exercise market power, because they suffer from a durable-goods monopoly problem; that is, they cannot credibly commit to keep a gap between their marginal costs and the allowance price.

on the initial allocation, independence does not hold, and cost-effectiveness is not achieved.

These results hold in the case of multiple dominant firms acting as Cournot players (Westskog 1996) and in a market with few buyer and sellers, all of which have some ability to set prices (Malueg and Yates 2009).[13] In both cases, the permit market achieves the cost-effective allocation only if the dominant firms initially receive the number of permits that leave them with no incentive to trade.[14]

The interaction of the permit market with a product market complicates the influence of market power on the final allocation of permits. For example, a dominant firm may utilize its power in the permit market to gain an advantage in its product market (Misiolek and Elder 1989) and thereby can manipulate the price of emissions permits to drive up the production costs of its rivals by forcing them to purchase less cost-effective abatement technology. In this case, the dominant firm would tend to purchase more permits (or sell less) than it would in a case in which there was no exclusionary effect in the product market.

Thus, when the dominant firm receives no permits and is a buyer on the permit market, the final distribution of permits may be closer to the efficient allocation than would obtain in a competitive market. But if the dominant firm receives a disproportionately large share of permits and is a seller, then the degree to which market power leads the ultimate allocation to depart from the cost-effective allocation is exacerbated.[15] In contrast, if the dominant firm exercises market power only in the product market and not in the permit market, the final allocation of permits will be independent of the initial allocation, but this allocation may not be cost-effective.[16]

Market structure can be important in other ways. One interesting case is

[13] Malueg and Yates (2009) also examine a market in which participants have market power and private information; the result is that the ultimate allocation is less predictable than it would be under typical circumstances, but the final allocation of permits remains a function of the initial distribution.

[14] Other variations of these basic models have been examined. Malik (2002) expands on Hahn (1984) to consider what would happen if some price-taking firms are noncompliant and suffer a penalty for noncompliance. This situation creates a greater elasticity of demand for permits, which reduces the extent to which the dominant firm can profit from its market power. Still, the cost-effective allocation is not achieved, because the dominant firm continues to exercise some market power and because noncompliant price-taking firms fail to equate their marginal abatement costs with the allowance price. Schwartz (2007) demonstrates that if permits are defined in terms of air quality space at a number of different receptor points with a different market pertaining to each receptor point, then, even with one or more dominant firms, the final allocation of permits may be independent of the initial allocation if the number of firms equals the number of markets in the ambient permit system.

[15] Eshel (2005) extends this model to encompass a firm that has market power in both the permit market and the product market and has found similar results.

[16] This result occurs because rival firms engaging in Cournot competition may be more aggressive in the output market if they have higher marginal abatement costs (and will therefore buy permits) but may decrease output production if they are more efficient (and will therefore sell permits). In equilibrium, the less efficient firms will hold a larger share of emissions permits than would be welfare maximizing (Malueg 1990; Sartzetakis 2004).

imperfections in capital markets that cause firms to face liquidity constraints. In this case, initial allowance allocations may affect the ultimate allocation of permits. Liquidity constraints may be one reason that smaller firms tend to adopt new technologies more slowly than do larger firms (Saloner and Sheppard 1995). More broadly, lack of liquidity can prevent firms from being able to minimize costs by investing in either permits or abatement technology. A firm initially allocated a relatively large number of permits will have greater liquidity, in the form of readily salable permits. Thus, allocation can affect entry and exit behavior.

3.3. Uncertainty

Uncertainty regarding future allowance prices can lead to a violation of the independence property under two conditions: limits to transferability of allowances (transaction costs) and risk aversion on the part of regulated firms. Without limits on transferability, an appropriate price for permits that incorporates their expected future value would emerge, and, absent other market imperfections, the market would clear so that the posttrading allocation of allowances would be independent of the initial distribution.

With uncertainty, firms with small initial allocations may be expected to overinvest in abatement technology in order to hedge against possible high future allowance prices, and firms with large initial allocations may be expected to underinvest in abatement technology in order to hedge against possible low future allowance prices (Baldursson and von dehr Fehr 2004; see also Ben-David et al. 2000). Laboratory experiments have provided some empirical support for these results (Betz and Gunnthorsdottir 2009).

Another way in which uncertainty may affect the final allocation of permits is also related to transaction costs. Building off the work of Stavins (1995), Montero (1998) incorporates uncertainty regarding whether a transaction, once initiated and in which costs have been sunk, can be completed. The results are similar to those described by Stavins (1995), but with uncertainty increasing the extent to which the final equilibrium will depart from the cost-effective one. Essentially, uncertainty increases the degree to which transaction costs impede mutually beneficial trades.

More broadly, if such uncertainty is endogenous and is a function of the number of permits demanded and the effort spent to attain trade approval, then for all types of transaction costs—constant, increasing, and decreasing—the tendency will be for an initial allocation further from the least-cost equilibrium to increase the departure of the ultimate allocation from the ideal one (Montero 1998). This result contrasts with that of Stavins (1995), who found that, under decreasing marginal transaction costs, economies of scale would lead an initial allocation further from the least-cost one to result in a more cost-effective ultimate allocation.

3.4. Conditional Allowance Allocations

Under a system of conditional allowance allocations, a firm's allocation in the current period is affected by its behavior in the previous period. This situation gives rise to inefficiencies, because firms no longer simply minimize their abatement costs but also take into account the profits they could obtain by receiving a larger allocation of allowances in the next period.

Although conditional allowance allocations do not strictly fit within the realm of the independence property because the government determines an allocation rule and not a specific allocation, this is an important case in which the initial allocation affects the outcome. For example, the government may assign allowances on the basis of a measure of output (such as gasoline production or electricity output) in the previous period.[17] We consider these rules because they determine the initial allocation and because they are proposed in a variety of applications.

An output-based updating-allocation rule serves as a production subsidy (Fischer 2001), which affects the posttrading allocation and thus can affect efficiency. Firms can be expected to increase production, because they can receive more permits in the subsequent period, and these increases contribute to their profits. The emissions price drives reductions in emissions intensity, but the prospect of allocations for additional production offsets (in part or in full) increases the prospect of additional costs from the emissions embodied in that production. Output-based allocation, therefore, means that emissions-intensive products have lower marginal costs than they would under ordinary free allocations, in which the full value of the embodied emissions would be embedded in the firm's marginal costs. Consequently, with higher output, to meet an equivalent emissions target, carbon prices must rise to bring about additional reductions in emissions intensity and shift abatement to other covered sectors (Fischer and Fox 2007).

The use of this allocation mechanism may be motivated by a desire to protect certain industries from adverse competitiveness effects (Houser et al. 2008) or by a desire to limit price impacts on consumers. For example, the American Clean Energy and Security Act of 2009 (H.R. 2454, 11th Cong. [2009])—the so-called Waxman-Markey legislation—employs this device for firms in energy-intensive trade-exposed sectors to limit the cost disadvantage that would otherwise shift competitiveness, economic activity, and emissions to countries without comparable regulation.

Unlike attempts to use ordinary free allocations to compensate the shareholders of firms in regulated sectors (which have no effects on behavior and no impacts on competitiveness), output-based updating allocations compensate firms in a way that lowers their marginal production costs. The allocations in H.R. 2454 to local distribution companies, which are regulated entities with the

[17] An extreme but not uncommon type of output-based updating-allocation mechanism is one in which a source loses its allocation if it closes down or receives a free allocation if it enters.

mandate to pass on the cost savings to consumers, have effects similar to those of output-based allocations.

Because product prices are kept lower than would otherwise occur, output-based allocations discourage reliance on conservation and substitution to less emitting alternative products as a means for emissions reductions. Instead, to meet the cap, there is greater reliance on technological solutions and conservation in other sectors that are better able to pass on the costs (Fischer and Fox 2007). This reliance means higher compliance costs.[18] In the absence of preexisting distortions, or if the allocations are poorly designed and inappropriately targeted, output-based allocations can reduce efficiency.[19]

Conditional allocations may become increasingly relevant to the climate change problem. Countries such as China and India may see it in their interest to build coal plants now if they believe it will be more costly to do so later or, alternatively, if they believe that their allocation of permits under a cap-and-trade regime would be increased (Kaplow 2009).

3.5. Non-cost-Minimizing Behavior

If some market participants are not cost minimizing (not equating their marginal abatement costs with the prevailing allowance price), then the final allocation of allowances will likely be a function of the initial allocation. There are a variety of potential sources of such non-cost-minimizing behavior.

An interesting case is that of public entities as market participants. The international emissions quota regime established under Article 17 of the Kyoto Protocol (United Nations Framework Convention on Climate Change, UN Doc. FCCC/CP/1997/7/Add. 1, December 10, 1997; 37 ILM 22 [1998]) allows trades in national targets (assigned amount units) among nations. But there is no reason to assume that nations are simple cost minimizers or even if they sought to be that they would have the necessary information regarding their national abatement costs to carry out cost-effective international exchanges (Hahn and Stavins 1999).

Behavioral economics provides several potential explanations for other sorts of non-cost-minimizing behavior. The endowment effect describes situations in which firms or individuals overvalue items already in their possession (Thaler 1980; Kahneman, Knetsch, and Thaler 1990). If the firms granted emissions allowances tend to value them more highly than do other firms simply because

[18] Evaluating the cost-effectiveness of output-based allocations requires an assessment of other distortions that we have not emphasized here, namely, tax interactions and the potential for emissions leakage (Fischer and Fox 2009; Bernard, Fischer, and Fox 2007).

[19] An alternative policy for addressing emissions leakage is border adjustment. In theory, border adjustment is generally superior to output-based allocation, since border adjustment enables the preservation of the conservation signal to consumers by ensuring that comparable carbon costs are imposed on all products. Many economists, however, would nevertheless prefer the option of output-based allocation, because border adjustments may be more likely than other policies to invite retaliation, which would lead to reduced trade and consequent welfare losses. But, in principal, production subsidies can invite retaliation as well, so it is not clear which approach is better in practice.

they happen to hold them, then fewer transactions will occur than would otherwise be predicted by market forces. If firms initially endowed with permits tend not to sell them, independence will not hold. Murphy and Stranlund (2007) suggest that the endowment effect may explain the low trading volumes in some cap-and-trade systems.

Status quo bias, a variant of the endowment effect, would lead to similar results. Like the endowment effect, status quo bias is a manifestation of loss aversion—in which actors consider the disutility of parting with an item as being greater than the utility associated with obtaining it (Kahneman, Knetsch, and Thaler 1991). A firm operating with such status quo bias might be expected to hold on to its initial allotment of permits simply because it has a preference for the status quo, whatever its mix of permits and abatement costs (Samuelson and Zeckhauser 1988).

Other sources of non-cost-minimizing behavior can also affect the performance of cap-and-trade markets. Even if managers wished to maximize profits, there may be such a high degree of organizational complexity in a firm that they cannot effectively do so (Tschirhart 1984). Related to this point is that principal-agent problems may cause firms not to minimize costs. A firm may continue to hold permits even if it would be more efficient to sell the permits and invest in abatement technology simply because it is administratively easier (Oates and Stassmann 1984).

In some scenarios, firms that are not generally cost minimizing might nevertheless minimize costs with respect to emissions abatement. Oates and Strassmann (1984) argue that this may be the case with the sort of managerial model of firms set forth by Williamson (1963). In this model, managerial utility includes a preference for expenditures on staff, salaries and benefits for management, and discretionary profits—the excess of actual profits over the minimum profits required. If abatement contributes nothing positive or negative to staff expenditures or managerial salaries, then firm managers will seek to maximize discretionary profits by minimizing expenditures on abatement—and will therefore act to equalize the permit price with marginal abatement costs. The independent, cost-effective allocation of permits will result.

Another reason for a firm's departure from cost-minimizing behavior is associated with regulatory compliance. Proponents of responsive regulation have argued that firms often do not make decisions on whether to comply with a regulatory scheme on the basis of simple cost-benefit calculations but rather are motivated—at least partly—by factors such as the shame associated with violating a regulation perceived to be morally valid (Parker 2006; Braithwaite 2002). Firms may comply with regulations out of a sense of civic duty and because of the pressure imposed by other actors not to deviate from established norms (May 2005). Thus, if the initial allocation of permits affects firms' perceptions of the moral validity of a cap-and-trade scheme, it may influence their decisions about whether to play by the rules and thereby affect the ultimate allocation of permits.

Finally, firms may be reluctant to engage in trading in the beginning stages of a market. For example, if a firm's managers are risk averse, they may try to meet the emission target that is implicitly defined by the initial allocations. So they would minimize costs across different plants that are part of their companies, subject to the constraint that emissions do not exceed the initial allocations. Ellerman (2000) calls this "autarkic compliance." In the early years of the SO_2 trading system, some utilities appear to have pursued this strategy.

3.6. Regulatory Treatment

If firms receive differential regulatory treatment, then initial allocations of allowances can affect equilibrium allocations, performance, and costs. State-level regulation of electricity producers, such as rate-of-return regulation, can discourage or even prevent firms from cost minimizing with respect to emissions (Hahn and Noll 1983; Tschirhart 1984; Oates and Strassman 1984). In addition, regulators may discourage trading, because of concerns about local pollution (Fullerton, McDermott, and Caulkins 1997). Finally, if a cap-and-trade system is interstate, then jurisdictions may be regulated differently. In all of these cases, the equilibrium allocation will not be independent of the initial allocation, and the outcome will not be cost-effective.[20]

Under typical rate-of-return regulation, utilities are permitted to extract only those revenues that are linked with certain costs determined by the regulator to be recoverable. Under the original cost method of rate making, a utility is allowed to retain as revenue its annual depreciation plus a certain proportion of its original expenditure on capital facilities. If permits are given out rather than sold, then the utility will not have spent anything to acquire its permit holdings, and the value of these permits will not be incorporated into its rate base. This problem is compounded by the common requirement that any gains from the sale of assets must be factored into utility prices—thus these utilities will effectively be taxed 100 percent on any revenue from the sale of permits. Utilities will, therefore, have incentives to spend money on other abatement mechanisms, a variation on Averch and Johnson's (1962) observation that regulated utilities will tend to overinvest in capital goods.

In states that calculate rates on the basis of the replacement cost method—adjusting the original cost of capital goods on the basis of inflation and technological changes—the above problems would not arise, as the market price of permits would be used to adjust the value of the utility's rate base, and thus the utility might be expected to minimize abatement costs (Hahn and Noll 1983). If the permit market spans several states, and some participants are subject to the original cost method regulation, then the cost-effective allocation will not be attained, as the lack of incentives to minimize cost will prevent the market from clearing efficiently. Further distortions are introduced if expenditures on abatement technologies are permitted to earn higher rates of return than amounts

[20] Obviously, a cap-and-trade system that restricts certain trades will violate the independence property if the restrictions are binding.

expended on emissions permits or if some abatement assets can be depreciated at a faster rate than others or treated as operating costs (Bohi and Burtraw 1992).

Problems can arise with respect to efficiency even if regulatory treatment is the same across utilities. Suppose there is free allocation for utilities in all states. This situation could lead to lower product prices (for example, for electricity) because of regulation, as noted above. These product prices will not generally reflect incremental costs (because of the nature of regulation) and will thus give rise to allocative inefficiencies (Spulber 1985). In particular, the equilibrium would be affected by how many allowances the utility sector is given directly. These problems can be addressed by allowing utilities to purchase allowances from local distribution companies, as suggested in the Waxman-Markey legislation.

Finally, regulators might conceivably take proactive steps that prevent the cost-effective allocation from being reached. If concerned about local pollution, a state regulator might implement rules that discourage the purchase of emissions permits, thus limiting the extent to which permits granted to utilities in other regions may cross political boundaries (Fullerton, McDermott, and Caulkins 1997).[21] In this scenario, the express purpose of regulation would be to ensure that independence does not hold and that firms are not allowed to increase their emissions to the amount that would be cost-effective for them and for the system as a whole.

4. Assessment of Effects of Allowance Allocation on Performance of Actual Cap-and-Trade Systems

In this section, we assess the independence property in existing and planned cap-and-trade systems. Our analysis proceeds in two ways: descriptively, by assessing the presence of the various conditions that could lead to a violation of the independence property, and statistically, by reviewing studies that test whether the final allocation of permits is affected by the initial allocation.

Market imperfections are present in most real-world markets. For example, in most markets, there are transaction costs in bringing buyers and sellers together. The question for our purposes is whether these market imperfections turn out to be important in terms of leading to significant violations of the independence property.

We review seven cap-and-trade systems: lead trading, CFCs under the Montreal Protocol, the SO_2 allowance trading program, RECLAIM in Southern California,

[21] Of course, there may be efficiency rationales for limiting trading when there is nonuniform mixing of pollutants or hot spots.

the eastern ozone transport NO_X market, EU ETS, and Article 17 of the Kyoto Protocol.[22]

4.1. Lead Trading

In 1973, the U.S. Environmental Protection Agency (EPA) began to impose regulations requiring refineries to reduce the content of lead in gasoline. In 1982, in conjunction with new, lower limits on lead content, the EPA authorized interrefinery trading of lead credits. If refiners produced gasoline with a lower lead content than was required, they earned lead credits, which could then be traded to other refineries. A major purpose of the program was to lessen the financial burden on smaller refineries, which were believed to have significantly higher compliance costs than larger ones had. In 1985, the EPA initiated a program to allow refineries to bank lead credits. In 1987, the program was terminated when the lead phasedown was completed (Stavins 2003).

The leaded gasoline phasedown is generally considered to have been a success (Kerr and Mare 1998). Part of this success has been attributed to the lack of regulatory restrictions on trading lead credits, the relatively low costs of trading, and the prior existence of markets among refineries for products such as gasoline additives (U.S. EPA 1985; Hahn and Hester 1989; Kerr and Newell 2003).

There is some evidence that transaction costs may have reduced the cost-effectiveness of the program. In a sample of 87 refineries in the period 1983–84, 20 percent of the refineries had not participated in the market at all, and this finding suggests that the market was probably not achieving the competitive equilibrium in the absence of transaction costs (Kerr and Mare 1998). Empirical analysis suggested that larger firms, and firms with a larger number of refineries, faced lower transaction costs than did smaller ones.[23]

Regulatory uncertainty may have added to the transaction costs. Refineries calculated whether they would have excess lead credits to transfer and traded these credits before they were verified by the EPA. In some cases, permits that were traded were later revealed to be invalid. Refineries, therefore, incurred additional costs through assessing the reputation of trading partners and investigating the validity of permits being sold. These costs led to a decrease in the cost-effectiveness of the program on the order of 10–20 percent (Kerr and Mare 1998).

There are two pieces of circumstantial evidence that suggest the independence property may not be supported in this application: differences in transaction

[22] In addition to these systems, we examined the Regional Greenhouse Gas Initiative but did not include it because of a lack of reliable data. We also examined four cap-and-trade systems that may be modified or implemented in the future: the EU Emission Trading Scheme as amended for 2012, the New Zealand Emissions Trading Scheme, the Australian cap-and-trade system, and the American Clean Energy and Security Act (H.R. 2454). At this point, it is too early to say whether these programs, if implemented, would likely support the independence property.

[23] Given that property rights were allocated on the basis of production, this system had features of an output-based allocation. See above.

costs across firms and absence of participation in the market by a significant fraction of firms. Still, it is not possible to reach a definitive conclusion on the basis of the data.

4.2. Chlorofluorocarbon Trading under the Montreal Protocol

The Montreal Protocol (1522 UNTS 3; 26 ILM 1550)—established in 1987 to limit emissions that damaged Earth's stratospheric ozone layer—required reductions in the use of CFCs and halons. The protocol called for a 50 percent reduction in the production of particular CFCs at 1986 levels by 1998 and froze halon production and consumption at 1986 levels beginning in 1992. Each country that signed the agreement was allowed to choose its own mechanism for limiting emissions. The United States implemented a cap-and-trade system in 1988 to help it comply with the protocol. The system placed limitations on both the production and the use of CFCs by issuing allowances that limited these activities. Because different types of CFCs have different effects on ozone depletion, each CFC was assigned a different weight on the basis of its depletion potential.

The production of CFCs was highly concentrated, with two producers—Du Pont and Allied Signal—accounting for 75 percent of domestic production and Du Pont alone holding a 49 percent market share. Because the protocol did not permit allowance trading across countries, these producers were shielded from potential international competition. There was, therefore, a real possibility that these firms might use their market power to manipulate the CFC allowance price. The EPA considered trying to dilute this potential price-setting ability by distributing production permits to importers as well as producers, but it ultimately decided to limit the allowance market to producers to minimize transaction and administrative costs (Hahn and McGartland 1989).

While the cap-and-trade program was a success (in that the limits imposed by the protocol were reached), the overall efficiency of the market is difficult to determine, because no studies have been conducted to estimate the cost savings (Stavins 2007). Because the market for CFC production was highly concentrated, however, it is possible that market power led to a violation of the independence property.

4.3. Sulfur Dioxide Allowance Trading under the Clean Air Act Amendments of 1990

The most important application in the United States of a market-based instrument for environmental protection is arguably the cap-and-trade system for regulating emissions of SO_2. Sulfur dioxide is the primary precursor of acid rain and a local pollutant with significant negative human health effects. This cap-and-trade system, which was established under Title IV-A of the U.S. Clean Air Act Amendments of 1990 (104 Stat. 2468; Pub. L. No. 101-549), was intended to reduce SO_2 emissions by 10 million tons (approximately 50 percent) below

1980 levels by the year 2000. Phase 1 of SO_2 emissions reductions was started in 1995, with phase 2 of reduction initiated in 2000 (Stavins 2003).

The program resulted in substantial cost savings relative to command-and-control regulation, with estimated annual savings ranging between $250 million and $1.6 billion, compared with an estimated $3 billion baseline (Carlson et al. 2000).[24] Of course, substantial cost savings do not imply that costs were necessarily minimized. It has been estimated that actual abatement costs exceeded the least-cost solution by $280 million in 1995 and by $339 million in 1996 (Carlson et al. 2000).[25] More than half of sources appeared not to have switched to using low-sulfur coal when it was economical.[26]

One possible explanation for deviations from the cost-minimizing outcome is the existence of transaction costs, particularly at the beginning of the program.[27] Firms initially confronted a very thin allowance market (Burtraw 1996). Transactions increased significantly over time, with the number of allowances traded doubling nearly every year during the first 3 years, an increase that is consistent with transaction costs decreasing as firms gained experience trading (Burtraw et al. 2005). By 1996, transaction costs were relatively low, with the average commission charged per trade at less than 2 percent (Joskow, Schmalensee, and Bailey 1998).[28]

Regulation may have brought about another distortion in the allowance market. Concerns were expressed that state regulatory authorities would limit trading to protect their domestic coal industries, and some research indicates that some state cost recovery rules were inefficient. One study examined the 11 states that would receive the most allowances and found that in six of these states there was a regulatory bias for utilities to pursue other means of compliance (for example, installing scrubbers or switching fuel types) in lieu of purchasing allowances (Bohi 1994). But other analysis concluded that public utility commission regulations did not substantially affect the allowance market (Bailey 1998). Subsequent research found a statistically significant relationship between a utility being subject to public utility commission regulations and its likelihood of choosing certain abatement methods in phase 1 (Amiru 2002). Regulated firms were more likely to switch to using low-sulfur coal, which was relatively costly in comparison with purchasing allowances.

Utility regulations may have contributed to uncertainty. Although the Federal

[24] Ellerman et al. (2000) report similar estimates, finding that costs of compliance were 55 percent of those that would be incurred under a command-and-control regime.

[25] Ellerman et al. (2000), in contrast, estimate that the cost of the program exceeded the least-cost solution by significantly less in phase 1 (between 3 and 15 percent in their estimate).

[26] Similar findings were reported by Swinton (2002), who examined the actions of seven coal-powered power plants in Florida from 1990 to 1998.

[27] Another possible explanation of the deviation observed by Carlson et al. (2000) may be that it is due to the use of an ex ante analysis, not an ex post analysis (Ellerman et al. 2000).

[28] Transaction costs may have been relatively low because the auctions the Environmental Protection Agency conducted in order to distribute allowances helped facilitate price discovery (Schmalensee et al. 1998).

Energy Regulatory Commission (FERC) established guidelines for accounting rules in 1993, FERC and the various state public utility commissions did not immediately set cost recovery rules, and many utilities were unsure of how allowance transactions would be treated (Burtraw 1996). Many utilities apparently felt comfortable trading allowances even in the absence of any sort of ruling (Bailey 1998). And the lack of formal guidance at the outset of the program does not appear to have impeded the development of the allowance market (Ellerman et al. 2000).

The presence of market power does not appear to have been a problem. From 1995 through 2003, the path of emissions was consistent with that expected from perfect competition (Ellerman and Montero 2007). In addition, the compliance decisions of the four largest firms in the market in the period 1995–99, which accounted for 43 percent of the total allowances granted, exhibit no evidence of market manipulation (Liski and Montero 2011).

In summary, it appears that, at least in the early years of the SO_2 cap-and-trade program, a perfectly cost-effective allocation of permits was not achieved, in part because of high transaction costs and utility regulations. But later, as trading became more frequent, transaction costs fell, and utility regulations were clarified. The subsequent market was likely consistent with the independence property.

4.4. The Regional Clean Air Incentives Market

The South Coast Air Quality Management District (SCAQMD), which is responsible for controlling emissions in a four-county area of Southern California, launched a cap-and-trade program in 1994 to reduce NO_x and SO_2 emissions in the Los Angeles area. This Regional Clean Air Incentives Market (RECLAIM) set an aggregate cap on NO_x and SO_2 emissions for all power plants, cement factories, refineries, and other industrial sources with emissions greater than 4 tons per year and employed a trading program that incorporates zonal restrictions, whereby trades are not permitted from downwind to upwind sources. The program also incorporates temporal restrictions, with the banking of permits limited (Stavins 2008).

In an analysis directly examining whether RECLAIM exhibited the independence property, the evidence indicated that it did (Fowlie and Perloff 2008). Regulators had randomly assigned firms in RECLAIM to one of two permit allocation cycles. Permit allocations were generally decreasing over time, and depending on which cycle a firm was randomly assigned to, the decline in a firm's allocation might occur either in January or in July. After an analysis employing instrumental variable procedures—and using random assignment into one of the two cycles as the instrument—there was no evidence of a statistically significant relationship between the initial and the final permit allocation (Fowlie and Perloff 2008).

At the same time, evidence suggests that RECLAIM has not achieved all of its cost-saving potential.[29] Transaction costs were present, in part because of the heterogeneous nature of firms participating in the market, with the result that the average facility had only a 20 percent probability of engaging in trade. The reduction in trading due to transaction costs has been estimated to be 32 percent (Gangadharan 2000).

These transaction costs likely decreased in subsequent years, as brokers and other intermediaries entered the market (Ellerman, Joskow, and Harrison 2003). Whereas only 28 percent of permit transactions in 1994 used brokers, by 2001 brokers were involved in 75 percent of trades (Fowlie and Perloff 2008). Transaction costs appear to be modest, as is evidenced by the high volume of trading in the market[30] and by SCAQMD's decision in 2002 to reject as unnecessary a proposal that would have established a centralized market for all RECLAIM permit transactions.

Uncertainty may also have created distortions in the permit market. One source of uncertainty resulted from the actions of a major private broker, which has been sued more than once for failing to deliver purchased emissions permits (Fowlie and Perloff 2008). The 17 amendments to the program rules between 1994 and 2006 likely contributed to firms' uncertainty regarding the permit market (Fowlie and Perloff 2008). Furthermore, there has been price volatility, with spikes in the price of NO_X permits in 2000 due to a spike in California energy prices, which ultimately required SCAQMD to remove electricity generators from the market. This price spike does not reflect a failing of RECLAIM design itself but rather flaws in California's deregulation of its electricity market (Harrison 2003).

It appears that RECLAIM functioned reasonably well. A direct statistical test of the independence property cannot reject the claim that the final allocation is independent of the initial allocation, but high initial transaction costs could have led to a violation of the independence property in the early years of the program.

4.5. *Eastern Ozone Transport Nitrogen Oxide Market*

Under EPA guidance, 12 northeastern states and the District of Columbia implemented a regional NO_X cap-and-trade system in 1999 to reduce compliance costs associated with the Ozone Transport Commission regulations of the 1990 amendments to the Clean Air Act.[31] Under the program, the EPA distributes NO_X allowances to each state, and states allocate allowances to sources in their jurisdictions. Sources may buy, sell, and bank allowances, although a system of

[29] One prospective analysis predicted cost savings of approximately 40 percent from the emissions trading program as compared with existing command-and-control regulation (Johnson and Pekelney 1996).

[30] By the end of 1997, permits corresponding to a total of 244,000 tons of emissions had been transferred (Nash and Revesz 2001), while as of the end of 2001, the corresponding figure was over 400,000 (Ellerman, Joskow, and Harrison 2003).

[31] Only nine of the 12 states actively participated in the program (Farrell 2001).

progressive flow controls limits the total number of banked allowances that can be used during the ozone season (Stavins 2003).

This program appears to have been successful. Nitrogen oxide emissions fell from 1.9 million tons in 1990 to 500,000 tons in 2006 (Stavins 2007). An ex ante examination estimated cost savings of between 40 percent and 47 percent in the period 1999–2003, compared with the costs of command-and-control regulation (Farrell, Carter, and Raufer 1999). At least the early phases of the market were broad and active (Farrell 2000).[32]

There is, however, some evidence that distortions may have affected the workings of the allowance market. Interviews with participants in the Ozone Transport Region revealed that firms had generally been conservative in their trading of allowances (Farrell 2003). First, they believed that there was a great deal of uncertainty in the market and were unsure of their ability to purchase and sell allowances. Second, because of market thinness, some firms thought they had some price-setting power and thus abstained from purchasing or selling allowances if doing so would affect the price (Farrell 2003). In summary, despite some concerns that arise from survey results, this market has appeared to work reasonably well and has reduced the costs of limiting emissions.

4.6. European Union Emission Trading System

In order to meet its commitments under the Kyoto Protocol, the EU adopted in 2003 the EU ETS, a cap-and-trade system for CO_2 emissions reduction. This downstream system covers about half of EU emissions from some 11,500 sources, including large sources such as oil refineries, combustion installations over 20 megawatts, coke ovens, cement factories, ferrous metal production, glass and ceramics production, and pulp and paper production. The program does not cover sources in the transportation, commercial, or residential sectors. The EU ETS was designed to be implemented in phases: a pilot or learning phase from 2005 to 2007, a Kyoto commitment period phase from 2008 to 2012, and a series of subsequent phases (Stavins 2008).

Direct statistical evidence exists in support of the independence property (Reguant and Ellerman 2008). In Spain, EU allowances (EUAs) were allocated on the basis of a function that rewards certain cleaner sources. Because of this allocation method, the endogeneity problems normally present in testing for independence could be avoided. No evidence was found that the initial allocation of EUAs had an effect on firms' production decisions.

Other evidence suggests that there have been relatively few distortions in the market for EUAs. Given the large number of sources covered by the program, few if any firms are capable of exercising market power. The largest player in the EUA market (in the trial phase) was allocated less than 6 percent of allowances. Even if the electricity-generating sector—which contained seven of the

[32] Including reductions achieved under the NO_X State Implementation Plan, nitrogen oxide emissions fell from 1.86 million tons in 1990 to .49 million tons in 2006 (Stavins 2008).

top 10 firms in terms of initial EUA holdings—were considered as a whole, there is no empirical evidence of market power (Convery and Redmond 2007).

Transaction costs also have been low. Although the EU ETS involves several nations, trading between firms in different countries has been frequent, in large part because of the presence of intermediaries in the EUA market (Ellerman 2008). Brokers entered the market even before the EU ETS went online, with futures trading beginning in 2004. And several companies, such as Point Carbon, facilitated transactions by tracking and reporting developments in the EUA market, providing information on prices for both buyers and sellers (Convery and Redmond 2007).

On the basis of these factors, we believe that this market's performance has been broadly consistent with the independence property. Distortions were introduced, however, by provisions that grant allowances freely to new entrants to the market and that require that unused allowances be forfeited by any facility that shuts down. Similar to output-based updating allocations, these provisions subsidize production and could lead to the independence property being violated (Ellerman and Buchner 2007).

4.7. Kyoto Protocol Article 17

In 1990, negotiations commenced that led to an agreement in 1992 on the United Nations Framework Convention on Climate Change (UNFCCC), which entered into force in 1994 with 190 countries as parties and established a long-term goal of stabilizing greenhouse gas concentrations in the atmosphere at a level that would prevent dangerous anthropogenic interference with the climate system. In Kyoto, Japan, in December 1997, the parties to the UNFCCC agreed on the terms of what came to be known as the Kyoto Protocol. The protocol includes a provision for cost-effective implementation through a set of tradable permit mechanisms, two of which are credit programs (Joint Implementation and the Clean Development Mechanism) and one of which is a cap-and-trade system (the international trading provision in Article 17). These are provided as options that countries can employ.

The agreement was intended to result in industrialized countries' emissions declining in aggregate by 5.2 percent below 1990 levels by the year 2012. Despite the failure of the United States and Australia to ratify the protocol,[33] the agreement became effective in 2005, once the requirement that 55 Annex I (predominantly industrialized) countries, jointly accounting for 55 percent of 1990 Annex I emissions, ratify the agreement was met (Stavins 2008).

Early on, the emissions trading provision of Article 17 was hailed by many as the most effective of the three compliance mechanisms (Weyant and Hill 1999). But others noted reasons to be skeptical (Hahn and Stavins 1999). First, one might not expect the market created by Article 17 to exhibit independence and achieve cost-effectiveness, because the relevant market actors are nations,

[33] Australia subsequently ratified the Kyoto Protocol late in 2007.

not private firms. Nations, unlike firms, are unlikely to have the incentive or the capability to act in a cost-minimizing way. Hence, nations are unlikely to equate their marginal abatement costs with permit prices (Hahn and Stavins 1999). Moreover, the status of the participants in the markets as sovereign nations also provides a strong opportunity for noncompliance, as participation in the market is ultimately voluntary.

Another problem is that some participants in the Article 17 market may be large enough to manipulate permit prices to their advantage (Bernard et al. 2003). Such countries may be able to raise prices by subsidizing certain abatement technologies or lower prices by imposing a tariff on permit exports. Even if a nation were to distribute its allowances to firms within its jurisdiction, it may be able to implement policies that encourage these firms to coordinate their actions (Liski and Montero 2011).

Another concern raised with respect to Article 17 trading has been transaction costs. The protocol involves many countries with distinct characteristics and experiences with emissions abatement. One might, therefore, expect permit trade between these countries to feature significant search and information costs (Pan and Regemortor 2004). The EU ETS may be considered an efficient submarket of the overarching Article 17 permit market that helps lower transaction costs, although the existence and conduct of the EU ETS does not depend upon Article 17 of the Kyoto Protocol.

While markets established in response to the Kyoto Protocol targets, such as the EU ETS, may be relatively cost-effective, it is unlikely that final permit allocations are independent of initial allocations in the international permit market created by Article 17 itself. Problems with non-cost-minimizing behavior, strategic price manipulation, and transaction costs could be significant.

4.8. Summary of Findings

Table 1 summarizes our findings regarding empirical support for the independence property among the seven cap-and-trade systems we have reviewed. We examined both indirect (circumstantial) evidence regarding the presence and importance of these factors in cap-and-trade systems, as well as direct (statistical) evidence of the violation of the independence property. Overall, we find support for the independence property to be high in three of the systems (counting SO_2 allowance trading), moderate in two of the systems, and low in two.

5. Conclusions

What has come to be known as the Coase theorem—the assertion that under certain conditions, bilateral negotiation between the generator and the recipient of an externality will lead to the same efficient outcome regardless of the initial assignment of property rights—has had important impacts not only on environmental (and other areas of) economics but also on the development of public

Table 1

Summary of Empirical Findings Regarding the Independence Property

Cap-and-Trade System	Indirect Evidence of Possible Permit Market Distortions	Support for the Independence Property
Lead trading	Transaction costs	Moderate
CFC trading under Montreal Protocol	Market power	Low but no rigorous empirical analysis
SO_2 allowance trading	Transaction costs, differential regulatory treatment, uncertainty	Low at the outset, high subsequently
RECLAIM	Transaction costs, uncertainty	High[a]
Eastern ozone transport NO_X markets	Market power, uncertainty, non-cost-minimizing behavior	Moderate
EU ETS	Possible uncertainty	High[a]
Kyoto Protocol Article 17	Transaction costs, market power, non-cost-minimizing behavior	Low but no rigorous empirical analysis

Note. CFC = chlorofluorocarbon; EU ETS = European Union Emission Trading Scheme; NO_x = nitrogen oxides; RECLAIM = Regional Clean Air Incentives Market; SO_2 = sulfur dioxide.

[a] This conclusion is partly based on statistical tests. Only those applications of cap-and-trade considered in the text are included.

policy in the environmental domain. Its close relative, the independence property, has been particularly important in obtaining political support for the use of cap-and-trade systems to address environmental issues.

This property is of great relevance to the practical development of policy, because it allows equity and efficiency concerns to be separated. In particular, a government can set an overall cap of pollutant emissions (a pollution reduction goal) and leave it up to a legislature to construct a constituency in support of the program by allocating shares of the allowances to various interests, such as sectors and geographic regions, without affecting either the environmental performance of the system or its aggregate social costs.

Because of the importance of this property, we examined the conditions under which it is more or less likely to hold—both in theory and in practice. We find that in theory, a number of factors can lead to the independence property being violated. These are particular types of transaction costs in cap-and-trade markets; significant market power in the allowance market; uncertainty regarding the future price of allowances; conditional allowance allocations, such as output-based updating-allocation mechanisms; non-cost-minimizing behavior by firms; and specific kinds of regulatory treatment of participants in a cap-and-trade market.

Of course, that these factors can lead to the violation of the independence property does not mean that in practice they do so in quantitatively significant ways. Overall, we find modest support for the independence property in the seven cases we examine, although it would surely be useful to have more empirical research in this realm.

That the independence property appears to be broadly validated provides support for the efficacy of past political judgments regarding constituency building through legislatures' allowance allocations in cap-and-trade systems. Governments have repeatedly set the overall emissions cap and then left it up to the political process to allocate the available number of allowances among sources to build support for an initiative without reducing the system's environmental performance or driving up its cost. This success with environmental cap-and-trade systems should be contrasted with many other public policy proposals for which the normal course of events is that the political bargaining that is necessary to develop support reduces the effectiveness of the policy or drives up its overall cost.

References

Arimura, Toshi H. 2002. An Empirical Study of the SO_2 Allowance Market: Effects of PUC Regulations. *Journal of Environmental Economics and Management* 44:271–89.

Arrow, Kenneth J., and Gerard Debreu. 1954. Existence of an Equilibrium for a Competitive Economy. *Econometrica* 22:265–90.

Averch, Harvey, and Leland Johnson. 1962. Behavior of the Firm under Regulatory Constraint. *American Economic Review* 52:1052–69.

Bailey, Elizabeth M. 1998. Allowance Trading Activity and State Regulatory Rulings: Evidence from the U.S. Acid Rain Program. Center for Energy and Environmental Policy Research Working Paper No. 98005. Massachusetts Institute of Technology, Cambridge, Mass.

Baldursson, Fridrick M., and Nils-Henrik M. von der Fehr. 2004. Price Volatility and Risk Exposure in Market-Based Environmental Policy Instruments. *Journal of Environmental Economics and Management* 48:682–704.

Becker, Gary S. 1983. A Theory of Competition among Pressure Groups for Political Influence. *Quarterly Journal of Economics* 98:371–400.

Ben-David, Shaul, David Brookshire, Stuart Burness, Michael McKee, and Christian Schmidt. 2000. Attitudes toward Risk and Compliance in Emission Permit Markets. *Land Economics* 76:590–600.

Bernard, Alain, Carolyn Fischer, and Alan K. Fox. 2007. Is There a Rationale for Output-Based Rebating of Environmental Levies? *Resource and Energy Economics* 29:83–101.

Bernard, Alain, Sergey Paltsev, John M. Reilly, Marc Vielle, and Laurent Viguier. 2003. Russia's Role in the Kyoto Protocol. Joint Program on the Science and Policy of Global Change Report No. 98. Massachusetts Institute of Technology, Cambridge, Mass.

Betz, R., and A. Gunnthorsdottir. 2009. Modeling Emissions Markets Experimentally: The Impact of Price Uncertainty. Unpublished manuscript. University of New South Wales, Department of Economics, Sydney.

Bohi, Douglas R. 1994. Utilities and State Regulators Are Failing to Take Advantage of Emission Allowance Trading. *Electricity Journal* 7:20–27.

Bohi, Douglas R., and Dallas Burtraw. 1992. Utility Investment Behavior and the Emission Trading Market. *Resources and Energy* 14:129–53.

Braithwaite, John. 2002. *Restorative Justice and Responsive Regulation*. Oxford: Oxford University Press.

Buchanan, James M., and Gordon Tullock. 1975. Polluters' Profits and Political Response: Direct Control versus Taxes. *American Economic Review* 65:139–47.

Burtraw, Dallas. 1996. The SO_2 Emissions Trading Program: Cost Savings without Allowance Trades. *Contemporary Economic Policy* 14:79–94.

Burtraw, Dallas, David A. Evans, Alan Krupnick, Karen Palmer, and Russell Toth. 2005. Economics of Pollution Trading for SO_2 and NO_x. *Annual Review of Environment and Resources* 30:253–89.

Carlson, Curtis, Dallas Burtraw, Maureen Cropper, and Karen L. Palmer. 2000. Sulfur Dioxide Control by Electric Utilities: What Are the Gains from Trade? *Journal of Political Economy* 108:1292–1326.

Cason, Timothy N., and Lata Gangadharan. 2003. Transactions Costs in Tradable Permit Markets: An Experimental Study of Pollution Market Designs. *Journal of Regulatory Economics* 23:145–65.

Coase, R. H. 1960. The Problem of Social Cost. *Journal of Law and Economics* 3:1–44.

Convery, Frank J., and Luke Redmond. 2007. Market and Price Developments in the European Union Emissions Trading Scheme. *Review of Environmental Economics and Policy* 1:88–111.

Crocker, Thomas D. 1966. The Structuring of Atmospheric Pollution Control Systems. Pp. 61–68 in *The Economics of Air Pollution*, edited by Harold Wolozin. New York: W. W. Norton.

Dahlman, Carl J. 1979. The Problem of Externality. *Journal of Law and Economics* 22: 141–62.

Dales, John H. 1968. *Pollution, Property, and Prices*. Toronto: University of Toronto Press.

Ellerman, A. Denny. 2000. From Autarkic to Market-Based Compliance: Learning from Our Mistakes. Pp. 190–215 in *Emissions Trading: Environmental Policy's New Approach*, edited by Richard F. Kosobud. New York: John Wiley & Sons.

———. 2008. The EU Emission Trading Scheme: A Prototype Global System? Discussion Paper No. 2008-02. Harvard Project on International Climate Agreements, Cambridge, Mass.

Ellerman, A. Denny, and Barbara K. Buchner. 2007. The European Union Emissions Trading Scheme: Origins, Allocation, and Early Results. *Review of Environmental Economics and Policy* 1:66–87.

Ellerman, A. Denny, Paul L. Joskow, and David Harrison Jr. 2003. Emissions Trading in the U.S.: Experience, Lessons, and Considerations for Greenhouse Gases. Unpublished manuscript. PEW Center for Global Climate Change, Arlington, Va.

Ellerman, A. Denny, Paul L. Joskow, Richard Schmalensee, Juan-Pablo Montero, and Elizabeth M. Bailey. 2000. *Markets for Clean Air: The U.S. Acid Rain Program*. New York: Cambridge University Press.

Ellerman, A. Denny, and Juan-Pablo Montero. 2007. The Efficiency and Robustness of Allowance Banking in the U.S. Acid Rain Program. *Energy Journal* 28:47–71.

Eshel, Dafna. 2005. Optimal Allocation of Tradable Pollution Rights and Market Structures. *Journal of Regulatory Economics* 28:205–23.

Farrell, Alex. 2000. The NO_x Budget: A Look at the First Year. *Electricity Journal* 13:83–93.

———. 2001. Multi-lateral Emission Trading: Lessons from Inter-state No_x Control in the United States. *Energy Policy* 29:1061–72.

———. 2003. Temporal Hotspots in Emission Trading Programs: Evidence from the Ozone-Transport Commission's NO_x Budget. Paper presented at Market Mechanisms and Incentives: Applications to Environmental Policy, Washington, D.C., May 1–2.

Farrell, Alex, Robert Carter, and Roger Raufer. 1999. The NO_x Budget: Market-based Control of Tropospheric Ozone in the Northeastern United States. *Resource and Energy Economics* 21:103–24.

Fischer, Carolyn. 2001. *Rebating Environmental Policy Revenues: Output-Based Allocations and Tradable Performance Standards*. Washington, D.C.: Resources for the Future.

Fischer, Carolyn, and Alan K. Fox. 2007. Output-Based Allocation of Emissions Permits for Mitigating Tax and Trade Interactions. *Land Economics* 83:575–99.

———. 2009. Combining Rebates with Carbon Taxes: Optimal Strategies for Coping with Emissions Leakage and Tax Interactions. Resources for the Future Discussion Paper No. 09-12. Resources for the Future, Washington, D.C.

Ford, Melanie, Andrew Gurney, Catherine Tulloh, Todd McInnis, Raymond Mi, and Helal Ahammad. 2009. Agriculture and the Carbon Pollution Reduction Scheme (CPRS): Economic Issues and Implications. ABARE Project 3372. *Issues Insights* 9.2.

Fowlie, Meredith, and Jeffrey M. Perloff. 2008. Distributing Pollution Rights in Cap-and-Trade Programs: Are Outcomes Independent of Allocation? Unpublished manuscript. University of California, Berkeley, Department of Agricultural and Resource Economics, Berkeley.

Fullerton, Don, Shaun P. McDermott, and Jonathan P. Caulkins. 1997. Sulfur Dioxide Compliance of a Regulated Utility. *Journal of Environmental Economics and Management* 34:32–53.

Gangadharan, Lata. 2000. Transaction Costs in Pollution Markets: An Empirical Study. *Land Economics* 76:601–14.

Goulder, Lawrence H., and Ian W. H. Parry. 2008. Instrument Choice in Environmental Policy. *Review of Environmental Economics and Policy* 2:152–74.

Hahn, Robert W. 1984. Market Power and Transferable Property Rights. *Quarterly Journal of Economics* 99:753–65.

Hahn, Robert W., and Gordon L. Hester. 1989. Marketable Permits: Lessons for Theory and Practice. *Ecology Law Quarterly* 16:361–406.

Hahn, Robert W., and Albert M. McGartland. 1989. The Political Economy of Instrument Choice: An Examination of the U.S. Role in Implementing the Montreal Protocol. *Northwestern University Law Review* 83:592–611.

Hahn, Robert W., and Roger G. Noll. 1983. Barriers to Implementing Tradable Air Pollution Permits: Problems of Regulatory Interactions. *Yale Journal on Regulation* 1:63–91.

Hahn, Robert W., and Robert N. Stavins. 1999. *What Has the Kyoto Protocol Wrought?* Washington, D.C.: American Enterprise Institute.

Harrison, David, Jr. 2003. Ex Post Evaluation of the Reclaim Emissions Trading Program for the Los Angeles Air Basin. Unpublished manuscript. National Policies Division, Organisation for Economic Co-operation and Development Environment Directorate, Paris.

Hazlett, Thomas. 2009. Ronald H. Coase. Chap. 1 in *Pioneers in Law and Economics,* edited by Lloyd R. Cohen and Joshua D. Wright. Northampton, Mass.: Edward Elgar.

Houser, Trevor, Rob Bradley, Britt Childs, Jacob Werksman, and Robert Heilmayr. 2008. *Leveling the Playing Field: International Competition and U.S. Climate Policy Design.* Washington, D.C.: Peterson Institute for International Economics and World Resources Institute.

Johnson, Scott Lee, and David M. Pekelney. 1996. Economic Assessment of the Regional Clean Air Incentives Market: A New Emissions Trading Program for Los Angeles. *Land Economics* 72:277–97.

Joskow, Paul L., and Richard Schmalensee. 1998. The Political Economy of Market-Based Environmental Policy: The U.S. Acid Rain Program. *Journal of Law and Economics* 41: 81–135.

Joskow, Paul L., Richard Schmalensee, and Elizabeth M. Bailey. 1998. The Market for Sulfur Dioxide Emissions. *American Economic Review* 88:669–85.

Kahneman, Daniel, Jack L. Knetsch, and Richard H. Thaler. 1990. Experimental Tests of the Endowment Effect and the Coase Theorem. *Journal of Political Economy* 98:1325–48.

———. 1991. Anomalies: The Endowment Effect, Loss Aversion, and Status Quo Bias. *Journal of Economic Perspectives* 5:193–206.

Kaplow, Louis. 2011. Taxes, Permits, and Climate Change. Pp. 168–92 in *U.S. Energy Tax Policy,* edited by Gilbert E. Metcalf. Cambridge: Cambridge University Press.

Kerr, Suzi, and David Mare. 1998. Transaction Costs and Tradable Permit Markets: The United States Lead Phasedown. Unpublished manuscript. Motu Economic and Public Policy Research, Wellington.

Kerr, Suzi, and Richard G. Newell. 2003. Policy-Induced Technology Adoption: Evidence from the U.S. Lead Phasedown. *Journal of Industrial Economics* 51:317–43.

Krupnick, Alan J., Wallace E. Oates, and Eric Van De Verg. 1983. On Marketable Air-Pollution Permits: The Case for a System of Pollution Offsets. *Journal of Environmental Economics and Management* 10:233–47.

Liski, Matti. 2001. Thin versus Thick CO_2 Market. *Journal of Environmental Economics and Management* 41:295–311.

Liski, Matti, and Juan-Pablo Montero. 2011. Market Power in an Exhaustible Resource Market: The Case of Storable Pollution Permits. *Economic Journal* 121:116–44.

Malik, Arun S. 2002. Further Results on Permit Markets with Market Power and Cheating. *Journal of Environmental Economics and Management* 44:371–90.

Malueg, David A. 1990. Welfare Consequences of Emission Credit Trading Programs. *Journal of Environmental Economics and Management* 18:66–77.

Malueg, David A., and Andrew J. Yates. 2009. Bilateral Oligopoly, Private Information, and Pollution Permit Markets. *Environmental and Resource Economics* 43:553–72.

May, Peter J. 2005. Compliance Motivations: Perspectives of Farmers, Homebuilders, and Marine Facilities. *Law and Policy* 27:317–47.

Misiolek, Walter S., and Harold W. Elder. 1989. Exclusionary Manipulation of Markets for Pollution Rights. *Journal of Environmental Economics and Management* 16:156–66.

Montero, Juan-Pablo. 1998. Marketable Pollution Permits with Uncertainty and Transaction Costs. *Resource and Energy Economics* 20:27–50.

Montgomery, W. David. 1972. Markets in Licenses and Efficient Pollution Control Programs. *Journal of Economic Theory* 5:395–418.

Murphy, James J., and John K. Stranlund. 2007. A Laboratory Investigation of Compliance Behavior under Tradable Emissions Rights: Implications for Targeted Enforcement. *Journal of Environmental Economics and Management* 53:196–212.

Nash, Jonathan R., and Richard L. Revesz. 2001. Markets and Geography: Designing Marketable Permit Schemes to Control Local and Regional Pollutants. *Ecology Law Quarterly* 28:569–662.

Oates, Wallace E., and Diana L. Strassmann. 1984. Effluent Fees and Market Structure. *Journal of Public Economics* 24:29–46.

Pan, Haoran, and Denise Van Regemortor. 2004. The Costs and Benefits of Early Action before Kyoto Compliance. *Energy Policy* 32:1477–86.

Parker, Christine. 2006. The "Compliance" Trap: The Moral Message in Responsive Regulatory Enforcement. *Law and Society Review* 40:591–622.

Pigou, Arthur C. 1920. *The Economics of Welfare*. London: Macmillan.

Reguant, Mar, and A. Denny Ellerman. 2008. Grandfathering and the Endowment Effect: An Assessment in the Context of the Spanish National Allocation Plan. Center for Energy and Environmental Policy Research, Cambridge, Mass.

Saloner, Garth, and Andrea Shepard. 1995. Adoption of Technologies with Network Effects: An Empirical Examination of the Adoption of Automated Teller Machines. *Rand Journal of Economics* 26:479–501.

Samuelson, William, and Richard Zeckhauser. 1988. Status Quo Bias in Decision Making. *Journal of Risk and Uncertainty* 1:7–59.

Sartzetakis, Eftichios S. 2004. On the Efficiency of Competitive Markets for Emission Permits. *Environmental and Resource Economics* 27:1–19.

Schmalensee, Richard, Paul L. Joskow, A. Denny Ellerman, Juan-Pablo Montero, and Elizabeth M. Bailey. 1998. An Interim Evaluation of Sulfur Dioxide Emissions Trading. *Journal of Economic Perspective* 12:53–68.

Schwartz, Sonia. 2007. Market Power Effects on Market Equilibrium in Ambient Permit Markets. *B. E. Journal of Economic Analysis and Policy* 7(1), art. 13, pp. 1–27.

Spulber, Daniel F. 1985. Effluent Regulation and Long-Run Optimality. *Journal of Environmental Economics and Management* 12:103–16.

Stavins, Robert N. 1995. Transaction Costs and Tradeable Permits. *Journal of Environmental Economics and Management* 29:133–48.

———. 1998. What Can We Learn from the Grand Policy Experiment? Lessons from SO_2 Allowance Trading. *Journal of Economic Perspectives* 12:69–88.

———. 2003. Experience with Market-Based Environmental Policy Instruments. *Handbook of Environmental Economics* 1:355–435.

———. 2007. Market-Based Environmental Policies: What Can We Learn from U.S. Experience (and Related Research)? Pp. 19–47 in *Moving to Markets in Environmental Regulation: Lessons from Twenty Years of Experience,* edited by Jody Freeman and Charles Kolstad. New York: Oxford University Press.

———. 2008. A Meaningful U.S. Cap-and-Trade System to Address Global Climate Change. *Harvard Environmental Law Review* 32:293–371.

Stigler, George J. 1966. *The Theory of Price.* 3d ed. New York: Macmillan.

Swinton, John R. 2002. The Potential for Cost Savings in the Sulfur Dioxide Allowance Market: Empirical Evidence from Florida. *Land Economics* 78:390–404.

Thaler, Richard. 1980. Toward a Positive Theory of Consumer Choice. *Journal of Economic Behavior and Organization* 1:39–60.

Tietenberg, Thomas H. 1985. *Emissions Trading: An Exercise in Reforming Pollution Policy.* Washington, D.C.: Resources for the Future.

Tschirhart, John T. 1984. Transferable Discharge Permits and Profit-Maximizing Behavior. Pp. 157–71 in *Economic Perspectives on Acid Deposition Control,* edited by Thomas D. Crocker. Stoneham, Mass.: Butterworth.

U.S. EPA (Environmental Protection Agency), Office of Policy Analysis. 1985. Costs and Benefits of Reducing Lead in Gasoline, Final Regulatory Impact Analysis. Washington, D.C.: Government Printing Office.

Westskog, Hege. 1996. Market Power in a System of Tradable CO_2 Quotas. *Energy Journal* 17:85–103.

Weyant, John, and Jennifer Hill. 1999. Introduction and Overview to Special Issue. *Energy Journal* 120:vii–xliv.

Williamson, Oliver E. 1963. Managerial Discretion and Business Behavior. *American Economic Review* 53:1032–57.

[11]

Environ Resource Econ (2007) 37:111–129
DOI 10.1007/s10640-007-9110-y

Second-best theory and the use of multiple policy instruments

Lori Snyder Bennear · Robert N. Stavins

Published online: 16 May 2007

Abstract In many cases policy makers employ multiple instruments to address a single environmental problem, but much of the economics literature on instrument choice focuses on comparing properties of single policy instruments. We argue that under a fairly broad set of circumstances the use of multiple policy instruments can be justified as optimal in a second-best world. We examine two broad categories of second-best policy making: cases with multiple market failures only some of which can be corrected at any one time; and cases with exogenous (often political) constraints that cannot be removed. The fact that the use of multiple policy instruments can be justified economically in these two cases does not imply, however, that all multiple instruments employed in actual practice are economically justified.

1 Introduction

A major issue for policy makers in the environmental domain concerns identifying an appropriate policy instrument, an issue that was central to David Pearce's interests, in terms of both his scholarly research and his participation in the policy process (Pearce 1998a, b). Regulators may have at their disposal a portfolio of potential policy instruments, including product or process bans, technology standards, performance standards, taxes, subsidies, tradable permits, tort liability, information disclosure, industry self-regulation, and management-based regulation. Different instruments are appropriate for different types of problems in different

L. S. Bennear
Nicholas School of Environment and Earth Sciences, Duke University, P.O. Box 90328, Durham, NC 27708, USA
e-mail: lori.bennear@duke.edu

R. N. Stavins (✉)
John F. Kennedy School of Government, Harvard University, 79 John F. Kennedy Street, Cambridge, MA 02138, USA
e-mail: robert_stavins@harvard.edu

circumstances. The challenge is to determine the conditions under which each instrument, or set of instruments, is the appropriate choice.

Several criteria can be brought to bear on this choice, including: (1) the efficiency criterion, which assesses an instrument's ability to attain a level of pollution control that maximizes net benefits; (2) the cost effectiveness criterion, which assesses an instrument's ability to achieve a given level of pollution reduction at the lowest cost; and (3) other economic and non-economic criteria, including distributional equity and political process transparency (Pearce 1998c, d, 1997, 1975; Helfand, et al. 2003; Baumol and Oates 1988; Revesz and Stavins 2005).

The normative literature on instrument choice is substantial, with the bulk focusing on the use of one instrument or comparisons of two or more instruments. More recently the literature has explored the economic properties of combinations of multiple instruments, particularly the use of hybrid instruments that combine a quantity and a price instrument. In contrast, in the policy world the use of combinations of multiple policy instruments is common and seldom takes the form of a hybrid instrument. To economists, such use of multiple instruments often appears ad hoc and unrelated to economic efficiency or cost effectiveness.

We postulate that the use of multiple policy instruments can be justified as optimal in a second-best world. An efficient general equilibrium, in the sense of the first theorem of welfare economics, requires a universal set of perfectly competitive markets. A general definition of the second-best problem is that if there is some constraint within the general equilibrium system that prevents attainment of at least one of the conditions of Pareto optimality, then attainment of the other Pareto optimal conditions is no longer necessarily welfare improving (Lipsey and Lancaster 1956).

There are two ways of thinking about the second-best problem in the context of our investigation of multiple policy instruments. One is that there are multiple constraints—these could be political constraints, market failures, or policy failures—but only some of these constraints can be addressed at any given time. What effect does the continued existence of the other constraints have on instrument choice? The other way to think about the second-best problem is that there is a single exogenous constraint that cannot be overcome, again this could be a political constraint, a market failure, or a policy failure. What effect does this exogenous constraint have on optimal instrument choice? We demonstrate that the second-best nature of problems addressed by policy makers justifies policy coordination and *can* justify the use of multiple policy instruments in a wide range of settings. Much research remains to be done, however, to understand the actual set of instruments that should be employed in such broad second-best settings.

We begin in Sect. 2 by describing some of the cases in which multiple policy instruments have been used in practice. This section illustrates the broad range of circumstances in the environmental and natural resource management fields where multiple policy instruments have been employed. In Sect. 3, we provide an overview of the second-best theory and one of its most studied applications in environmental policy—environmental policy in the presence of pre-existing distortionary taxes. This application highlights the importance of policy coordination in second-best settings. It also illustrates how the use of multiple policy instruments can be optimal in a second-best world. These findings turn out to be broadly applicable in environmental and resource management.

The remainder of the paper highlights several areas where multiple policy instruments can be justified using second-best theory. In Sect. 4, we examine the case where there are multiple constraints, but only a few can be addressed at a given time. This section focuses primarily on the existence of multiple market failures and demonstrates that these multiple constraints on the general equilibrium system also justify policy coordination and can often justify the use of multiple policy instruments. In Sect. 5, we examine the case where there is one exogenous constraint that affects the optimal choice of policy instruments for environmental policy. We conclude in Sect. 6.

2 The Use of Multiple Policy Instruments in Practice

The use of multiple policy instruments is the norm, rather than the exception in environmental and natural resource management. Here we describe the actual use of multiple policy instruments in three major areas: energy efficiency, toxic chemical control, and fisheries management. A full discourse on all of the combinations of policy instruments used throughout the world in these policy areas is well beyond the scope of this paper.[1] Rather, we intend to illustrate the prevalence of multiple instrument use over a broad range of environmental and resource issues.

2.1 Energy efficiency

A wide range of policy instruments has been used in pursuit of greater energy efficiency. For purposes of illustration we focus on the combinations used to regulate energy efficiency in the United States, but many of these combinations are used throughout the world (OECD 2004, 2005).

In the United States, national energy efficiency standards have been developed for commercial and residential appliances. Congress first mandated the development of these standards in the National Appliance Energy Conservation Act (NAECA) of 1987.[2] The U.S. Department of Energy develops and updates energy efficiency standards for household and commercial appliances, including: refrigerators and freezers, room air conditioners, fluorescent lamp ballasts, incandescent reflector lamps, clothes washers and dryers, dishwashers, kitchen ranges and ovens, pool heaters, and water heaters. Products cannot be sold in the United States unless they meet these minimum energy efficiency standards.

In addition to these standards, an information disclosure instrument is also used to promote energy efficiency. The Federal Trade Commission issued the Appliance Labeling Rule that became effective in 1980. This rule requires that all new refrigerators and freezers, water heaters, dishwashers, clothes washers and dryers, room air conditions, heat pumps and furnaces, and boilers sold must display an "EnergyGuide" label. These labels describe the estimated annual energy use of the appliance and how this energy use compares with other brands of appliances in the same category.

An additional labeling program, the Energy Star program, is jointly run by the U.S. Environmental Protection Agency and the U.S. Department of Energy. This program encourages the development and consumption of energy efficiency products by developing energy efficiency qualifications for numerous household and commercial appliances. Appliances that meet the qualifications can bear the Energy Star label. The Energy Star program also works with retailers to promote and market Energy Star labeled products and provides information to consumers on cost savings associated with Energy Star products.

There are also numerous voluntary programs used to promote energy efficiency. In the 1990s, the Green Lights program was a prominent voluntary program that encouraged

---

1 For more examples of the use of multiple policy instruments in practice see OECD 2004 and OECD 2005.

2 This was not the first attempt to develop national energy efficiency standards. The National Energy Policy and Conservation Act of 1975 required the U.S. Department of Energy to develop energy efficiency standards for appliances, but these standards were never finalized.

businesses to adopt more efficient lighting. Participating businesses signed a Memorandum of Understanding with EPA and agreed to survey their facilities and upgrade where profitable.[3] In exchange, EPA provided information on energy-efficient lighting technology, financing options, and public recognition opportunities. More recently, EPA has established the Climate Leaders program—a voluntary program designed to encourage businesses to reduce their greenhouse gas emissions (often through increased energy efficiency).[4] Participants set greenhouse gas reduction goals and conduct regular inventories to measure their progress toward achieving the goals. In exchange, EPA provides guidance on both setting and attaining goals and public recognition.

2.2 Toxic chemical control

There are at least six different policy instruments used in the United States to control pollution and minimize hazards from toxic chemicals—technology standards, information disclosure requirements, management-based regulations, government-industry partnerships, voluntary programs, and industry self-regulatory programs. While we illustrate the use of these multiple instruments with examples from the United States, many of these same combinations of instruments are used throughout the world for toxic chemical control.

There are technology-based standards regulating air releases of toxic chemicals in some industries in the United States under the Clean Air Act (U.S. Environmental Protection Agency 2000). Maximum attainable control technology standards are set at the industry level for hazardous air pollutants. The industries that are regulated (or scheduled to be regulated) range from electric utilities, metal smelters, dry cleaners, and numerous manufacturing industries, to publicly owned treatment works.

At the same time, most manufacturing facilities, electric utilities, federal facilities, and mining facilities are subject to information disclosure regulations. With the Emergency Planning and Community Right-To-Know Act (EPCRA) Congress required manufacturing facilities to publicly disclose their releases of nearly 300 different toxic chemicals as part of the Toxic Release Inventory program.[5] In addition to reporting releases of these chemicals to air, water, landfills, and underground injection, facilities must report amounts of listed substances that are sent off-site for treatment, disposal, or recycling. Facilities' annual reports on their releases, offsite-transfers, and recycling are made available via the Internet through the U.S. Environmental Protection Agency (EPA) and through third party sites, such as Environmental Defense's "Scorecard."[6]

Many facilities face additional regulation of toxic chemicals at the state level. Twelve states have adopted management-based regulations of toxic chemicals (Bennear 2006). These regulations require facilities to review their production processes and develop a set of goals and procedures that will reduce toxic chemical use and release. Similar management-based regulations have been imposed at the federal level for chemicals that can create industrial safety hazards. EPA requires manufacturers that use certain chemicals to develop risk management

[3] The exact agreement was to upgrade 60% of the square footage where upgrading was profitable without affecting lighting quality.

[4] See Smith and Swierzbinski (2007) and Ulph and Ulph (2007) in this collection for European and theoretical perspectives on this issue.

[5] Facilities are required to disclose releases if they have more than 10 full-time equivalent employees and either manufacture or process more than 25,000 pounds or otherwise use more than 10,000 pounds of a listed chemical. For lead, mercury, and persistent bioaccumulating toxins, the reporting thresholds are lower. The program has since been expanded to include other industry sectors and to include an additional 300 chemicals.

[6] The scorecard can be accessed at www.scorecard.org.

plans that assess the potential for accidents within their plant and evaluate alternatives for reducing or eliminating this risk (Kleindorfer et al. 2000; Coglianese and Lazer 2003).

Both EPA and the states have developed industry-government partnership programs to address toxic chemical use and release. The most widely known partnership program is EPA's Design for the Environment (DfE) program, which began after Congress passed the Pollution Prevention Act in 1990 (Lenox et al. 2000, Lenox and Ehrenfeld 1997, 1995) and involves government industry collaboration in assessing currently available technologies, identifying environmentally-friendly alternatives, and assessing the financial payoffs of these alternatives.

EPA also developed a voluntary program to address toxic chemical releases. The Industrial Toxics Program, generally referred to as the 33/50 program, was a voluntary program begun in 1991 where participants agreed to reduce their releases of 17 priority toxic chemicals by 33% by 1992 and 50% by 1995 (Khanna and Damon 1999; Gamper-Rabindrin 2006). In exchange companies received public recognition benefits.

2.3 Fisheries management

Multiple policy instruments are also used extensively in natural resource management. For example, combinations of instruments are used to manage fisheries in New Zealand. As with the other policy cases, some of the combinations of policy instruments used in New Zealand are used in other countries as well.

Since 1986, the primary tool for managing fisheries in New Zealand has been an individual transferable quota (ITQ) program. Commercial fishers were originally granted quotas that entitled them to a share of the total allowable catch of a species. These quotas can be bought and sold.[7] Currently, over 90 species of fish are regulated using the ITQ system.

While economists often argue that a successful ITQ system can be used in lieu of more traditional fisheries management methods such as gear restrictions and season closures, in New Zealand, several of these traditional management methods are used in conjunction with the ITQ system.

New Zealand established closed fishing areas for both commercial and recreational fishing. Quotas cannot be used to cover fish caught in closed areas. Similarly, many fisheries are subject to season closures, often during the breeding season. Quotas cannot be used to cover catch during the closed season. Size limits are imposed on fish caught by both recreational and commercial fishers. For example, blue cod must be at least 30 cm long to be considered allowable catch. Finally, New Zealand still imposes gear restrictions on some fisheries. These restrictions include things like the size of nets and the size of mesh in nets.

New Zealand also imposes restrictions on the trading of the quotas in the ITQ system. For example, there is a ceiling on the total number of permits any one firm can hold (Connor 2001; Runolfsson and Arnason 2001). In addition, there are limits on foreign-ownership of quotas and a requirement that quotas be associated with a fishing vessel to prevent absentee ownership (Connor 2001; Runolfsson and Arnason 2001).

These three cases illustrate that the use of multiple policy instruments is common in environmental and resource management. But is such use justified on economic grounds, and, if so, under what circumstances? It is to those questions that we now turn.

[7] Every month the total amount caught by a commercial fisher is compared to his quota. If a fisher exceeds his quota for the month he is assessed a penalty. The penalty is such that it eliminates any profit the fisher has from obtaining the excess fish.

3 Multiple instruments and policy making in a second-best setting—an overview

The second-best problem is generally defined as follows. If a constraint exists within the general equilibrium system that prevents attainment of one Pareto optimal condition, then attainment of other Pareto optimal conditions is no longer necessarily desirable, i.e., welfare improving (Lipsey and Lancaster 1956). The constraint may be a market failure, a policy failure, or a political constraint. For example, imagine a country that decides to reduce trade tariffs in a world where other countries utilize trade tariffs. A basic economic analysis would suggest that the decision to reduce trade tariffs by one country is welfare improving. However, given the existence of the exogenous constraint(s)—namely that other countries continue to use tariffs—this reduction in tariffs can actually be welfare reducing (Ozga 1955).

Another way to frame the second-best problem is that if there are multiple constraints that prevent the attainment of multiple Pareto optimal conditions, the elimination of only one of the constraints does not necessarily lead to a welfare improvement (Lipsey and Lancaster 1956). If the constraints are market failures, then second-best theory suggests that the elimination of one market failure, in a world with many market failures, may not be welfare enhancing. In fact, market failures can be *jointly ameliorating* (correction of one market failure ameliorates welfare losses from the other), *jointly reinforcing* (correction of one market failure exacerbates welfare losses from the other), or *neutral* (correction of one market failure does not affect the welfare losses from the other).

How does the theory of the second-best affect instrument choice for environmental and resource policy? The implications have been well studied for a few particular applications. For example, there is a substantial body of theoretical and empirical research that examines the interactions of environmental policy and "pre-existing" distortions that emanate from labor and capital taxes. If one ignores the pre-existing tax distortions and compares pollution taxes and tradable permits, under a broad set of circumstances, the cost-effectiveness and efficiency properties of these two instruments would be the same. However, this equivalency does not hold when one examines pollution control policy together with tax policy.

One reason for the difference between the first-best and second-best outcomes is that the use of a revenue-generating policy instrument for pollution control provides the opportunity to reduce the levels of other distortionary taxes. In particular, revenue from pollution taxes can be used to offset the marginal rate of other distortionary taxes, such as income and capital taxes (Lee and Misiolek 1986; Oates and Schwab 1988; Oates 1993; Repetto et al. 1992; and Terkla 1984). Such "revenue recycling" can lower the overall cost of environmental protection in the presence of distortionary taxes, generating a so-called "double-dividend."[8]

The use of revenue-generating instruments also leads to a tax interaction effect, whereby pollution taxes act as implicit taxes on labor and capital, and thus increase costs of goods and services and reduce after-tax returns (Bovenberg and de Mooij 1994; Bovenberg and van der Poeg 1994; Bovenberg and Goulder 1996, 1997; Goulder 1998; and Parry 1995, 1997). This tax interaction effect offsets some of the welfare gains attained by the revenue recycling effect. Theoretical models that combine both the tax interaction effect and the revenue recycling effect suggest that the tax interaction effect dominates. Hence, the revenue recycling effect will offset some, but not all of the costs of pollution control (Goulder 1998).

Because of the revenue recycling effect, the general equilibrium costs of environmental policy can be lower with the use of multiple policy instruments. In particular, the costs are

[8] The strong version of the double-dividend hypotheses asserts that the efficiency gains from reducing distortionary taxes can more than offset the actual costs of the pollution taxes themselves. Hence, pollution control policy with taxes (or auctioned permits) is costless in welfare terms. A weaker version of the double-dividend asserts that the efficiency gains from reducing distortionary taxes offsets part of the costs of pollution taxes.

lowest when policy makers use a revenue-neutral approach that *combines* a pollution tax with a proportionate cut in particularly distortionary taxes. The cost of this combined tax and tax-cut policy are lower than with the use of a single instrument, whether that single instrument be a pollution tax (with no revenue recycling) or a freely allocated permit system (Bovenberg and Goulder 1997; Goulder et al. 1997; Parry et al. 1999).[9]

In a second-best world, coordination of policies is required to attain an efficient outcome. As it turns out, the efficient coordination of policies in this case also involves the use of multiple policy instruments. In the next section, we demonstrate that this finding applies broadly to other cases of policy making in a second-best setting and that the need for coordinated response in a second-best world can often serve as an economic justification for the use of multiple policy instruments.

4 Second-best policy making with multiple constraints

One way to think about the second-best problem is that there are multiple constraints, any one of which causes the general equilibrium system to fail to reach a Pareto optimum. What are the interaction effects of these various constraints? To what extent does policy designed to address one constraint need to be coordinated or changed in light of the other constraints? In this section we focus on five pairs of constraints: (1) imperfect property rights and other externalities, (2) multiple externalities, (3) market power and externalities, (4) unobservable behavior and externalities, and (5) imperfect information and externalities. In each case, the constraints are market failures, but the general findings are likely to apply whenever there are multiple constraints, regardless of the nature of those constraints.

4.1 Imperfect property rights & other externalities

Most natural resource management problems are the direct results of poorly defined property rights. Open access fisheries are a classic example. If the oceans were privately held, the optimal management strategy would be to harvest fish until the marginal benefit of harvesting equals the marginal cost of harvesting (or net benefits are maximized).[10] But since no one owns the oceans, fishing continues as long as each fisher experiences a positive profit. Thus, fishing continues until total benefits of harvesting equal total costs (net benefits are zero), which generally results in higher initial harvest rates and lower long-run equilibrium stock and harvest rates. Many instruments for fisheries management are designed to address this imperfect property rights issue.

In multi-species fisheries, there is potentially a second market failure, namely a negative externality from fishing effort. There are two primary types of externalities that result from fishing activity. The first results from either incidental catch of non-targeted species or high-grading of targeted species.[11] High-profile examples of incidental catch include dolphins caught during the process of harvesting tuna, and turtles caught in the process of harvesting shrimp. The second type of externality results from degradation of habitat caused by fishing effort. For example, bottom-trawling fishing vessels can degrade habitat for benthic species.

[9] This finding has been demonstrated for several policy applications, including carbon dioxide and sulfur dioxide regulation. Goulder (1998) provides a summary of these findings.

[10] In the standard dynamic fisheries model, this point is where the growth rate of the value of the fish left in the ocean is exactly equal to the growth rate of the value of money (interest rate) (Conrad 1999:45).

[11] Highgrading occurs when fishers discard smaller fish so that they do not count against their quota. The survival rates of the discarded fish vary widely by species and fishing method (Squires et al. 1998).

Instruments currently employed to manage the excess fishing effort that results from open access fisheries include: fisheries closures, seasonal limits, licensing requirements, gear restrictions, landing fees, user fees, and individual transferable quota (ITQ) systems. Fee systems and ITQ systems can achieve cost-effective reductions in fishing effort (Conrad 1999: 52–53)[12] but employing ITQs to correct the property rights market failure can increase the incentives for incidental take and highgrading (Arnason 1994; Anderson 1994; Boyd and Dewees 1991; Sampson 1994; Vestergaard 1996). The typical ITQ system employed issues quotas for landed fish as opposed to quotas for fish that are caught. Under this type of ITQ system the net profit of a particular fish that is caught must be compared to the alternative benefit of discarding the fish and selling the quota. This alternative benefit will include the price of the quota less any cost of sorting and discarding lower valued fish. In comparison, in an open access fishery the incentive to discard simply compares the net profit from selling the fish to cost of sorting and discarding the fish (Arnason 1994; Anderson 1994). This *can* lead to an increased incentive for highgrading under an ITQ system.[13] Thus, imperfect property rights and externalities can be *jointly reinforcing* market failures. The optimal policy to correct the property rights problem can exacerbate the externality problem.

Does the mere existence of multiple market failures necessarily warrant the use of multiple policy instruments? The answer is no. Sometimes one instrument can be manipulated to address both market failures. Some of the externalities generated by ITQ systems can be corrected within the ITQ system. For example, when the incidental catch is also a species of market value and is regulated with an ITQ system, then banking and trading of quotas can help insure that fishers' quotas for different species match their catch (Squires et al. 1998). Similarly, an ITQ system can be designed over catch rather than landings, which would eliminate the increased incentive for highgrading (Arnason 1994).

It may not be feasible, however, to rely on a single instrument—an ITQ system—to correct for the externality in the case of highgrading or bycatch of ecologically but not economically valuable species or in cases where habitat externalities are significant. In such cases, additional policy instruments may be required to address the negative externalities. Some have suggested that combining ITQs with taxes on high grade fish or subsidies for low grade fish can correct the incentive problem (Anderson 1994). Alternatively, landing restrictions that requires a certain amount of low grade fish be landed for each high grade fish, could help ameliorate the incentive for high-grading (Anderson 1994).

The New Zealand fisheries management case (described in Sect. 2.3) highlights how traditional management instruments are often combined with an ITQ system. The important insight from the theory of the second-best as applied in this case is not necessarily the optimal number of instruments to be used, but rather that the existence of multiple market failures sometimes requires coordination of instruments to address these failures. However, no theoretical or empirical studies have examined the properties of fisheries management with alternative combinations of instruments.

4.2 Multiple externalities

Environmental externalities may interact with other types of externalities in ways that are jointly ameliorating or jointly reinforcing. A prime example is the interaction between positive

[12] Fishers who are more efficient tend to catch a higher share of the total allowable catch under an ITQ, and less efficient fishers are encourage to exit the fishery by selling their quotas.

[13] However, this result is not given as a corner solution, one with no discarding, is possible. In this case the open access and ITQ discarding rates would be the same. But for an interior solution, the discarding rate will be higher under the ITQ.

externalities (spillovers) from technological innovation and negative externalities from pollution. Policies that target only one of these externalities may have indirect effects on the other. Multiple policy instruments may be better suited to address the joint externality than one instrument alone. Indeed, this particular case of combined market failures provide a rationale for a portfolio of public policies that foster both emissions reductions and the development and adoption of environmentally beneficial technologies (Jaffe et al. 2005).

Research and development (R&D) activities can result in technological change that affects environmental quality. While it is frequently said that technological change improves the environment, in fact, technological change may be pollution-intensive or pollution-saving. Firms invest in R&D activity to maximize the expected returns of this activity. The returns to R&D investment, however, are highly uncertain, and often involve low probabilities of high payoffs. These two features of R&D investment can make it difficult to raise capital in traditional capital markets (Scherer and Harhoff 2000; Jaffe et al. 2003). In addition, if the outcome of R&D investment is advancement in general knowledge or technology, it may be difficult to exclude others from utilizing the results of R&D investments. These positive externalities of R&D efforts result in the under-provision of such investment (Arrow 1962; Spence 1984; Griliches 1979, 1992; Jaffe 1986, 1998).[14]

The interaction of the negative pollution and positive technology externality argues for policy coordination. For example, policies to reduce pollution should be designed in such a way as to promote, rather than hinder technological change (Jaffe et al. 2003). Market-based instruments, such as tradable permits and pollution taxes, increase the cost of pollution to firms and thereby create incentives for increased diffusion of existing pollution control technologies and may increase incentives for investment in innovation of new technologies. In contrast, technology-based standards inhibit innovation and diffusion of new technologies (Jaffe et al. 2004).

While the use of a single well-crafted environmental policy instrument may increase the incentives for technological change, evidence suggests that environmental policy alone does not create sufficient incentives to overcome completely the technological market failures (Jaffe et al. 2005). Combining environmental policies with policies to promote both innovation and diffusion of technology can enhance welfare. These additional policy instruments include promoting investment in technology in the public sector, changing the costs of R&D via tax incentives and subsidies, and changing the benefits of R&D through the use of patents (Jaffe et al. 2003). Again, the theory of the second-best suggests that coordination is required and that some combination of these instruments is likely to be optimal.

4.3 Market power and externalities

Environmental externalities and market power are another pair of jointly reinforcing market failures that can necessitate the use of multiple policy instruments. A firm exhibits market power if it holds such a large share of the overall market that its behavior as either a buyer or seller affects market price and there are barriers to entry that prevent additional firms from entering the industry. The combination of market power and negative externalities can yield problematic outcomes. In the case of a market with a negative externality, the socially

[14] It is worth noting that the fact that individual firms cannot fully appropriate the gains from research can theoretically lead to an oversupply of private R&D activity. This theoretical result stems from the fact that R&D is financed by the rents it produces, and hence investment in R&D effectively "steals" rents from competing firms by lowering the expected rents they will receive from their own investment (Mankiw and Whinston 1986). Empirically, this rent-stealing effect does not appear to be significant, but rather the data support the more traditional argument for under-provision of R&D (Griliches 1992).

optimal outcome results in less output being produced and being sold at a higher price than what would occur under perfect competition. But a monopolist in her efforts to maximize profits chooses to produce less and sell at a higher price than what would result under perfect competition. Thus market power may reduce the unwanted welfare effects of a negative externality (Lipsey and Lancaster 1956; Hammer 2000).[15] Policies implemented to rectify one market failure may affect the nature of policies required to rectify the other market failure.

The tradeoff between social welfare losses that result from decreases in production in markets with market power, and social welfare gains that result from those same decreases in production in markets with negative externalities are well-known. Does there exist an optimal number of firms such that these welfare losses and gains exactly offset one another (Cornes et al. 1986; Mason and Polasky 1997; Gopinath and Wu 1999)? The answer is that in general, there is an optimal number of firms that would lead – in theory – to welfare neutrality, suggesting that coordination of anti-trust and environmental policy could be welfare improving. But no studies have attempted to identify the instrument or set of instruments that would be likely to yield such a welfare-neutral second-best optimum.[16]

4.4 Unobservable behavior and externalities

Some environmental problems are characterized by regulators' inability to observe, measure, or monitor relevant behavior. For example, a regulator may wish to penalize improper disposal of a waste product, but improper disposal may be clandestine and difficult to detect. Likewise, emissions from mobile sources are costly to measure.

Under such circumstances, two-part instruments—instruments that combine a tax and a subsidy—can be equivalent to a Pigovian tax on the unobservable and unmeasurable activity (Fullerton and Wolverton 2000). Two-part instruments are generalizations of deposit-refund systems, where buyers pay a deposit on purchase and receive a refund of the deposit upon proper disposal (Sigman 1995).

The generalized two-part instrument does not require the tax and subsidy to be of the same magnitude or involve the same entities. In the case of recyclables, the efficient Pigouvian tax on improper disposal of recyclable material (for example, aluminum cans) can be replicated with a tax upon purchase of the cans (paid by the consumer) and a subsidy for recycling, which may be paid to the household or to a recycling company (Fullerton and Wolverton 2000). The use of such tax-subsidy combinations is not limited to waste disposal, but has potentially broad applications to any area where the key activity to be restricted is unobservable (Fullerton and Mohr 2002).

The optimal use of multiple instruments when undesirable behavior is unobservable is not restricted to two instruments nor to combinations of taxes and subsidies. In some cases, combinations of taxes on various inputs and outputs may be more efficient than a single tax. For example, mobile source emissions cannot be effectively taxed, due to the cost of measuring these emissions. Traditional approaches to mobile source regulation have focused on fuel efficiency restrictions and input taxes on gasoline. Given heterogeneity in preferences and driving styles, a combination of taxes on gasoline, engine size, and age of car can better

[15] This is not restricted to the static case. In the case of extraction of non-renewable resources, monopolists tend to extract more slowly over time, which is consistent (at least in terms of the direction if not the magnitude) with what would be socially optimal if there are negative externalities associated with the resource's extraction (Conrad 1999: 86–87).

[16] There is some legal scholarship on whether the courts should be allowed to consider welfare losses and gains from second-best situations in anti-trust cases (Hammer 2000)

approximate the Pigovian tax on emissions than the use of any of these taxes in isolation (Fullerton and West 2000).

Once again, in the presence of multiple externalities a coordinated policy that combines multiple instruments can be more efficient or cost-effective than the single first-best policy. This is an area that has received considerable attention from economists, who have identified optimal combinations of instruments, but outside of one policy area, namely waste management and the use of deposit-refund systems, these cost-effective combinations of instruments have not been widely used.

4.5 Imperfect information and externalities

The final combination of market failures that we examine is imperfect information combined with externalities. Consider the standard consumer utility maximization problem with multiple goods. Some of these goods may entail a negative externality in either production or consumption. Further assume that the consumer has preferences regarding the environmental impacts of specific consumption choices, but that the impacts of different consumption choices are not perfectly known. Two market failures are involved: there is production and consumption of goods that involves external costs; and there is imperfect information that impedes consumption choices.

How do these two market failures interact? In many cases, correcting the information problem can help reduce the externality problem. That is, the two market failures are jointly ameliorating.[17] To see how they might be jointly ameliorating, consider labeling programs, such as the EnergyGuide or Energy Star programs discussed in Sect. 2.1. These programs increase consumers' information about energy costs. For the purely private calculation about whether the additional costs of the good upfront are justified by the longer-term energy savings, labels can facilitate better decision-making. These decisions may in turn result in lower energy use. Empirical evidence on the effects of the Green Lights and Energy Star programs suggests that the programs have successfully lowered energy consumption (DeCanio 1998; DeCanio and Watkins 1998; Howarth et al. 2000).

A mix of instruments, in particular, some form of standards (which could be in the form of tradable permits or a tax) plus information disclosure requirements, can yield a more efficient solution in the presence of these potentially jointly ameliorating market failures. In practice, this coupling of instruments is frequently observed, but current practice appears to be to use one instrument to address one failure and another instrument to address the other failure without consideration of their potential interaction.

5 Second-best policy making with an exogenous constraint

There are several situations in which the existence of an exogenous constraint can lead to a second-best optimum that involves multiple policy instruments. We examine three cases: (1) uncertainty; (2) stakeholder support, and (3) administrative capacity constraints. In all three cases, a single policy instrument is desirable in the absence of the constraint, but multiple policy instruments may be the second-best solution in the presence of the constraint.

[17] The existence of markets for "green goods" which are jointly private consumption goods and public goods does not necessarily lead to environmental improvements, because private provision of the public good can be crowded out (Kotchen 2006). For an economy that is sufficiently large or where the public good (environmental quality) is a gross complement to private consumption, however, the existence of differentiated green products is likely to result in environmental improvement (Kotchen 2006).

5.1 Uncertainty

When government regulators are uncertain about pollution abatement costs, which is typically the case, how can the government seek to achieve the economically efficient degree of abatement? One approach, in theory, would be for the government to iteratively search for the optimum by successively announcing and revising its policies in the light of responses by firms. But much of the investment that firms carry out for pollution abatement takes years to accomplish and is essentially irreversible, convergence may be impossible because the optimal solution is itself changing over time, due to technological change and other factors.

Most analysis of policy choice under uncertainty has implicitly assumed that the government cannot alter a policy once it has been announced. This eliminates an iterative approach. What is the second-best policy in light of this political constraint? In comparing a choice between two distinct instruments—a price and quantity instrument—the optimal policy to choose *ex ante* is the instrument that is *expected* to exhibit the smaller degree of deadweight loss *ex post* (Weitzman 1974). Thus, the price instrument will be more efficient if the slope of the marginal abatement cost function is greater than the (absolute value of the) slope of the marginal benefit function (the inverse of the marginal damage function), and the quantity instrument will be more efficient if the slope of the marginal cost function is less than the slope of the marginal benefit function (Weitzman 1974).[18]

If we broaden government's choice set to include multiple instruments, then it turns out that under a range of realistic conditions the most efficient policy *ex ante* will combine a price and quantity instrument.[19] That is, a mixed system — involving charges (and possibly subsidies) and quantity restrictions via tradable permits will be preferable to either pollution charges or tradable permits used separately (Roberts and Spence 1976).[20]

The logic behind this theoretical result is as follows. Taxes (pollution charges) and permits (tradable permits) can be used to protect against the failings of the other. Permits can protect society from extremely high levels of pollution, whereas taxes can provide incentives to clean up more than required by permits if costs are low. This suggest a three-part policy mechanism: a fixed quantity of tradable emission (abatement) permits; a subsidy (per unit of emissions) paid by the government to any firm whose permit holdings exceed its emissions; and a tax (per unit of emissions) paid by firms if their emissions exceed their permit holdings (Roberts and Spence 1976).

This system has three key properties: first, it is efficient; second, it is superior to either a pure tax or pure permit system; and third, the system creates a piecewise linear penalty function faced by firms, which approximates the non-linear damage function assumed to be linked with pollution. So, under this system, emissions are limited cost-effectively through the tradable permit allocation. If abatement costs turn out to be less than anticipated, a residual incentive for additional abatement is provided by the subsidy.[21] On the other hand, if abatement costs turn out to be greater than anticipated, a "safety valve" from the restriction imposed by the permit allocation is provided by the possibility of a tax being paid for emissions exceeding permitted levels.

[18] Things also become more complex if there is uncertainty regarding both costs and benefits, and there is statistical dependence between the marginal benefit and marginal cost functions (Stavins 1996).

[19] This is frequently referred to as a hybrid instrument.

[20] Weitzman (1978) also examined the potential efficiency of hybrid instruments in the presence of uncertainty. In a related response to the dichotomous choice of a linear tax versus a permit system analyzed by Weitzman (1974) and others, Kaplow and Shavell (2002) argue for a non-linear tax.

[21] A hybrid policy that includes such a "price floor" through a government promise to buy back permits can foster dynamic inefficiencies (Baumol and Oates 1988).

The use of a combined tax and tradable permit system has been examined in recent years in the context of addressing global climate change (Jacoby and Ellerman 2004; Kopp et al. 2000; Pizer 1999, 2002; McKibbin and Wilcoxen 2002). The proposed systems combine an ordinary tradable permit system with a government promise to sell additional permits at a stated price, thereby creating a price (and cost) ceiling. These approaches have been labeled "safety-valve systems." This is a case where policy making in a second-best setting would recommend the use of multiple policy instruments. To date, despite substantial enthusiasm from economists, no such hybrid instrument has been implemented.

5.2 Stakeholder support

The choice of policy instrument can be limited by political participation constraints. The first-best policy may not be feasible because incidence of the policy's cost and benefits are such that key stakeholders are unwilling to provide sufficient political support.

Consider our previous example of a two-part hybrid policy combining an initial distribution of tradable permits with a provision that additional permits can be purchased at a specified "trigger" price (in essence, a tax). Empirically, such a policy may be only slightly more efficient than a pure tax regime, such as in the case of climate change (Pizer 2002), but it achieves its efficiency while preserving the political appeal of tradable permit systems (in which the permits are allocated without charge). In particular, the government authority has great flexibility to distribute the rents associated with the scarce emission permits as it chooses, and can thereby build a political constituency for the overall policy. Thus, such a two-part policy instrument can be motivated by a desire to generate stakeholder support.

There has been considerable attention given to the political attractions of a tradable permit system in which the permits are allocated without charge (Stavins 1989; Goulder et al. 1997; Parry and Williams 1999). It is well-known that when valuable permits are allocated gratis to those otherwise bearing the greatest cost burden, as with the SO_2 allowance trading program in the United States, political support for a program can be increased without an increase in the policy's social cost. Opposition to pollution taxes is great, because they fail to distribute such rents and because they force firms to pay not only the cost of abatement but also the tax penalty associated with residual emissions. Hybrid systems—tradable permits with a safety valve—have welfare properties equal to or better than a pure price-based mechanism (tax system) while retaining some of the politically desirable characteristics of a pure permit system (Pizer 2002).

The stakeholder support argument for hybrid policy instruments also operates in the presence of other distortionary taxes. A pollution tax may be efficient but politically unacceptable because of its adverse impacts on particular industries. These adverse impacts can be reduced or virtually eliminated by replacing the simple tax with an appropriate hybrid policy. For example, Goulder (2000) reports simulations of CO_2 abatement policies in which the efficiency costs of a standard carbon tax are greatly reduced by combining the carbon tax with industry-specific corporate tax cuts. This contrasts with a system of gratis tradable permits, which has considerably greater efficiency costs. Most striking is Goulder's finding that a permit system with only a small degree of gratis distribution is only slightly more costly than a standard carbon tax program, but can be designed to eliminate virtually all of the cost burden on the hardest hit sectors.

Stakeholder support constraints arise frequently in international environmental and resource policy. The constraint is often that the policy should be perceived to be equitable (Cazorla and Toman 2000). While definitions of what constitutes an equitable outcome vary (Cazorla and Toman 2000), perception of fairness is as important as efficiency or cost-effectiveness in many international negotiations. Unfortunately, there is often a tradeoff between cost-effectiveness (or efficiency) and distributional equity. In these circumstances, the second-best optimum may be to use a combination of instruments to better attain both goals.

A key example of how the stakeholder support constraint can affect policy choice arose in negotiations over developing country participation in the Kyoto Protocol. Economists have argued for developing country participation on cost-effectiveness grounds (Stavins 2001, 2005). But targets for developing countries, even if indexed by economic performance, are widely viewed by developing countries themselves as inequitable (Aldy et al. 2003). The policy eventually established in the Kyoto Protocol combined specific targets for developed countries with optional provisions in the form of the Clean Development Mechanism for developing countries.

5.3 Administrative capacity constraints

Sometimes the constraint that forces policy to deviate from it's first-best solution has to do with the administrative capacity of the state. A prime example of this comes from policies to regulate non-uniformly mixed pollution.

It is well known that cost-effectiveness can be achieved—in principle—with market-based systems, such as tradable emission permits, when pollutants are *uniformly mixed* in receiving airsheds or watersheds. With *non-uniformly mixed* pollutants, where location matters, things may become more complex. In theory, an "ambient tradable permit" system will be cost-effective in such circumstances (Montgomery 1972).

How might such an instrument be designed? In principle, if the government can define a vector of transfer coefficients for each pollutant emitting source, linking emissions from each source with concentrations (ambient conditions) at specified receptor locations, then a trading system can be designed that will achieve a cost-effective allocation of the abatement burden among sources (Tietenberg 1995). Although the design is relatively straight-forward, the implementation of such as system is not, and no such systems exist in practice. Under such a system, each emitter would need to obtain separate permits for each and every affected receptor. With a large number of receptor sites, the set of necessary transactions could be overwhelming.

A practical, second-best alternative may be to use multiple instruments. A market-based instrument—such as tradable emission permits or an emission tax (in the interest of cost-effectiveness)—can be combined with localized ambient standards to prevent concentrations from exceeding accepted levels in particular localities (so-called "hot spots"). This can be done by combining tradeable permits with: (1) ambient concentration limits (Krupnick et al. 1983; McGartland and Oates 1985), or (2) restrictions on both ambient concentrations and emissions (Atkinson and Tietenberg 1982).

There are several examples of the use of market-based instruments in concert with ambient air standards in the United States. One is the sulfur dioxide (SO_2) allowance-trading program carried out in the United States under the Clean Air Act amendments of 1990. This permit trading program exists in the context of ambient standards that limit local concentrations of sulfur dioxide and correlated pollutants. This combination of permits and standards promotes cost-effectiveness while preventing the existence of hot spots (Stavins 2003).

Under the Regional Clean Air Markets (RECLAIM) program in the Los Angeles metropolitan area, a tradable permit system for NO_x and SO_2 was combined with trading restrictions to reduce hot-spots. The Los Angeles basin was divided into two zones, inland and coastal. Inland facilities are allowed to purchase permits from any facility in the Los Angeles

Basin, but coastal facilities are allowed to purchase permits only from other coastal facilities (Harrison 1999).

Water quality regulation provides another example of how multiple policy instruments can lead to a better second-best outcome than a single policy instrument. In water quality regulation, not only do marginal abatement costs vary widely among pollution sources, but marginal damages can also vary widely, depending on source location, pollutant transport characteristics, weather, and other factors. Under these conditions, taxes targeted to address individual facilities' effects on ambient water quality are more cost-effective than quantity-based standards. But the typical proxy for site-specific taxes—a uniform tax—may actually be more costly than quantity-based standards if high abatement cost facilities are also high marginal damage facilities (Boyd 2003; Brown and Johnson 1984). This has been demonstrated empirically in the Delaware River basin the United States (Spofford 1983).

While multiple instruments are used in these policy situations, and this use may be justified on the bases of maximizing social welfare subject to exogenous constraints on administrative capacity, such cap-and-trade policies could benefit from further research on their second-best properties. For example, what is the optimal number of trading zones in a cap-and-trade within zone system? How should standards be set if standards will be combined with a cap-and-trade system? For water regulation, a hybrid approach, combining a quantity-based approach and a fee for exceeding the quantity limit may ameliorate the problems raised by uniform taxes or permit systems, but no such hybrid approach has been studied in this context.

6 Conclusion

We began by highlighting the frequency with which policy makers use multiple instruments to address environmental and natural resource problems, and we described three areas—energy efficiency, toxic chemical control, and fisheries management—where policy makers routinely use a mix of several instruments. Yet with few exceptions, economic research has focused on the economic properties of single policy instruments or comparative properties of individual instruments.

We find that under a fairly broad set of circumstances the use of multiple policy instruments *can* be justified on economic grounds. This justification comes from recognition that environmental policy is usually developed and executed in second-best settings. While some applications of second-best theory to environmental policy have been well examined, in particular, the interaction of environmental policy and pre-existing distortionary taxes, many other applications of second-best theory exist within the environmental domain. We highlighted several cases. A first group consisted of cases where an environmental market failure (imperfect property rights or negative externality) co-exists simultaneously with other market failures. A second group consisted of cases where environmental externalities exist in the presence of an additional (often political) constraint that cannot be removed.

The examination of policy making in these second-best settings reveals two commonalities. First, in the case of multiple market failures, coordination of policies to address the interaction of these failures will often be more efficient that policies that address these failures separately. Second, in all cases, the use of multiple policy instruments may be more efficient than the use of a single policy instrument.

Do these findings eliminate the apparent disconnect between the use of multiple policy instruments in practice and the relative paucity of economic research on multiple instruments? As with many economics questions, the answer is "both yes and no." It is true that the use of multiple policy instruments may be justified when one examines environmental policy in a second-best setting. But there is little research to suggest that the actual mix of instruments employed in practice is the economically efficient mix. Similarly, where there is good economic research to justify the use of multiple policy instruments—particularly research on hybrid instruments—there is little use in the policy world. Thus, the disconnect between instrument mixes used in practice and instrument mixes recommended by economists is only partially explained by the second-best nature of environmental problems. A particularly fruitful area for future research with direct policy implications will be to examine optimal instrument mixes in a second-best setting.

Acknowledgements We are grateful to Scott Barrett, Nathanial Keohane, Sheila Olmstead, Tom Tietenberg, and two anonymous referees for their helpful comments on earlier drafts. All views expressed here, as well as any errors, are the responsibility of the authors. This paper is based on work commissioned and owned by the OECD. The original work can be found in OECD [The Use of Multiple Policy Instruments For Environmental Protection: An Economic Perspective, ENV/EPOC/WPNEP(2005)5].

References

Aldy JE, Barrett S, Stavins RN (2003) Thirteen plus one: a comparison of global climate policy architectures. Climate Policy 3(4):373–397

Anderson L (1994) Highgrading in ITQ fisheries. Marine Resour Econ 9:209–226

Arnason R (1994) On catch discarding in fisheries. Marine Resour Econ 9:189–208

Arrow KJ (1962) Economic welfare and the allocation of resources for invention. In: Nelson R (ed) The rate and direction of inventive activity. Princeton University Press, Princeton, NJ

Atkinson SE, Tietenberg TH (1982) The empirical properties of two classes of designs for transferable discharge permit markets. J Environ Econ Manage 9:101–121

Baumol WJ, Oates WE (1988) The theory of environmental policy, 2nd edn. Cambridge University Press, Cambridge, United Kingdom

Bennear LS (2006) Are management-based regulations effective: evidence from state pollution prevention programs. Working Paper, Duke University

Bovenberg AL, de Mooij RA (1994) Environmental levies and distortionary taxation. Am Econ Rev 84(4):1085–1089

Bovenberg AL, Goulder LH (1996) Optimal environmental taxation in the presence of other taxes: general equilibrium analyses. Am Econ Rev 86(4):985–1000

Bovenberg AL, Goulder LH (1997) Costs of environmentally motivated taxes in the presence of other taxes: general equilibrium analyses. National Tax J 50(1):59–87

Bovenberg AL, van der Ploeg F (1994) Environmental policy, public finance, and the labour market in a second-best world. J Public Econ 66:127–144

Boyd J (2003) Water pollution taxes: a good idea doomed to failure?. Public Finance Manage 3(1):34–66

Boyd R, Dewees C (1991) Putting theory in practice: individual transferable quotas in New Zealand's fisheries. Soc Nat Resour 5:179–198

Brown G, Johnson R (1984) Pollution control by effluent charges: it works in the federal republic of Germany, why not in the U.S.?. Nat Resour J 24:929–966

Cazorla M, Toman M (2000) International equity and climate change policy. Resources for the Future Climate Issue Brief No. 27, Resources for the Future, Washington, D.C

Coglianese C, Lazer D (2003) Management-based regulation: prescribing private management to achieve public goals. Law Soc Rev 37(4):691–729

Connor R (2001) Changes in fleet capacity and ownership rights in New Zealand Fisheries: Case Studies on the Effects of Transferable Fishing Rights on Fleet Capacity and Concentration of Quota Ownership, FAO Fisheries Technical Paper 412 ed. Ross Shotton. Rome: Food and Agriculture Organization of the United Nations

Conrad JM (1999) Resource economics. Cambridge University Press, Cambridge, U.K

Cornes R, Mason CF, Sandler T (1986) The commons and the optimal number of firms. Q J Econ 101(3):641–646

DeCanio SJ (1998) The efficiency paradox: bureaucratic and organizational barriers to profitable energy-saving investments. Energy Policy 26(5):441–454

DeCanio SJ, Watkins WE (1998) Investments in energy efficiency: do the characteristics of firms matter?. Rev Econ Stat 80(1):95–107

Fullerton D, Mohr RD (2002) Suggested subsidies are suboptimal unless combined with an output tax. National Bureau of Economic Research Working Paper #w8723

Fullerton D, West S (2000) Tax and subsidy combinations for the control of car pollution. National Bureau of Economic Research Working Paper #w7774

Fullerton D, Wolverton A (2000) Two generalizations of a deposit refund system. National Bureau of Economic Research Working Paper #w7505

Gamper-Rabindran S (2006) Did the EPA's voluntary industrial toxics program reduce emissions?: a GIS analysis of distributional impacts and by-media analysis of substitution. J Environ Econ Manage 52(1):391–410

Gopinath M, Wu J (1999) Environmental externalities and the optimal level of market power. Am J Agric Econ 81(4):825–833

Goulder LH (1998) Environmental policy making in a second-best setting. J Appl Econ 1(2):279–328

Goulder LH (2000) Confronting the adverse industry impacts of CO_2 Abatement policies: what does it cost?. Climate Issues Brief, No. 23, Resources for the Future, Washington, D.C

Goulder LH, Parry IW, Burtraw D (1997) Revenue-raising vs. Other approaches to environmental protection: the critical significance of pre-existing tax distortions. RAND J Econ 28:708–731

Griliches Z (1979) Issues in assessing the contribution of research and development to productivity growth. Bell J Econ 10:92–116

Griliches Z (1992) The search for R&D spillovers. Scandinavian J Econ 94:S29–S47

Hammer PJ (2000) Antitrust beyond competition: market failures, total welfare, and the challenge of intramarket second-best tradeoffs. Michigan Law Rev 98:849–925

Harrison D Jr (1999) Turning theory into practice for emissions trading in the Los Angeles air basin. In: Steve S, Jim S (eds) Pollution for sale: emissions trading and joint implementation. Edward Elgar, Cheltenham, UK, pp 63–79

Helfand GE, Berck P, Maull T (2003) The theory of pollution policy. In: Karl-Göran M, Jeffrey V (eds) Handbook of Environmental Economics, vol I. Elsevier Science, Amsterdam, pp 249–303

Howarth RB, Haddad BM, Paton B (2000) The economics of energy efficiency: insights from voluntary participation programs. Energy Policy 28(6–7):477–486

Jacoby HD, Ellerman AD (2004) The safety valve and climate policy. Energy Policy 32(4):481–491

Jaffe AB, Newell RG, Stavins RN (2003) Technological change and the Environment. In: Karl-Göran M, Jeffrey V (eds) Handbook of environmental economics, vol I. Elsevier Science, Amsterdam, pp 461–516

Jaffe AB, Newell RG, Stavins RN (2004) Technology policy for energy and the environment. In: Jaffe AB, Lerner J, Stern S (eds) Innovation policy and the economy, vol 4. The MIT Press for the National Bureau of Economic Research, Cambridge, Massachusetts, pp 35–68

Jaffe AB (1986) Technological opportunity and spillovers of R&D: evidence from firms' patents, profits and market value. Am Econ Rev 76:984–1001

Jaffe AB (1998) The importance of 'spillovers' in the policy mission of the advanced technology program. J Technol Transfer. Summer 1998:11–19

Jaffe AB, Newell RG, Stavins RN (2005) A tale of two market failures: technology and environmental policy. Ecol Econ 54(2005):164–174

Kaplow L, Shavell S (2002) On the superiority of corrective taxes to quantity regulation. Am Law Econ Rev 4(1):1–17

Khanna M, Damon LA (1999) EPA's Voluntary 33/50 Program: impact on toxic releases and economic performance of firms. J Environ Econ Manage 37(1):1–25

Kleindorfer PR, Feldman H, Lowe RA (2000) Accident Epidemiology and the U.S. Chemical Industry: Preliminary Results from RMP*Info. Working Paper 00-01-15. (Philadelphia, PA: Center for Risk Management and Decision Processes, The Wharton School, University of Pennsylvania.)

Kopp RJ, Morgenstern RD, Pizer WA, Toman MA (2000) A proposal for credible early action in US climate policy. In: Brockmann KL, Stonzik M (eds) Flexible mechanisms for efficient climate policy: cost saving policies and business opportunities.. Physica-Verlag, Heidelberg, Germany

Kotchen MJ (2006) Green markets and private provision of public goods. J Pol Econ 114:816–834

Krupnick AJ, Oates WE, Van De Verg (1983) On marketable air pollution permits: the case for a system of pollution offsets. J Environ Econ Manage 10:233–247

Ledyard JO, Szakaly-Moore K (1994) Designing organizations for trading pollution rights. J Econ Behav Org 25:167–196

Lee DR, Misiolek WS (1986) Substituting pollution taxation for general taxation: some implications for efficiency in pollution taxation. J Environ Econ Manage 13:1338–1347

Lenox M, Ehrenfeld J (1995) Design for environment: a new framework for strategic decisions. Environ Qual Manage 4(4):37–52

Lenox M, Ehrenfeld J (1997) The development & implementation of DfE programs. J Sustain Prod Design 1(1):17–27

Lenox M, King A, Ehrenfeld J (2000) An assessment of design-for-environment practices in leading U.S. electronics firms. Interfaces 30(3):83–94

Lipsey RG, Lancaster K (1956) The general theory of second best. Rev Econ Stud 24(1):11–32

Mankiw NG, Whinston MD (1986) Free entry and social inefficiency. RAND J Econ 17:48–58

Mason CF, Polasky S (1997) The optimal number of firms in the commons: a dynamic approach. Can J Econ 30(4):1143–1160

McGartland AM, Oates WE (1985) Marketable permits for the prevention of environmental deterioration. J Environ Econ Manage 12(3):207–228

McKibbin W, Wilcoxen P (2002) The role of economics in climate change policy. J Econ Perspect 16(2):107–129

Montgomery WD (1972) Markets in licenses and efficient pollution control programs. J Econ Theor 5:395–418

Oates WE (1993) Pollution charges as a source of public revenues. In: Herbert Giersch (ed) Economic progress and environmental concerns. Springer-Verlag, Berlin, pp 135–152

Oates WE, Schwab RM (1988) Economic competition among jurisdictions: efficiency enhancing or distortion inducing. J Public Econ 35:333–354

Organization for Economic Cooperation and Development (OECD) Environment Directorate, Environment Policy Committee (2004) Instrument Mixes Used For Environmental Policy: Further Analyses and Additional Case Studies, OECD Working Paper [ENV/EPOC/WPNEP(2004)18/REV1]

Organization for Economic Cooperation and Development (OECD) Environment Directorate, Environment Policy Committee (2005) Instrument Mixes Used For Environmental Policy: Preliminary cross-cutting findings and some issues for discussion, OECD Working Paper [ENV/EPOC/WPNEP/RD(2005)1]

Ozga SA (1955) An essay in the theory of tariffs. J Political Econ 63(6):498–499

Parry IWH (1995) Pollution taxes and revenue recycling. J Environ Econ Manage 29:S64–S77

Parry IWH (1997) Environmental taxes and quotas in the presence of distortionary taxes in factor markets. Resour Energy Econ 19:203–220

Parry IW, Williams RC (1999) A second-best evaluation of eight policy instruments to reduce carbon emissions. Resour Energy Econ 21:347–373

Parry IWH, Williams R, Goulder LH (1999) When can carbon abatement policies increase welfare? The fundamental role of distorted factor markets. J Environ Econ Manage 37:52–84

Pearce DW (1975) An evaluation of cost-benefit analysis criteria. Scottish J Political Econ 22(2):121–134

Pearce DW (1997) Benefit-cost analysis, environment, and health in the developed and developing world. Environ Dev Econ 2(2):210–214

Pearce DW (1998a) Packaging waste and the polluter pays principle: a taxation solution. In: Pearce DW (ed) Economics and environment: essays on ecological economics and sustainable development (Cheltenham, U.K. and Northampton, Mass.: Edward Elgar), pp 231–240

Pearce DW (1998b) The role of carbon taxes in adjusting to global warming. In: Pearce D W (ed) Economics and environment: essays on ecological economics and sustainable development (Cheltenham, U.K. and Northampton, Mass.: Edward Elgar), pp 337–349

Pearce DW (1998c) Limits of cost-benefit analysis as a guide to environmental policy. In: Pearce DW (ed) Economics and environment: essays on ecological economics and sustainable development (Cheltenham, U.K. and Northampton, Mass.: Edward Elgar): 55–66

Pearce DW (1998d) The social incidence of environmental costs and benefits. In: Pearce DW (ed) Economics and environment: essays on ecological economics and sustainable development (Cheltenham, U.K. and Northampton, Mass.: Edward Elgar): 241–267

Pizer WA (1999) Choosing price or quantity controls for greenhouse gases. Climate Issues Brief, No. 17, Resources for the Future, Washington, D.C

Pizer WA (2002) Combining price and quantity controls to mitigate global climate change. J Public Econ 85:409–434

Repetto R, Dower RC, Jenkins R, Geoghegan J (1992) Green fees: how a tax shift can work for the environment and the economy. World Resources Institute, Washington, D.C

Revesz RL, Stavins RN (2005) Environmental law and policy. In: Mitchell Polinsky A, Shavell S (eds) Handbook of law and economics. Elsevier Science, Amesterdam (forthcoming)

Roberts MJ, Spence M (1976) Effluent charges and licenses under uncertainty. J Public Econ 5:193–208

Runolfsson B, Arnason R (2001) The effects of introducting transferable property rights on fleet capacity and ownership of harvesting rights in Iceland's fisheries: Case Studies on the Effects of Transferable Fishing Rights on Fleet Capacity and Concentration of Quota Ownership, FAO Fisheries Technical Paper 412 ed. Ross Shotton. Rome: Food and Agriculture Organization of the United Nations

Sampson D (1994) Fishing tactics in a two species fisheries model: the bioeconomics of bycatch and discarding. Can J Fish Aqua Sci 51:2688–2964

Scherer FM, Harhoff D (2000) Technology policy for a world of skew-distributed outcomes. Res Policy 29:559–566

Sigman H (1995) A comparison of public policies for lead recycling. RAND J Econ 26(3):452–478

Smith S, Swierzbinski J (2007) Assessing the performance of the UK Emissions Trading Scheme. Environ Resour Econ 37(1):137–158 (this issue)

Spence AM (1984) Cost reduction, competition and industry performance. Econometrica 52(1):101–121

Spofford W (1983) Properties of alternative source control policies: case study of the lower delaware valley. Resources for the Future, Washington, D.C

Squires D, Campbell H, Cunningham S, Dewees C, Quentin Grafton R, Herrick SF Jr, Kirkley J, Pascoe S, Salvanes K, Shallard B, Turris B, Vestergaard N (1998) Individual transferable quotas in multispecies fisheries. Marine Policy 22(2):135–159

Stavins RN (1989) Harnessing market forces to protect the environment. Environ 31(1):5–7, 28–35

Stavins RN (1996) Correlated uncertainty and policy instrument choice. J Environ Econ Manage 30:218–232

Stavins RN (2001) Economic analysis of global climate change policy: a primer. In: Claussen E, Cochran VA, Davis DP (eds) Climate change: science, strategies, and solutions. Brill Publishing, Boston, pp 177–192

Stavins RN (2003) Experience with market-based environmental policy instruments. In: Mäler K-G, Vincent J V (eds) Handbook of Environmental Economics, vol I. Chapter 9, pp 355–435, Elsevier Science, Amsterdam

Stavins RN (2005) Beyond Kyoto: getting serious about climate change. Milken Inst Rev 7(1):28–37

Terkla D (1984) The efficiency value of effluent tax revenues. J Environ Econ Manage 11:107–123

Tietenberg TH (1995) Tradable permits for pollution control when emission location matters: what have we learned?. Environ Resour Econ 5(2):95–113

Ulph A, Ulph D (2007) Climate change – environmental policies and technology policies in a strategic context. Environ Resour Econ 37(1):159–180 (this issue)

United States Environmental Protection Agency (2000) Taking Toxics Out of the Air. Research Triangle Park, NC: United States Environmental Protection Agency, Office of Air Quality Planning and Standards, EPA-452/K-00-002

Vestergaard N (1996) Discard behavior, high-grading and regulation: the case of the Greenland shrimp Fishery. Marine Resour Econ 11:247–266

Weitzman ML (1978) Optimal rewards for economic regulation. Am Econ Rev 68:683–691

Weitzman ML (1974) Prices vs. Quantities. Rev Econ Stud 41:477–491

PART IV

ECONOMICS AND TECHNOLOGICAL CHANGE

[12]

Environmental and Resource Economics **22:** 41–69, 2002.

Environmental Policy and Technological Change

ADAM B. JAFFE[1], RICHARD G. NEWELL[2] and ROBERT N. STAVINS[3*]
[1]*Department of Economics, Brandeis University and National Bureau of Economic Research;* [2]*Resources for the Future;* [3]*John F. Kennedy School of Government, Harvard University and Resources for the Future (*author for correspondence, e-mail: robert_stavins@harvard.edu)*

Abstract. The relationship between technological change and environmental policy has received increasing attention from scholars and policy makers alike over the past ten years. This is partly because the environmental impacts of social activity are significantly affected by technological change, and partly because environmental policy interventions themselves create new constraints and incentives that affect the process of technological developments. Our central purpose in this article is to provide environmental economists with a useful guide to research on technological change and the analytical tools that can be used to explore further the interaction between technology and the environment. In Part 1 of the article, we provide an overview of analytical frameworks for investigating the economics of technological change, highlighting key issues for the researcher. In Part 2, we turn our attention to theoretical analysis of the effects of environmental policy on technological change, and in Part 3, we focus on issues related to the empirical analysis of technology innovation and diffusion. Finally, we conclude in Part 4 with some additional suggestions for research.

Key words: diffusion, environmental policy, innovation, invention, technological change

1. Economic Frameworks and Issues in Technological Change

Economists have examined a diverse set of issues associated with technological change that go well beyond those analyses that have focused directly on implications for environmental policy, including: the theory of incentives for research and development (Tirole 1988; Reinganum 1989; Geroski 1995); the measurement of innovative inputs and outputs (Griliches 1984 and Griliches 1998); analysis and measurement of externalities resulting from the research process (Griliches 1992; Jaffe 1998); the measurement and analysis of productivity growth (Jorgenson 1990; Griliches 1998; Jorgenson and Stiroh 2000); diffusion of new technology (Karshenas and Stoneman 1995; Geroski 2000); the effect of market structure on innovation (Scherer 1986; Sutton 1998); market failures related to innovation and appropriate policy responses (Martin and Scott 2000); the economic effects of publicly funded research (David et al. 2000); the economic effects of the patent system (Jaffe 2000); and the role of technological change in endogenous macroeconomic growth (Romer 1994; Grossman and Helpman 1994). In this part of the article, we provide a very brief guide to some of this research. In particular, we introduce approaches for measuring technological change, we examine critical aspects of the process of technological change, and we describe modeling approaches and potential market failures relating to technology innovation and diffusion.

1.1. MEASUREMENT OF TECHNOLOGICAL CHANGE

The measurement of the rate and direction of technological change rests fundamentally on the concept of the transformation function,

$$T(Y, I, t) \leq 0, \tag{1}$$

where Y represents a vector of outputs, I represents a vector of inputs, and t is time. Equation (1) describes a production possibility frontier, that is, a set of combinations of inputs and outputs that are technically feasible at a point in time. Technological change is represented by movement of this frontier that makes it possible over time to use given input vectors to produce output vectors that were not previously feasible.

In most applications, separability and aggregation assumptions are made that make it possible to represent the economy's production technology with a production function,

$$Y = f(K, L, E; t), \tag{2}$$

where Y is now a scalar measure of aggregate output (for example, gross domestic product), and the list of inputs on the right-hand side of the production function can be made arbitrarily long. For illustrative purposes, we conceive of output as being made from a single composite of capital goods, K, a single composite of labor inputs, L, and a single composite of environmental inputs, E (for example, waste assimilation). Again, technological change means that the relationship between these inputs and possible output levels changes over time.

Logarithmic differentiation of the production function (Equation (2)) with respect to time yields

$$y_t = A_t + \beta_{Lt} l_t + \beta_{Kt} k_t + \beta_{Et} e_t, \tag{3}$$

in which lower case letters represent the percentage growth rates of the corresponding upper case variable; the β's represent the corresponding logarithmic partial derivatives from Equation (2); and the t indicate that all quantities and parameters may change over time.[1] The term A_t corresponds to "neutral" technological change, in the sense that it represents the rate of growth of output if the growth rates of all inputs were zero. But the possibility that the β's can change over time allows for "biased" technological change, that is, changes over time in *relative* productivity of the various inputs.

Equations (2) and (3) are most easily interpreted in the case of process innovation, in which firms figure out more efficient ways to make existing products, allowing output to grow at a rate faster than inputs are growing. In principle, these equations also apply to product innovation. Y is a composite or aggregate output measure, in which the distinct outputs of the economy are each weighted by their relative value, as measured by their market price. Improved products will typically sell at a price premium, relative to lower quality products, meaning that their introduction will increase measured output even if the physical quantity of the new goods does not exceed the physical quantity of the old goods they replaced. In practice, however, product improvement will be included in measured productivity only to the extent that the price indices used to convert nominal GDP or other nominal output measures to real output measures are purged of the effects of product innovation. In general, official price indices and the corresponding real output measures achieve this objective only to a limited extent.

On its face, Equation (3) says nothing about the *source* of the productivity improvement associated with the neutral technological change term, A_t. If, however, all inputs and outputs are properly measured, and inputs (including R&D) yield only normal investment returns, then all endogenous contributions to output should be captured by returns to inputs, and there should be no "residual" difference between the weighted growth rates of inputs and the growth rate of output. The observation that the residual has been typically positive is therefore interpreted as evidence of exogenous technological change.

1.2. PROCESS OF TECHNOLOGICAL CHANGE

Economic theories of the process of technological change can be traced to the ideas of Josef Schumpeter (1942), who distinguished three stages in the process by which a new, superior technology permeates the marketplace. *Invention* constitutes the first development of a scientifically or technically new product or process. Inventions may be patented, though many are not. Either way, most inventions never actually develop into an *innovation*, which is accomplished only when the new product or process is commercialized, that is, made available on the market. A firm can innovate without ever inventing, if it identifies a previously existing technical idea that was never commercialized, and brings a product or process based on that idea to market. The invention and innovation stages are carried out primarily in private firms through a process that is broadly characterized as "research and development" (R&D).[2] Finally, a successful innovation gradually comes to be widely available for use in relevant applications through adoption by firms or individuals, a process labeled *diffusion*. The cumulative economic or environmental impact of new technology results from all three of these stages,[3] which we refer to collectively as the process of technological change.

1.3. INDUCED INNOVATION AND EVOLUTIONARY APPROACHES

If the imposition of environmental requirements can stimulate invention and innovation that reduces the cost of complying with those requirements, this has profound implications for both the setting of environmental policy goals and the choice of policy instruments. It is useful to identify two major strands of thought regarding the determinants of innovative activity. We call these two broad categories of modeling approaches the "induced innovation" approach and the "evolutionary" approach.

Induced innovation. Within the induced innovation approach, firms undertake an investment activity called "R&D" with the intention of producing profitable new products and processes. Decisions regarding the magnitude and nature of R&D activities are governed by firms' efforts to maximize their value, or, equivalently, to maximize the expected discounted present value of cash flows. In some applications, the output of R&D is explicitly modeled as "knowledge capital," an intangible asset that firms use together with other assets and other inputs to generate revenues (Griliches 1979; Hall et al. 2000).

When viewed as an investment activity, R&D has important characteristics that distinguish it from investment in equipment or other tangible assets. First, although the outcome of any investment is uncertain to some extent, R&D investment appears to be qualitatively different. Not only is the variance of the distribution of expected returns much larger than for other investments, but much or even most of the value may be associated with very low-probability but very high value outcomes (Scherer et al. 2000). This skewness in the distribution of the outcomes of the research process has important implications for modeling firms' R&D decision making (Scherer and Harhoff 2000). In addition, the asset produced by the R&D investment process is specialized, sunk and intangible, so that it cannot be mortgaged or used as collateral. The combination of great uncertainty and intangible outcomes makes financing of research through capital market mechanisms much more difficult than for traditional investment. The difficulty of securing financing for research from outside sources may lead to under-investment in research, particularly for small firms that have less internally generated cash and/or less access to financial markets.

In addition to these financing difficulties, research investment differs from physical investment because the asset produced by the research process – new knowledge about how to make and do things – is difficult to exclude others from using. As first noted in the classic paper by Arrow (1962), this means that the creator of this asset will typically fail to appropriate all or perhaps most of the social returns it generates. Much of this social return will accrue as "spillovers" to competing firms, to downstream firms that purchase the innovator's products, or to consumers (Griliches 1979, 1992; Jaffe 1986, 1998). This "appropriability problem" is likely to lead to significant underinvestment by private firms in R&D, relative to the social optimum (Spence 1984).

The recognition that R&D is a profit-motivated investment activity also leads to the hypothesis that the rate and direction of innovation are likely to respond to changes in relative prices. Since environmental policy implicitly or explicitly makes environmental inputs more expensive, the "induced innovation" hypothesis suggests an important pathway for the interaction of environmental policy and technology, and for the introduction of impacts on technological change as a criterion for evaluation of different policy instruments. We consider empirical approaches and evidence on induced innovation in Section 3.1 below.

The evolutionary perspective. While viewing R&D as a profit-motivated investment activity comes naturally to most economists, the large uncertainties surrounding the outcomes of R&D investments make it very difficult for firms to make optimizing R&D decisions. Accordingly, Nelson and Winter (1982) used Herbert Simon's idea of boundedly rational firms that engage in "satisficing" rather than optimizing behavior (Simon 1947) to build an alternative model of the R&D process. In this "evolutionary" model, firms use "rules of thumb" and "routines" to determine how much to invest in R&D, and how to search for new technologies. The empirical predictions of this model depend on the nature of the rules of thumb that firms actually use (Nelson and Winter 1982; Winter et al. 2000).

If firms are not optimizing, a logical consequence of the evolutionary model is that it cannot be presumed that the imposition of a new external constraint (for example, a new environmental rule) necessarily reduces profits. There is at least the theoretical possibility that the imposition of such a constraint could be an event that forces a satisficing firm to rethink its strategy, with the possible outcome being the discovery of a new way of operating that is actually more profitable for the firm. This raises the possibility that environmental regulation can lead to a "win-win" outcome in which pollution is reduced and profits increased.

Porter and other "win-win" theorists have argued that in a non-optimizing world, regulation may lead to "innovation offsets" that "can not only lower the net cost of meeting environmental regulations, but can even lead to absolute advantages over firms in foreign countries not subject to similar regulations" (Porter and van der Linde 1995: 98). Of course, the fact that firms engage in non-optimizing behavior creates a *possibility* for profit improvements, without suggesting that such improvements would be the norm, would be systematic, or even likely.

Porter and van der Linde (1995) provided case studies of firms which adopted new technology in response to regulation, and appear to have benefited, but win-win theorists do not claim that all environmental regulations generate significant innovation offsets. Indeed, they emphasize that regulation must be properly designed in order to maximize the chances for encouraging innovation. Quantitative evidence is limited – much of it from a large related literature on the impact of environmental regulation on productivity and investment[4] – and results seem to be industry and methodology dependent.

Boyd and McClelland (1999) and Boyd and Pang (2000) employ data envelopment analysis to evaluate the potential at paper and glass plants for "win-win" improvements that increase productivity and reduce energy use or pollution. They suggest that the paper industry could reduce inputs and pollution by 2–8% without reducing productivity. Berman and Bui (2001) found significant productivity increases associated with air pollution regulation in the oil refining industry, but Gray and Shadbegian (1998) found that pollution abatement investment "crowds out" productive investment almost entirely in the pulp and paper industry. Greenstone (1998) found that air pollution regulation has a statistically significant but very small impact on overall costs, implying a small negative productivity impact.

Generally, economists have been skeptical of the win-win theory (Palmer et al. 1995; Oates et al. 1993). From a theoretical perspective, it is possible to model apparently inefficient firm behavior as the (second-best) efficient outcome of imperfect information and divergent incentives among managers or between owners and managers in a principal/agent framework.[5] From this perspective, the *apparent* inefficiency does not have normative implications. Since firms are doing the best they can given their information environment, it is unlikely that the additional constraints represented by environmental policy interventions would be beneficial. On a more concrete level, Palmer et al. (1995) surveyed firms affected by regulation – including those cited by Porter and van der Linde as success stories – and found that most firms say that the net cost to them of regulation is, in fact, positive.

For regulation to have important informational effects, the government must have better information than firms have about the nature of environmental problems and their potential solutions. Furthermore, while it seems likely that environmental regulation will stimulate the innovation and diffusion of technologies that facilitate compliance, creation and adoption of new technology will typically require real resources, and have significant opportunity costs. Overall, the evidence on induced innovation and the win-win hypothesis seems to be a case of a "partially full glass" that analysts see as mostly full or mostly empty, depending on their perspective.

1.4. MICROECONOMICS OF TECHNOLOGY DIFFUSION

From the mechanical reaper of the nineteenth century (David 1966), through hybrid corn seed (Griliches 1957), steel furnaces (Oster 1982), optical scanners (Levin et al. 1987) and industrial robots (Mansfield 1989), research has consistently shown that the diffusion of new, economically superior technologies is a gradual process. Typically, the fraction of potential users that has adopted a new technology follows a sigmoid or "S-shaped" path over time, rising only slowly at first, then entering a period of very rapid growth, followed by a slowdown in growth as the technology reaches maturity and most potential adopters have switched (Geroski 2000).

The explanation for the apparent slowness of the technology diffusion process has been a subject of considerable study. Two main forces have been emphasized. First, potential technology adopters are heterogeneous, so that a technology that is generally superior will not be equally superior for all potential users, and may remain inferior to existing technology for some users for an extended period of time after its introduction. Second, adopting a new technology is a risky undertaking, requiring considerable information, both about the generic attributes of the new technology and about the details of its use in the particular application being considered. It takes time for information to diffuse sufficiently, and the diffusion of the technology is limited by this process of diffusion of information.

The two main models of the diffusion process each emphasize one of these two aspects of the process. The *probit* or *rank* model, first articulated in an unpublished paper by David (1969), posits that potential adopters are characterized by a distribution of returns associated with the new technology. Because adoption is costly, at any moment in time there is a threshold point on this distribution, such that potential users with values above this threshold will want to adopt, and users for whom the value of the new technology is at or below this threshold will not want to adopt. Because the new technology will typically get cheaper and better as time passes, this threshold will gradually move to the left, and eventually sweep out the entire distribution. If the distribution of underlying values is normal (or another single-peaked distribution with similar shape), this gradual movement of the threshold across the distribution will produce the typical S-shaped diffusion curve.

The other widely-used model is the *epidemic* model (Griliches 1957; Stoneman 1983). The epidemic model presumes that the primary factor limiting diffusion is information, and that the most important *source* of information about a new technology is people or firms who have tried it. Thus technology spreads like a disease, with the instigation of adoption being contact between the "infected" population (people who have already adopted) and the uninfected population. Denoting the fraction of the potential using population that has adopted as f, this leads to the differential equation $\frac{df}{dt} = \beta f(1 - f)$. Solution of this equation yields a logistic function, which has the characteristic S-shape. The parameter β captures the "contagiousness" of the disease, presumably related to the cost of the new technology and the degree of its superiority over the technology it replaces (Griliches 1957).[6]

Both of the models discussed above predict that the present value of benefits from adoption and the initial adoption cost enter into decisions affecting the diffusion rate. In the probit model, this net present value comparison determines the location of the adoption threshold that determines what fraction of potential adopters will adopt at a moment in time. In the epidemic model, this net present value comparison determines the magnitude of the "contagiousness" parameter, which in turn determines the speed at which the technology spreads from adopters to previous non-adopters.

While the induced innovation literature focuses on the potential for environmental policy to bring forth new technology through innovation, there is also a widely-held view that significant reductions in environmental impacts could be achieved through more widespread diffusion of *existing* economically-attractive technologies, particularly ones that increase energy efficiency and thereby reduce emissions associated with fossil fuel combustion. For example, the report of the

Interlaboratory Working Group (1997) compiled an analysis of technologies that reportedly could reduce energy use and hence CO_2 emissions at low or even negative net cost to users. The observation that energy-efficient technologies that are cost-effective at current prices are diffusing only slowly dates to the 1970s, having been identified as a "paradox" at least as far back as Shama (1983).

The apparent potential for emissions reductions associated with faster diffusion of existing technology raises two important questions. First, what is the theoretical and empirical potential for "induced diffusion" of lower-emissions technologies? Specifically, how do environmental policy instruments that implicitly or explicitly increase the economic incentive to reduce emissions affect the diffusion rate of these technologies? A second and related question is the degree to which historical diffusion rates have been limited by market failures in the energy and equipment markets themselves (Jaffe and Stavins 1994). To the extent that diffusion has been and is limited by market failures, it is less clear that policies that operate by increasing the economic incentive to adopt such technology will be effective. On the other hand, if such market failures are important, then policies focused directly on correction of such market failures provide, at least in principle, opportunities for policy interventions that are social-welfare increasing, even without regard to any environmental benefit. Potential sources of market failure include problems regarding inadequate information and uncertainty, principal-agent problems, constrained capital financing, and positive adoption spillovers.

Information plays a particularly important role in the technology diffusion process. First, information is a public good that may be expected in general to be underprovided by markets. Second, to the extent that the adoption of the technology by some users is itself an important mode of information transfer to other parties, adoption creates a positive externality and is therefore likely to proceed at a socially suboptimal rate. As discussed further in Section 3.2, Howarth et al. (2000) explored the significance of inadequate information in inhibiting the diffusion of more efficient lighting equipment. Metcalf and Hassett (1999) compared available estimates of energy savings from new equipment to actual savings realized by users who have installed the equipment. They found that actual savings, while significant, were less than those promised by engineers and product manufacturers.

Also related to imperfect information are a variety of agency problems that can inhibit the adoption of superior technology. An example of an external agency problem would be a landlord/tenant relationship, in which a tenant pays for utilities but the landlord makes decisions regarding which appliances to purchase, or vice versa. Internal agency problems can arise in organizations where the individual or department responsible for equipment purchase or maintenance differs from the individual or department whose budget covers utility costs.[7] DeCanio (1998) explored the significance of organizational factors in explaining firms' perceived returns to installation of energy-efficient lighting.

Uncertainty is another factor that may limit the adoption of new technology (Geroski 2000). Such uncertainty is not a market failure, merely a fact of economic life. Uncertainty can be inherent in the technology itself, in the sense that its newness means that users are not sure how it will perform (Mansfield 1968). For resource-saving technology, there is the additional uncertainty that the economic value of such savings depends on future resource prices, which are themselves uncertain. This uncertainty about future returns means that there is an "option value" associated with postponing the adoption of new technology (Pindyck 1991; Hassett and Metcalf 1995, 1996).

Closely related to the issue of uncertainty is the issue of the discount rate or investment hurdle rate used by purchasers in evaluating the desirability of new technology. A large body of research demonstrates that purchasers *appear* to use relatively high discount rates in evaluating energy-efficiency investments (Hausman 1979; Ruderman et al. 1987; Ross 1990). The implicit or explicit use of relatively high discount rates for energy savings does not represent a market failure in itself; it is rather the manifestation of underlying aspects of the decision process including those just discussed. At least some portion of the discount rate premium is likely to be related to uncertainty, although the extent to which the premium can be explained by uncertainty and option value is subject to debate (Hassett and Metcalf 1995, 1996; Sanstad et al. 1995).

Capital market failures that make it difficult to secure external financing for these investments may also play a role (Shrestha and Karmacharya 1998). For households and small firms, adoption of new technologies with significant capital costs may be constrained by inadequate access to financing. And in some countries, import barriers may inhibit the adoption of technology embodied in foreign-produced goods (Reppelin-Hill 1999). It is impossible to generalize, however, particularly across countries.

Finally, the presence of increasing returns in the form of learning effects, network externalities, or other positive adoption externalities suggests that market outcomes for technologies exhibiting these features may be inefficient. For example, the idea that we are "locked into" a fossil-fuel-based energy system is a recurring theme in policy discussions regarding climate change and other energy-related environmental problems. At a more aggregate level, there has been much discussion of the question of whether it is possible for developing countries to take less environmentally-damaging paths of development than have industrialized countries (Evenson 1995).

2. Theory of the Effects of Environmental Policy on Technological Change

The effects of environmental policies on the development and spread of new technologies may, in the long run, be among the most important determinants of success or failure of environmental protection efforts (Kneese and Schultze 1975). It has long been recognized that alternative types of environmental policy instruments can have significantly different effects on the rate and direction of technological change (Orr 1976). Environmental policies, particularly those with large economic impacts (for example, those intended to address global climate change) can be designed to foster rather than inhibit technological invention, innovation, and diffusion (Kempe and Soete 1990).

For purposes of examining the link between environmental policy instruments and technological change, policies can be characterized as either command-and-control or market-based approaches. Market-based instruments – such as pollution charges, subsidies, tradeable permits, and some types of information programs – can encourage firms or individuals to undertake pollution control efforts that are in their own interests and that collectively meet policy goals (Stavins 2001). Command-and-control regulations tend to force firms to take on similar shares of the pollution-control burden, regardless of the cost. They often do this by setting uniform standards for firms, the most prevalent of which are performance- and technology-based standards. But holding all firms to the same target can be expensive and, in some circumstances, counterproductive, since standards typically exact relatively high costs by forcing some firms to resort to unduly expensive means of controlling pollution. Because the costs of controlling emissions may vary greatly among firms, and even among sources within the same firm, the appropriate technology in one situation may not be cost-effective in another.

All of these forms of intervention have the potential for inducing or forcing some amount of technological change, because by their very nature they induce or require firms to do things they would not otherwise do. Performance and technology standards can be explicitly designed to be "technology forcing," mandating performance levels that are not currently viewed as technologically feasible or mandating technologies that are not fully developed. One problem with these approaches, however, is that while regulators can typically assume that *some* amount of improvement over existing technology will always be feasible, it is impossible to know how much. Standards must either be made unambitious, or else run the risk of being ultimately unachievable, leading to political and economic disruption (Freeman and Haveman 1972).

Technology standards are particularly problematic, since they tend to freeze the development of technologies that might otherwise result in greater levels of control. Under regulations that are targeted at technologies, as opposed to emissions levels, no financial incentive exists for businesses to exceed control targets, and the adoption of new technologies is discouraged. Under a "Best Available Control Technology" (BACT) standard, a business that adopts a new method of pollution abatement may be "rewarded" by being held to a higher standard of performance and thereby not benefit financially from its investment, except to the extent that its competitors have even more difficulty reaching the new standard (Hahn and Stavins 1991). On the other hand, if third parties can invent and patent better equipment, they can – in theory – have a ready market. Under such conditions, a BACT type of standard can provide a positive incentive for technology innovation. Unfortunately, as we note below, there has been very little theoretical or empirical analysis of such technology-forcing regulations.

In contrast with such command-and-control regulations, market-based instruments can provide powerful incentives for companies to adopt cheaper and better pollution-control technologies. This is because with market-based instruments, it pays firms to clean up a bit more if a sufficiently low-cost technology or process for doing so can be identified and adopted.[8]

There are two principal ways in which environmental policy instruments can be compared with regard to their effects on technological change. First, one can ask – both with theoretical models and with empirical analyses – what effects alternative instruments have on the rate and direction of relevant technological change. Second, one can ask whether environmental policies encourage an *efficient* rate and direction of technological change, or more broadly, whether such policies result in overall economic efficiency (that is, whether the efficient degree of environmental protection is achieved). We consider both sets of criteria.

2.1. TECHNOLOGY INVENTION AND INNOVATION

Although decisions about technology invention and commercialization are partly a demand-side function of anticipated sales (adoption), the relevant literature comparing the effects of alternative environmental policy instruments has given greater attention to the supply side, focusing on incentives for firm-level decisions to incur R&D costs in the face of uncertain outcomes. Such R&D can be either inventive or innovative, but the theoretical literature in this area typically makes no particular distinction.

The earliest work that is directly relevant was by Magat (1978), who compared effluent taxes and CAC standards using an innovation possibilities frontier (IPF) model of induced innovation, where research can be used to augment capital or labor in a standard production function. Subsequently, Magat (1979) compared taxes, subsidies, permits, effluent standards, and technology standards, and showed that all but technology standards would induce innovation biased toward emissions reduction.[9]

Taking a somewhat broader view than most economic studies, Carraro and Siniscalco (1994) suggested that environmental policy instruments should be viewed jointly with traditional industrial policy instruments in determining the optimal way to attain a given degree of pollution abatement. They showed that innovation subsidies can be used to attain the same environmental target, but without the output reductions that result from pollution taxes. Laffont and Tirole (1996) examined how a tradeable permit system could – in theory – be modified to achieve desired incentive effects for technological change. They demonstrated that although spot markets for permits cannot induce the socially optimal degree of innovation, futures markets can improve the situation (Laffont and Tirole 1996).[10]

Cadot and Sinclair-Desgagne (1996) posed the following question: if a regulated industry has private information on the costs of technological advances in pollution control (frequently a reasonable assumption), then since the industry has an incentive to claim that such technologies are prohibitively expensive, can the government design an incentive scheme that will avoid the problems posed by this information asymmetry? The authors developed a solution to this game-theoretic problem. Not surprisingly, the scheme involves government issued threats of regulation (which diminish over time as the firm completes stages of technology development).

It was only recently that theoretical work followed up on Magat's attempt in the late 1970s to rank policy instruments according to their innovation-stimulating effects. Fischer et al. (1998) found that an unambiguous ranking of policy instruments was not possible. Rather, the ranking of policy instruments depended on the innovator's ability to appropriate spillover benefits of new technologies to other firms, the costs of innovation, environmental benefit functions, and the number of firms producing emissions.

In an analysis that is quite similar in its results to the study by Fischer et al. (1998), Ulph (1998) compared the effects of pollution taxes and command-and-control standards, and found that increases in the stringency of the standard or tax had ambiguous effects on the level of R&D, because environmental regulations have two competing effects: a direct effect of increasing costs, which increases the incentives to invest in R&D in order to develop cost-saving pollution-abatement methods; and an indirect effect of reducing product output, which reduces the incentive to engage in R&D. Carraro and Soubeyran (1996) compared an emission tax and an R&D subsidy, and found that an R&D subsidy is desirable if the output contractions induced by the tax are small or if the government finds output contractions undesirable for other reasons. Addressing the same trade-off, Katsoulacos and Xepapadeas (1996) found that a simultaneous tax on pollution emissions and subsidy to environmental R&D may be better suited to overcoming the joint market failure (negative externality from pollution and positive externality or spillover effects of R&D).[11] Finally, Montero (2000) compared instruments under non-competitive circumstances, and found that the results are less clear than when perfect competition is assumed. Standards and taxes yield higher incentives for R&D when the market is characterized by Cournot competition, but the opposite holds when the market is characterized by Bertrand competition.

2.2. TECHNOLOGY DIFFUSION

The predominant theoretical framework for analyses of diffusion effects has been what could be called the "discrete technology choice" model: firms contemplate the use of a certain technology which reduces marginal costs of pollution abatement and which has a known fixed cost associated with it. While some authors have presented this approach as a model of "innovation," it is more appropriately viewed as a model of adoption. With such models, several theoretical studies have found that the incentive for the adoption of new technologies is greater under market-based instruments than under direct regulation (Zerbe 1970; Downing and White

1986; Milliman and Prince 1989; Jung et al. 1996). With the exception of Downing and White (1986), all of these studies examined the gross impacts of alternative policy instruments on the quantity of technology adoption.

Theoretical comparisons *among* market-based instruments have produced only limited agreement. In a frequently-cited article, Milliman and Prince (1989) examined firm-level incentives for technology diffusion provided by five instruments: command-and-control; emission taxes; abatement subsidies; freely-allocated emission permits, and auctioned emission permits. Firm-level incentives for adoption in this representative-firm model were pictured as the anticipated change in producer surplus. They found that auctioned permits would provide the largest adoption incentive of any instrument, with emissions taxes and subsidies second, and freely allocated permits and direct controls last. The Milliman and Prince (1989) study was criticized by Marin (1991) because of its assumption of identical firms, but it was subsequently shown that the results remain largely unchanged with heterogeneous abatement costs (Milliman and Prince 1992).

In 1996, Jung et al. built on Milliman and Prince's basic framework for comparing the effects of alternative policy instruments, but rather than focusing on firm-level changes in producer surplus, they considered heterogeneous firms, and modeled the "market-level incentive" created by various instruments. Their rankings echoed those of Milliman and Prince (1989): auctioned permits provided the greatest incentive, followed by taxes and subsidies, free permits, and performance standards.

Subsequent theoretical analyses (Parry 1998; Denicolò 1999; Keohane 1999) clarified several aspects of these rankings. First, there is the question of relative firm-level incentives to adopt a new, cost-saving technology when the price of pollution (permit price or tax level) is endogenous. Milliman and Prince (1989), as well as Jung et al. (1996), argued that auctioned permits would provide greater incentives for diffusion than freely-allocated permits, because technology diffusion lowers the equilibrium permit price, bringing greater aggregate benefits of adoption in a regime where all sources are permit buyers. But when technology diffusion lowers the market price for tradeable permits, all firms benefit from this lower price regardless of whether they adopt the given technology (Keohane 1999). Thus, if firms are price takers in the permit market, auctioned permits provide no more adoption incentive than freely-allocated permits.

The overall result is that both auctioned and freely-allocated permits are inferior in their diffusion incentives to emission tax systems (but superior to command-and-control instruments). Under tradeable permits, technology diffusion lowers the equilibrium permit price, thereby reducing the incentive for participating firms to adopt. Thus, a permit system provides a lower adoption incentive than a tax, assuming the two instruments are equivalent before diffusion occurs (Denicolò 1999; Keohane 1999).

More broadly, it appears that an unambiguous exhaustive ranking of instruments is not possible on the basis of theory alone. Parry (1998) found that the welfare gain induced by an emissions tax is *significantly* greater than that induced by tradable permits only in the case of very major innovations. Similarly, Requate (1998) included an explicit model of the final output market, and finds that whether (auctioned) permits or taxes provide stronger incentives to adopt an improved technology depends upon empirical values of relevant parameters.

Furthermore, complete theoretical analysis of the effects of alternative policy instruments on the rate of technological change must include modeling of the government's response to technological change, because the degree to which regulators respond to technologically-induced changes in abatement costs affects the magnitude of the adoption incentive associated with alternative policy instruments. Because technology diffusion presumably lowers the aggregate marginal abatement cost function, it results in a change in the efficient level of control. Hence, following diffusion, the optimal agency response is to set a more ambitious target. Milliman and Prince (1989) examined the incentives facing private industry, under alternative policy instruments, to oppose such policy changes. Their conclusion was that firms would oppose optimal agency adjustment of the policy under all instruments *except taxes*. Under an emissions tax, the optimal agency response to cost-reducing technological change is to lower the tax rate (assuming convex damages); under a subsidy, the optimal response is to lower the subsidy; under tradeable permit systems, the optimal response is to decrease the number of available permits, and thereby drive up the permit price. Thus, firms have clear incentives to support the optimal agency response only under an emissions tax regime.

In a comparison of tradeable permits and pollution taxes, Biglaiser et al. (1995) examined these instruments' ability to achieve the first-best outcome in a dynamic setting. They found that effluent taxes can do so, but permits cannot, but that this result depends on an assumption of constant marginal damages. If marginal damages are not constant, the optimal policy is determined by the interaction of marginal damages and marginal abatement costs for both taxes and permits. The result is analogous to Weitzman's (1974) rule: if the marginal damage curve is relatively flat and there is uncertainty in marginal costs (from the regulator's perspective) due to potential innovation at the firm level, then a price instrument is more efficient.

2.3. INDUCED INNOVATION AND OPTIMAL ENVIRONMENTAL POLICY

It seems logical that if environmental policy intervention induces innovation, this reduces the social cost of environmental regulation, suggesting that the optimal policy is more stringent than it would be if there were no induced innovation. This intuition contains an element of truth, but a number of complexities arise. First, one has to be careful what is meant by reducing the cost of regulation. If the policy intervention induces a reduction in the marginal cost of abatement, then any given policy target (for example, a particular aggregate emission rate or a particular ambient concentration) will be achieved at lower cost than it would without induced innovation. On the other hand, the lower marginal abatement cost schedule arising from induced innovation makes it socially optimal to achieve a greater level of pollution abatement. For a flat marginal social benefit function evaluated at the social optimum, or for any emission tax, this results in greater *total* expenditure on abatement even as the *marginal* abatement cost falls.

Another important issue is the general equilibrium effect of induced environmental innovation on innovation elsewhere in the economy (Schmalensee 1994). If inducement operates through increased R&D expenditure, then an issue arises as to the elasticity of supply of R&D inputs. To the extent that this supply is inelastic, then any induced innovation must come at the expense of other forms of innovation, creating an opportunity cost that may negate the effects observed in the regulated portion of the economy. The general equilibrium consequences of these effects for welfare analysis depend on the extent of R&D spillovers or other market failures, and the magnitude of these distortions in the regulated firms or sectors relative to the rest of the economy (Goulder and Schneider 1999).

In an application to global climate policy, Goulder and Mathai (2000) looked at optimal carbon abatement in a dynamic setting, considering not only the optimal overall amount of abatement but also its timing. In addition to R&D-induced innovation, they considered (in a separate model) reductions in abatement costs that come about via learning-by-doing. Induced innovation reduces marginal abatement costs, which increases the optimal amount of abatement, but it also increases the cost of abatement today relative to the future, because of lower abatement costs in the future, implying that with R&D-induced innovation, optimal abatement is lower in early years and higher in later years than it would otherwise be. In the learning-by-doing model, there is a third effect: abatement today lowers the cost of abatement in the future. This reinforces the tendency for cumulative optimal abatement to be higher in the presence of induced innovation, but makes the effect on optimal near-term abatement ambiguous.[12]

3. Empirics of the Innovation and Diffusion of Green Technology

3.1. EMPIRICAL ANALYSIS OF INNOVATION

There has been exceptionally little empirical analysis directly of the effects of alternative policy instruments on technology innovation in pollution abatement, principally because of the paucity of available data. One study by Bellas (1998) carried out a statistical analysis of the costs of flue gas desulfurization (scrubbing) installed at coal-fired power plants in the United States under the new-source performance standards of the 1970 and 1977 Clean Air Acts, but failed to find any evidence of effects of scrubber vintage on cost, suggesting little technological innovation had taken place under this regulatory regime.

Although there has been very little analysis in the context of pollution-abatement technologies, there is a more extensive literature on the effects of alternative policy instruments on the innovation of energy-efficiency technologies, because data have been available. The greatest challenge in testing the induced innovation hypothesis specifically with respect to environmental inducement is the difficulty of measuring the extent or intensity of inducement across firms or industries (Jaffe et al. 1995). Ideally, one would like to look at the relationship between innovation and the shadow price of pollution or environmental inputs, but such shadow prices are not easily observed. Instead, one must use proxies, such as expenditures on pollution abatement, prices of polluting inputs, and characteristics of environmental regulations.[13] We consider studies that have used each of these approaches.

Lanjouw and Mody (1996) showed a strong association between pollution abatement expenditures and the rate of patenting in related technology fields. Jaffe and Palmer (1997) examined the correlation between pollution expenditures by industry and indicators of innovation more broadly. They found that there is a significant correlation within industries over time between the rate of expenditure on pollution abatement and the level of R&D spending. They did not, however, find evidence of an effect of pollution control expenditure on overall patenting.

Evidence of inducement has also been sought by examining the response to changing energy prices. Newell et al. (1999) examined the extent to which the energy efficiency of the menu of home appliances available for sale changed in response to energy prices between 1958 and 1993, using a model of induced innovation as changing characteristics of capital goods. Newell et al. (1999) generalized Hicks' (1932) concept of induced innovation (in terms of factor prices) to include inducement by regulatory standards, such as labeling requirements that might increase the value of certain product characteristics by making consumers more aware of them. More generally, non-price regulatory constraints can fit within the inducement framework if they can be modeled as changing the shadow or implicit price that firms face in emitting pollutants. In their framework, the existing technology for making a given type of equipment at a point in time is identified in terms of vectors of characteristics (including cost of manufacture) that are feasible. The process of invention makes it possible to manufacture "models" (characteristics vectors) that were previously infeasible. Innovation means the offering for commercial sale of a model that was not previously offered for sale. Induced innovation is then represented as movements in the frontier of feasible models that reduce the cost of energy efficiency in terms of other attributes.

With this product-characteristic approach, Newell et al. (1999) assessed the effects of changes in energy prices and in energy-efficiency standards in stimulating innovation, and found that energy price changes induced both commercialization of new models and elimination of old models. Regulations, however, worked largely through energy-inefficient models being dropped, since that is the intended effect of the energy-efficiency standards (models below a certain energy efficiency level may not be offered for sale). Through econometric estimation and a series of dynamic simulations, Newell et al. (1999) examined the effects of energy price changes and efficiency standards on average efficiency of the menu of products over time. They found that a substantial amount of the improvement was what may be described as autonomous (that is, not otherwise explained by the model and associated with the passage of time), but significant amounts of innovation were also due to changes in energy prices and changes in energy-efficiency standards. They found that technological change in air conditioners was actually biased against energy efficiency in the 1960s (when real energy prices were falling), but that this bias was reversed after the two energy shocks of the 1970s. In terms of the efficiency of the average model offered, they found that energy efficiency in 1993 would have been about one-quarter to one-half lower in air conditioners and gas water heaters, if energy prices had stayed at their 1973 levels, rather than following their historical path. Most of the response to energy price changes came within less than five years of those changes.

A closely related approach to investigating the same phenomena is that of hedonic price functions. One hedonic study examined the effects of public policies in the context of home appliances. Greening et al. (1997) estimated the impacts of the 1990 and 1993 national efficiency standards on the quality-adjusted price of household refrigerator/freezer units, and found that quality-adjusted prices fell after the implementation of the energy efficiency standards. However, such quality-adjusted price decreases are consistent with historical trends in refrigerator/freezer prices, and hence, one cannot rule out the possibility that the imposition of efficiency standards slowed the rate of quality-adjusted price decline.

Given the attention paid to automobile fuel economy over the past two decades, it is not surprising that several hedonic studies of automobiles have addressed or focused on energy-efficiency, including Ohta and Griliches (1976) and Goodman (1983). Atkinson and Halvorsen (1984) found that the fuel efficiency of the new car fleet responds more than proportionally to changes in expected fuel prices. Using an analogue to the hedonic price technique, Wilcox (1984) constructed a quality-adjusted measure of automobile fuel economy over the period 1952–1980, finding that it was positively related to oil prices. Ohta and Griliches (1986) found that gasoline price changes over the period 1970–1981 could alone explain much of the observed change in related automobile characteristics.

More recently, Pakes et al. (1993) investigated the effects of gasoline prices on the fuel economy of motor vehicles offered for sale, and found that the observed increase in miles per gallon (mpg) from 1977 onward was largely due to the consequent change in the mix of vehicles on the market. Fewer low-mpg cars were marketed, and more high-mpg cars were marketed. Subsequently, Berry et al. (1996) combined plant-level cost data for the automobile industry and information on the characteristics of models that were produced at each plant to estimate a hedonic cost function – the supply-side component of the hedonic price function – finding that quality-adjusted costs generally *increased* over the period 1972–1982, thus coinciding with rising gasoline prices and emission standards.

Goldberg (1998) combined a demand-side model of discrete vehicle choice and utilization with a supply-side model of oligopoly and product differentiation to estimate the effects of CAFE (Corporate Average Fuel Economy) standards on the fuel economy of the new car fleet. She found that automobile fuel operating costs have had a significant effect, although a gasoline tax of a magnitude that could match the effect of CAFE on fuel economy would have to be very large.

Finally, Popp (2001a, b) looked more broadly at energy prices and energy-related innovation. In the first paper, he found that patenting in energy-related fields increases in response to increased energy prices, with most of the effect occurring within a few years, and then fading over time. Popp attributed this fading to diminishing returns to R&D. In the second paper, he attempted to decompose the overall reduction in energy use that is associated with changing energy prices between the substitution effect – movements along a given production frontier – and the induced innovation effect – movement *of* the production frontier itself induced by the change in energy prices. Using energy-related patents as a proxy for energy innovation, he found that approximately one-third of the overall response of energy use to prices is associated with induced innovation, with the remaining two-thirds associated with factor substitution. Because energy patents are likely to measure energy innovation only with substantial error, one might interpret this result as placing a lower bound on the fraction of the overall response of energy use to changing prices that is associated with innovation.

3.2. EMPIRICAL ANALYSIS OF DIFFUSION

One of the great successes during the modern era of environmental policy was the phasedown of lead in gasoline, which took place in the United States principally during the decade of the 1980s. The phasedown was accomplished through a tradeable permit system among refineries, whereby lead rights could be exchanged and/or banked for later use. Kerr and Newell (2000) used a duration model to assess the effects of the phasedown program on technology diffusion. They found that increased stringency (which raised the effective price of lead) encouraged greater adoption of technology that substitutes for lead in increasing octane. They also found that larger and more technically sophisticated refineries, which had lower costs of adoption, were more likely to adopt the new technology. As theory suggests (Malueg 1989), they also found that the tradeable permit system provided incentives for more *efficient* technology adoption decisions, as evidenced by a significant divergence in the adoption behavior of refineries with low versus high compliance costs. Namely, the positive differential in the adoption propensity of expected permit sellers (i.e., low-cost refineries) relative to expected permit buyers (i.e., high-cost refineries) was significantly greater under market-based lead regulation compared to under individually binding performance standards.

Another prominent application of tradeable permit systems which has provided an opportunity for empirical analysis of the effects of policy instruments on technology diffusion is the sulfur dioxide allowance trading program, initiated under the U.S. Clean Air Act amendments of 1990. In an econometric analysis, Keohane (2001) found evidence that the increased flexibility of the market-based instrument provided greater incentives for technology adoption. In particular, he found that the choice of whether or not to adopt a "scrubber" to remove sulfur dioxide – rather than purchasing (more costly) low-sulfur coal – was more sensitive to cost differences (between scrubbing and fuel-switching) under the tradeable permit system than under the earlier emissions rate standard.[14]

Turning from pollution abatement to energy efficiency, Jaffe and Stavins (1995) carried out econometric analyses of the factors affecting the adoption of thermal insulation technologies in new residential construction in the United States between 1979 and 1988. They examined the dynamic effects of energy prices and technology adoption costs on average residential energy-efficiency technologies in new home construction, and found that the response of mean energy efficiency to energy price changes was positive and significant, both statistically and economically. Interestingly, they also found that equivalent percentage adoption cost changes were about three times as effective as energy price changes in encouraging adoption, although standard financial analysis would suggest they ought to be about equal in percentage terms. This finding offers confirmation for the conventional wisdom that technology adoption decisions are more sensitive to up-front cost considerations than to longer-term operating expenses.

Hassett and Metcalf (1995) found an even larger discrepancy between the effect of changes in installation cost and changes in energy prices. There are three possible explanations for this. One possibility is a behavioral bias that causes purchasers to focus more on up-front cost than they do on the lifetime operating costs of an investment. An alternative view is that purchasers focus equally on both, but uncertainty about future energy prices makes them give less weight to energy prices than they do to capital cost, which is known. A final interpretation might be that consumers have reasonably accurate expectations about future energy prices, and their decisions reflect those expectations, but our proxies for their expectations are not correct.

Although empirical evidence from these two studies indicate that subsidies may be more effective than "equivalent" taxes in encouraging technology diffusion, it is important to recognize some disadvantages of such subsidy approaches. First, unlike energy prices, adoption subsidies do not provide incentives to reduce utilization. Second, technology subsidies and tax credits can require large public expenditures per unit of effect, since consumers who would have purchased the product even in the absence of the subsidy still receive it. In the presence of fiscal constraints on public spending, this raises questions about the feasibility of subsidies that would be sizable enough to have desired effects.

Rose and Joskow (1990) also found a positive effect of fuel price increases on the adoption of a new fuel-saving technology in the U.S. electricity-generation sector; and in a tobit analysis of steel plant adoption of different furnace technologies, Boyd and Karlson (1993) found a significant positive effect of increases in a fuel's price on the adoption of technology that saves that fuel, although the magnitude of the effect was modest. For a sample of industrial plants in four heavily polluting sectors (petroleum refining, plastics, pulp and paper, and steel), Pizer et al. (2001) found that both energy prices and financial health were positively related to the adoption of energy-saving technologies.

Greene (1990) used data on fuel prices and fuel economy of automobiles from 1978 to 1989 to test the relative effectiveness of CAFE standards and gasoline prices in increasing fuel economy. He found that the big three U.S. firms faced a binding CAFE constraint, and for these firms compliance with CAFE standards had roughly twice the impact on fuel economy as did fuel prices. Japanese firms, however, did not face a binding CAFE constraint, and fuel prices had only a small effect. Luxury European manufactures seemed to base their fuel efficiency largely on market demand and often exceeded CAFE requirements. For these firms, neither the standards nor prices seemed to have much effect.

Another body of research has examined the effects on technology diffusion of command-and-control environmental standards when they are combined with "differential environmental regulations." In many situations where command-and-control standards have been used, the required level of pollution abatement has been set at a far more stringent level for new sources than for existing ones. There is empirical evidence that such differential environmental regulations have lengthened the time before plants were retired (Maloney and Brady 1988; Nelson et al. 1993). Further, this dual system can actually worsen pollution by encouraging firms to keep older, dirtier plants in operation (Stewart 1981; Gollop and Roberts 1983; McCubbins et al. 1989).

What about conventional command-and-control approaches? Jaffe and Stavins (1995) also examined the effects of more conventional regulations on technology diffusion, in the form of state building codes. They found no discernable effects. It is unclear to what extent this is due to inability to measure the true variation across states in the effectiveness of codes, or to codes that were in many cases not binding relative to typical practice. This is a reminder, however, that although price-based policies will always have some effect, typical command-and-control may have little effect if they are set below existing standards of practice.[15]

Attention has also been given to the effects on energy-efficiency technology diffusion of voluntary environmental programs. Howarth et al. (2000) examined two voluntary programs of the U.S. Environmental Protection Agency, the Green Lights and Energy Star programs, both of which are intended to encourage greater private industry use of energy-saving technologies. A natural question from economics is why would firms carry out *additional* technology investments as part of a voluntary agreement? The authors respond that there are a set of agency problems that inhibit economically wise adoption of some technologies. For example, most energy-saving investments are small, and senior staff may rationally choose to restrict funds for small projects that cannot be perfectly monitored. The Green

Lights program may be said to attempt to address this type of agency problem by providing information on savings opportunities at the level of the firm where decisions are made.

Although the empirical literature on the effects of policy instruments on technology diffusion by no means settles all of the issues that emerge from the related theoretical studies, a consistent theme that runs through both the pollution-abatement and energy-efficiency empirical analyses is that market-based instruments are decidedly more effective than command-and-control instruments in encouraging the cost-effective adoption and diffusion of relevant new technologies.

4. Conclusions

Virtually all research on the relationship between technological change and environmental policy has been linked with one of two underlying realities: first, the environmental impacts of social and economic activity is greatly affected by the rate and direction of technological change; and second, environmental policy interventions themselves create new constraints and incentives that affect the process of technological developments.

One important research need, linked with the first reality, is the frequent necessity of determining the economic and environmental baseline against which to measure the impacts of proposed policies. Forecasts based on historical experience depend on the relative magnitude of the effects of price-induced technological change, learning-by-doing, public sector R&D, and exogenous technical progress. Sorting out these influences with respect to environmentally relevant technologies and sectors poses a major challenge.

There has also been much debate surrounding the "win-win" hypothesis. Much of this debate has been explicitly or implicitly ideological or political. More useful would be detailed examinations regarding the kinds of policies and the kinds of private-sector institutions that are most likely to generate innovative, low-cost solutions to environmental problems.

More research is also needed on the second broad linkage between technology and environment, the effect of environmental policy interventions on the process of technological change. The empirical evidence is generally consistent with theoretical findings that market-based instruments for environmental protection are likely to have significantly greater, positive impacts over time than command-and-control approaches on the invention, innovation, and diffusion of desirable, environmentally-friendly technologies. But empirical studies have also produced some results that appear not to be consistent with theoretical expectations, such as the finding from two independent analyses that the diffusion of energy-efficiency technologies is more sensitive to variation in adoption-cost than to commensurate energy price changes. Further theoretical and/or empirical work may resolve this apparent anomaly.

Refutable hypotheses have emerged from theoretical models of alternative policy instruments, but most have not been tested rigorously with empirical data. Whereas the predictions from theory regarding the ranking of alternative environmental policy instruments is relatively consistent, most of the empirical analysis has focused on energy-efficient technologies, rather than pollution abatement technologies *per se*. The increased use of market-based instruments and performance-based standards brings with it information with which hypotheses regarding the effects of policy instruments on technology innovation and diffusion can be tested.

Finally, the long-term nature of policy challenges such as that posed by the threat of global climate change makes it all the more important that we improve our understanding of the effects of environmental policy on innovation and diffusion of new technology. What is clear is that many relevant issues cannot be resolved at a purely theoretical level or on the basis of aggregate empirical analysis alone. Serious investigation of induced technological change and its consequences for environmental policy requires going beyond studies that examine whether or not such effects exist, to carry out detailed analyses in a variety of sectors in order to understand the circumstances under which the effects are large or small. This will inevitably require research from multiple methodological viewpoints over an extended period of time.

Acknowledgements

We are grateful for valuable research assistance from Lori Snyder and helpful comments from Ernst Berndt, Karl-Göran Mäler, Lawrence Goulder, Nathaniel Keohane, Charles Kolstad, Ian Parry, Steven Polasky, David Popp, Vernon Ruttan, Manuel Trajtenberg, Jeffrey Vincent, and David Zilberman, but the authors alone are responsible for all remaining errors.

Notes

1. This formulation can be considered a first-order approximation to an arbitrary functional form for Equation (2). Higher-order approximations can also be implemented.
2. Data regarding R&D expenditures of firms are available from the financial statements of publicly traded firms, if the expenditure is deemed "material" by the firm's auditors, or if the firm chooses for strategic reasons to report the expenditure (Bound et al. 1984). In the United States, the government carries out a "census" of R&D activity, and reports totals for broad industry groups (National Science Board 1998). Many industrialized countries now collect similar statistics, which are available through the Organization of Economic Cooperation and Development (OECD 2000).
3. Typically, for there to be environmental impacts of a new technology, a fourth step is required – utilization, but that is not part of the process of technological change *per se*. Thus, for example, a new type of hybrid motor vehicle engine might be *invented*, which emits fewer pollutants per mile; the same or another firm might commercialize this engine and place the *innovation* in new cars available for purchase on the market; individuals might purchase (or adopt) these cars,

leading to *diffusion* of the new technology; and finally, by driving these cars instead of others (*utilization*), aggregate pollutant emissions might be reduced. Conversely, if higher efficiency and the resulting reduced marginal cost causes users to increase utilization, then the emissions reduction associated with higher efficiency may be partially or totally offset by higher utilization.

4. See, for example, Gollop and Roberts (1983), Kolstad and Turnovsky (1998) and Yaisawarng and Klein (1994).
5. For a survey, see Holmström and Tirole (1987).
6. Both the probit and epidemic models typically focus on the fraction of the population that had adopted at a point in time. If one has individual-level data on adopters, one can take as the dependent variable the individual time until adoption. This leads to a duration or hazard model (Rose and Joskow 1990). Kerr and Newell (2000) employed a duration model to analyze technology adoption decisions by petroleum refineries during the phasedown of lead in gasoline.
7. For a discussion of the implications of the separation of environmental decision-making in major firms from relevant economic signals, see: Hockenstein et al. (1997). A series of related case studies are provided by Reinhardt (2000).
8. In theory, the relative importance of the dynamic effects of alternative policy instruments on technological change (and hence long-term compliance costs) is greater in the case of those environmental problems which are of great magnitude (in terms of anticipated abatement costs) and/or very long time horizon. Hence, the increased attention being given by scholars and by policy makers to the problem of global climate change has greatly increased the prominence of the issues that are considered in this article.
9. A considerable amount of theoretical work followed in the 1980s. Although much of that work characterized its topic as the effects of alternative policy instruments on technology innovation, the focus was in fact on effects of policy on technology diffusion. Hence, we defer consideration of those studies to the next section.
10. In a subsequent analysis, Laffont and Tirole (1996) examined the government's ability to influence the degree of innovative activity by setting the number of permits (and permit prices) in various ways in a dynamic setting.
11. See, also, Conrad (2000).
12. Nordhaus (2000) introduced induced technological change into the "DICE" model of global climate change and associated economic activities, and found in that case that the impact of induced innovation was modest.
13. In the literature on the relationship between environmental regulation and productivity, discussed in section 1.3, to measure the characteristics of environmental regulations studies have used expert judgements about relative regulatory stringency in different states (Gray and Shadbegian 1998), number of enforcement actions (Gray and Shadbegian 1995), attainment status with respect to environmental laws and regulations (Greenstone 1998), and specific regulatory events (Berman and Bui 2001).
14. In an examination of the effects of alternative policy instruments for reducing oxygen-demanding water pollutants, Kemp (1998) found that effluent charges were a significant predictor of adoption of biological treatment by facilities. In earlier work, Purvis and Outlaw (1995) carried out a case study of EPA's permitting process for acceptable water-pollution control technologies in the U.S. livestock production sector.
15. In a separate analysis of thermal home insulation, this one in the Netherlands, Kemp (1997) found that a threshold model of diffusion (based on a rational choice approach) could not explain observed diffusion patterns. Instead, epidemic models provided a better fit to the data. Kemp also found that there was no significant effect of government subsidies on the adoption of thermal insulation by households.

References

Arrow, K. J. (1962), 'Economic Welfare and the Allocation of Resources for Invention', in R. Nelson, ed., *The Rate and Direction of Inventive Activity*. Princeton NJ: Princeton University Press.

Atkinson, S. E. and R. Halvorsen (1984), 'A New Hedonic Technique for Estimating Attribute Demand: An Application to the Demand for Automobile Fuel Efficiency', *Review of Economics and Statistics* **66**, 417–426.

Bellas, A. S. (1998), 'Empirical Evidence of Advances in Scrubber Technology', *Resource and Energy Economics* **20**, 327–343.

Berman, E. and L. Bui (2001), 'Environmental Regulation and Productivity: Evidence from Oil Refineries', *Review of Economics and Statistics* **83**(3), 498–510.

Berry, S., S. Kortum and A. Pakes (1996), 'Environmental Change and Hedonic Cost Functions for Automobiles', *Proceedings of the National Academy of Sciences, U.S.A.* **93**, 12731–12738.

Biglaiser, G., J. K. Horowitz and J. Quiggin (1995), 'Dynamic Pollution Regulation', *Journal of Regulatory Economics* **8**, 33–44.

Bound, J., C. Cummins, Z. Griliches, B. Hall and A. Jaffe (1984), 'Who Does R&D and Who Patents?', in Z. Griliches, ed., *R&D, Patents and Productivity*. Chicago: University of Chicago Press.

Boyd, G. A. and S. H. Karlson (1993), 'The Impact of Energy Prices on Technology Choice in the United States Steel Industry', *The Energy Journal* **14**(2), 47–56.

Boyd, G. and J. McClelland (1999), 'The Impact of Environmental Constraints on Productivity Improvement and Energy Efficiency an Integrated Paper and Steel Plants', *The Journal of Economics and Environmental Management* **38**, 121–146.

Boyd, G. and J. Pang (2000), 'Estimating the Linkage between Energy Efficiency and Productivity', *Energy Policy* **28**, 289–296.

Cadot, O. and B. Sinclair-Desgagne (1996), 'Innovation Under the Threat of Stricter Environmental Standards', in C. Carraro et al., eds., *Environmental Policy and Market Structure*. Dordrecht: Kluwer Academic Publishers, pp. 131–141.

Carraro, C. and A. Soubeyran (1996), 'Environmental Policy and the Choice of Production Technology', in C. Carraro et al., eds., *Environmental Policy and Market Structure*. Dordrecht: Kluwer Academic Publishers, pp. 151–180.

Carraro, C. and D. Siniscalaco (1994), 'Environmental Policy Reconsidered: The Role of Technology Innovation', *European Economic Review* **38**, 545–555.

Conrad, K (2000), 'Energy Tax and Competition in Energy Efficiency: The Case of Consumer Durables', *Environmental and Resource Economics* **15**, 159–177.

David, P. (1966), 'The mechanization of reaping in the ante-bellum Midwest', in H. Rosovsky, ed., *Industrialization in Two Systems*. Cambridge, MA: Harvard University Press, pp. 3–39.

David, P. (1969), 'A Contribution to the Theory of Diffusion', mimeo. Forthcoming as Chapters 1–3 of David, P., *Behind the Diffusion Curve*. Boulder: Westview Press.

David, P., B. Hall and A. Toole (2000), 'Is Public R&D a Complement or Substitute for Private R&D? A Review of the Econometric Evidence', *Research Policy* **29**, 497–529.

DeCanio, S. J. (1998), 'The Efficiency Paradox: Bureaucratic and Organizational Barriers to Profitable Energy-Saving Investments', *Energy Policy* **26**, 441–454.

Denicolo, V. (1999), 'Pollution-reducing Innovations Under Taxes or Permits', *Oxford Economic Papers* **51**, 184–199.

Downing, P. B. and L. J. White (1986), 'Innovation in Pollution Control', *Journal of Environmental Economics and Management* **13**, 18–29.

Evenson, R. (1995), 'Technology Change and Technology Strategy', in J. Behrman and T.N. Srinivasan, eds., *Handbook of Development Economics*, Vol. 3A. Amsterdam: North-Holland.

Fischer, C., I. W. H. Parry and W. A. Pizer (1998), 'Instrument Choice for Environmental Protection When Technological Innovation is Endogenous', Resources for the Future Discussion Paper 99-04. Resources for the Future, Washington, D.C.

Freeman, A. M. and R. H. Haveman (1972), 'Clean Rhetoric and Dirty Water', *Public Interest* **28**, 51–65.

Geroski, P. A. (1995), 'Markets for Technology: Knowledge, Innovation and Appropriability', in P. Stoneman, ed., *Handbook of the Economics of Innovation and Technological Change*. Oxford: Blackwell Publishers, pp. 90–131.

Geroski, P. A. (2000), 'Models of Technology Diffusion', *Research Policy* **29**, 603–626.

Goldberg, P. K. (1998), 'The Effects of the Corporate Average Fuel Efficiency Standards in the U.S.', *Journal of Industrial Economics* **46**, 1–3.

Gollop, F. M. and M. J. Roberts (1983), 'Environmental Regulations and Productivity Growth: The Case of Fossil-Fueled Electric Power Generation', *Journal of Political Economy* **91**, 654–674.

Goodman, A. C. (1983), 'Willingness to Pay for Car Efficiency: A Hedonic Price Approach', *Journal of Transport Economics* **17**, 247–266.

Goulder, L. H. and K. Mathai (2000), 'Optimal CO2 Abatement in the Presence of Induced Technological Change', *Journal of Environmental Economics and Management* **39**, 1–38.

Goulder, L. H. and S. L. Schneider (1999), 'Induced Technological Change and the Attractiveness of CO_2 Emissions Abatement Policies', *Resource and Energy Economics* **21**, 211–253.

Gray, W. B. and R. J. Shadbegian (1995), 'Pollution Abatement Costs, Regulation, and Plant-Level Productivity', National Bureau of Economic Research Working Paper No. 4994.

Gray, W. B. and R. J. Shadbegian (1998), 'Environmental Regulation, Investment Timing, and Technology Choice', *Journal of Industrial Economics* **46**, 235–256.

Greene, D. L. (1990), 'CAFE or Price?: An Analysis of the Effects of Federal Fuel Economy Regulations and Gasoline Price on New Car MPG, 1978–89', *The Energy Journal* **11**(3), 37–57.

Greening, L. A., A. H. Sanstad and J. E. McMahon (1997), 'Effects of Appliance Standards on Product Price and Attributes: An Hedonic Pricing Model', *Journal of Regulatory Economics* **11**, 181–194.

Greenstone, M. (1998), 'The Marginal Effects of Environmental Regulations of the Manufacturing Sector: Evidence from the 1970 and 1977 Clean Air Act Amendments', mimeo Princeton University.

Griliches, Z. (1957), 'Hybrid Corn: An Exploration in the Economics of Technical Change', *Econometrica* **48**, 501–522.

Griliches, Z. (1979), 'Issues in Assessing the Contribution of Research and Development to Productivity Growth', *Bell Journal of Economics* **10**, 92–116.

Griliches, Z. (1984), *R&D, Patents and Productivity*. Chicago: University of Chicago Press.

Griliches, Z. (1992), 'The Search for R&D Spillovers', *Scandinavian Journal of Economics* **94**, S29–S47.

Griliches, Z. (1998), *R&D and Productivity, The Econometric Evidence*. Chicago: University of Chicago Press.

Grossman, G. and E. Helpman (1994), 'Endogenous Innovation in the Theory of Growth', *Journal of Economic Perspectives* **8**, 23–44.

Hahn, R. W. and R. N. Stavins (1991), 'Incentive-based Environmental Regulation: a New Era from an Old Idea?', *Ecology Law Quarterly* **18**, 1–42.

Hall, B., A. Jaffe and M. Trajtenberg (2000), 'Market Value and Patent Citations: A First Look', National Bureau of Economic Research Working Paper No. 7741.

Hassett, K. A. and G. E. Metcalf (1995), 'Energy Tax Credits and Residential Conservation Investment: Evidence form Panel Data', *Journal of Public Economics* **57**, 201–217.

Hassett, K. A. and G. E. Metcalf (1996), 'Can Irreversibility Explain the Slow Diffusion of Energy Saving Technologies?', *Energy Policy* **24**, 7–8.

Hausman, J. A. (1979), 'Individual Discount Rates and the Purchase and Utilization of Energy-Using Durables', *Bell Journal of Economics* 10, 33.

Hicks, J. (1932), *The Theory of Wages* London: Macmillan.

Hockenstein, J. B., R. N. Stavins and B. W. Whitehead (1997), 'Creating the Next Generation of Market-Based Environmental Tools', *Environment* **39**(4), 12–20, 30–33.

Holmström, B. R. and J. Tirole (1987), 'The Theory of the Firm', in R. Schmalensee and R. Willig, eds., *Handbook of Industrial Organization*, vol. 1. Amsterdam: North-Holland, pp. 61–133.

Howarth, R. B., B. M. Haddad, and B. Paton (2000), 'The Economics of Energy Efficiency: Insights from Voluntary Participation Programs', *Energy Policy* **28**, 477–486.

Interlaboratory Working Group (1997), *Scenarios of Carbon Reductions: Potential Impacts of Energy Technologies by 2010 and Beyond* Washington: Office of Energy Efficiency and Renewable Technologies, U.S. Department of Energy.

Jaffe, A. B. (1986), 'Technological Opportunity and Spillovers of R&D: Evidence from Firms' Patents, Profits and Market Value', *American Economic Review* **76**, 984–1001.

Jaffe, A. B. (1998), 'The Importance of 'Spillovers' in the Policy Mission of the Advanced Technology Program', *Journal of Technology Transfer*, Summer, 11–19.

Jaffe, A. B. (2000), 'The U.S. Patent System in Transition: Policy Innovation and the Innovation Process', *Research Policy* **29**, 531–558.

Jaffe, A. B. and K. Palmer (1997), 'Environmental Regulation and Innovation: A Panel Data Study', *Review of Economics and Statistics* **79**, 610–619.

Jaffe, A. B. and R. N. Stavins (1994), 'The Energy Paradox and the Diffusion of Conservation Technology', *Resource and Energy Economics* **16**, 91–122.

Jaffe, A. B. and R. N. Stavins (1995), 'Dynamic Incentives of Environmental Regulations: The Effects of Alternative Policy Instruments on Technology Diffusion', *Journal of Environmental Economics & Management* **29**, S43–S63.

Jaffe, A. B., S. Peterson, P. Portney and R. N. Stavins (1995), 'Environmental Regulation and the Competitiveness of U.S. Manufacturing: What Does the Evidence Tell Us?', *Journal of Economic Literature* **33**, 132–163.

Jorgenson, D. (1990), 'Productivity and Economic Growth', in E. Berndt and J. Triplett, eds., *Fifty Years of Economic Measurement*. Chicago: University of Chicago Press, pp. 19–118.

Jorgenson, D. W. and K. J. Stiroh (2000), 'Raising the Speed Limit: U.S. Economic Growth in the Information Age', *Brookings Papers on Economic Activity*, 125–211.

Jung, C. H., K. Krutilla and R. Boyd (1996), 'Incentives for Advanced Pollution Abatement Technology at the Industry Level: An Evaluation of Policy Alternatives', *Journal of Environmental Economics and Management* **30**, 95–111.

Karshenas, M. and P. Stoneman (1995), 'Technological Diffusion', in P. Stoneman, ed., *Handbook of the Economics of Innovation and Technological Change*. Oxford: Blackwell Publishers.

Katsoulacos, Y. and A. Xepapadeas (1996), 'Environmental Innovation, Spillovers and Optimal Policy Rules', in C. Carraro et al., eds., *Environmental Policy and Market Structure*. Dordrecht: Kluwer Academic Publishers, pp. 143–150.

Kemp, René (1997), *Environmental Policy and Technical Change*. Cheltenham, UK: Edward Elgar.

Kemp, René (1998), 'The Diffusion of Biological Waste-water Treatment Plants in the Dutch Food and Beverage Industry', *Environmental and Resource Economics* **12**, 113–136.

Kemp, R. and L. Soete (1990), 'Inside the "Green Box": on the Economics of Technological Change and the Environment', in Freeman, C. and L. Soete, eds., *New Explorations in the Economics of Technological Change*. London: Pinter.

Keohane, N. O. (1999), 'Policy Instruments and the Diffusion of Pollution Abatement Technology', mimeo, Harvard University.

Keohane, N. O. (2001), 'Essays in the Economics of Environmental Policy', unpublished Ph.D. dissertation, Harvard University.

Kerr, S. and R. G. Newell (2000), 'Policy-Induced Technology Adoption: Evidence from the U.S. Lead Phasedown', Resources for the Future Discussion Paper 01-14, (Resources for the Future, Washington, DC).

Kneese, A. and C. Schultze (1975), *Pollution, Prices, and Public Policy*. Washington, D.C.: Brookings Institution.

Kolstad, C. D. and M. H. L. Turnovsky (1998), 'Cost Functions and Nonlinear Prices: Estimating a Technology with Quality-Differentiated Inputs', *Review of Economics & Statistics* **80**, 444–453.

Laffont, J. and J. Tirole (1996), 'Pollution Permits and Compliance Strategies', *Journal of Public Economics* **62**, 85–125.

Landjouw, J. O. and A. Mody (1996), 'Innovation and the International Diffusion of Environmentally Responsive Technology', *Research Policy* **25**, 549–571.

Levin, S., S. Levin and J. Meisel (1987), 'A Dynamic Analysis of the Adoption of a New Technology: The Case of Optical Scanners', *Review of Economics and Statistics* **69**, 12–17.

Magat, W. A. (1978), 'Pollution Control and Technological Advance: A Dynamic Model of the Firm', *Journal of Environmental Economics and Management* **5**, 1–25.

Magat, W. A. (1979), 'The Effects of Environmental Regulation on Innovation', *Law and Contemporary Problems* **43**, 3–25.

Maloney, M. T. and G. L. Brady (1988), 'Capital Turnover and Marketable Pollution Rights', *Journal of Law and Economics* **31**, 203–226.

Malueg, D. A. (1989), 'Emission Credit Trading and the Incentive to Adopt New Pollution Abatement Technology', *Journal of Environmental Economics and Management* **16**, 52–57.

Mansfield, E. (1968), *Industrial Research and Technological Innovation*. New York: Norton.

Mansfield, E. (1989), 'Industrial robots in Japan and the USA', *Research Policy* **18**, 183–192.

Marin, A. (1991), 'Firm Incentives to Promote Technological Change in Pollution Control: Comment', *Journal of Environmental Economics and Management* **21**, 297–300.

Martin, S. and J. Scott (2000), 'The Nature of Innovation Market Failure and the Design of Public Support for Private Innovation', *Research Policy* **29**, 437–448.

McCubbins, M. D., R. G. Noll and B. R. Weingast (1989), 'Structure and Process, Politics and Policy: Administrative Arrangements and the Political Control of Agencies', *Virginia Law Review* **75**, 431–482.

Metcalf, G. E. and K. A. Hassett (1999), 'Measuring the Energy Savings from Home Improvement Investments: Evidence from Monthly Billing Data', *The Review of Economics and Statistics* **81**, 516–528.

Milliman, S. R. and R. Prince (1989), 'Firm Incentives to Promote Technological Change in Pollution Control', *Journal of Environmental Economics and Management* **17**, 247–265.

Milliman, S. R. and R. Prince (1992), 'Firm Incentives to Promote Technological Change in Pollution Control: Reply', *Journal of Environmental Economics and Management* **22**, 292–296.

Montero, J. P. (2000), 'Market Structure and Environmental Innovation', mimeo Catholic University of Chile.

National Science Board (1998), *Science and Engineering Indicators – 1998*. Washington: National Science Foundation, available online at http://www.nsf.gov/sbe/srs/seind98/frames.htm

Nelson, R. and S. Winter (1982), *An Evolutionary Theory of Economic Change* Cambridge, MA: The Belknap Press of Harvard University Press.

Nelson, R., T. Tietenberg and M. Donihue (1993), 'Differential Environmental Regulation: Effects on Electric Utility Capital Turnover and Emissions', *Review of Economics and Statistics* **75**, 368–373.

Newell, R. G., A. B. Jaffe and R. N. Stavins (1999), 'The Induced Innovation Hypothesis and Energy-Saving Technological Change', *The Quarterly Journal of Economics* **114**, 941–975.

Nordhaus, W. D. (2000), 'Modeling Induced Innovation in Climate-Change Policy', mimeo, Yale University.

Oates, W. E., K. Palmer and P. R. Portney (1993), 'Environmental Regulation and International Competitiveness: Thinking About the Porter Hypothesis', Resources for the Future Discussion Paper 94-02, Resources for the Future, Washington, D.C.

Ohta, M. and Z. Griliches (1976), 'Automobile prices revisited: Extensions of the hedonic hypothesis', in N.E. Terleckyj, ed., *Household Production and Consumption*. New York: National Bureau of Economic Research.

Ohta, M. and Z. Griliches (1986), 'Automobile Prices and Quality: Did the Gasoline Price Increases Change Consumer Tastes in the U.S.?', *Journal of Business & Economic Statistics* **4**, 187–198.

Organization for Economic Cooperation and Development (2000), *Basic Science and Technology Statistics: 1999 Edition*. Paris: Organization for Economic Cooperation and Development.

Orr, L. (1976), 'Incentive for Innovation as the Basis for Effluent Charge Strategy', *American Economic Review* **66**, 441–447.

Oster, S. (1982), 'The Diffusion of Innovation Among Steel Firms: The Basic Oxygen Furnace', *The Bell Journal of Economics* **13**, 45–56.

Pakes, A., S. Berry and J. A. Levinsohn (1993), 'Applications and Limitations of Some Recent Advances in Empirical Industrial Organization: Prices Indexes and the Analysis of Environmental Change', *American Economic Review* **83**, 240–246.

Palmer, K., W. E. Oates and P. R. Portney (1995), 'Tightening Environmental Standards: The Benefit-Cost or the No-Cost Paradigm?' *Journal of Economic Perspectives* **9**, 119–132.

Parry, I. W. H. (1998), 'Pollution Regulation and the Efficiency Gains from Technological Innovation', *Journal of Regulatory Economics* **14**, 229–254.

Pindyck, R. (1991), 'Irreversibility, Uncertainty, and Investment', *Journal of Economic Literature* **29**, 1110–1152.

Pizer, W. A., W. Harrington, R. J. Kopp, R. D. Morgenstern and J. Shih (2001), 'Technology Adoption and Aggregate Energy Efficiency', Resources for the Future Discussion Paper 01-21. Resources for the Future, Washington, DC.

Popp, D. (2001a), 'Induced Innovation and Energy Prices', *American Economic Review*, forthcoming.

Popp, D. (2001b), 'The Effect of New Technology on Energy Consumption', *Resource and Energy Economics* **23**(4), 215–239.

Porter, M. E. and C. van der Linde (1995), 'Toward a New Conception of the Environment-Competitiveness Relationship', *Journal of Economic Perspectives* **9**, 97–118.

Purvis, A. and J. Outlaw (1995), 'What We Know About Technological Innovation to Achieve Environmental Compliance: Policy Issues for an Industrializing Animal Agriculture Sector', *American Journal of Agricultural Economics* **77**, 1237–1243.

Reinganum, J. (1989), 'The Timing of Innovation: Research, Development and Diffusion', in R. Schmalensee and R. Willig, eds., *Handbook of Industrial Organization*, vol. 1. Amsterdam: North-Holland, pp. 850–908.

Reinhardt, F. L. (2000), *Down to Earth: Applying Business Principles to Environmental Management*. Boston: Harvard Business School Press.

Reppelin-Hill, V. (1999), 'Trade and Environment: An Empirical Analysis of the Technology Effect in the Steel Industry', *Journal of Environmental Economics and Management* **38**, 283–301.

Requate, T. (1998), 'Incentives to Innovate under Emission Taxes and Tradeable Permits', *European Journal of Political Economy* **14**, 139–165.

Romer, P. M. (1994), 'The Origins of Endogenous Growth', *Journal of Economic Perspectives* **8**, 3–22.

Rose, N. and P. Joskow (1990), 'The Diffusion of New Technologies: Evidence from the Electric Utility Industry', *Rand Journal of Economics* **21**, 354–373.

Ross, M. (1990), 'Capital Budgeting Practices of Twelve Large Manufacturers', in P. Cooley, ed., *Advances in Business Financial Management*. Chicago: Dryden Press, pp. 157–170.

Ruderman, H., M. D. Levine and J. E. McMahon (1987), 'The Behavior of the Market for Energy Efficiency in Residential Appliances Including Heating and Cooling Equipment', *The Energy Journal* **8**, 101–123.

Sanstad, A., C. Blumstein and S. Stoft (1995), 'How High are Option Values in Energy-efficiency Investments', *Energy Policy* **23**, 739–743.

Scherer, F. (1986), *Innovation and Growth, Schumpeterian Perspectives*. Cambridge, MA: The MIT Press.

Scherer, F. M. and D. Harhoff (2000), 'Technology Policy for a World of Skew-Distributed Outcomes', *Research Policy* **29**, 559–566.

Scherer, F., D. Harhoff and J. Kukies (2000), 'Uncertainty and the Size Distribution of Rewards from Technological Innovation', *Journal of Evolutionary Economics* **10**, 175–200.

Schmalensee, R. (1994), 'The Costs of Environmental Protection', in M.B. Kotowski, ed., *Balancing Economic Growth and Environmental Goals*. Washington, D.C.: American Council for Capital Formation, Center for Policy Research, pp. 55–75.

Schumpeter, J. (1942), *Capitalism, Socialism and Democracy*. New York: Harper.

Shama, A. (1983), 'Energy Conservation in U.S. Buildings, Solving the High Potential/Low Adoption Paradox from a Behavioral Perspective', *Energy Policy* **11**, 148–168.

Shrestha, R. M. and B. K. Karmacharya (1998), 'Testing of Barriers to the Adoption of Energy-Efficient Lamps in Nepal', *The Journal of Energy and Development* **23**(1), 71–82.

Simon, H. A. (1947), *Administrative Behavior: A Study of Decision-making Processes in Administrative Organization*. New York: Macmillan Company.

Spence, A. M. (1984), 'Cost Reduction, Competition and Industry Performance', *Econometrica* **52**(1), 101–121.

Stavins, R. N. (2001), 'Experience with Market-Based Environmental Policy Instruments', in K. Mäler and J. Vincent, eds., *Handbook of Environmental Economics*. Amsterdam: Elsevier Science, forthcoming.

Stewart, R. B. (1981), 'Regulation, Innovation, and Administrative Law: A Conceptual Framework', *California Law Review* **69**, 1256–1270.

Stoneman, P. (1983), *The Economic Analysis of Technological Change*. Oxford: Oxford University Press.

Sutton, J. (1998), *Technology and Market Structure*. Cambridge, MA: The MIT Press.

Tirole, J. (1988), *The Theory of Industrial Organization*. Cambridge, MA: The MIT Press.

Ulph, D. (1998), 'Environmental Policy and Technological Innovation', forthcoming in C. Carraro and D. Siniscalaco, eds., *Frontiers of Environmental Economics* Cheltenhsam: Edward Elgar.

Weitzman, M. L. (1974), 'Prices vs. Quantities', *Review of Economic Studies* **41**, 477–491.

Wilcox, J. (1984), 'Automobile Fuel Efficiency: Measurement and Explanation', *Economic Inquiry* **22**, 375–385.

Winter, S. G., Y. M. Kaniovski and G. Dosi (2000), 'Modeling Industrial Dynamics with Innovative Entrants', *Structural Change and Economic Dynamics* **11**, 255–293.

Zerbe, R. O. (1970), 'Theoretical Efficiency in Pollution Control', *Western Economic Journal* **8**, 364–376.

Available online at www.sciencedirect.com

SCIENCE @ DIRECT®

ELSEVIER

Ecological Economics 54 (2005) 164–174

ECOLOGICAL ECONOMICS

www.elsevier.com/locate/ecolecon

ANALYSIS

A tale of two market failures: Technology and environmental policy

Adam B. Jaffe[a], Richard G. Newell[b,*], Robert N. Stavins[c,b]

[a]*Brandeis University and National Bureau of Economic Research, United States*
[b]*Resources for the Future, 1616 P St NW, Washington, DC 20036, United States*
[c]*Harvard University, United States*

Received 1 May 2004; accepted 1 December 2004
Available online 9 March 2005

Abstract

Market failures associated with environmental pollution interact with market failures associated with the innovation and diffusion of new technologies. These combined market failures provide a strong rationale for a portfolio of public policies that foster emissions reduction as well as the development and adoption of environmentally beneficial technology. Both theory and empirical evidence suggest that the rate and direction of technological advance is influenced by market and regulatory incentives, and can be cost-effectively harnessed through the use of economic-incentive based policy. In the presence of weak or nonexistent environmental policies, investments in the development and diffusion of new environmentally beneficial technologies are very likely to be less than would be socially desirable. Positive knowledge and adoption spillovers and information problems can further weaken innovation incentives. While environmental technology policy is fraught with difficulties, a long-term view suggests a strategy of experimenting with policy approaches and systematically evaluating their success.

Keywords: Technology; Research and development; Environment; Externality; Policy

JEL classification: O38; Q28; H23

1. Introduction

The influence of technological development on energy and environmental systems permeates discussions of energy and environmental policy. New technology has been credited with solving environmental problems by mitigating the effects of pollutants, and has been maligned as a source of increased pollution. For modeling long-term environmental problems such as global climate changes, the effects of technological change compounded over long time horizons will likely be large. Thus, the single largest source of difference among modelers' predictions of the cost of climate policy is often differences in assumptions about the future rate and

* Corresponding author. Tel.: +1 202 328 5111.
E-mail address: newell@rff.org (R.G. Newell).

0921-8009/$ - see front matter
doi:10.1016/j.ecolecon.2004.12.027

direction of technological change (Clarke and Weyant, 2002; Carraro et al., 2003; Energy Modeling Forum, 1996).

But technological change does not exist in a vacuum. Environmental policy interventions, such as carbon cap and trade systems and carbon taxes, generate incentives that will affect which new technologies will be developed and how rapidly and deeply they will diffuse. The induced effects of environmental policy on technology can therefore have substantial implications for the normative analysis of policy. While researchers dispute the extent to which environmental policy-induced technological change reduces the social cost of environmental compliance, there is little dispute among economists that flexible, incentive-oriented policy approaches are more likely to foster low-cost compliance paths than prescriptive regulatory approaches.[1]

The realization that the process of technological change is itself characterized by market failures complicates policy analysis, and increases the likelihood that a portfolio of policies, rather than policy directed at emissions reduction alone, will offer a more complete response to environmental problems. The seeming intractability of some energy and environmental problems, such as global climate change, combined with considerable uncertainties and the long time frame over which their ultimate consequences will play out, may make the development and deployment of new technologies attractive as a major policy response. That is, policies whose purpose is generating technological change are likely to be important parts of the policy portfolio for addressing certain environmental problems, in addition to the rules and regulations we normally think of as environmental policies. Technology policy can be a costly approach, however, if it is used as a substitute for, rather than complement, to environmental policy. Environmental policy targeted directly at emissions (for example through an emissions tax or cap-and-trade system) will still typically provide the most important single element of a cost-effective environmental policy strategy.

[1] For a detailed survey of the influence of environmental policies on innovation and diffusion see Jaffe et al. (2003).

This paper provides background for consideration of these issues. We begin by discussing the key analytic issues that permeate policy discussions occurring at the nexus between technology and environmental policy. Section 3 discusses the possibilities for policies designed to operate directly on technology to improve our ability to cope with environmental problems. We offer concluding observations in Section 4.

2. Key analytical issues

2.1. Fundamentals of environmental economics

Economic analysis of environmental policy is based on the idea that the potentially harmful consequences of economic activities on the environment constitute an "externality," an economically significant effect of an activity, the consequences of which are borne (at least in part) by a party or parties other than the party that controls the externality-producing activity. A factory that pollutes the air, water, or land imposes a cost on society. The firm that owns the factory has an economic incentive to use only as much labor or steel as it can productively employ, because those inputs are costly to the firm. The cost to society of having some of its labor and steel used up in a given factory is "internalized" by the firm, because it has to pay for those inputs. But the firm does not have an economic incentive to minimize the "external" costs of pollution.

Environmental policies attempt to equalize this imbalance by raising the incentive for a firm to minimize these externalities. Policy choices accomplish this in one of two general ways—either by internalizing the environmental costs so polluters make their own decisions regarding their consumption of environmental inputs, or by imposing a limit on the level of environmental pollution.

The cost of environmental policies could be in the form of decreased output of desired products (for example, a scrubber on an electric power plant reduces its electricity production from a given quantity of fuel), increased use of other variable inputs (for example, eliminating certain gases from the waste stream in a smokestack may require more fuel to be burned), purchase of specialized pollution control equipment (for example, catalytic converters on automobiles), or substitution of inferior or more expensive products or production methods to avoid pollution-causing products or methods (for example, less effective pesticides used when DDT was banned).

In the short run, setting an efficient environmental policy requires a comparison of the marginal cost of reducing pollution with the marginal benefit of a cleaner environment. All else being equal, emissions of pollutants that are very harmful should be greatly restricted, because the pollutants otherwise produce large marginal costs to society. But, all else being equal, emissions of pollutants that are very costly to eliminate should be tolerated, because the marginal cost of reducing them is high.

When technology enters the equation, the terms of the tradeoff between the marginal cost of pollution control and its marginal social benefit are altered. In particular, technology innovations–such as new pollution control equipment, cleaner production methods, or new substitutes for environmentally harmful products–typically reduce the marginal cost of achieving a given unit of pollution reduction. This means that a specified level of environmental cleanup can be achieved at lower total cost to society, and it also means that a lower total level of pollution can be attained more efficiently than would be expected if the cost of cleanup were higher. Thus, in this simple static picture, technology improvements can be good for the environment and good for the firm that must meet environmental mandates.

2.2. Fundamentals of the economics of technology

In this simple analytic scenario, the technology innovation results in greater overall social benefit because the cost of reducing pollution has decreased and environmental health has improved. If this were the end of this static story, than the only effect would be to convert the analysis of environmental policy from a static cost/benefit tradeoff to a dynamic one. Policies to reduce pollution have two effects, however—they reduce pollution today, and they also typically change the incentives that firms face with regard to investing resources in developing new technology for the future. In particular, when firms face an incentive to reduce their emissions, this simultaneously creates an incentive for them to find ways to reduce pollution at lower cost. The fact that the development of such technology will, over time, change the pollution benefit/cost calculus means that choosing efficient environmental policy requires an analysis of this dynamic interaction. The simple static model does not take into account the fact that new technology is itself not free.

To reach the point where pollution is being reduced or some other benefit is realized, two things must happen, both of which require the investment of resources. The first step–innovation–involves scientific or engineering research to establish a new technical idea and to develop that idea into a commercial product or process.[2] The second step–adoption (or diffusion)–is the process by which a new product or process gradually replaces older technology throughout many firms and applications. Adoption is also costly, because firms must learn about new technology, purchase new equipment, and adapt it to their particular circumstances. If technological change is not free, can we expect Adam Smith's "invisible hand" to choose the right level of investment in both innovation and diffusion of new technology?

The problem compounds, because independent of the externality associated with pollution, innovation and diffusion are both characterized by externalities as well as other market failures.

2.2.1. Knowledge externalities

In the case of pollution as an externality, the polluter reaps the benefits derived from polluting while imposing the pollution costs on others. The polluter therefore lacks an incentive to reduce those costs. However, in the case of technology, the problem is reversed. A firm that invests in or implements a new technology typically creates benefits for others while incurring all the costs. The firm therefore lacks the incentive to increase those benefits by investing in technology. Pollution creates a negative externality, and so the invisible hand allows too much of it. Technology creates

[2] Schumpeter (1942) identified three steps in technological change: invention, innovation and diffusion. In the Schumpeterian trichotomy, invention is the first technical development, and innovation the first commercial introduction. For simplicity, we have collapsed these two steps into one and labeled it innovation.

positive externalities, and so the invisible hand produces too little of it.[3]

The positive externality of innovation comes from the public–good nature of new knowledge—innovating firms cannot keep other firms from also benefiting from their new knowledge and therefore cannot capture for themselves all the benefits of the innovation. In addition, the process of competition will typically drive a firm to sell a new device at a price that captures only a portion of its full value, which means that consumers also reap some of the benefits from new technology. While patents and other institutions are employed to protect firms' investments in innovation, such protection is inherently imperfect. A successful innovator will capture some rewards, but those rewards will always be only a fraction–and sometimes a very small fraction–of the overall benefits to society of the innovation. Hence innovation creates positive externalities in the form of "knowledge spillovers" for other firms, and spillovers of value or consumer surplus for the users of the new technology.

2.2.2. Adoption externalities

The environmental and knowledge externalities discussed above have long been at the center of economic debates about technology policy. More recently, we have come to understand some additional market failures that may operate in the adoption and diffusion of new technology. For a number of reasons, the cost or value of a new technology to one user may depend on how many other users have adopted the technology. In general, users will be better off the more other people use the same technology. This benefit associated with the overall scale of technology adoption has sometimes been referred to as "dynamic increasing returns."

Dynamic increasing returns can be generated by learning-by-using, learning-by-doing, or network externalities.[4] While the image of the world beating a path to the door of the successful innovator may seem compelling, the diffusion of a new technology is typically gradual. It takes time for potential users to learn of the new technology, try it, adapt it to their circumstances, and become convinced of its superiority. An important mechanism in this learning process is the observation of the adoption of the new technology by others. Hence the adopter of a new technology creates a positive externality for others in the form of the generation of information about the existence, characteristics, and success of the new technology. This phenomenon is often called "learning-by-using."

The supply-side counterpart, "learning-by-doing," describes how production costs tend to fall as manufacturers gain production experience. If this learning spills over to benefit other manufacturers without compensation it can represent an additional adoption externality. Finally, network externalities exist if a product becomes technologically more valuable to an individual user as other users adopt a compatible product (as with telephone and computer networks, for example). These phenomena can be critical to understanding the existing technological system, forecasting how that system might evolve, and predicting the potential effect of some policy or event.

2.2.3. Incomplete information

Both innovation and diffusion of new technology are characterized by additional market failures related to incomplete information. While all investment is characterized by uncertainty, the uncertainty associated with the returns to investment in innovation is often particularly large. Further, information about the prospects for success of given technology research investments is asymmetric, in the sense that the developer of the technology is in a better position to assess its potential than outsiders. A firm attempting to raise investment capital to fund the development of new technology will therefore find such investors skeptical about promised returns, and likely to

[3] There is, however, an offsetting negative externality because R&D is a fixed cost that must, in equilibrium, be financed by the stream of quasi-rents it produces. The entry of another R&D competitor, or an increase in the R&D investment level of a competitor, reduces the expected quasi-rents earned by other R&D firms. This "rent-stealing" effect (Mankiw and Whinston, 1986) could, as a theoretical matter, lead to over-investment in R&D. The empirical evidence suggests, however, that positive externalities associated with knowledge spillovers dominate the rent-stealing effect, leading to social rates of return to R&D substantially in excess of the private rates of return (Griliches, 1992).

[4] See Jaffe et al. (2003, pp. 491–494) for a review of the literature on dynamic increasing returns.

demand a premium for investment that carries such risks. This likely imperfection in the market for capital for funding technology development exacerbates the "spillover" problem and therefore contributes to our expectation that the invisible hand encourages too little research and development.

In the context of environmental problems such as climate change, the huge uncertainties surrounding the future impacts of climate change, the magnitude of the policy response, and thus the likely returns to R&D investment, would seem to exacerbate this problem further. In the extreme, for example, it is difficult to see how the technological solutions that would be required to address the possibility of catastrophic effects of climate change would be provided for by the market even if environmental policies sent appropriate signals about expected costs. In this sense, there may be considerable option value to the development of certain environmental technologies that would be difficult to capture solely through emissions policy.

With respect to technology adoption and diffusion, we have already noted that imperfect information can slow the diffusion of new technology. First, information has important "public good" attributes: once created it can be used by many people at little or no additional cost. It may be difficult or impossible for an individual or firm that invests in information creation to prevent others who do not pay for the information from using it. It is well known that such public goods will tend to be underprovided by ordinary market activity. Incomplete information can also foster principal–agent problems, as when a builder or landlord chooses the level of investment in energy efficiency in a building, but the energy bills are paid by a later purchaser or a tenant. If the purchaser has incomplete information about the magnitude of the resulting energy savings, the builder or landlord may not be able to recover the cost of such investments, and hence might not undertake them. These market failures with respect to adoption of new technology are part of the explanation for the apparent "paradox" of underinvestment in energy-saving technologies that appear cost-effective but are not widely utilized (Jaffe and Stavins, 1994).

Thus the interplay of technology and the environment involves the interaction of two analytically distinct but linked sets of market failures. The consequences of this interaction can be complex. The fact that markets underinvest in new technology strengthens the case for making sure that environmental policy is designed to foster, rather than inhibit innovation. It may mean that the social cost of environmental policy is less than it would otherwise appear, because part of the cost is in the form of investments in innovation that yield positive externalities outside the environmental arena. Whether this is true or not depends on, among other things, whether the increased investment in environmental innovation brought forth by environmental policy comes at the expense of innovation in other areas. If it does, the net effect on the costs of environmental policy will depend on the relative spillovers of environmental innovation compared to innovation that is displaced.[5] In practice, it may be difficult to sort out all of these effects, and very difficult to do so with quantitative reliability.

Thus, technological change is important for environmental policy, and analysis of energy and environmental policy can benefit from the perspective of the economics of technological change. Our general approach is to view technological change relative to the environment as occurring at the nexus of two distinct and important market failures: pollution represents a negative externality, and new technology generates positive externalities. Hence, in the absence of public policy, new technology for pollution reduction is, from an analytical perspective, doubly underprovided by markets. This suggests that the efficiency of environmental policy depends on its consequences for technological change, and also that there is a potential role for policy aimed directly at the stimulation of environmentally beneficial technological change.

3. Environmental technology policy

Given that the development of environmentally beneficial technology is subject to two interacting market failures, in cases where environmental externalities have not been fully internalized it is likely that the rate of investment in such technology is below the

[5] See, for example, Goulder and Schneider (1999) on the importance of spillover assumptions for the cost of climate mitigation.

socially optimal level. And it is unlikely that environmental policy alone creates sufficient incentives. Hence the optimal set of public policies likely also includes instruments designed explicitly to foster innovation and possibly technology diffusion, as distinct from environmental policies that stimulate new technology as a side effect of internalizing environmental externalities.

Of course, one way to foster environmental technology is to foster technology in general, and allow the market to determine what portion of the stimulated development will be in the environment area. The arguments for generally greater public investment in technology infrastructure are well known and have been the subject of numerous studies. We focus instead on the potential for policies aimed explicitly at the development and diffusion of environmentally benign and/or energy-saving technology.

There is a strong strain in the economic analysis of technology policy of *avoiding* choosing particular technical areas for support, that is "picking winners." There are, however, several interrelated reasons why technology policy narrowly focused on energy and environment is likely to be socially desirable under certain circumstances. First is the public good nature of the environment itself, which makes environment, in effect, an area of government procurement like defense and space, and hence a suitable area for focused governmental technology efforts.

Another is a second-best argument related to the practical limitations of environmental policy. Most economists, present authors included, would argue that the most efficient single policy for addressing global climate change is an emissions policy that places a price on greenhouse gases (for example, through an emissions tax or cap-and-trade system). However, in the area of global climate change–arguably the most significant long-run environmental threat–the United States and much of the world has largely put off, for the moment, significant environmental policy intervention. Hence there is little environmental policy-induced incentive to develop technologies that reduce greenhouse gas emissions. In this second-best setting, policy to foster greenhouse-gas-reducing technology may be one of the main policy levers available and can be justified on economic grounds so long as it has positive net benefits. Technology policy can be a costly approach, however, if it is used as a substitute for, rather than complement, to environmental policy.[6]

One might reasonably ask, however, why governments who are unwilling to impose costs on the economy to reduce greenhouse gas emissions directly should be more willing to invest resources in improving energy technologies to reduce greenhouse gases? One reason is purely political: policies subsidizing technology do in fact receive considerable political support in most countries (see, for example, Section 3.3). An explanation is that the benefits of such policies tend to be focused and the costs dispersed, giving rise to favorable political–economic conditions.

Another possible explanation may be found in the limitations of feasible environmental policies to address the dynamic nature of the problem.[7] If the social costs of climate change are considered modest at the present but are expected to rise considerably in the future, this may warrant current investment in R&D at the same time as it explains tempered interest in current mitigation. In principle, governments could announce a time path for future environmental policies that might induce the appropriate level of R&D investment in anticipation of future emissions policy. But there are many reasons, both practical and theoretical, why such advance policy commitments are unlikely to be forthcoming, and why they may not represent credible commitments if they were announced. This difficulty of setting appropriate dynamic environmental policies may warrant more reliance on technology policy, to which governments can commit now.

In the context of environmental problems having important international dimensions, such as global climate change, there can be additional reasons related to international cooperation for considering environmental and technology policies simultaneously. The nature of global environmental problems, technological diffusion, and international trade can provide

[6] See, for example, Fischer and Newell (2004) on the increasing cost of using certain focused technology policies to achieve greenhouse gas reduction goals.

[7] We thank an anonymous referee for emphasizing these additional features of the problem.

arguments for issues linkage where more countries may participate and comply with international agreements on environmental policy and technology policy if they are linked than if they are treated separately (see Folmer and van Mouche, 1993; Cesar and de Zeeuw, 1996; Carraro and Egenhofer, 2002; Kemfert, 2004).

3.1. Innovation policies

Policies that internalize the cost of environmental harm stimulate the creation of environment-friendly technology by increasing the demand for low-cost pollution-reduction methods. Thus, so-called "demand-pull" increases the return to developing such technologies. The spillover problem implies that firms can expect to capture only a portion of that return, but a portion of a large return is still more of an incentive than a portion of a small return. Government can also stimulate innovation through the supply side, either by making it less expensive for firms to undertake research in this area, or by performing the research in public institutions.

Where research produces potentially large social benefits, but is so prone to the spillover problem that firms will not view it as profitable, there is an analytical basis for performing that research in the public sector or through direct private research contracts. In the United States, for example, there is a long tradition of performing such basic research at the U.S. National Energy Laboratories. The National Labs, such as Lawrence Berkeley, Brookhaven, Oak Ridge, Sandia, Lawrence Livermore, and the National Energy Technology Lab are owned by the U.S. Department of Energy (DOE), but operated by either a private firm or a university. Most of the research on energy and environment that is performed at these facilities is funded by the DOE and other U.S. federal agencies.

So long as firms see some potential for private return, public policy can counter-balance the spillover problem by subsidizing research in the private sector rather than performing it in the public sector. The advantage of this approach is that private firms may have better information than the government about the likely commercial feasibility of technologies, and hence be more successful at choosing which technologies to pursue. Subsidies can take the form of fairly general tax credits, or matching funds provided to firms for specific research proposals. In areas where the public research institutions have specific expertise, joint industry–government research can be undertaken via vehicles such as Cooperative Research and Development Agreements. Finally, because the supply of appropriately trained scientists and engineers is relatively inelastic in the short run, there is a danger that any increased expenditure on research in a given area will be at least partly consumed by an increase in wages (Goolsbee, 1998), rather than going to more research effort. This tendency can be offset if subsidies to or expenditures on research are complemented by subsidies for education and training in the appropriate areas.[8]

It is generally the case with science and technology programs that systematic assessment efforts are woefully lacking. Because success is uncertain and difficult to measure, most agencies engaged in support of research and technology adoption have resisted efforts to measure their output against quantitative benchmarks, as is required in the United States by the Government Performance and Results Act (Jaffe, 1998, 2002). Granted, such evaluation is very difficult, and there is a real danger that imperfect assessment methods will distort policy by encouraging efforts that "look good" on evaluation, even if such efforts are not ideally suited to the program's mission. But continuous, systematic, quantitative assessment is the only way that the relative effectiveness of alternative policy approaches can be compared over time. In particular, collecting information in a standardized way as projects are begun, implemented, and terminated is the only way to amass the data necessary for a rigorous retrospective analysis.

Retrospective assessment must of course account for the considerable uncertainties that exist when projects are undertaken, so that projects that turn out to have low value ex post may still have been good investments given ex ante possibilities. Consideration of such uncertainties may be particularly problematic for problems such as climate mitigation, where timeframes are especially long, and considerable uncertainties exist with respect to environmental benefits and with respect to baseline energy market

[8] For a general discussion of support for training and education as a complement to research subsidies, see Romer (2000).

conditions (for example, natural gas prices). Such considerations underscore the need to go beyond simplistic ex post rate-of-return analyses that ignore ex ante information sets, changing conditions, and option value. It also reinforces the value of taking a portfolio approach to R&D investments, whereby investments are made in a set of projects that are likely to pay off under differing future conditions.

The analyses that have been conducted of U.S. federal research relating to energy and the environment have come to mixed conclusions. Cohen and Noll (1991) documented the monumental waste represented by the breeder reactor and synthetic fuel programs in the 1970s, but in the same volume Pegram (1991) concluded that the photovoltaics research program undertaken in the same time frame had significant benefits. More recently, the U.S. National Research Council attempted a fairly comprehensive overview of energy efficiency and fossil energy research at DOE over the last two decades (National Research Council, 2001). Using both estimates of overall return and case studies, they concluded, as one might expect, that there were only a handful of programs that proved highly valuable. Their estimates of returns suggest, however, that the benefits of these successes justified the overall portfolio investment.

Perhaps more important than the question of the overall rate of return is what distinguishes the successful programs from the failures. While the small numbers and inherent randomness make it difficult to draw strong conclusions, it seems that the successful programs are ones in which significant participation by industry–in the form of many firms or consortia rather than individual contractors–helped to ensure that the photovoltaics, building energy efficiency, and advanced engine programs produced outputs that are actually or potentially of real commercial value.

3.2. Adoption policies

There is a long history of public support for research in the United States and other industrialized countries. There has been less policy consensus regarding the desirability of using public policy to speed the adoption of new technology. Because of the positive information externality associated with technology adoption, there is a valid analytical basis for considering such policy. Further, if learning curves or other sources of dynamic increasing returns are important, there could be social benefits associated with speeding diffusion of new energy-saving or otherwise environmentally beneficial technologies.

On the other hand, the possibility of technology "lock-in" makes this a potentially two-edged sword. If the government encourages the diffusion of a particular technology, it is possible that it could become so entrenched in the market place that it stifles, at least for a time, the development of some other, superior technology. This danger creates a tension in the design of policies to encourage adoption. To maximize the exploitation of dynamic increasing returns, it is desirable to focus on the development of a small number of promising technologies. Yet to avoid accidentally helping to entrench the wrong technology, it is desirable for policy to be "technology neutral," encouraging all efforts that achieve specified objectives *without* focusing on a particular approach.

Given limited public resources, the government clearly cannot subsidize all new technologies, so there is a need to focus scarce resources on commercialization opportunities for which there is the clearest need for a public role. As stated earlier, this case will be more compelling the lower are the private incentives for adoption, as in the case of environmental problems that have not otherwise been fully priced into private decisions.

As with research, the government can encourage adoption both in its own operations and by subsidizing the efforts of others. As the government is a very large landlord, vehicle operator, and user of many other kinds of equipment, its decision to purchase certain technologies for its own use can have significant effects on the rate of diffusion.

Technology diffusion, and achievement of any associated benefits of dynamic increasing returns, can also be encouraged with tax credits that reduce the effective purchase price of new equipment that meets specified criteria. There is some literature analyzing the effectiveness of energy conservation tax credits at inducing conservation investment. The empirical evidence from this literature is mixed, with some early studies suggesting that tax credits are a very ineffective policy (for example, Dubin and Henson, 1988; Walsh, 1989), while some later evidence points to some effectiveness (for example, Hassett and Metcalf, 1995). Nonetheless, it is important to recognize some disadvantages of subsidy approaches. First, unlike policies that raise the price of emissions, adoption subsidies do not provide incentives to reduce utilization of polluting technology (Fischer and New-eell, 2004). Second, technology subsidies and tax credits can require large public expenditures per unit of effect, since consumers who would have purchased the product even in the absence of the subsidy (i.e., free-riders) still receive it. This free-rider effect is likely to be more of a problem for technologies that have already penetrated to a significant extent, and less so for very new technologies that are expensive compared with substitutes.

Since a major aspect of market failure in technology diffusion is imperfect information, another category of policy to encourage diffusion is information provision. With respect to technologies that appear cost-effective, but are not yet widely utilized, this kind of policy can help overcome the apparent market failure without putting the government in the position of betting on particular technologies. While analysis is limited, there is some evidence of success for information programs (Anderson and Newell, 2004). For example, the DOE provides free energy audits to small and medium-sized companies using university-based engineering teams that recommend energy-saving projects that appear to be desirable. DOE has maintained an extensive database on the technology costs, projected energy savings, and which recommendations were adopted. Overall, about 50% of recommended projects are adopted. These programs are relatively inexpensive, and so are probably earning a reasonable social return. But firms' decisions to adopt only 50% of the projects recommended by engineering experts suggest that imperfect information is not the only reason for non-adoption; rate-of-return requirements and other economic considerations remain.

The early years of utility demand side management in the 1970s, which emphasized information and low-interest loans, also demonstrated to utilities that education alone produced limited energy savings. Thus utilities were led to consider programs that contained stronger financial incentives to convince consumers to make energy-saving choices (Gillingham et al., 2004).

Finally, command and control regulations can also be used to try to force the diffusion of particular technologies, often by removing less expensive and less environmentally beneficial competing technologies from the market. The Corporate Average Fuel Economy ("CAFE") standards have been designed to force an improvement in auto efficiency.[9] Energy efficiency standards have also been implemented for major home appliances. Such standards can in principle be beneficial by conserving on the need for every individual to undertake the information and assessment process inherent in trading off capital and energy operating costs. However, they also raise the risk of going beyond an economically justified minimum, at which point they can impose limits to product choice and undesirable costs on what is a very heterogeneous population of adopters.

3.3. Current U.S. climate technology policy efforts

As an illustration of the range of technology policy initiatives related to energy and environment, Table 1 summarizes current U.S. Federal climate change initiatives. Based on the 2004 fiscal year budget request (OMB, 2003), about $1.3 billion dollars will be spent on research, and a similar amount on technology diffusion. On the research side, about 40% relates to energy conservation, one-third for sources of renewable energy, and most of the remainder going to "clean coal" and other forms of carbon reduction from fossil fuels. On the diffusion side, a little over half is proposed to go toward tax credits for renewable energy production, cogeneration, hybrid/fuel cell autos, landfill gas production, and solar homes. About one-third of the money goes for state energy efficiency grants, and 9% for EPA information and voluntary initiatives, such as the "Green Lights" program.

While the level of proposed climate technology R&D funding is very close to past funding levels (for example, $1.4 billion in 2002), the proposed funding for adoption incentives reflects $288 million in proposed new tax incentives for hybrid/fuel

[9] For an assessment of the cost-effectiveness of the CAFE standards see National Research Council (2002) and the research it reviews.

Table 1
Overview of U.S. federal climate technology R&D and adoption initiatives (proposed for FY04)

Climate-related technology R&D spending ($1.3 billion/year)	
Energy conservation	41%
Renewables supply	33%
Fossil fuel GHG reductions	14%
Carbon capture and sequestration	8%
Nuclear	2%
EPA science and technology	1%
Forest, range, agriculture	1%
Energy Information Admin.	<1%
Climate-related technology adoption spending ($1.0 billion/year)	
State energy efficiency grants	34%
Renewables production tax incentives	27%
Hybrid/fuel cell car tax credits	16%
Cogen. production tax incentives	10%
EPA information/voluntary programs	9%
Landfill gas production tax credit	3%
Solar homes tax credits	1%

cell cars, cogeneration of electricity, landfill gas, residential solar, and expansion of the existing renewables tax credit to include certain forest-related resources, agricultural, and other sources. While this paper has argued that there may be sound economic reasons for pursuing some of these programs, the net benefits of individual initiatives remain an open question and is the subject of more detailed analysis.

4. Concluding remarks

When economists evaluate public policies that intervene in the market economy, they generally view it from the analytical perspective of market failure. When it comes to green technology, two mutually reinforcing sets of market failures are at work—which decrease the likelihood that the rate of investment in the development and diffusion of such technology would occur at the socially optimal level. The solutions fall into two categories of approaches. One approach is to foster the development and diffusion of new technology by designing environmental policies to increase the perceived market payoff and maximize flexibility in compliance. The other approach is to implement policies aimed directly at encouraging the development and diffusion of environmentally friendly technologies. Theory suggests and empirical research confirms that innovation and technology diffusion do respond to the incentives of the market, and that properly designed regulation can create such incentives.

The double market failure further clarifies the case for broad-based public support of technology innovation and diffusion. And for cases in which private incentives do not reflect the full costs of environmental externalities, for whatever reason, the efficiency of the policy mix will likely be improved by including public policies aimed directly at stimulating the development and diffusion of new environmentally benign technology. This argument is particularly strong with respect to those aspects of technology development that are most subject to market failure in the form of difficulty by private firms in appropriating the returns to innovation and adoption. Technology "infrastructure" such as data collection and dissemination, and training of scientists and engineers is likely to be seriously underprovided by market incentives alone.

Technology policy that goes beyond basic scientific research, toward the development and diffusion of specific technologies is politically controversial. There are good reasons for this controversy, including the question of whether the government is the appropriate arbiter for determining which aspects of technology should be supported, as well as concern over the effect of political momentum forming behind ill-advised initiatives, which then became difficult to stop. But problems such as global climate change are too important–and the potential positive technological externalities are too clear–to abandon policy efforts simply because they are difficult. Government must remain engaged in technology policy, but it should try a variety of ways to structure policy in this area to minimize the known policy problems. Models are already working, such as public–private partnerships that subsidize research but retain significant elements of market forces in determining which technologies to pursue. Failure of some policy initiatives should be expected, and those failures should be used to terminate or improve particular programs, not to rationalize total inaction.

Policy experimentation would logically work hand-in-hand with systematic policy evaluation. On the ground, however, policy success is very difficult to measure, because the output or effect is often intangible, the expected benefits of technologies change with changing conditions, and the evaluation period must take place over a long time period. This leads some advocates of public investment in technology to resist quantitative evaluation of technology programs on the grounds that measurements of such intangible outputs will understate the benefits and hence undermine political support for such programs. The danger of not even attempting to evaluate policies is that we perpetuate our ignorance in solving the problem, and thereby consign technology policy forever to the realm of ideology. Rather, we should embrace the fact that technological change is a long-term process, and we ought to be willing to take a long-term view. We should remain hopeful that on the time scale of years and decades, systematic evaluation will eventually allow the creation of a solid empirical base for the design of technology policy to maximize its social returns.

References

Anderson, S.T., Newell, R.G., 2004. Information programs for technology adoption: the case of energy-efficiency audits. Resource and Energy Economics 26 (1), 27–50.

Carraro, C., Egenhofer, C., 2002. Firms, governments, and climate policy: incentive-based policies for long-term climate change. Edward Elgar, Cheltenham.

Carraro, C., van der Zwaan, B.C.C., Gerlagh, R., 2003. Endogenous technological change in economy–environment modeling, an introduction. Resource and Energy Economics 25, 1–10.

Cesar, H., de Zeeuw, A., 1996. Issue linkage in global environmental problems. In: Xepapadeas, A. (Ed.), Economic Policy for the Environment and Natural Resources (197–216). Edward Elgar, Cheltenham.

Cohen, Linda R., Noll, R.G., 1991. The technology pork barrel. Brookings, Washington, DC.

Clarke, Leon, Weyant, John, 2002. Modeling induced technological change: an overview. In: Grübler, A., Nakicenovic, N., Nordhaus, W. (Eds.), Technological Change and the Environment. Resources for the Future Press, Washington, DC.

Dubin, J.A., Henson, S.D., 1988. The distributional effects of the federal energy tax act. Resources and Energy 10, 191–212.

Energy Modeling Forum, Stanford University, 1996. Markets for energy efficiency, EMF report 13 vol. I.

Fischer, Carolyn, and Newell Richard, 2004. "Environmental and Technology Policies for Climate Change," RFF Discussion Paper 04-05. Washington, DC: Resources for the Future.

Folmer, H., van Mouche, P., 1993. Interconnected games and international environmental problems. Environmental and Resource Economics 3, 313–335.

Gillingham, K., Wewell, R.G., Palmer, K., 2004. Retrospective Examination of Demand-Side Energy Efficiency Policies. RFF Discussion Paper 04-19rev. Washington, DC: Resources for the Future.

Goolsbee, A., 1998. Does government R&D policy mainly benefit scientists and engineers? American Economic Review 88, 298–302.

Goulder, Lawrence H., Schneider, Stephen, H., 1999. Induced technological change and the attractiveness of CO2 emissions abatement. Resource and Energy Economics 21, 211–253.

Griliches, Z., 1992. The search for R&D spillovers. Scandinavian Journal of Economics 94, S29–S47.

Hassett, K.A., Metcalf, G.E., 1995. Energy tax credits and residential conservation investment: evidence form panel data. Journal of Public Economics 57, 201–217.

Jaffe, Adam B., 1998. Measurement issues. In: Branscomb, L.M., Keller, J. (Eds.), Investing in Innovation. MIT Press, Cambridge.

Jaffe, Adam B., 2002. Building programme evaluation into the design of public research-support programmes. Oxford Review of Economic Policy 18, 22–34.

Jaffe, A.B., Stavins, R.N., 1994. The energy paradox and the diffusion of conservation technology. Resource and Energy Economics 16, 91–122.

Jaffe, Adam B., Newell, Richard G., Stavins, Robert, N., 2003. Technological change and the environment Chapter 11. In: Mäler, K.-G., Vincent, J.R. (Eds.), Handbook of Environmental Economics, vol. 1. Elsevier Science, Amsterdam, pp. 461–516.

Kemfert, C., 2004. International climate coalitions and trade: assessment of cooperation incentives by issue linkage. Energy Policy 32, 455–465.

Mankiw, N.G., Whinston, M.D., 1986. Free entry and social inefficiency. Rand Journal of Economics 17, 48–58.

National Research Council, 2001. Energy research at DOE: was it worth it? National Academy Press.

National Research Council, 2002. Effectiveness and impact of Corporate Average Fuel Economy (CAFE) standards. National Academy Press.

OMB, 2003. Federal climate change expenditures report to congress. Executive Office of the President, Washington, DC.

Pegram, William M., "The Photovoltaics Commercialization Program," in Cohen and Noll, op cit, 1991.

Romer, P., 2000. Should the government subsidize supply or demand for scientists and engineers. In: Jaffe, A., et al., (Eds.), Innovation Policy and the Economy. MIT Press, Cambridge, MA.

Schumpeter, J., 1942. Capitalism, socialism and democracy. Harper, New York.

Walsh, Michael J., 1989. Energy tax credits and housing improvement. Energy Economics 11 (2), 274–284.

[14]

The Effects of Environmental Regulation on Technology Diffusion: The Case of Chlorine Manufacturing

By LORI D. SNYDER, NOLAN H. MILLER, AND ROBERT N. STAVINS*

In order to assess the effects of regulation on the environmental and economic performance of firms, it is important to understand the interaction between regulation and technological change. The costs of compliance with environmental regulations are determined, in part, by the cost and availability of alternative production and abatement technologies, and regulations themselves can affect the nature and rate of technological change (Adam B. Jaffe et al., 2003). As a result, the economic and environmental impacts of regulation are affected by the technology choices made by individual plants.

We study the effect of regulation on technological change in the chlorine manufacturing industry by focusing on the diffusion of membrane-cell technology, widely viewed as environmentally superior to both mercury-cell and diaphragm-cell technologies. The chlorine manufacturing industry has experienced a substantial shift over time toward the membrane technology. This diffusion is the result of three different processes: adoption of cleaner technologies at existing plants (adoption),[1] the use of membrane cells by newly constructed facilities (entry), and the closing of facilities using diaphragm and mercury cells (exit).

In this paper, we examine econometrically two of these aspects of diffusion (adoption and exit behavior) to assess the effects of environmental regulation.[2] Employing plant-level data on technology choice, economic variables, and regulatory variables from 1976 to 2001, we examine these adoption and exit decisions with a hazard model, considering the effects of both direct regulation of chlorine manufacturing and regulation of downstream users of chlorine.

I. The Chlorine Manufacturing Industry

Over 95 percent of the world's chlorine is produced by an electrolytic process in which electric current is introduced to a salt-water brine, resulting in the separation of chlorine, hydrogen gas, and caustic soda (sodium hydroxide). Three different types of cells have been employed in this electrolytic process: the mercury cell, the diaphragm cell, and the membrane cell.

The mercury cell is widely viewed as having the greatest potential for environmental damage, due to the potential release of mercury, a highly persistent and bio-accumulative toxin. The diaphragm-cell technology, which accounts

† *Discussants:* Arik Levinson, Georgetown University; John List, University of Maryland; Randy Becker, U.S. Bureau of the Census.

* John F. Kennedy School of Government, Harvard University, 79 JFK St., Cambridge, MA 02138 (affiliation and address for all authors); Stavins is also affiliated with Resources for the Future. We are grateful for helpful comments from Angelique Augereau, Christopher Forman, Michael Greenstone, Arik Levinson, Richard Newell, and David Popp, but the authors alone are responsible for all remaining errors. This research was partially funded by grant number R-82968901-0 from the National Center for Environmental Research (NCER) STAR Program, at the U.S. Environmental Protection Agency and by a grant from the Heinz Scholars for Environmental Research.

[1] In this context, adoption by an existing plant can mean both conversion of all of an existing plant's capacity (retrofitting) or a combination of retrofitting and new capacity expansion at an existing plant. In the latter case, some capacity is maintained using the older technology.

[2] The entry aspect is empirically difficult to estimate because one only has data on facilities that choose to enter and cannot observe or infer all the possible firms that could have entered during a time period but did not. This renders probabilistic empirical analysis of the entry decision infeasible.

for two-thirds of chlorine produced in the United States (Chlorine Institute, 2001) is considered to be more environmentally benign than the mercury-cell technology, although not without its own environmental risks, since the diaphragm cell is composed of layers of asbestos. The newer membrane process is the most environmentally benign.

Over the past 25 years, there has been a gradual movement from mercury and diaphragm cells to membrane technology. In 1975, plants using mercury cells accounted for 22 percent of total chlorine capacity, plants using diaphragm cells accounted for 73 percent, and membrane-cell plant capacity was less than 1 percent of the total. By 2001, mercury-cell capacity had fallen to 10 percent, diaphragm cells accounted for 67 percent, and membrane cells accounted for 20 percent (Chlorine Institute, 2001).[3] Some of the significant increase in membrane-cell capacity has come from adoption of membrane cells at existing facilities, but the bulk of the diffusion has taken place through exit of facilities using mercury and diaphragm cells.

II. The Regulation of Chlorine Manufacturing

In 1972, a widely publicized incident of mercury poisoning in Minamata Bay, Japan, led the Japanese government to prohibit the use of mercury cells for chlorine production. The United States did not follow suit, but it did impose more stringent environmental constraints on mercury-cell units during the early 1970's. Subsequently, chlorine manufacturing became subject to increased regulation under the Clean Air Act, the Clean Water Act, the Resource Conservation and Recovery Act (RCRA), and the Comprehensive Environmental Response, Compensation, and Liability Act (CERCLA), popularly known as Superfund. In addition, chlorine manufacturing became subject to public-disclosure requirements under the Toxics Release Inventory (TRI), which has required large manufacturing facilities to make public their annual releases of over 300 different toxic chemicals since 1987.

[3] The remaining 4 percent of chlorine in 1975 and 3 percent in 2001 was produced at facilities that used a method other than electrolysis.

In addition to regulation of the chlorine manufacturing process, there has also been increased environmental pressure on industries that use chlorine as an input. This indirect regulation is potentially important for choices of chlorine manufacturing technology because a large share of chlorine is manufactured for on-site use in the production of other products. Changes in regulations in these downstream industries can have substantial impacts on the demand for chlorine and can affect the rate of entry and exit of chlorine production plants.

Two major (indirect) regulations may have altered the demand for chlorine. One is the Montreal Protocol, which regulates the production of ozone-depleting chemicals, such as chloroflorocarbons (CFCs), for which chlorine is a key ingredient. In 1987, the Montreal Protocol imposed a timeline for phasing out CFC production in the United States and other industrialized countries. The other potentially important indirect regulation is the so-called "Cluster Rule" regulation of releases from pulp and paper mills. Finalized in 1998, the Cluster Rule tightened restrictions on the release of chlorinated compounds to both water and air. This led to increased interest by the industry in non-chlorine bleaching agents, which in turn may have affected the economic viability of some chlorine plants.

III. Empirical Model

We utilize a proportional hazard model to analyze the effects of economic and regulatory variables on adoption and exit decisions by chlorine manufacturing plants from 1976 to 2001. The hazard model is appropriate because our focus is on the *timing* of technology decisions (Nicholas Kiefer, 1988; Suzi Kerr and Richard G. Newell, 2001). Plants face different anticipated returns to adoption of membrane technology. Economic and regulatory conditions can affect expected costs and returns and, hence, affect the timing of adoption.

We employ the proportional hazard model, which separates the hazard rate into two components: a baseline hazard rate which is a function of time, $h_0(t)$; and a function of the covariates, $\exp(\mathbf{X}'\boldsymbol{\beta})$. In a proportional hazard model, changes in the explanatory variables shift the baseline hazard. The baseline hazard function can be left unspecified, as in a Cox proportional hazard model, or parameterized. While the Cox proportional hazard model has the advantage of not requiring assumptions about the form of the underlying hazard function, it also has a major shortcoming for this analysis: the effects of any pure time-series variables cannot be estimated. Since these effects are potentially important, we adopt the parameterized-baseline hazard approach.

Use of a parameterized-baseline hazard model requires a choice of specification. An exponential density function yields a constant baseline hazard function. Other specifications yield baseline hazard functions that are functions of time, allowing for the possibility that the baseline hazard increases or decreases over time (Peter Kennedy, 1998 pp. 249–62). Without strong a priori notions regarding the shape of the underlying baseline hazard, we begin by assuming a constant baseline hazard function (i.e., an exponential density function) and estimate the hazard model for adoption and exit decisions. We conduct sensitivity analyses to determine the robustness of the results to the specification of the baseline hazard function.

IV. Data

Data on cell technology use at U.S. chlorine manufacturing plants were obtained from the Chlorine Institute (2001) and from the *Directory of Chemical Producers*. One of our two dependent variables, CONVERT, is an indicator of membrane adoption at an existing plant and takes a value of 1 in the last year a facility used the old technology. The other dependent variable, SHUTDOWN, takes a value of 1 if a facility closed during a given year.

We construct variables that reflect both economic and regulatory determinants of adoption or exit timing:

Size.—Empirical research on technology diffusion has frequently found that larger plants are more likely to adopt new technologies or adopt more quickly (Zvi Griliches, 1957; Kerr and Newell, 2001). We employ two measures of size: plant capacity and net sales of the parent company. These measures reflect two different notions of size. Plant capacity refers to the share of the individual plant in the chlorine market. In contrast, net sales are measured at the parent company, not the individual production unit, and hence serve as a proxy for access to capital, risk-aversion, and other firm-specific factors.

Complexity.—Chlorine is often produced and used on-site as an input in the production of other products. Such "complexity" of operation might affect both exit and adoption decisions, because the more integrated is chlorine production with other high-value-added goods, the more costly and difficult it might be to shut down a plant either temporarily or permanently. This is particularly true for chlorine manufacturing because transportation of chlorine is dangerous and is itself highly regulated. The variable COMPLEX is an integer variable that indicates the number of other production processes, as captured in three-digit SIC codes, that occur at the manufacturing site. We also use a series of dummy variables representing the industry or industries co-located with a given chlorine plant. The variables CO-PAPER, CO-PLASTICS, and CO-ORGANICS take a value of 1 if the plant also produces pulp and paper products, plastics, and organic chemicals respectively.

General Economic Conditions.—Chlorine production is a pro-cyclical industry. We employ two measures of business activity: lagged values of GDP in the chemical manufacturing industry and lagged values of the real price of chlorine.[4]

Regulatory Variables.—As described above, several regulatory regimes may have affected the rate and direction of technology diffusion in the chlorine industry. We construct dummy variables that capture whether a facility was affected by a specific regulatory regime, including Superfund, the Montreal Protocol, the pulp and paper cluster rule, and the Toxics Release Inventory. The pulp and paper and Montreal Protocol indicators take a value of 1 if the respective rule was in effect during that year *and* the chlorine plant was co-located with a

[4] We tried alternative lag structures to account for different expectations mechanisms. The results were robust to changes in the lag structure over 1–5 years.

pulp and paper mill or plant producing ozone-depleting chemicals. We also include a dummy variable that indicates whether the plant uses mercury cells, because these cells have been more heavily regulated under various regulatory regimes.

V. Results

For the adoption model, we have complete data on 51 facilities, eight of which adopted the membrane technology during the sample period. We estimate the hazard model under a variety of specifications.[5] Due to the nonlinear nature of the hazard function, the coefficients do not have a direct interpretation. In order to provide information on the relative importance of the explanatory variables on the hazard rate, we calculate the hazard rate when all continuous variables are evaluated at their mean and indicator variables are set to zero. The mean hazard rate for the adoption model is 1.28 percent. The impact on the mean hazard rate of increasing a continuous variable by 10 percent or changing an indicator variable to 1, ceteris paribus, is reported in Table 1.

The analysis shows that larger plants and plants owned by firms with larger sales volumes were slightly more likely to switch to the membrane technology. Facilities co-located with paper mills and plastics plants were substantially less likely to switch. Increases in the lagged real price of chlorine (a proxy for expected future prices) made technology changes less likely. A potential explanation for this result is that the opportunity cost of the downtime necessary to change technologies is greater when real prices are higher. When GDP in the chemical industry was used as an indicator of business activity, the sign and magnitude of the effect were similar, but the results were not statistically significant.

The effects of the regulatory variables on the likelihood of adopting membrane technology are not statistically significant in any of the specifications. Mercury plants, which were subject to stringent regulation for water, air, and hazardous-waste removal, were no more likely to switch to the membrane technology than diaphragm plants. Similarly, TRI reporting appears to have had no statistically significant effect on adoption decisions, whether TRI is measured as a simple indicator variable, by rank of total releases, or by magnitude of reported releases.

We also estimate a hazard model for the exit decision. We have complete data for 55 facilities, 21 of which ceased operations between 1976 and 2001. We test a variety of specifications, controlling for economic and regulatory determinants. The mean hazard rate for the exit model is 2.1 percent. The incremental effect of the covariates relative to the mean hazard can be found in Table 1.

Despite the fact that the results of the exit analysis are not as robust to changes in specification as are the results of the adoption analysis, some interesting and quite striking patterns emerge. Although the economic covariates

TABLE 1—EFFECTS OF COVARIATES ON THE MEAN HAZARD RATE

	Change from mean hazard (percentage points)	
Covariate	Adoption	Exit
Economic Variables:		
Capacity	**0.08**	**−0.34**
Net sales	**0.10**	0.02
Real price of chlorine	**−0.26**	0.08
COMPLEX		−0.02
CO-PAPER	**−1.28**	
CO-PLASTICS	**−1.27**	
CO-ORGANICS	0.00	
Regulations on Chlorine Production:		
Mercury	0.00	−0.40
Superfund	0.71	−0.28
TRI (Disclosure)	1.10	**−1.69**
Regulations on End-Use of Chlorine:		
Pulp and paper cluster		**11.18**
Montreal		**15.36**

Note: Bold type indicates that the corresponding coefficients are statistically different from zero at the 10-percent level of significance.

[5] Due to space constraints, only a single specification is reported here. See Snyder et al. (2002) for full regression results. Because the likelihood ratio is estimated using repeated observations on the same plant, the assumption of independent identically distributed error terms is suspect. We instead estimate and report clustered standard errors that are adjusted for correlation among observations from the same plant, but we assume independence across plants.

seem to have had little effect on the timing of exit decisions (with the exception of facility-level capacity, which retarded shutdown slightly), regulations do explain some of the variation in exit decisions. In particular, indirect regulation of the end-uses of chlorine accelerated shutdowns in certain industries. Facilities affected by the pulp and paper cluster rule and the Montreal Protocol were substantially more likely to shut down than were other facilities.

It may appear from the results that facilities that report to TRI were less likely to shut down than facilities that did not report, but it is important to note that nearly all facilities that were still in operation in 1986 reported to TRI, and so this parameter may be picking up a simple time effect, rather than a true regulatory effect.

VI. Conclusions

Diffusion of new technology is the result of a combination of adoption at existing facilities and entry and exit of facilities with various technologies in place. In chlorine manufacturing, total capacity using the cleaner membrane technology has been increasing over the past 25 years. This increase is partly a result of adoption of membrane cells at existing plants, but mainly a consequence of plant shutdowns.

Our results indicate that regulatory factors have not had a statistically significant effect on the decision to adopt membrane-cell technology at existing plants. On the other hand, indirect regulation of the end-uses of chlorine appears to have accelerated facility closures significantly, and thereby to have increased the share of plants using the cleaner, membrane technology for chlorine production.

Environmental regulation, in this study, did affect technological change. It did so not by encouraging the adoption of membrane cells by existing facilities, but by reducing the demand for chlorine and hence encouraging the shutdown of facilities using environmentally inferior options.

REFERENCES

Chlorine Institute. *North American chlor-alkali industry plants and production data report.* Washington, DC: Chlorine Institute, 2001.

Directory of chemical producers, United States. Menlo Park, CA: SRI Consulting, 1976–2001.

Griliches, Zvi. "Hybrid Corn: An Exploration in the Economics of Technical Change." *Econometrica*, October 1957, *25*(4), pp. 501–22.

Jaffe, Adam B.; Newell, Richard G. and Stavins, Robert N. "Technological Change and the Environment," in K. G. Maler and J. Vincent, eds., *Handbook of environmental economics.* Amsterdam: North-Holland, 2003, pp. 461–516.

Kennedy, Peter. *A guide to econometrics.* Cambridge, MA: MIT Press, 1998.

Kerr, Suzi and Newell, Richard G. "Policy-Induced Technology Adoption: Evidence from the U.S. Lead Phasedown." Resources for the Future (Washington, DC) Discussion Paper No. 01-14, May 2001.

Kiefer, Nicholas. "Economic Duration Data and Hazard Functions." *Journal of Economic Literature*, June 1988, *26*(2), pp. 646–79.

Snyder, Lori D.; Miller, Nolan H. and Stavins, Robert N. "The Effects of Environmental Regulation on Technology Diffusion in the Chlorine Manufacturing Industry." Mimeo, Harvard University, 2002.

[15]

Available online at www.sciencedirect.com
ScienceDirect
ELSEVIER
Energy Economics 28 (2006) 563 – 578
Energy Economics
www.elsevier.com/locate/eneco

The effects of economic and policy incentives on carbon mitigation technologies

Richard G. Newell [a,*], Adam B. Jaffe [b,c], Robert N. Stavins [a,d]

[a] *Resource for the Future, United States*
[b] *Economics, Brandeis University, United States*
[c] *National Bureau of Economic Research, United States*
[d] *John F. Kennedy School of Government, Harvard University, United States*

Received 29 July 2006; accepted 29 July 2006

Abstract

The ability to estimate the likely effects of potential climate change policies on energy use and greenhouse gas (GHG) emissions requires an improved understanding of the relationship between different policy alternatives and energy-saving and GHG-reducing changes in technology. A particularly important and understudied aspect of this set of issues is the conceptual and empirical modeling of how the various stages of technological change are interrelated, how they unfold over time in response to market forces, and the differential impact of various policies (for example, R&D subsidies, environmental taxes, information programs). We summarize several contributions to this literature and suggest promising areas for continued research on empirical analysis and modeling of induced technological change.

Keywords: Carbon dioxide; Emission reduction; Incentives; Technology; Policy

1. Introduction

Serious consideration is currently being given to a range of international and domestic policy actions to reduce carbon dioxide emissions due to their potentially damaging effects on the climate. Changes in carbon emissions can be driven by a number of factors, including changes in economic activity, energy use per unit of economic activity (energy efficiency), and the carbon intensity of energy used (carbon efficiency). Obviously, limiting economic activity as a means of reducing

* Corresponding author.
E-mail address: newell@rff.org (R.G. Newell).

0140-9883/$ - see front matter
doi:10.1016/j.eneco.2006.07.004

carbon emissions has scant political appeal for rich countries, let alone poor ones. Technological improvements that generate enhanced energy and carbon efficiency have therefore been the principal means discussed for addressing climate change. Recent policy proposals have included R&D funding, tax credits for the purchase of energy-efficient equipment, public–private partnerships aimed at developing and deploying energy-efficient technologies, carbon cap-and-trade systems, renewable-energy performance standards for electricity generation, and energy-efficiency standards for products. Broader policies under discussion–including a tradable carbon permit system or carbon emission fees–could have even greater effects on energy-related technologies.

An improved understanding of the relationships between market incentives, government policies, and technological change is therefore crucial for forecasting the baseline level of technology, its associated carbon emissions, and the likely magnitude, cost, and timing of emission reductions associated with alternative policies. In fact, different assumptions about baseline improvements in energy efficiency are often the single largest source of difference in predictions about the cost of achieving given climate policy objectives (Gaskins and Weyant, 1996).

In addition, public policies themselves can affect the rate and/or direction of technological change: by directly mandating characteristics of products that can be offered for sale (for example, through appliance efficiency standards) or that can be purchased and installed (local zoning requirements); by directly carrying out research and development activities at government laboratories or sponsoring them in the private sector; or by providing economic incentives through taxes, tradable permits, and other instruments that indirectly encourage the development and use of technologies with specific characteristics. In the present context, this last option refers to the possibility that higher (expected) energy prices will lead to the development and adoption of energy-consuming capital goods that are less energy-intensive in their use, a notion that is directly related to Hicks' (1932) induced innovation hypothesis.[1]

Economists who have studied technological change agree that the rate and direction of innovation is affected both by *exogenous* "technological opportunity" and by the *endogenous* expected rate of return to particular innovations (Schmookler, 1966; Rosenberg, 1982; Jaffe, 1988).[2] In the climate change context, much analysis of these issues has revolved around the extent to which changes in energy efficiency are induced by energy prices, and the extent to which energy efficiency improves "autonomously", or purely as a function of time.[3]

One can distinguish three steps or stages in the process by which a new, superior technology permeates the marketplace. *Invention* constitutes the first development of a scientifically or technically new product or process. Inventions may be patented, though many are not. Either way, most inventions never actually develop into an *innovation*, which is accomplished only when the

[1] Discussions regarding global climate change policy reinforce the importance of the relationship between public policy and technological change. In U.S. Senate testimony, Janet Yellen, then Chair of the Council of Economic Advisers, commented on the difficulty of carrying out economic analysis of the Kyoto Protocol: "One area in which the uncertainty is particularly large is the pace of technological progress–including diffusion of existing energy-efficient technologies, as well as research and development of new technologies–and the extent to which the pace will accelerate in response to government programs" (Yellen, 1998).

[2] In some cases, the non-endogenous component of technological change has been characterized as an "autonomous" component. See, for example, Manne and Richels (1992).

[3] There has been much debate regarding the rate of "autonomous energy efficiency improvement" or AEEI (Manne and Richels, 1992) in integrated assessment models of climate change. The AEEI is measured as a residual; that is, it is defined to be the rate of improvement in energy efficiency that is not explained by other factors contained within a given model. Hence, the magnitude of the AEEI should be, by definition, a function of the particular model being employed. As several authors have noted, the results produced by integrated assessment models tend to be highly sensitive to the magnitude of the assumed AEEI (Gaskins and Weyant, 1996).

new product or process is commercialized, that is, made available on the market. A firm can innovate without ever inventing, if it identifies a previously existing technical idea that was never commercialized, and brings a product or process based on that idea to market. The invention and innovation stages are carried out primarily in private firms through a process that is broadly characterized as "research and development" (R&D). A successful innovation gradually earns a significant share of the purchases of firms or individuals, a process labeled *diffusion*. The cumulative economic or environmental impact of new technology results from all three of these stages, which we refer to collectively as the process of technological change.

For the purpose of modeling the interaction of energy use and climate change, however, there are two additional steps that mediate the impact of technological change on emission of greenhouse gases. If sales of new equipment were increasingly dominated by more efficient models, this would gradually increase the efficiency of the stock of equipment in use, a process we label *stock turnover*.[4] Finally, the efficiency of the stock of equipment in use will in turn affect how intensely people utilize the equipment. The (greenhouse-gas) efficiency of the existing stock, combined with the intensity with which it is used, determines the emissions of greenhouse gases.

Each of the elements of this process is limited by steps further back in the chain of technological change and also subject to technical limitations. Current greenhouse-gas efficiency is to a great extent constrained by the efficiency of the existing stock of equipment, subject to the level of utilization.[5] The efficiency of this year's stock is significantly determined by the efficiency of last year's stock, and only gradually affected by changes in the efficiency of new models being purchased. Diffusion of more efficient equipment into the sales of new models is limited by the menu of available products from which purchasers can chose. Innovation can change these offerings, but innovation is in turn limited by the feasible set of product models (that is, characteristics bundles) given by invention at a point in time. Invention is in turn limited by invention possibilities based on the costs of research and development, the availability of technical knowledge, and technical constraints such as physical limits to energy efficiency. Table 1 provides an overview of these relationships.

As an example, consider tradable carbon permits or taxes on carbon content which raise the price of fossil fuels.[6] Over a sufficiently long time horizon, such policies may be anticipated to have a number of effects. First, for any given stock of equipment or technologies, they may cause less energy to be consumed. Second, among the menu of equipment that is available, such policies will cause more efficient technologies (capital) to be chosen. Third, it is also possible that sufficiently strong carbon policies would induce greater investment of resources into the research and design of efficient machines, so that the efficiency of the *menu* of options from which people can choose in the market place is improved. These last two linkages refer to the potential endogeneity of technological change, and over the course of several decades or more, these invention, innovation, and diffusion effects are likely to dwarf any utilization effects.[7]

[4] In many studies of technology diffusion, the word "diffusion" is used to encompass both the increasing sales of the new technology (what we have labeled diffusion), and its increasing share of equipment in use (what we have labeled stock turnover). This approach makes sense if the new technology is so superior to existing equipment that its availability causes many users to scrap existing equipment sooner than they otherwise would have and replace it with new equipment. In the case of much energy-using equipment, it seems more plausible that more efficient models will gradually replace existing models as those older models wear out and are eventually replaced.

[5] Short-run price elasticities of demand for energy are typically quite low, being estimated at about −0.1 to −0.2 (Bohi, 1981).

[6] See, for example, Nordhaus (1991), Jorgenson and Wilcoxen (1992), Goulder (1995), and Pizer (1999).

[7] Hogan and Jorgenson (1991) provide evidence that technological change could indeed overwhelm short-run substitution effects, even over the span of a few years.

Table 1
Interrelationships among technological change stages and incentives

Stage and key decision maker	Key influences	Taken as given	How measured?
Invention (inventing firm)	R&D costs, expected revenues, market share, royalties, non-pecuniary rewards	Invention/innovation possibilities and tradeoffs; expected diffusion	Patents, patent citations, characteristics of available products, tradeoffs among characteristics
Innovation (commercializing firm)	R&D costs, manufacturing costs, expected revenues, market share		
Diffusion (purchases) (user)	Capital and operating costs, product characteristics, exogenous "environmental" variables	Menu of available products, their characteristics, and tradeoffs; expected usage	Average characteristicsof new purchases, number or proportion of adopters
Stock turnover (user)	Growth in demand for energy services; depreciation of existing equipment	Distribution of characteristics in new purchases; previous stock	Average characteristics of stock of equipment, number or proportion of stock that incorporates new technology
Use (user)	Energy price, weather, final product demand	Characteristics of equipment stock	Units of energy use (e.g., hours, kW)

Large-scale simulation models used to model energy and carbon policies do not adequately treat the potential for endogenous energy-saving technological change. Rather, technological change is typically treated simply as a function of calendar time, which–assuming that the direction of technological change may in fact be price-responsive–makes the estimation results from such models sensitive to the time period studied, among other potential problems. As a result, policy-analytic forecasts from models that do not properly treat endogenous technological change are likely to misestimate the timing, cost, and extent of energy and carbon reductions associated with various public policies such as carbon taxes or tradable carbon permits (Popp, 2003).

The purpose of this paper is to summarize some recent research on the effects of government policies and economic incentives on the invention, innovation, and diffusion of mitigation technologies. In Section 2 we review empirical research on the influence of energy prices and regulatory policies on energy-efficient innovation. In Section 3 we summarize research on the influence of market and policy incentives on technology diffusion. Section 4 reviews the prospects for and integrated assessment modeling results regarding carbon capture and storage technologies. Section 5 offers some concluding remarks and suggestions for future research.

2. Price- and policy-induced innovation

A natural way to model induced technical change at the microeconomic level is to recognize that energy-saving technological change comes about largely through the introduction of new capital goods that embody improved energy efficiency, or energy input per unit of output. Energy efficiency can then be thought of as an attribute or characteristic of the capital goods. Newell (1997) and Newell, Jaffe, and Stavins (1999) examined the extent to which the energy efficiency of the menu of home appliances available for sale changed in response to energy prices between 1958 and 1993, using a model of induced innovation as changing characteristics of capital goods. Data were for hundreds of product models from the Sears catalogue and other publicly available data sources. Hicks formulated the induced innovation hypothesis in terms of factor prices. Newell et al. (1999) generalized this concept to include inducement by regulatory standards, such as labeling requirements that might increase the value of certain product characteristics by making consumers more aware of them. More generally, non-price regulatory constraints can fit within the inducement framework if they can be modeled as changing the shadow or implicit price that firms face in emitting pollutants. In their framework, the existing technology for making a given type of equipment at a point in time is identified in terms of vectors of characteristics (including cost of manufacture) that are feasible. The process of invention makes it possible to manufacture "models" (characteristics vectors) that were previously infeasible. Innovation means the offering for commercial sale of a model that was not previously offered for sale. Induced innovation is then represented as movements in the frontier of feasible models that reduce the cost of energy efficiency in terms of other attributes, as shown in Fig. 1.

By constructing a series of dynamic simulations, they examined the effects of energy price changes and efficiency standards on average efficiency of the menu of products over time. They found that a substantial amount of the improvement was what may be described as autonomous (that is, associated with the passage of time), but significant amounts of innovation were also due to changes in energy prices and changes in energy-efficiency standards. They found that technological change in air conditioners was actually biased against energy efficiency in the 1960s (when real energy prices were falling), but that this bias was reversed after the two energy shocks of the 1970s. In terms of the efficiency of the average model offered, they found that energy efficiency in 1993 would have been about one-quarter to one-half lower in air conditioners and gas water heaters, if energy prices had stayed at their 1973 levels, rather than following their historical path. Most of the response to energy price changes came within less than 5 years of those changes.

Newell, Jaffe, and Stavins have applied the same approach to investigate innovation in other important energy-using technologies, including farm tractors, which are representative of the general purpose internal combustion engine, and jet aircraft. There is an exceptionally rich array of data for

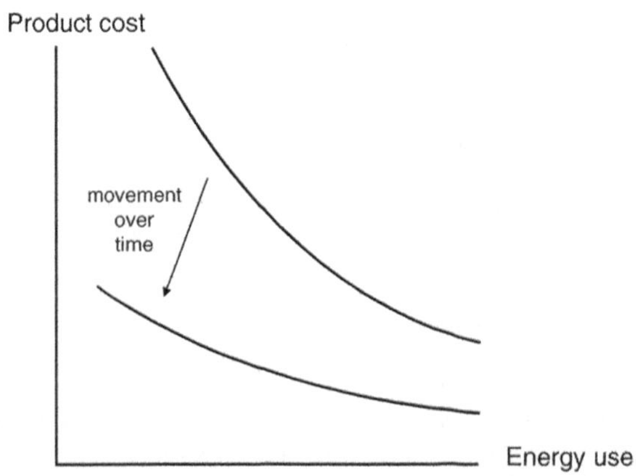

Fig. 1. Innovation in product characteristics.

farm tractors that spans over 80 years and across thousands of different product models and dozens of manufacturers. In prior research we found that the key information requirements for an investigation of induced energy-saving product innovation are data on product characteristics (e.g., horsepower and energy efficiency), purchase price, production costs, product output, and characteristics "prices" such as the price of energy (for energy efficiency) and the wage rate (for horsepower).

For the case of farm tractors, we constructed a database on product characteristics from the Nebraska Tractor Test Data reports (1920–1996), which have been conducted by the University of Nebraska for virtually every tractor model sold in the U.S. since 1920. The data set contains 1631 different models of wheel tractors from seven major manufacturers (Allis Chalmers, Case, Deere, Ford, International Harvester, Massie Ferguson, and White) and several other minor manufacturers spanning 78 years (1918–1995). The full data set has 5724 complete observations, with information on each tractor's test number, year tested, make, model, fuel type, energy efficiency, horsepower, and whether the tractor had a cab or four-wheel drive. We also employed data on the cost of tractor production, product output, energy prices, and farm wages.

Overall, the results are consistent with the economic interpretation of the parameters. The estimated elasticities for the various characteristics all have the expected signs and reasonable magnitudes; and the coefficient on cumulative production is negative, indicating positive technological change. The results confirm that the cost of durable goods increases with increasing energy efficiency, capacity, and other desirable characteristics, and that the cost of producing a given bundle of characteristics tends to fall with increased cumulative production experience as a result of technological change in production techniques and product design.

In addition, we found support for induced innovation in the tradeoffs between product cost, energy efficiency, and horsepower. That is, the slope of the technological frontier with respect to fuel efficiency was less negative during periods of higher energy prices, meaning that the elasticity of product cost with respect to energy flow is lower, or, equivalently, that the tradeoff at a point in time between production cost and energy efficiency has shifted so that energy efficiency is less expensive on the margin. The same was found to be true for horsepower with respect to changes in the real farm wage.

In addition, we found support for the existence of a "learning curve" in tractor production, as represented by a strong positive relationship between product cost reductions and cumulative output. Specifically, a 10% increase in cumulative production was associated with about a 7% decrease in quality-adjusted product cost. In fact, cumulative output proved to bear a clearer relationship to product cost reductions than did time, which is typically employed in analyses of technological change. Given the pervasive association between production experience and product costs found in the empirical literature, an important area of further research is the possible incorporation of learning effects into aggregate modeling efforts. This raises significant analytical and empirical challenges.

3. The influence of market and policy incentives on technology diffusion

3.1. Technology information programs

As policies that would entail significant energy price increases are unlikely to be politically attractive in the near term, the near-term focus in the United States has been on the development and diffusion of technology through other means. Thus, policy proposals have tended to emphasize programs that foster research, development, and deployment of technologies, government-industry partnerships, tax credits and other financial incentives, minimum appliance efficiency standards, voluntary agreements, and information programs.

Information programs, which seek to encourage energy efficiency by increasing awareness of conservation opportunities and offering technical assistance with their implementation, are an important element of this energy-efficiency policy portfolio. These programs take a variety of forms, including educational workshops and training programs for professionals, advertising, product labeling, and energy audits of manufacturing plants. In addition to alerting firms to profitable conservation opportunities, access to more accurate performance information can reduce the uncertainty and risk associated with adopting technologies that are new, or that receive differing reviews from equipment vendors, utilities, or consultants. The economic rationale for these programs lies primarily in public good aspects of knowledge and information provision. Although these public information programs are not free, the cumulative benefit of educating many users with similar information can greatly exceed the costs. Such information, however, tends to be under-provided by the private sector. Concerns about environmental externalities associated with energy production and use provide additional justification for these programs.

Despite the role that information programs play in existing and proposed energy-efficiency policy portfolios, surprisingly little is known about how participants respond to such programs. Although a reasonably large literature surveys various potential market barriers and market failures in energy-efficiency investment, few analyses have focused specifically on information programs. This is in part due to a lack of adequate data for analysis. One exception is Morgenstern and Al-Jurf (1999), who analyze data from the Department of Energy's 1992 Commercial Buildings Energy Consumption Survey. They find that information provided through demand-side-management utility programs appears to make a significant contribution to the diffusion of high-efficiency lighting in commercial buildings. Although not the focus of their examination of energy-saving product innovation, Newell et al. (1999) find that the responsiveness of energy-efficient innovation in home appliances to energy price changes increased substantially during the period after energy-efficiency product labeling was required.

Anderson and Newell (2004) focused on actions taken by manufacturing plants in response to energy audits offered through the U.S. Department of Energy's Industrial Assessment Centers (IAC) program, which has been providing energy assessments at no financial cost to small- and medium-sized manufacturers since 1976. This program is of interest for several reasons. First, significant opportunities to conserve energy may exist in the industrial sector, which represents 37% of total national energy consumption. Second, the opportunity to focus on the behavior of small- and medium-sized firms is rare due to data constraints, even though these firms represent over 98% of all manufacturing firms and more than 42% of total manufacturing energy consumption. This focus is particularly appropriate given that smaller firms are more likely to benefit from access to information and expertise, which tend to be more readily available to larger firms. Finally, the IAC program has generated an unusually extensive set of data on the characteristics of conservation opportunities identified and actions taken under the program. One attractive aspect of these data is that there are multiple observations available for each firm, allowing the use of a fixed effects model to control for unobserved differences in firms' propensities to adopt technology.

Because of their detail, these data provide a unique opportunity to quantify the factors that encourage small- and medium-sized industrial firms to invest in energy-conserving technologies. After summarizing the general character of projects adopted under the IAC program, Anderson and Newell (2004) explore the influence of technology costs, expected energy savings, and individual firm characteristics on the likelihood of adopting projects. They employ models of varying flexibility to examine and compare the degree of response to differences in capital costs and operating cost savings, as well as the energy price and quantity differences that underlie savings. The results strengthen our understanding of how certain factors influence technology adoption decisions, and whether this behavior is consistent with economic expectations. In addition, the results offer evidence on the likely relative effectiveness of policies aimed at increasing energy efficiency, such as energy or carbon price increases, technology subsidies, and policies that directly alter the energy use of technologies.

Another important aspect of this type of investment decision-making is the "payback cutoff," "hurdle rate," or other discounting factor that firms employ when measuring current costs against future benefits. The rate of discount used for climate policy analysis has huge implications on the results. There is a substantial literature that suggests that "implicit discount rates," which one can calculate based on the capital cost versus operating cost savings of various implemented and unimplemented projects, can in practice be quite high relative to market interest rates (Hausman, 1979; Train, 1985). A related literature further contends that these high implicit discount rates are attributable to various market barriers and market failures–including information problems–and that these problems can be ameliorated by appropriate policies (Ruderman et al., 1987; also see Jaffe and Stavins, 1994).

Accordingly, several analyses of carbon mitigation costs have modeled the effect of information programs and other policies by significantly lowering the discount rate used for energy conservation decisions. The Clean Energy Futures study (Brown et al., 2001; Interlaboratory Working Group, 2000), for example, lowered investment hurdle rates to 15% in the industrial sector (and 7% in the residential sector) to capture the effect of information programs and other energy conservation policies. Such lowering of hurdle rates has the intended effect of decreasing estimated energy use in the model, but modeling the effect of information programs in this way also leads to a number of side effects. Lower hurdle rates also increase the rate at which energy use declines in response to energy price increases resulting, for example, from a carbon permit system or carbon tax. This implies a reduction in the cost of carbon mitigation efforts through carbon price policies.

By expanding the perceived range of investment opportunities available to firms, information programs may indeed lead to the adoption of profitable but previously unimplemented technologies, associated energy use reduction, and lower observed *implied* hurdle rates. But this does not imply an across-the-board reduction in the *actual* investment hurdle rate, which is unobserved and could remain at pre-policy levels. In other words, it is entirely possible that managers continue to apply hurdle rates well above market interest rates to the new set of possibilities brought forth by an information program. On the other hand, it is possible that information programs actually do significantly alter the way in which firms trade off the current costs and future benefits of all energy conservation opportunities, for example, by educating managers to focus more on the operating cost savings of projects.

Anderson and Newell explore these issues by examining the rates of return for potential projects faced by firms that participated in the IAC program to determine whether the level of implicit discounting used by plants that received information assistance may have decreased to levels that some studies suggest. Finally, they analyze the reasons given by firms for not adopting recommended projects in order to determine whether this decision is due to the economic undesirability of the projects, or to some remaining type of market barrier or failure.

They find that about half of the projects recommended by energy assessment teams are actually adopted by the plants receiving these recommendations, although they cannot say how many of these projects might have been adopted in the absence of the energy audit. They find that that firms respond as expected to marginal changes in the financial characteristics of projects (i.e., technology costs, energy prices, the quantity of energy saved, energy operating cost savings, and the payback period) (Fig. 2). Firms are about 40% more responsive to investment costs than to energy savings, suggesting that policies to reduce implementation costs may be somewhat more

R.G. Newell et al. / Energy Economics 28 (2006) 563–578 571

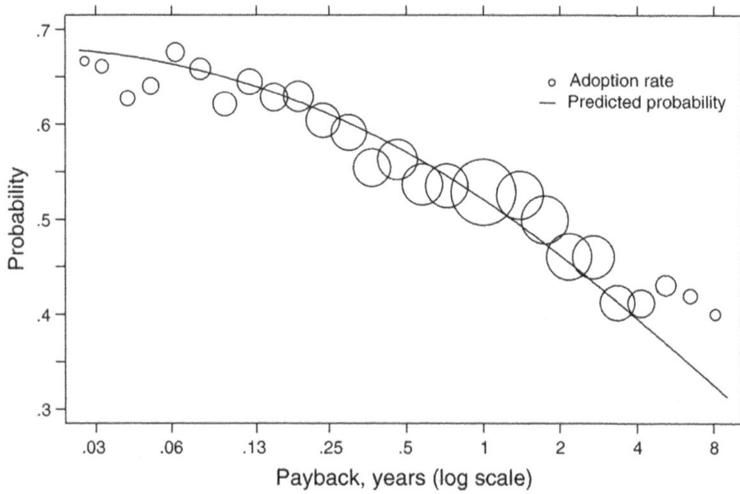

Fig. 2. Probability of adoption of energy efficient technologies versus payback. See Anderson and Newell (2004). Circles represent the observed adoption rates for fixed intervals of payback in log scale. The areas of the circles are proportional to the number of observations in each interval. The solid line is the predicted probability of adoption for estimated econometric model.

effective than various mechanisms that raise energy prices. Although the financial characteristics of projects are clearly important, there also appear to be other, unmeasured project-specific factors (e.g., individual project lifetimes, unmeasured costs and benefits, uncertainty regarding costs and benefits, or project complexity and risks) that influence the investment decision.

Anderson and Newell estimate that the investment threshold typically used by the plants in evaluating which energy audit recommendations to adopt was about a 1 to 2-year payback, which corresponds to an implicit hurdle rate of 50% to 100% for projects lasting 10 years or more. Although they are unable to determine whether participation in the IAC program actually lowered investment hurdle rates, these payback thresholds are consistent with what many surveys of plant managers suggest that they deliberately use for many types of investments, including those for energy conservation. In any event, these hurdle rates are many times higher than those assumed in many analyses of the effects of various climate policies.

Finally, the reasons given by program participants for not adopting certain project recommendations suggest that most of these disregarded projects may have been economically undesirable. Many of these reasons hint at various unmeasured costs, project risks, and uncertainty that are unlikely to be reflected in estimated implementation costs and projected annual savings. On the other hand, many projects were likely rejected because of institutional or bureaucratic barriers within firms, and most of the reasons are sufficiently vague that they cannot rule them out as indicative of institutional or bureaucratic barriers.

Overall, Anderson and Newell (2004) found that one can view the glass as either half full or half empty. Although the results suggest that the IAC program has led to the adoption of many financially attractive energy conservation projects, plants found about half of the projects recommended by assessment teams to be unattractive. This suggests that other, more costly policies targeted at increasing the financial attractiveness of these projects (e.g., energy/carbon taxes, or tax breaks/

subsidies for implementation) may be needed to further promote energy efficiency in these firms. Furthermore, it would seem that policies that could lengthen the short paybacks that firms routinely demand from all types of projects (not just those for energy efficiency) would have implications that extend well beyond the realm of energy and climate policy.

3.2. The influence of alternative policy instruments on technology diffusion

The theoretical literature has long recognized that alternative types of policy instruments can have significantly different effects on the rate and direction of technological change, typically finding that economic incentive-based instruments (e.g., pollution taxes and tradable pollution permits) can provide more efficient incentives for technology adoption than conventional regulations (e.g., technology and performance standards). Despite a reasonable amount of theoretical attention, little empirical evidence exists on the dynamic effects of environmental regulation, particularly with respect to the relative effects of alternative policy instruments. Kerr and Newell (2004) provide some of the first such evidence.

That paper reports a detailed empirical study of these issues for an important industry undergoing technological responses to a dramatic decrease in allowed pollution levels. While many are familiar with the success of the tradable permit system for sulfur dioxide, the phasedown of lead in gasoline by U.S. petroleum refineries during the 1970s and 1980s was the first major success in implementing a market-based environmental policy. Historically, lead was added to gasoline to inexpensively boost octane levels, but it also has serious side effects on human health. They assess the pattern of technology adoption by refineries during the lead phasedown, both across refineries and across time, with the intent of understanding how various economic incentives, market factors, and the stringency and form of regulation influenced this process.

Toward this end, Kerr and Newell (2004) develop a model of the technology adoption decision in the presence of regulation and derive an econometrically testable duration model. Their econometric approach is related to that taken by several applied industrial organization studies of technology adoption, although those studies do not assess the influence of regulation on the process of technological change. The model suggests that firms will gradually adopt the technology as its costs fall and increased regulatory stringency increases the value of adoption; firms with lower benefits or higher costs will adopt more slowly. They also test the proposition that there will be a divergence in the adoption propensities of low versus high compliance cost plants during periods with a tradable permit system versus an individually binding performance standard (Malueg, 1989). Plants with relatively low costs of compliance (i.e., sellers in a permit market) will have greater incentives for cost-saving technology adoption within a trading regime. At the same time, relatively high-cost plants (i.e., permit buyers) will have decreased adoption incentives under the permit system.

The intuition behind this latter proposition is that the tradable permit system encourages all plants to take action until their marginal cost of pollutant reduction equals the permit price, while the individual standard forces all plants to attain a fixed target. Sellers' incentives to adopt are higher under the permit system because they can undertake additional reductions and get extra profits by selling permits. The incentives to adopt would be lower for buyers under the permit system, however, since they can buy permits rather than being forced to self-comply with relatively expensive reductions. Thus, the tradable permit system provides incentives for *more efficient* adoption, but it can lower adoption incentives for some plants with high compliance costs. Under a nontradable performance standard, such opportunities for flexibility do not exist to the same degree. If plants face individually binding standards, they will be forced to take individual action–such as technology adoption–regardless of the cost, with the resultant inefficiency reflected in a divergence across plants in the marginal costs of pollution control.

They employ a unique panel data set on petroleum refineries covering the full period of the U.S. lead phasedown, which began with a requirement that new cars after 1974 use unleaded gasoline. This was followed by performance standards on lead in gasoline, a tradable permit market controlling the lead in leaded gasoline (1983–1987), ending with a more stringent performance standard and ultimately a ban in 1996. The adoption of pentane–hexane isomerization technology–a substitute for lead as a source of octane–was one of the major responses to the increased severity of regulation.

Kerr and Newell (2004) find that increased regulatory stringency (which raised the effective price of lead) encouraged greater adoption of lead-reducing technology. They also show that larger and more technically sophisticated refineries were more likely to adopt the new technology. Importantly, they further find that the tradable permit system provided incentives for more efficient technology adoption decisions. The relative adoption propensity of refineries with low versus high compliance costs was significantly greater under the tradable permit regime than under a nontradable performance standard.

4. Invention and innovation of unconventional technologies: the case of carbon capture and storage

Energy efficiency and low-carbon or renewable energy sources were once seen as the only realistic means to reduce carbon dioxide (CO_2) emissions. In recent years, however, analysts and policymakers have begun to recognize the potential for a third option. Carbon capture and storage (CCS) technologies would remove CO_2 emissions from large, stationary sources (e.g., power plant flue gases). Captured emissions would then be compressed and transported for storage in geologic reservoirs (e.g., saline aquifers or depleted oil wells) or the deep ocean. In contrast to indirect forms of sequestration (e.g., forestation or enhanced ocean uptake of CO_2), which absorb CO_2 from the atmosphere, CCS seeks to avoid atmospheric emissions altogether.

Recognizing this, policymakers have recently begun to view the potential for CCS with increased seriousness. For instance, the budget devoted by the U.S. Department of Energy to CCS research has increased from about $1 million in 1998 to $44 million in 2003, just 5 years later. Internationally, the IPCC recently convened a group of policymakers and experts to outline the structure of a future IPCC special report on CCS technologies. And just this February, the U.S. Department of Energy announced plans for FutureGen, a $1 billion project to build a prototype power plant that integrates hydrogen and electricity production from gasified coal with geologic storage of captured CO_2 emissions.

In a review of CCS technologies and related modeling results, Anderson and Newell (2004) examine the prospects for CCS in terms of its technical feasibility, cost, timing, ancillary environmental effects, and potential contribution to an overall climate policy portfolio. Given current experience with these technologies in the oil, gas, and other industries, they find that their application to carbon mitigation is already technically feasible. Experience with these technologies in the oil, gas, and other niche industries shows that their application to carbon mitigation is technically feasible.

The existing evidence also suggests that these technologies could be economically attractive, given sufficiently stringent climate policies. Niche industries such as natural gas and hydrogen production already produce pure streams of CO_2, which could be compressed and diverted to storage sites at relatively low costs (i.e., under $50/t C). In fact, natural gas production is the only known case in which CCS technologies have been applied for the sole purpose of emissions reduction. This and similar opportunities are, however, quite small. Recent estimates suggest that the application of CCS in the electric power and industrial sectors could significantly reduce total U.S. emissions at a current cost of about \$200/t C to \$250/t C avoided. This is within the range of estimated costs for domestic U.S. compliance with the Kyoto Protocol, and many expect that these costs could fall substantially with time and technological development. In addition, a rise in natural gas prices–as would likely occur with the onset of a price on carbon emissions–could also lower the carbon price at which CCS technologies become competitive.

Although CCS may be economic under stringent climate policies, a number of technical, environmental, and political issues arise with regard to transportation and storage of captured CO_2. Despite significant experience with storage of CO_2 and other substances in underground reservoirs, there is substantial uncertainty regarding how much CO_2 such reservoirs can hold, how long injected CO_2 would remain trapped, and whether injected CO_2 would escape from storage reservoirs to other formations. The effects of ocean storage are even more uncertain, raise additional environmental concerns, and are more likely to generate controversy. Storage of CO_2 as carbonates could lessen many of the concerns related to ocean storage but would generate other environmental concerns and would entail substantially higher storage costs. Finally, leakage from storage facilities would weaken CCS as a source of permanent emissions reductions, though CCS could still provide valuable temporary storage while less costly permanent means of mitigation are being developed (e.g., renewable energy sources).

Recent modeling efforts at MIT (McFarland et al., 2004), Carnegie Mellon University (Johnson and Keith, 2004), and Pacific Northwest National Laboratory (Edmonds et al., 2002) have examined the role of CCS technologies under various carbon policies. Although these models differ significantly in methodology and geographic scope, their CCS results are fairly consistent. CCS technologies are typically found to enter into significant use after 20 to 35 years at carbon prices of \$50/t C to \$100/t C (see Table 2). Also notable is the finding that IGCC (Integrated Gasification Combined Cycle) plants with CCS are surprisingly competitive in these models, eventually surpassing NGCC (Natural Gas Combined Cycle) as the dominant fossil fuel technology. These results appear to differ markedly from those presented above, which indicate that when NGCC plants without CCS are the relevant reference technology, CCS plants only become competitive at \$200/t C or higher.

There are two primary sources of this apparent inconsistency. First, these models assume that technological change will lower the cost of CCS by as much as one-third compared with our estimates above (David and Herzog, 2000). Second, and more importantly, these models all

Table 2
Key results for CCS costs in electricity sector from integrated modeling studies

Study, scenario, and timeframe	CCS technology	CCS price (\$/t C)	CCS entry year	Max share of electricity production, year
EPPA, global price of \$50/t C in 2010,	NGCC	100	2020	16%, 2040
rising to \$200/t C by 2040, 1995–2095	IGCC	100	2020	50%, 2100
CMU, \$150/t C applied across MAAC region,	NGCC	175	–	–
2000–2040	IGCC	75	Immediate	35%, 2040
	PC retrofit	50	Immediate	10%, 2040
MiniCAM, stabilization at 550 ppmv, 1995–2095	NGCC	90	2020–2035	15%, 2095
	New PC	90	2020–2035	6%, 2095

MIT Emissions Prediction and Policy Analysis (EPPA) results are from McFarland et al. (2004). Carnegie Mellon University (CMU) results are from Johnson and Keith (2004). MiniCAM results from Pacific Northwest National Laboratory are from Edmonds et al. (2002). The \$175/t C entry price for NGCC plants in the CMU results represents the level at which NGCC would penetrate, where the 2000–2040 tax is higher than \$150/t C.

predict–either through endogenous modeling or exogenous assumptions–that gas prices will rise relative to coal during this century. As we discuss above, if the price of gas rises sufficiently, the relevant reference technology actually shifts from NGCC plants without capture to IGCC plants without capture, leading to a drop in the carbon price at which CCS technologies for IGCC plants become competitive.

Overall, the results of these studies suggest that fuel switching from coal to natural gas and energy efficiency improvements would provide the least costly options for moderate reductions in carbon emissions. Given sufficiently stringent carbon reduction targets, however, CCS technologies could play an important role in lowering mitigation costs. By allowing for the continued use of cheap coal at low effective rates of emissions, CCS technologies would provide a competitive alternative to costly renewable energy sources and fuel switching in the face of rising gas prices. Finally, CCS could capture significant quantities of CO_2 without exceeding most current estimates of probable storage capacity.

Several integrated assessment modeling studies suggest that CCS could play an important role in mitigating carbon emissions, conditional on policies that impose a sufficiently high implicit or explicit price on such emissions. The results indicate that fuel switching from coal to natural gas and energy efficiency improvements would be the least costly options for moderate reductions in emissions. For larger reductions and higher carbon prices, however, CCS substantially lowers mitigation costs. Assuming no barriers to implementation other than cost (i.e., ignoring political and environmental issues) and given certain assumptions (e.g., regarding fuel prices and energy demand), these studies suggest that a significant number of new plants with CCS would enter the power supply sector within the next few decades, though CCS retrofits could enter in just a few years given a sufficiently high price on emissions. The availability and use of CCS technologies would decrease reliance on renewable energy sources while encouraging electricity production to shift from natural gas to coal power. CCS would significantly reduce the present value of the cost of mitigation over time. Finally, CCS would result in the capture of significant quantities of CO_2 without exceeding most current storage capacity estimates.

5. Concluding remarks

An understanding of the process of technological change is important for economic analysis of climate change policy for two broad reasons. First, the environmental impact of social and economic activity is greatly affected by the rate and direction of technological change. This linkage occurs because new technologies may either create or mitigate pollution, and because many environmental problems and policy responses are evaluated over timeframes in which the cumulative impact of technological changes is likely to be large.

The importance of the first link is manifest in determining the economic and environmental "baseline" against which to measure the impacts of proposed policies. That is, before we can discuss what we should or should not do about some environmental problem, we need to forecast how severe the problem will be in the absence of any action. Such forecasts are always based, in some way, on extrapolation of historical experience. Within that historical experience, the processes of technological change have been operating, often with significant consequences for the severity of environmental impacts. Forecasts for the future based on this historical experience depend profoundly on the relative magnitude of the effects of price-induced technological change, learning-by-doing and learning-by-using, public sector R&D, and exogenous technical progress. Sorting out these influences with respect to environmentally relevant technologies and sectors poses a major modeling and empirical challenge.

A particularly important and understudied aspect of this set of issues is the conceptual and empirical modeling of how the various stages of technological change are interrelated, how they unfold over time, and the differential impact that various policies (for example, public-sector R&D, R&D subsidies to the private sector, environmental taxes, information programs) may have on each phase of technological change. There has been relatively little empirical analysis of these policy options directed specifically at the development of environmentally beneficial technology.

Looking forward, there are several promising areas for continued research on the modeling of induced technological change in CGE models:

- Further empirical analysis of the effects of alternative policy instruments on technological change, including non-carbon-price policies such as subsidies for technology adoption, R&D funding and tax incentives, information programs, renewable portfolio standards, and energy efficiency standards.
- Improved ability of assessment models to assess the effects of both price and non-price technology policies. This may require relaxing assumptions that the only relevant market failure is that related to the climate externality. In this regard, attention to knowledge spillovers and informational market failures may be necessary.
- Critical assessment and synthesis of the existing empirical literature on learning curves in energy technologies and new empirical research to avoid any pitfalls of previous studies and that can potentially be incorporated into computational models including those of an aggregate nature.
- Improved ability of assessment models to incorporate learning effects in an economically coherent manner.
- Empirical research on the degree of knowledge spillovers across different sectors and across national boundaries.
- Improved treatment by assessment models of technology spillovers and the opportunity cost of R&D directed toward carbon mitigation technology.
- Empirical research and improved incorporation into aggregate model of "nonconventional" mitigation technologies such as carbon capture and storage and renewable energy technologies.
- Empirical, analytical and numerical research on the treatment of long-term discounting in assessment models and its influence on technological change.

The potential long-run consequences of today's policy choices create a high priority for broadening and deepening our understanding of the effects of environmental policy on innovation and diffusion of new technology. Unfortunately, these issues cannot be resolved at a purely theoretical level, or on the basis of aggregate empirical analyses. For both benefit-cost and cost-effectiveness analysis, we need to know the *magnitudes* of these effects, and these magnitudes are likely to differ across markets, technologies, and institutional settings. Thus, taking seriously the notion of induced technological change and its consequences for environmental policy requires going beyond demonstration studies that test whether or not such effects exist, to carry out detailed analyses in a variety of sectors in order to understand the circumstances under which they are large or small. This will require significant research attention from multiple methodological viewpoints over an extended period of time. But the alternative is continuing to formulate public policies with significant economic and environmental consequences without being able to take into account what is going on "inside the black box" of technological change.

Acknowledgement

We acknowledge financial support from U.S. DOE Grant DE-FG02-98ER62702.

References

Anderson, Soren T., Newell, Richard G., 2004. Information programs for technology adoption: the case of energy-efficiency audits. Resource and Energy Economics 26 (1), 27–50.

Bohi, D.R., 1981. Analyzing Demand Behavior: A Study of Energy Elasticities. Resources for the Future, Washington, DC.

Brown, M.A., Levine, M.D., Short, W., Koomey, J.G., 2001. Scenarios for a clean energy future. Energy Policy 29 (14), 1179–1196.

David, Jeremy, Herzog, Howard J., 2000. The cost of carbon capture. Paper read at Fifth International Conference on Greenhouse Gas Control Technologies, August 13–16, at Cairns, Australia.

Edmond, James A., Clarke, John, Dooley, James J., Kim, Son H., Smith, Steven J., 2002. Modeling greenhouse gas energy technology responses to climate change. Paper Read at Sixth International Conference on Greenhouse Gas Control Technologies, September 30–October 4, at Kyoto, Japan.

Gaskins, Darius, Weyant, John (Eds.), 1996. Reducing Global Carbon Dioxide Emissions: Costs and Policy Options. Stanford University Press, Stanford.

Goulder, Lawrence, 1995. Effects of carbon taxes in an economy with prior tax distortions. Journal of Environmental Economics and Management 29 (3), 271–297.

Hausman, J.A., 1979. Individual discount rates and the purchase and utilization of energy—using durables. Bell Journal of Economics 10, 33B54.

Hicks, John R., 1932. The Theory of Wages. MacMillan, London.

Hogan, William W., Jorgenson, Dale W., 1991. Productivity trends and the cost of reducing CO_2 emissions. Energy Journal 12 (1), 67–85.

Interlaboratory Working Group, 2000. Scenarios for a Clean Energy Future. Oak Ridge National Laboratory, Oak Ridge, TN.

Jaffe, A.B., 1988. Demand and supply influences in R&D intensity and productivity growth. Review of Economics and Statistics 70, 431–437.

Jaffe, A.B., Stavins, R.N., 1994. The energy-efficiency gap: what does it mean? Energy Policy 22 (1), 804–810.

Johnson, Timothy L., David W. Keith, 2004. Fossil electricity and CO_2 sequestration: how natural gas prices, initial conditions and retrofits determine the cost of controlling CO_2 emissions. Energy Policy 32, 367–382.

Jorgenson, Dale, Wilcoxen, Peter, 1992. Reducing U.S. carbon dioxide emissions: an assessment of different instruments. Journal of Policy Modeling 15 (5).

Kerr, Suzi, Newell, Richard G., 2004. Policy-induced technology adoption: evidence from the US lead phasedown. Journal of Industrial Economics 51 (3), 317–343.

Malueg, D.A., 1989. Emission credit trading and the incentive to adopt new pollution abatement technology. Journal of Environmental Economics and Management 16 (1), 52–57.

Manne, Alan S., Richels, Richard G., 1992. Buying Greenhouse Insurance—The Economic Costs of Carbon Dioxide Emission Limits. MIT Press, Cambridge, Massachusetts.

McFarland, Jim, John, Reilly, Howard, Herzog, 2004. Representing energy technologies in top-down economic models using bottom-up information. Energy Economics 26, 685–707.

Morgenstern, R.D., Al-Jurf, S., 1999. Can free information really accelerate technology diffusion. Technological Forecasting & Social Change 61, 13–24.

Newell, Richard G. 1997. Environmental policy and technological change: the effects of economic incentives and direct regulation on energy-saving innovation. Ph.D. thesis, Harvard University, Cambridge, MA.

Newell, Richard G., Jaffe, Adam B., Stavins, Robert N., 1999. The induced innovation hypothesis and energy-saving technological change. Quarterly Journal of Economics 114 (3), 941–975.

Nordhaus, W.D., 1991. The costs of slowing climate change: a survey. The Energy Journal 12 (1), 37–64.

Pizer, William A., 1999. Optimal choice of climate change policy in the presence of uncertainty. Resource and Energy Economics 21, 255–287.

Popp, D., 2003. ENTICE: endogenous technological change in the DICE model of global warming. NBER Working Paper, vol. 9762.

Rosenberg, N., 1982. Inside the Black Box. Cambridge University Press, Cambridge, U.K.

Ruderman, H., Levine, M.D., McMahon, J., 1987. The behavior of the market for energy efficiency in residential appliances including heating and cooling equipment. The Energy Journal 8 (1), 101–124.
Schmookler, J., 1966. Invention and Economic Growth. Harvard University Press, Cambridge, MA.
Train, K., 1985. Discount rates in consumers' energy-related decisions: a review of the literature. Energy 10, 1243–1253.
Yellen, Janet, 1998. Testimony Before the House of Representatives Commerce Committee on the Economics of the Kyoto Protocol, March 4, 1998.

PART V

NATURAL RESOURCE ECONOMICS: LAND AND WATER

[16]

What Drives Land-Use Change in the United States? A National Analysis of Landowner Decisions

Ruben N. Lubowski, Andrew J. Plantinga, and Robert N. Stavins

ABSTRACT. *Land-use changes involve important economic and environmental effects with implications for international trade, global climate change, wildlife, and other policy issues. We use an econometric model to identify factors driving land-use change in the United States between 1982 and 1997. We quantify the effects of net returns to alternative land uses on private landowners' decisions to allocate land among six major uses, drawing on detailed micro-data on land use and land quality that are comprehensive of the contiguous United States. This analysis provides the first evidence of the relative historical importance of markets and federal farm policies affecting land-use changes nationally.* (JEL R14, Q15)

I. INTRODUCTION

Over the past quarter century, the United States has experienced dramatic land-use changes—including declines in cropland, increases in forests, and accelerated expansion of urban areas—that depart from trends over the previous decades.[1] Broad-scale changes in land use produce important effects with implications for international trade, global climate change, wildlife habitat and a host of other policy issues. For example, land use has figured prominently in international negotiations on climate change. Under the Kyoto Protocol, countries can claim credit for offsetting carbon dioxide (CO_2) emissions through changes in land use and land management. Also, in communities across the United States, "urban sprawl" has been a concern voiced by many citizens (Pew 2000). And the current Doha Round of the World Trade Organization (WTO) has stalled over the issue of reducing agricultural support programs in the United States and Europe (e.g., WTO 2006). Developing nations oppose these programs on the grounds that they increase production and lower global crop prices (e.g., Oxfam 2003).

What has caused the changes in land use that have occurred in the United States over the past twenty-five years? What has been the role of major federal farm policies? We investigate the influences of different economic and policy factors on national land use from 1982 to 1997 by estimating an econometric model of the revealed preferences of landowners drawing on detailed micro-data on land use and land quality. Our model measures the effects of estimated net returns to alternative land uses on decisions by private landowners to allocate their land among six major uses: crops, pasture, forest, urban, range, and a federally financed use, the Conservation Reserve Program (CRP).[2] Be-

[1] Between 1982 and 2002, total cropland in the 50 United States, including land in conservation programs, declined by 27 million acres (6%) after rising 25 million acres over 1964–1982 (Lubowski et al. 2006b). Forest area grew by 11 million acres (2%) over 1987–2002 after falling 24 million acres over 1963–1987 (Smith et al. 2004). The average urbanization rate rose from 1.0 to 1.4 million acres annually from 1960–1990 to 1990–1997 (Vesterby and Krupa 1995).

The authors are, respectively, economist, Environmental Defense Fund; associate professor, Department of Agricultural and Resource Economics, Oregon State University; and the Albert Pratt Professor of Business and Government, John F. Kennedy School of Government, Harvard University and University Fellow, Resources for the Future. This paper is based, in part, on Ruben N. Lubowski's Ph.D. dissertation at Harvard University. The authors thank Michael Roberts, Patrick Sullivan, and two anonymous reviewers for very helpful comments and acknowledge financial support from the U.S. Department of Energy. The authors alone are responsible for any errors and for all opinions expressed in this paper.

[2] The CRP, established by the Food Security Act of 1985, is the largest Federal program targeting land use. The program offers rental payments to landowners who voluntarily retire environmentally sensitive cropland under 10- to 15-year contracts. CRP presently pays about $1.8 billion per year to retire 36.1 million acres (about the size of Iowa) (FSA 2006).

Land Economics • November 2008 • 84 (4): 529–550
ISSN 0023-7639; E-ISSN 1543-8325

cause individual landowners are likely to take prices and production technologies as given, we focus on net returns as the drivers of land-use change, abstracting from the various underlying factors affecting the net returns for land. These fundamental drivers include population growth, consumer tastes, international trade, and other factors affecting the demand for land in different uses. Underlying drivers also include weather, technology, local rules, and other factors determining production possibilities from different land-use alternatives.

Our national-level model allows us to estimate the impacts on land-use decisions of federal policies, such as CRP, and other factors affecting the land base nationwide. Previous econometric land-use studies have focused on relatively small geographic areas, such as single regions or states. The few studies using national-level data have examined changes in a single land-use category, without modeling competition among land-use alternatives.[3] Modeling this competition is critical for measuring the impact of factors affecting land-use choices when multiple land-use options are economically viable. We estimate a national model that accounts for transitions among a comprehensive set of major private land-use categories.[4] Detailed data on the same points of land over time allows us to model transitions among different land uses—rather than just net changes in particular category—and to account for variation in land quality that affects the profitability and corresponding choice of alternative uses.

Most econometric analyses of land use examine the relationship between observed land-use patterns and measures or proxies for land rents (see Bockstael and Irwin 2000 for a review), but there has been little analysis of the relative importance of various factors in driving land-use changes. One exception was provided by Stavins and Jaffe (1990), who used an econometric model based on county data to examine the relative importance of federal flood control projects and other factors in driving the depletion of forested wetlands in the Mississippi Delta from 1935 to 1984. Our study extends their analysis by examining changes among a comprehensive set of major private land-use categories on a national scale. We use econometric estimates to simulate national changes in landowners' willingness to supply land in various uses from 1982 to 1997 under a series of scenarios that isolate the impact of particular market and policy factors.

Changes in national land use will influence the production of crops, timber, and other land-based commodities, affecting the prices for these products and, in turn, the incentives to allocate land among alternate uses. While we do not explicitly model these feedbacks, our simulations suggest the direction of impacts and relative importance of different factors affecting national land use. We also compare results from simulations using smaller changes in market and policy factors for which price feedbacks are likely to be minor.

To model both the supply and demand for land, other researchers have developed optimization models of the agriculture and forestry sectors. For example, Alig, Adams, and McCarl (1998) examined different future scenarios for CRP, farm program payments, and afforestation policies using a two-sector, multi-period simulation model, in which the forest and agricultural sectors are linked, and the welfare of producers and consumers in the two sectors is maximized.[5] While such models enable simulations of market or policy changes accounting for

[3] Previous national-level econometric studies examined urban area (Alig and Healy 1987), conservation programs (Parks and Kramer 1995; Plantinga, Alig and Cheng 2001), and timberland (Plantinga and Buongiorno 1990).

[4] While land-use categories can always be subdivided more specifically, our categories are more detailed than the "major" private uses reported in the USDA's *Major Land Uses* data series, the only accounting of land use for the entire nation (e.g., Lubowski et al. 2006b). In contrast to this series, we distinguish CRP from cropland; differentiate pasture and range; and exclude miscellaneous uses that are not identified.

[5] Other applications of this Forest and Agricultural Sector Optimization Model (FASOM) include Alig et al. (1997), Adams et al. (1999), McCarl and Schneider (2001), and EPA (2005).

both changes in land use and corresponding prices, optimization models are limited in their ability to measure the impacts of different factors on land use historically and, to our knowledge, have not been used for this purpose. Optimization models are governed by assumed decision rules and elasticities and cannot incorporate all factors that affect land use in reality. These factors include option values in the face of uncertainty and irreversible investments, liquidity constraints, and unobserved market or non-monetary costs or benefits (Stavins 1999). Existing models are also limited to changes among agricultural and forest uses, rather than the broad set of land uses considered in our study. Our revealed preference approach aims to address some of these limitations and, in doing so, could provide estimates for calibrating or cross-validating optimization analyses.

We find that private land-use decisions have depended critically on land quality and have been steered by anticipated economic returns to alternative uses, which in some cases have been affected significantly by public policies, sometimes intentionally and sometimes unintentionally. Our results have implications for a number of policy issues related to land use. For example, with regard to potential market impacts from federal farm programs, we find that positive effects on crop acreage of all direct federal payments to farmers were more than offset by cropland retirement under the CRP.[6] Indeed, we estimate that, in total, these federal farm programs resulted in a modest decline in U.S. cropland area.

II. ECONOMETRIC LAND-USE MODEL

Consider a risk-neutral and price-taking landowner facing the choice of allocating a parcel of land of uniform quality among a set of alternative uses.[7] We posit that landowners choose uses to maximize the present discounted value of the stream of expected net benefits from the land, and base their expectations of future land-use profits on current and historic values of relevant variables. Landowners would expect current values to persist over time if land-use net returns follow a random walk. Limited empirical evidence suggests that this is approximately true (Schatzki 2003). For simplicity, our analysis also abstracts from the effects of uncertainty in the presence of fixed, irreversible costs (Dixit and Pindyck 1994). We thus model land-use choices as if landowners ignore the value of the option to delay irreversible costs and obtain additional information.[8]

Given these simplifying assumptions, the decision rule that emerges from the related dynamic optimization problem is to choose the use with the highest expected one-period return at time t minus the current one-period expected opportunity cost of undertaking conversion (Lubowski 2002). With K potential land uses (j, $k = 1,\ldots,K$), the owner of a unit of land in use j will choose the use k at time t that satisfies:

$$\arg\max_{k}\left(R_{kt} - rC_{jkt}\right) \geq R_{jt}, \qquad [1]$$

where R_{jt} and R_{kt} represent the instantaneous expected net benefits at time t from a unit of land in use j and k, respectively; C_{jkt}

[6] Federal funds paid directly to farmers include commodity programs, disaster assistance, land retirement and other conservation programs, and miscellaneous payments. Funds paid under commodity and disaster programs averaged $6.3 billion annually from 1982 to 1997 (ERS 2003).

[7] If net returns and the costs of converting land between different uses are approximately linear in land quantity, the size of parcels will not affect the relative profitability of land-use options, in which case land-use decisions for a heterogeneous parcel can be treated as the sum of land-use choices on constituent uniform-quality parcels. While studies show that parcel size affects the value of land for urban development due to costs of subdividing and assembling parcels (e.g., Thornes and Miller 1998), we do not model these effects for lack of data and because urban development is a small share of the land-use changes considered.

[8] Given the broad scope of our analysis, it would be impossible to model explicitly the landowners' stochastic dynamic optimization problem (Rust 1987). However, our econometric approach incorporates option values indirectly. If fixed costs and other irreversible choices introduce rigidities in land use, these will be reflected in the estimated coefficients.

is the expected marginal cost of converting one unit of land from use j to use k at time t ($C_{jjt} = 0$); and r is the discount rate. If the use k satisfying equation 1 equals j, then the land unit will remain in its current use at time t; otherwise, the landowner will reallocate the land to the use $k \neq j$ that maximizes expected net returns after conversion costs.

The landowner's profit function may be thought of as including both observed and unobserved components. Specific restrictions on the structure of the unobserved components yield alternative specifications of probabilistic models. An ordinary logit model is one obvious possibility, but the assumption of independent disturbances in the simple logit model implies that the ratio of the probabilities of any two choices is independent of the other alternatives.

We allow for differences in substitutability among alternatives using a *nested* logit specification, which imposes this property of "independence of irrelevant alternatives" within but not across specified subgroups ("nests") of choices. Dividing the choice set into mutually exclusive subgroups K_s ($s=1,\ldots,S$), we specify U_{ijkt}, the landowner's utility from converting land parcel i from use j to use k at time t, as the sum of a component, V_{ijkt}, that is unique to the alternative k and another component, V_{ijst}, that is common to all the alternatives in K_s. Each of these components, in turn, includes an observed component plus an unobserved component characterized as a random error.

Under assumptions analogous to the standard logit model, the probability of choosing alternative k that is grouped in K_s can be expressed as the product of two terms: the probability, P_{ijst}, of choosing any of the alternatives within K_s; and the conditional probability, $P_{ijkt\|s}$, of choosing k given the choice of K_s.[9] For land parcel i starting in use j, the probability of choosing land use $k \in K_s$ between time t and $t + 1$ is thus

$$P_{ijkt} = P_{ijst} \cdot P_{ijkt|s} = \frac{\exp(V_{ijst} + \tau_{st} I_{ijst})}{\sum_{s=1}^{S} \exp(V_{ijst} + \tau_{st} I_{ijst})} \cdot \frac{\exp(V_{ijkt})}{\sum_{l=1}^{J_s} \exp(V_{ijlt})}, \qquad [2]$$

where τ_{st} are parameters, and $I_s = \ln \sum_{l=1}^{J_s} \exp(V_{ijlt})$. This "inclusive value" for nest K_s equals the expected utility for the choice of alternatives within a nest. The expression in equation [2] embodies the first-order Markov property since the probability of the parcel changing use depends only on exogenous covariates at time t.

Our chosen nesting structure is based on the premise that land uses with more similar land quality requirements are closer substitutes.[10] We expect land quality to affect land-use net returns principally in terms of agricultural yields, and we proxy land quality using the Land Capability Class (LCC), a summary measure of the suitability of land and associated climate for producing crops (USDA 1973).[11] Land in crops has the highest average quality as measured by the LCC, while pasture and CRP uses tend to be adopted on higher quality lands relative to forest and range uses.[12] In the case of CRP,

---

[9] Assumptions of the nested logit model imply that the two disturbances are independent and that their sum has the Gumbel distribution (Ben-Akiva and Lerman 2000).

[10] Land quality is only one potential determinant of substitutability among land uses. To the extent that farmers operate joint crop and livestock operations, farmers may have skills for pasture and range uses—rather than forestry, for example—so crops, pasture, and range uses may be closer substitutes for each other than for other uses. On the other hand, forest and pasture are similar in terms of lower labor requirements.

[11] The LCC measure was designed to target conservation programs, rather than to measure soil productivity, and allows local flexibility in rankings, producing some inconsistencies across regions (Helms 1992). Despite these limitations, we use LCC because estimated crop yields, the Soil Rating for Plant Growth (SRPG) (Soil Survey Staff 2000), and other explicit productivity measures are not currently available for all soil types and land uses.

[12] As a rough indicator, assigning values 1 through 4 to LCCs I–II, III–IV, V–VI, and VII–VIII, respectively, the NRI indicates that the average qualities in the contiguous United States over 1982–1997 were 1.5 for cropland, 1.9 for CRP, 2.0 for pasture, 2.2 for urban land, 2.7 for forests, and 3.0 for range, with lower values indicating higher quality.

program rules restrict which lands are eligible to participate; notably, eligibility is limited to lands that were planted to an agricultural commodity for four of the previous six crop years. We incorporate these differences in land quality requirements by specifying our nested logit model with three nests: K_1 (crops, CRP, and pasture); K_2 (forest and range); and K_3 (urban). We model urban land use as a unique nest, due to its greater degree of irreversibility, and because land quality, as measured by the LCC, is likely to be a much less important determinant of urban development returns.

Landowners presumably compare net returns to alternative uses on particular parcels. Although we have land-use data at the parcel level, we lack parcel-level observations of net returns. Instead, we observe county-level average returns and, to allow for parcel-level variation, we interact the average return variables for each land use with parcel-level indicators of land quality. The county average return for a particular use reflects the average return on those parcels of land in the county currently in that use. The estimated land quality adjustment factors scale this average net return to approximate the net returns to each use on every parcel of land, regardless of its current use. The magnitude of the land quality adjustment could be different for different land uses depending on the sensitivity of the different net returns to land quality as well as on the land quality distribution for the land in each use. In reality, in large parts of the country, forests, range, or other land uses in our choice set may not be a feasible alternative given climatic conditions or other biophysical, economic, or legal constraints. Given the national focus of our study, our model captures the average effects of LCC on land-use conversions but does not identify in what specific locations these effects might be more or less important.

We thus specify the component of utility that is unique to each alternative as

$$V_{ijkt} = \alpha^0_{jkt} + \alpha^q_{jkt}LCC^q_{it} + \beta^0_{jkt}R_{kc} + \beta^q_{jkt}LCC^q_{it}R_{kc} + \varepsilon_{ijkt}, \qquad [3]$$

where α^0_{jkt} is an alternative-specific intercept, α^q_{jkt}, β^0_{jkt}, and β^q_{jkt} are parameters, R_{kc} is the level of net returns to use k in county c, LCC^q_{it} is a dummy variable indicating whether parcel i is of quality q at time t, and ε_{ijkt} is the error term.[13] We lack data on the costs of changing land use, but we expect these costs to be closely related to land cover as well as slope and other factors affecting LCC. Accordingly, we model conversion costs with the terms α^0_{jkt} + $\alpha^q_{jkt}LCC^q_{it}$ in equation [3], which provide an intercept term varying with initial use and our land quality proxy.[14]

CRP participation depends on a different set of decisions than other land-use choices, because enrollment depends on both the landowner's bid, which includes a proposed rental rate, and the government's choice of whether to accept the bid, which depends on the environmental characteristics of a parcel as well as the cost. Because the program targets cropland, CRP rental rates are highly correlated with the profitability of cropping in a given locality. We account for the effect of crop net returns on the incentive to remain in cropland. Incentives to enroll in CRP are specified as a function of LCC, which was a consistent element of the program's eligibility criteria during our period of analysis.[15]

[13] The eight LCCs are merged into four groups: I–II, III–IV, V–VI, and VII–VIII.

[14] Costs of converting land to crops, such as clearing and drainage, have been estimated at state and regional levels (e.g., Heimlich et al. 1998). We proxy such costs with LCC as they are likely to be very site specific and because explicit cost estimates are not available for most of the land-use changes in our model. We expect conversion costs to increase with slope, erodibility, excess water, and stoniness, which are key factors proxied by higher LCC. An anonymous reviewer noted that conversion costs could inversely vary with LCC if more productive lands have more vegetation requiring clearing. The intercept term varying by initial use will partially proxy for vegetative cover.

[15] Until sign-up 13 in 1995, all lands with LCC ratings VI–VIII were eligible for CRP, while lands in LCC II–V were only eligible subject to erosion criteria and other requirements which changed over time. We did not include erosion factors in the NRI data as determinants of CRP participation because eligibility as measured by these variables was not correlated with CRP enrollments, possibly due to insufficient sampling of CRP parcels.

Also, because CRP rental rates are subject to caps, only lands earning returns below this cap are likely to be enrolled in the program. For these reasons, greater enrollment would be expected on lower quality croplands, as proxied by LCC. Equation [3] thus becomes $V_{ijkt} = \alpha^{0}_{jkt} + \alpha^{q}_{jkt}LCC^{q}_{it} + \varepsilon_{ijkt}$ for $k = CRP$.

For the component of utility that is constant across the alternatives within each nest, we include constant terms for the nest and interactions with the land quality indicator variables. For land parcel i in use j, the component of utility that is constant within each nest is thus

$$V_{ijst} = \gamma^{0}_{jst} + \gamma^{q}_{jst}LCC^{q}_{it} + \tau_{st}I_{ijst}, \qquad [4]$$

where γ^{0}_{jst} is a constant specific to nest s and γ^{q}_{jst} is a coefficient on the land quality indicators. Substituting equations [3] and [4] into [2] yields a complete nested logit model for estimation.

We estimate the model using repeated observations of land use on individual parcels from the National Resources Inventory (NRI) of the U.S. Department of Agriculture's Natural Resources Conservation Service (NRCS). The NRI is a panel survey of land use and land characteristics on non-Federal lands conducted at five-year intervals from 1982 to 1997 over the entire United States, excluding Alaska.[16] Data include about 844,000 sample "points," each representing a land area given by a sampling weight (Nusser and Goebel 1997). We observe land use at each point in four survey years, providing information on land-use changes over three transition periods: 1982–1987, 1987–1992, and 1992–1997. We focus on six land uses: crops, pasture, forest, urban, range, and CRP.[17] The land base in our analysis comprises 1.4 billion acres, representing about 74% of the total land area and 91% of non-federal land in the 48 contiguous states. Further details on the NRI data are provided in the Data Appendix.

The matrix of land-use changes in Table 1 illustrates the information provided by the NRI on the disposition of land across major uses over time. Land units generally remain in the same use. For example, of land parcels cropped in 1982, 84.2% remained in crops in 1997, while 7.3% had been converted to CRP, 4.6% to pasture, 1.6% to urban use, 1.3% to forest, and 0.9% to range. Of land parcels forested in 1982, 95.4% remained in forests in 1997, while 2.5% had been converted to urban use, 1% to pasture, 0.5% to crops, another 0.5% to range, and 0% to CRP. The urban land-use category appears to be an absorbing state, with lands almost never converted from urban to non-urban uses.

Our dependent variable is the choice of land use in year, $t + 5$, at each NRI point, and our covariates are the land use in year t, the land quality rating of the point, and proxies for the expected net returns from the land-use alternatives as of year t. To smooth temporary shocks from weather and other factors that affect net returns in particular years, we assume that landowners use an average of the annual net returns per acre to each land use over the most recent five-year period as the basis for their expectations of future net returns. Denoting each year as t, we specify land-use choices observed at time $t + 5$ as a function of the average land-use net returns between the years $t - 4$ and t, inclusive.[18]

Assembling data from a variety of private and public sources, we constructed county-level estimates of annual financial net

[16] Beginning in 2001, the NRI has been conducted annually on a smaller sample of points. Micro-data from the annual NRI are not publicly available to date and were not used in our analysis.

[17] Public lands and transportation infrastructure are excluded from the analysis, as changes in these uses are not affected directly by utility maximization by private landowners. We omit water bodies and barren lands as these uses are unlikely to vary over time. Finally, we exclude lands classified as marshlands and "miscellaneous," because data are not available to measure net returns to these uses.

[18] While some information after year t could be relevant for explaining subsequent changes, we only use information as of year t for our net returns proxy to avoid using data from future periods as predictors of past land-use decisions. Because observations of land use are at five-year intervals, our dependent variable includes land-use changes occurring at any time between t to $t + 5$, and we do not know when during this period these changes occurred.

TABLE 1

CHANGES IN MAJOR NON-FEDERAL LAND USES FROM 1982 TO 1997 IN THE CONTIGUOUS 48 STATES (1,000S OF ACRES)[a]

Land Use in 1982	Land Use in 1997						
	Cropland	CRP[b]	Forest Land	Pastureland	Rangeland	Urban Land	1982 Total[c]
Cropland	350,265 84.19%	30,412 7.31%	5,607 1.35%	19,269 4.63%	3,659 0.88%	6,816 1.64%	416,029 100%
Forest land	2,037 0.51%	4,168 1.05%	380,343 95.42%	9,803 2.46%	2,099 0.53%	129 0.03%	398,579 100%
Pastureland	15,347 11.85%	1,330 1.03%	14,091 10.88%	92,088 71.12%	2,568 1.98%	4,053 3.13%	129,478 100%
Rangeland	6,968 1.69%	729 0.18%	3,022 0.73%	3,037 0.74%	394,617 95.91%	3,055 0.74%	411,427 100%
Urban land	2 0.00%	0 0.00%	2 0.00%	1 0.00%	0 0.00%	51,946 99.99%	51,951 100%
1997 Total[c]	376,383 19.42%	32,696 1.69%	404,680 20.88%	119,513 6.17%	404,824 20.89%	75,924 3.92%	1,937,664 100.00%

[a] Read the table horizontally to see how land that was under a particular use in 1982 was subsequently allocated in 1997. Read the table vertically to see how land that that was in a particular land use in 1997 was previously allocated in 1982. Percentages are of 1982 totals (far right column).

[b] There is no corresponding row entry for the Conservation Reserve Program (CRP) as this Federal program was established in 1985.

[c] In addition to the land uses reported, totals include federal lands, water bodies, and other land-use categories not examined in this study.

returns (revenues less variable cash costs per acre) for crops, pasture, forest, range, and urban uses for all 3,014 counties in the 48 contiguous states (see Data Appendix). Net returns for crops, pasture, and range are computed in annual terms given annual yields from major crops and forage. On the other hand, returns to forests and urban uses are calculated as the net present values of a perpetual stream of forest and urban returns, respectively, and then converted to a constant annualized equivalent with an assumed private discount rate of 5%.[19] The estimates for cropland include net returns from market sales as well as direct farm program payments.[20] Our measure of payments is from the Census of Agriculture and includes all direct receipts from the Federal government, excluding payments for land retirement under the CRP and Wetlands Reserve Program (WRP). These payments are excluded because we model CRP as a distinct land-use category, and CRP and WRP payments are jointly reported in the Census. Table 2 provides summary statistics on land-use net returns.

III. ESTIMATION AND RESULTS

We use maximum likelihood methods to estimate the parameters of the model, drawing on cross-sectional variation in the independent variables to estimate parameters for the transition probabilities from each of four starting land uses (crops, pasture, forest, and range) to each of our six ending uses (crops, pasture, forest, range, urban and CRP) during each of the temporal transitions.[21] We use all of our observations on the NRI sample points in each respective land use at the start of each of the three transition periods (1982–1987, 1987–1992, and 1992–1997), and so in total,

[19] Annualized returns are thus $r{*}NPV$ where $NPV = \sum_{t=0}^{\infty} \frac{R_t}{(1+r)^t}$ and $r = 0.05$.

[20] We assume landowners are indifferent between a dollar from the government and a dollar from the market. We conjecture that farmers' expectations of future returns are approximately static with respect to the sum of market net returns and government payments, rather than to these income streams individually, given that federal farm program payments are highly countercyclical, partially insuring market risks (Roberts, Osteen, and Soule 2004).

[21] We also estimated models with pooled data and found that estimates based on separate cross-sections were superior at replicating land-use changes over these time periods (Lubowski 2002).

TABLE 2

ESTIMATED COUNTY-LEVEL ANNUAL NET RETURNS PER ACRE FROM THE MAJOR LAND USES FOR THE 48 CONTIGUOUS UNITED STATES, 1978–1992

	Mean and (Standard Deviation)				Change from 1978 (% Below)			
Variable ($/acre/year)[a]	1978	1978–1982	1983–1987	1988–1992	1978	1978–1982	1983–1987	1988–1992
Crop market net returns[b]	102.0	43.2	52.0	70.6	0	−58.8	−50	−31.4
	(69.1)	(34.4)	(39.5)	(46.6)	0%	−57.7%	−49.0%	−30.8%
Government payments[c]	8.7	15.2	21.8	12.2	0	6.5	13.1	3.5
	(4.7)	(7.2)	(8.8)	(6.0)	0%	74.3%	149.0%	39.6%
Crop market plus government net returns[d]	110.7	58.4	73.8	82.8	0	−52.3	−36.9	−27.9
	(69.8)	(38.4)	(45.0)	(48.2)	0%	−47.3%	−33.4%	−25.3%
Pasture net returns	32.6	16.1	7.9	12.7	0	−16.5	−24.7	−19.9
	(28.4)	(11.3)	(8.0)	(9.0)	0%	−50.5%	−75.6%	−61.1%
Forest net returns	6.4	6.0	9.0	17.2	0	−0.4	2.6	10.8
	(6.4)	(5.9)	(9.0)	(17.1)	0%	−7.2%	39.7%	166.2%
Urban net returns	1,809.0	1,946.3	2,389.5	2,348.8	0	137.3	580.5	539.8
	(1,808.9)	(1,946.3)	(2,389.4)	(2,348.8)	0%	7.6%	32.0%	29.8%
Range net returns	11.3	11.2	10.3	10.4	0	−0.1	−1	−0.9
	(10.5)	(10.1)	(8.9)	(9.3)	0%	−1.4%	−8.8%	−8.0%

[a] Values are averages over each five-year period of weighted annual county-level returns (see Data Appendix). Values are in 1990 dollars, deflated using the producer price index for all commodities from the Bureau of Labor Statistics.
[b] Includes only the market-component of crop net returns (price times yield minus variable costs).
[c] Includes direct federal farm program payments per acre, excluding the Conservation and Wetlands Reserve Programs.
[d] The sum of the market-component of crop net returns plus direct federal farm program payments.

we estimate twelve separate equations (four starting land uses in three time periods).[22]

The results indicate good fit of the model, and are consistent with the expected economic relationships.[23]

For brevity, Table 3 reports elasticities, rather than estimated parameters for the twelve equations.[24] For each starting use, we report the estimated elasticity for the probability of choosing each land-use alternative with respect to the net returns to that alternative ("own return elasticities"). These elasticities indicate the percentage change in the probability of a particular land-use change for a 1% change in the corresponding net returns. In 35 out of 60 cases, the own-return elasticities are positive and significant at the 0.05 level. In the seven cases where the own-return elasticities are negative, they are never significantly different from zero at the 0.05 level. The "cross-elasticities" are the elasticities of the probability of choosing a particular use *j* with respect to the profits of a different use *k*. In all but four of 300 cases, the cross-elasticities (not reported) are opposite in sign to the own-return elasticities and thus usually negative.

The elasticities indicate that landowners starting with lands in crops or pasture responded as anticipated to net returns from alternative land uses. For land starting in range, the own-return elasticities with respect to urban net returns are positive and significant, as well as the forest net return elasticity for 1982–1997. None of the other own-return elasticities for land starting in

[22] Because land virtually never transitions out of urban uses, we do not model these potential land-use changes and, in the simulations, assume land parcels remain urbanized with 100% probability. We estimated our model for points starting in CRP over 1992–1997 but not 1982–1987 or 1987–1992. Some land became eligible to exit the program in 1996 and changed to other uses by 1997. Other research models this post-CRP land use (Roberts and Lubowski 2007).

[23] Pseudo R^2 values range from 0.71 to 0.95 for the different equations. The positive and significant estimates of the inclusive value parameters and tests (Hausman and McFadden 1984) of a simpler, non-nested logit model with the same variables are consistent with violations of the "independence of irrelevant alternatives" hypothesis, supporting the use of the less restrictive nested logit specification. However, the nested specification was not critical for our overall findings, as the non-nested model yielded qualitatively similar results in our land-use simulations.

[24] Parameter estimates are reported in Lubowski (2002), and are available from the authors upon request.

TABLE 3

OWN-RETURN LAND-USE CHOICE ELASTICITIES[a]

	Final Land Use and Time Period[b]														
	Crops			Forest			Pasture			Range			Urban		
Initial Land Use	1982–1987	1987–1992	1992–1997	1982–1987	1987–1992	1992–1997	1982–1987	1987–1992	1992–1997	1982–1987	1987–1992	1992–1997	1982–1987	1987–1992	1992–1997
Crops	0.01**	0.03**	0.01**	0.88**	0.75**	0.31**	−0.00	0.08**	0.18*	0.68**	0.29**	0.38**	0.39**	0.25**	0.34**
	(0.00)	(0.00)	(0.00)	(0.05)	(0.07)	(0.04)	(0.03)	(0.02)	(0.03)	(0.087)	(0.07)	(0.05)	(0.02)	(0.01)	(0.02)
Pasture	0.23**	0.38**	0.34**	0.22**	0.08	0.00	0.03**	0.00**	−0.01	0.85**	0.71**	1.04**	0.43**	0.30**	0.33**
	(0.02)	(0.03)	(0.02)	(0.03)	(0.05)	(0.03)	(0.00)	(0.00)	(0.01)	(0.13)	(0.17)	(0.05)	(0.05)	(0.03)	(0.026)
Forest	0.21**	0.28**	0.29**	0.00	0.00	0.00	0.10	0.04	−0.01	0.28	−0.56	0.23	0.23**	0.30**	0.80**
	(0.07)	(0.05)	(0.06)	(0.00)	(0.00)	(0.05)	(0.06)	(0.09)	(0.06)	(0.23)	(3.46)	(0.33)	(0.02)	(0.07)	(0.06)
Range	−0.04	0.35	0.065	0.08	0.03	0.13	0.22	−0.15	0.40	−0.00	−0.00	−0.00	0.56**	0.40*	0.42**
	(0.05)	(0.18)	(0.23)	(0.36)	(0.42)	(0.91)	(0.17)	(0.53)	(0.42)	(0.00)	(0.00)	(0.97)	(0.05)	(0.18)	(0.03)

[a] Elasticities are evaluated at the means of the data and are the percentage change in the probability of choosing the final land use, conditional on being in the initial use, for a 1% change in the net returns to the final use. Standard errors, in parentheses, are estimated using the Delta Method (Greene 2000).

[b] There are no own-return elasticities for choosing CRP as net returns for CRP are not explicitly included in the econometric model.

* and ** denote significance at the 5% and 1% levels, respectively.

range are significantly different from zero, suggesting that range owners are relatively insensitive to the profitability of alternative uses, except for urban development. This is reasonable, given that range lands tend to be of the lowest quality and thus unsuitable for other agricultural uses.

A final econometric consideration is that land parcels located near one another may have unobserved characteristics correlated across space. If such characteristics influence land-use decisions or if local land-use choices are interdependent, error terms will be correlated across space, leading to inconsistent and inefficient parameter estimates in a logit model due to induced heteroskedasticity (McMillen 1992). A common approach for dealing with spatial autocorrelation is to estimate a model with a spatial weight matrix that relates the error terms in each location to error terms in all the other locations. In the context of discrete dependent variable models, this approach is very computationally intensive and has only been applied to small data sets (e.g., Carrion-Flores and Irwin 2004). We explored the potential importance of spatial dependence by eliminating observations near to one another, an alternative method adopted in previous discrete choice land-use models (Nelson and Hellerstein 1997; Nelson, Harris, and Stone 2001; Cropper, Puri, and Griffiths 2001). Estimates with samples that included only a single point within each NRI primary sampling area produced results similar to estimates including all points, suggesting that spatial dependence is not a critical concern.[25]

[25] Because the precise location of each point is not disclosed to preserve confidentiality, there is no way to directly test for spatial autocorrelation in a model with NRI point data. However, NRI's stratified sampling design provides some information on spatial relationships. Data on urban and water areas are collected for about 300,000 primary sampling areas varying from 40 to 640 acres in size. More detailed land characteristics and use data are collected at two to three points randomly selected within each of these areas (Nusser and Goebel 1997). Restricting our sample to one point within each of these clusters excludes points potentially on the same parcel or otherwise likely to share features related to location that might affect land-use choices.

IV. SIMULATIONS

In order to explore the relative importance of factors affecting land use, we simulate nationwide changes in landowners' willingness to supply land in different uses using fitted values of land-use change probabilities derived from the econometric model. These simulations estimate changes in land allocation at the parcel-level given hypothetical exogenous changes in land-use incentives. Aggregating these individual responses yields an estimate of changes in the national supply of land in alternative uses under various counterfactual scenarios for government payments and market net returns. These simulations suggest the direction of impacts and the relative importance of different factors affecting national land use.

Simulation Scenarios

We simulate a factual and eight counterfactual scenarios (Table 4). The "factual" simulation uses the historically observed values of all variables and provides a baseline with which to compare land-use changes under the counterfactual scenarios. The counterfactual scenarios differ in the levels of land-use net returns used to compute transition probabilities. In each counterfactual, we fix a particular variable at a hypothetical level, keeping all other variables at their actual historical values, so as to identify the effects of individual factors on the supply of land in each use. These simulations illustrate how landowners would have responded if equilibrium net returns incentives had been different as a result of an exogenous market or policy change.

Comparing the factual simulation with counterfactuals in which particular variables are held constant at 1978 levels allows us to isolate the effects on land-use decisions of changes in those variables that occurred over the period of analysis, assuming all else had remained constant.[26] For example, the

[26] Table 2 lists the estimated mean annual net returns per acre over the different time periods and the absolute and percentage changes subsequent to 1978. Landowners are assumed to consider the average of 1978–1982 values when choosing land-use over the first transition period (1982–1987).

TABLE 4
DESCRIPTION OF SIMULATION SCENARIOS

Scenario	Description
Factual	All returns at actual historical values
No Change in Crop Market Returns	Fix crop returns from market (but not government payments) at 1978 values
No Change in Government Payments	Fix government farm payments at 1978 values.
No Government Payments	Set government farm payments at zero
No Conservation Reserve Program (CRP)	Eliminate CRP as land-use option
No Government Payments and no CRP	Set government farm payments at zero and remove CRP as land-use option
No Change in Pasture Returns	Fix pasture returns at 1978 values
No Change in Forest Returns	Fix forest returns at 1978 values
No Change in Urban Returns	Fix urban returns at 1978 values
No Change in Range Returns	Fix range returns at 1978 values

scenario "No Change in Crop Market Returns" holds the market-component of cropland net returns at 1978 levels, while allowing government payments and all other variables to take on their historically observed values. By comparing the results with the factual simulation, we can estimate what impact changing crop market returns had on land allocation decisions. In addition to scenarios examining market-driven changes in net returns, we explore the influence of key government policies through additional scenarios: No Change in Government Payments; No Government Payments; No Conservation Reserve Program;[27] and No Government Payments and No CRP.

The simulations indicate the impact of each factor examined on the national supply of land in each use. Aggregating individual decisions to supply land for a particular use given an exogenous change in the net returns to that use yields an estimate of movements along the national supply curve for that land-use alternative. For example, our simulated changes in cropland under the scenario "No Change in Crop Market Returns" indicate movements along the national supply curve for cropland given a shift in cropland demand. At the same time, more or less cropland supplied will reduce or increase, respectively, the amount of land available for pasture, forests, and all other alternatives. Our estimates of these changes thus represent an inward or outward shift in the land supply curve for each of these other uses.

The estimated impacts on land supply from our simulations are not equivalent to estimated changes in equilibrium land use at the intersection of both supply and demand as determined simultaneously. If crop prices had remained at their relatively high 1978 levels, for instance, less land might have shifted from crops to alternative uses, lowering the production and raising the prices of pasture, range, and forest products. As a result, this would likely have produced a different set of incentives and resulting land-use allocation compared to our simulation.

If such price feedbacks are significant, our simulation results will overstate the absolute equilibrium impacts from the different market and policy factors, but still reflect the relative ranking of effects at a first order of approximation. Larger shifts in the supply of land in a particular use will induce greater movements along a downward sloping demand curve for land in that

[27] We keep all variables at historically observed values, but remove CRP as a land-use option by zeroing the terms in equation [3] for k = CRP. This yields a zero probability of choosing CRP and reallocates the probability to the remaining alternatives, accounting for differences in substitutability over choice sets as allowed by the nested model.

use, producing larger changes in the price for that land use. If limited to a single sector—that is, price feedbacks affecting land use A are mainly effects on the net returns to use A—these feedbacks would attenuate the equilibrium land-use impacts from each factor, but still preserve our estimates for the relative ranking of these effects. Large enough price feedbacks on the net returns to other uses, however, could potentially alter these rankings. Changes in the net returns to land use B (e.g., range) might have very small direct effects on the acreage and/or price of land use A (e.g., crops), but have effects on the price of land use C (e.g., pasture), which in turn have large impacts on land in use A. The likelihood of such indirect effects will depend on the relative elasticities of supply and demand for the different uses, namely the response of acreage (and thus net returns) from C to changes in net returns from B and the relative sensitivity of land use A to the net returns from C.

Although we do not model price feedbacks, our simulations reveal the major pathways along which market and policy changes have affected land supply in each use and the relative impacts on land supply in other uses. This information provides a gauge for when different feedback effects may potentially be significant, depending on the relevant elasticities of demand. To reduce potentially confounding influences from price feedbacks, we also examine changes in land allocation in response to smaller changes in our market and policy drivers.

As a sensitivity test, we simulate scenarios in which the changes in net returns and government programs are reduced by just 5% relative to historical levels. For example, we simulate "5% Less Change in Crop Market Return" where the absolute change since 1978 is reduced by just 5% in each period relative to the factual simulation. This change is one-twentieth compared to that in "No Change in Crop Market Returns" where, as already described, the change in crop market net returns is set at zero (a 100% reduction from the historical levels of change).[28] Because the nested logit land-use change probabilities are nonlinear functions of each exogenous variable, the relative magnitudes of the estimated effects can vary even as all variables are scaled down equally. Examining smaller levels of changes provide estimates of the marginal rather than overall impacts of the different factors affecting national land-use changes from 1982 to 1997.

Simulation Methods

Using the estimated coefficients from our econometric model, land-use change probabilities are calculated for each NRI point and each transition period, using point-level data on land use and land quality together with county-level data on net returns. We begin the simulations in 1982 with each NRI point in the land use reported for 1982. We multiply the acreage weight for the point by the estimated 1982–1987 transition probabilities for the corresponding initial land use to produce an estimate of the area that converts from 1982 to 1987 from the initial land use to each of the six potential land uses, including the status quo. We then multiply these predictions of land use in 1987 by the corresponding 1987–1992 probabilities to estimate land-use changes from 1987 to 1992; and we multiply the acreage predictions for 1992 by the estimated 1992–1997 probabilities to estimate changes from 1992 to 1997. Summing the predicted acres in each land use in 1997 across all the

[28] "Five Percent Less Government Payments" and "Five Percent Less CRP" are scenarios with total payments (in each period) and total CRP enrollment (in 1997) reduced by 5%, respectively To simulate 5% less land in CRP by 1997, we keep all variables at historically observed levels, but scale the component of the utility function for the CRP alternative so as to achieve the acreage target. This can be viewed as a reduction in total CRP incentives, net of conversion costs, applied in equal proportion across all land quality types and time periods. For the combined scenario "Five Percent Less Government Payments and Five Percent Less CRP," the utility from CRP needed to be further scaled back to reduce CRP acreage by 5% given the reduced incentives for cropland use.

TABLE 5
SIMULATED CHANGE IN CROP ACREAGE IN CONTIGUOUS UNITED STATES, 1982–1997

Scenario	Change in Crop Acres (1,000s)	Factually Simulated Change (%)	Maximum Acreage Change Attributable to Variable Fixed[a] (%)
Factual Simulation	−41,137	100.0	0.0
No Change in Crop Market Returns	−23,477	57.1	−42.9
No Change in Govt. Payments	−44,670	108.6	8.6
No Government Payments	−48,627	118.2	18.2
No CRP	−11,826	28.7	−71.3
No Government Payments and no CRP	−16,359	39.8	−60.2
No Change in Pasture Returns	−48,262	117.3	17.3
No Change in Forest Returns	−42,532	103.4	3.4
No Change in Urban Returns	−40,527	98.5	−1.5
No Change in Range Returns	−41,061	99.8	−0.2

[a] The difference between the counterfactual and factual simulation divided by the factual simulation. Positive (negative) values indicate that the crop acreage *decrease* was smaller (greater) in the factual versus counterfactual simulation. When the historical changes are reduced by 5%, rather than 100%, the results in descending order are 0.00%, −2.25%, 0.42%, 0.89%, −3.70%, −3.16%, 0.87%, 0.02%, −0.10%, and −0.01%.

points provides estimates of land-use totals for the 48 contiguous states in the final year of the simulations.

Simulation Results

The simulation model performs well in reproducing the direction and relative magnitudes of land-use changes during the period of analysis. The factually-simulated 1997 land uses are within 1% of actual totals for each land use, except in the case of urban land, for which the factual estimate is within 2% of the actual value.[29] The results in Tables 5 through 10 indicate the relative importance of the factors examined in the different scenarios in driving acreage changes for each land-use category across the 48 contiguous states over 1982–1997. The first column of each table contains the simulated acreage change under each scenario, providing an estimate of the impact on the national supply of land in that use. The second column indicates this acreage change as a share of the factually-simulated acreage change. The third column reports the difference between the counterfactual and factual acreage change as a percentage of the factual acreage change. This provides an estimate of the maximum share of the historically observed change that can be attributed to each factor in the absence of attenuating price feedbacks. Comparing the magnitudes of these values across rows provides an estimate of the relative importance of the different market and policy factors affecting national land use.

The ranking of these factors' importance is robust to smaller-sized changes which are unlikely to involve price feedbacks. As reported in the footnotes to Tables 5–10, the relative ranking is virtually unchanged if we scale down the historical changes by only 5%, rather than 100%. Thus, in general, results from our simulations based on the larger changes also convey the relative importance from more marginal changes in the different factors affecting national land use from 1982 to 1997. One exception is the relative importance of forest net returns in driving changes in forestland and, by extension, rangeland, as described further below.

Cropland. In the case of cropland (Table 5), the introduction of CRP followed by declining crop market conditions were the most important determinants of the decrease in cropland observed between

[29] Simulated land-use changes are also in line with the NRI. For 1982–1997, the NRI indicates that crop, pasture, and range areas decreased by 10.44%, 8.91%, and 2.59%, while forest and urban acres increased by 0.88% and 47.19%, respectively. Our corresponding factually-simulated changes are 9.82%, 8.30%, and 1.93%, and 0.46% and 49.70%.

TABLE 6
SIMULATED CHANGE IN CONSERVATION RESERVE PROGRAM (CRP) ACREAGE IN CONTIGUOUS UNITED STATES, 1982–1997

Scenario	Change in CRP Acres (1,000s)	Factually Simulated Change (%)	Maximum Acreage Change Attributable to Variable Fixed[a] (%)
Factual Simulation	32,437	100.0	0.0
No Change in Crop Market Returns	26,368	81.3	−18.7
No Change in Govt. Payments	34,048	105.0	5.0
No Government Payments	35,718	110.1	10.1
No CRP	0	0.0	−100.0
No Government Payments and no CRP	0	0.0	−100.0
No Change in Pasture Returns	31,540	97.2	−2.8
No Change in Forest Returns	32,374	99.8	−0.2
No Change in Urban Returns	32,475	100.1	0.1
No Change in Range Returns	32,439	100.0	0.0

[a] The difference between the counterfactual and factual simulation divided by the factual simulation. Positive (negative) values indicate that the CRP acreage *increase* was smaller (greater) in the factual versus counterfactual simulation. When the historical changes are reduced by 5%, rather than 100%, the results in descending order are 0.00%, −1.14%, 0.24%, 0.48%, −5.00%, −5.00%, 0.14%, 0.01%, −0.01%, and −0.00%.

1982 and 1997. Disaggregated results indicate that the decline in crop net returns chiefly induced land to leave crop production, while also discouraging the entry of new cropland, shifting the supply of cropland by an estimated 10.7 million and 6.9 million acres, respectively. The change in crop net returns principally affected land at the margins with pasture and the CRP. While falling crop net returns spurred landowners to leave crop production, government farm payments played a role in blunting the effect of market forces. The sum of all government payments offset an estimated 42% of the decline induced by crop markets, increasing 1997 cropland by a maximum of 7.5 million acres or 2% in the absence of price feedbacks. The disaggregated results indicate that government payments boosted crop acreage chiefly by increasing the retention of existing cropland, rather than by bringing new land into cultivation.

The land-use effects of government payments that raised the profitability of cropland were dwarfed by the success of the CRP in retiring cropland over 1982–1997. The estimated combined effect of CRP and direct federal farm program payments was to reduce 1997 cropland supply by 24.8 million acres or almost 7%. The CRP accounts for an estimated 29.3 million acre (8%) decline in cropland supply from 1982 to 1997 or as much as 71% of the decline in crop acres during this period without any price feedbacks. This suggests that approximately 91% of land in CRP in 1997 constituted additional reductions in crop acreage over and above the decrease that would have occurred in the absence of the program.[30]

The decline in pasture net returns after 1978 also appears to have been a significant factor restraining the crop area decrease, with an estimated effect on cropland supply equivalent to that of all direct government payments. Other factors had only minor estimated impacts on the supply of cropland. Changes in forest net returns restrained the decline in cropland by an estimated 3.4%, while changes in urban returns increased the decline by an estimated 1.5%.

CRP and pasture. The key determinants of CRP enrollment were changes in cropland net returns (Table 6), while government payments also reduced incentives for

[30] The fact that some fraction of CRP land may have exited crop production even without the program is distinct from the notion of "slippage," the possibility that crop acreage reductions in CRP might be offset by induced increases in cropland in other areas (Wu 2000; Roberts and Bucholtz 2005).

TABLE 7
SIMULATED CHANGE IN PASTURE ACREAGE IN CONTIGUOUS UNITED STATES, 1982–1997

Scenario	Change in Pasture Acres (1,000s)	Factually Simulated Change (%)	Maximum Acreage Change Attributable to Variable Fixed[a] (%)
Factual Simulation	−10,812	100.0	0.0
No Change in Crop Market Returns	−19,645	181.7	81.7
No Change in Govt. Payments	−9,257	85.6	−14.4
No Government Payments	−7,434	68.8	−31.2
No CRP	−8,205	75.9	−24.1
No Government Payments and no CRP	−4,546	42.0	−58.0
No Change in Pasture Returns	−2,826	26.1	−73.9
No Change in Forest Returns	−11,049	102.2	2.2
No Change in Urban Returns	−10,153	93.9	−6.1
No Change in Range Returns	−10,785	99.7	−0.3

[a] The difference between the counterfactual and factual simulation divided by the factual simulation. Positive (negative) values indicate that the pasture acreage *decrease* was smaller (greater) in the factual versus counterfactual simulation. When the historical changes are reduced by 5%, (rather than 100%, the results in descending order are 0.00%, 4.11%, −0.71%, −1.54%, −0.68%, −2.29%, −3.26%, 0.05%, −0.35%, and −0.01%.

TABLE 8
SIMULATED CHANGE IN FOREST ACREAGE IN CONTIGUOUS UNITED STATES, 1982–1997

Scenario	Change in Forest Acres (1,000s)	Factually Simulated Change (%)	Maximum Acreage Change Attributable to Variable Fixed[a] (%)
Factual Simulation	1,847	100.0	0.0
No Change in Crop Market Returns	293	15.9	−84.1
No Change in Govt. Payments	2,033	110.0	10.0
No Government Payments	2,259	122.3	22.3
No CRP	2,122	114.8	14.8
No Government Payments and no CRP	2,550	138.0	38.0
No Change in Pasture Returns	2,958	160.1	60.1
No Change in Forest Returns	−616	−33.4	−133.4
No Change in Urban Returns	3,181	172.2	72.2
No Change in Range Returns	1,861	100.7	0.7

[a] The difference between the counterfactual and factual simulation divided by the factual simulation. Positive (negative) values indicate that the forest acreage *increase* was smaller (greater) in the factual versus counterfactual simulation. When the historical changes are reduced by 5%, rather than 100%, the results in descending order are 0.00%, −3.33%, 0.51%, 1.12%, −0.67%, 1.85%, 2.07%, −0.36%, 4.18%, and 0.00%.

CRP enrollment.[31] The observed decline in pasture area was partly due to decreases in pasture net returns, but this was less important than the decrease in crop net returns, which restrained the decline in pasture acres (Table 7). Government policies also had significant impacts on pasture acreage. Both government payments and the CRP raised incentives for non-pasture land uses, promoting the decline in pasture acres. The CRP reduced pasture acres by diverting cropland that would have converted to pasture in the absence of the program. To the extent that CRP was a dominant factor behind the decline in cropland, it may have had a positive influence on cropland net returns, an important determinant of pasture acreage. Through this indirect feedback, CRP may have had an additional effect on reducing pastureland that we do not measure.

Forests and range. In the case of forest land, the rise in forest net returns after 1978 was the most important factor affecting

[31] Although illustrative of the direction and relative magnitude of the impacts of factors, our results should not be interpreted as realistic estimates of total CRP acres under the different scenarios. In the simulations, the policies governing the program are held constant and the government is assumed to accept acres of each quality into CRP with the same historically-observed probability, allowing total acreage to expand and contract according to the amount of land offered for enrollment.

TABLE 9

SIMULATED CHANGE IN RANGE ACREAGE IN CONTIGUOUS UNITED STATES, 1982–1997

Scenario	Change in Range Acres (1,000s)	Factually Simulated Change (%)	Maximum Acreage Change Attributable to Variable Fixed[a] (%)
Factual Simulation	−7,971	100.0	0.0
No Change in Crop Market Returns	−8,601	107.9	7.9
No Change in Govt. Payments	−7,890	99.0	−1.0
No Government Payments	−7,771	97.5	−2.5
No CRP	−8,134	102.1	2.1
No Government Payments and no CRP	−7,950	99.7	−0.3
No Change in Pasture Returns	−9,067	113.8	13.8
No Change in Forest Returns	−3,998	50.2	−49.8
No Change in Urban Returns	−7,311	91.7	−8.3
No Change in Range Returns	−8,157	102.3	2.3

[a] The difference between the counterfactual and factual simulation divided by the factual simulation. Positive (negative) values indicate that the range acreage *decrease* was smaller (greater) in the factual versus counterfactual simulation. When the historical changes are reduced by 5%, rather than 100%, the results in descending order are 0.00%, −1.30%, −0.05%, −0.13%, 0.05%, −0.08%, −0.15%, −0.11%, −0.67%, and 0.08%.

changes in supply (Table 8). The rise in forest net returns had a greater influence reducing acreage changes from forests rather than increasing conversions of land into new forests. The decline in crop net returns after 1978 also had an important influence on forestland change. In addition, the rise in urban net returns was an important factor affecting forest area changes, restraining forest increases. Government policies also influenced forestland changes by increasing incentives for alternative land uses. As noted in the case of pasture, to the extent that the CRP raised cropland net returns, the program may also have had an indirect effect on reducing forestland that is not captured by our estimates.

In the case of range acreage, the observed decline was mainly induced by increases in forest net returns after 1978, which accounted for up to 50% of the range area decrease (Table 9). Declines in pasture net returns and crop returns restrained the decline in rangeland, while the rise in urban net returns increased it. Other factors are estimated to have minor effects on rangeland. If changes in crop net returns and government payments reduced forest area enough to raise forest net returns, however, these factors could have had an additional negative effect on rangeland acres.

When we examine the smaller scenarios entailing a 5% reduction from the historical trend, the change in forest net returns declines from being the highest ranked factor driving the forest acreage change to second-to-last in importance behind the change in rangeland net returns. Because this smaller impact on forest acres means less land is diverted from range at the margin, the relative importance of forest net returns on the change in rangeland acreage also declines compared to the scenarios with the full-scale changes.

Urban areas. Urban areas increased by nearly 50% in the factual simulation (Table 10). The only substantial driver of this urban expansion was the increase in urban net returns, which accounted for as much as 13% of the increase in urban land area.[32] We estimate that the decline in crop net returns boosted the increase by a maximum of 2%, while other factors had trivial effects, affecting the change in urban areas by less than 1% overall.

V. CONCLUSIONS

The econometric and simulation analyses demonstrate that private land-use decisions have depended critically on land quality and have been steered by anticipated economic returns to alternative uses, which in some cases

[32] The relatively low explanatory power of the urban net return variable could be a result of the coarseness of our data. While our measure of urban net returns is at the county level, there is a great deal of within-county variation in development potential that our measures do not capture.

TABLE 10
SIMULATED CHANGE IN URBAN ACREAGE IN CONTIGUOUS UNITED STATES, 1982–1997

Scenario	Change in Urban Acres (1,000s)	Factually Simulated Change (%)	Maximum Acreage Change Attributable to Variable Fixed[a] (%)
Factual Simulation	25,635	100.0	0.0
No Change in Crop Market Returns	25,062	97.8	−2.2
No Change in Govt. Payments	25,736	100.4	0.4
No Government Payments	25,855	100.9	0.9
No CRP	26,043	101.6	1.6
No Government Payments and no CRP	26,304	102.6	2.6
No Change in Pasture Returns	25,657	100.1	0.1
No Change in Forest Returns	25,821	100.7	0.7
No Change in Urban Returns	22,336	87.1	−12.9
No Change in Range Returns	25,704	100.3	0.3

[a] The difference between the counterfactual and factual simulation divided by the factual simulation. Positive (negative) values indicate that the urban acreage *increase* was smaller (greater) in the factual versus counterfactual simulation. When the historical changes are reduced by 5%, rather than 100%, the results in descending order are 0.00%, −0.11%, 0.02%, 0.04%, 0.08%, 0.13%, 0.01%, 0.04%, −0.83%, and 0.01%.

have been affected significantly by public policies, sometimes intentionally and sometimes unintentionally. The evidence is strong that cropland declines over the past two decades have been due to falling crop net returns and the existence of the Conservation Reserve Program (CRP), as has been suggested by other research (Vesterby and Krupa 1995; Alig, Dicks, and Moulton 1998). We have also isolated the opposing influences of the CRP and other federal farm payments. By raising the profitability of cropland, the government increased acreage in crops and directly competed with itself in providing incentives for landowners to retire environmentally-sensitive cropland under the CRP. The increase in cropland area also came largely at the expense of land in pasture. Because cropping is a more intensive land use, typically involving greater application of agricultural chemicals, government payments are likely to have had unintended environmental impacts.

Apart from environmental impacts, the effect of government programs on aggregate crop production is a critical issue for on-going deliberations on global trade liberalization.[33] We estimated the contribution of CRP and direct Federal payments, two major elements of U.S. farm policy, to aggregate crop production.[34] Excluding CRP and WRP, direct federal payments to farmers increased land in crops by as much as 2% in 1997, partially offsetting the acreage decline due to the change in crop commodity markets. Our results also indicate that the positive effects on crop acres of direct federal payments to farmers were more than offset by cropland retirement under the CRP.[35]

For forest areas, we identified the rise in timber net returns as the most important factor driving the increase in forest areas between 1982 and 1997. This is consistent with reports that the increase in forests largely involved timberland acreage

[33] Federal farm payments under CRP and other environmental programs are currently permitted by WTO rules under the "green box" exemption if they are deemed "minimally trade distorting." On the other hand, trade-distorting impacts of direct payments are potentially actionable, regardless of countervailing impacts from other programs.

[34] We did not examine the effects of agricultural tariffs or the influence of indirect federal assistance to farmers, such as subsidized crop insurance through private firms, or federal funding of agricultural research or other third-party activities. Hoekman, Ng, and Olarreaga (2003) estimate that agricultural tariffs worldwide have had more than twice the impact on trade flows as domestic support policies. Lubowski et al. (2006a) estimate that crop insurance subsidy increases after 1994 raised cultivated acreage in 1997 by about 1%, with disproportionate impacts on wetlands and other environmentally sensitive lands.

[35] Our estimates for the historic impact of direct government payments on crop acreage lie the middle to upper range of other studies (Westcott and Price 2001; FAPRI 2001). Previous studies have also estimated offsetting effects of CRP and direct federal farm payments under recent Farm Acts (Gardner 2002; Westcott, Young, and Price 2002).

(Anderson and Magleby 1997; Alig, Dicks, and Moulton 1998). We also identified declining crop net returns as a major factor affecting forest area during this period, which is consistent with reports that forest areas have increased due to passive regrowth on abandoned agricultural lands (Alig and Wear 1992; Anderson and Magleby 1997; Alig, Dicks, and Moulton 1997; Flather, Brady, and Knowles 1999). Our findings suggest that policies targeting forest net returns, such as payments for carbon sequestration, are likely to be particularly effective at encouraging the retention of existing forests, rather than new forest establishment, as others have also suggested (Newell and Stavins 2000).[36] We also find that the marginal impact of forest net returns on forestland acreage is sensitive to the level of the simulated changes, suggesting that policies targeting forest net returns may be relatively ineffectual until the change in incentives exceeds a critical threshold. Urban net returns appear as the only significant driver of urban land increases, supporting the notion that the dramatic increase in urban land observed subsequent to 1982 was largely a response to increased housing demand driven by demographic changes and economic growth (Heimlich and Anderson 2001). Our findings suggest that efforts to protect open space by increasing net returns to agricultural uses are likely to have only limited impacts. Once urban development becomes feasible, development returns are so much greater than returns to other land uses that observed changes in non-urban returns are of insufficient magnitude to make a significant difference. This may explain why use-value assessments and other preferential tax policies, employed in virtually all states to encourage the retention of cropland, have had minimal effects in restraining urban development (Heimlich and Anderson 2001). This suggests that more targeted and aggressive policies would be needed to reduce urbanization, such as land purchases, acquisition of development rights, and substantial development charges.

This study demonstrates the importance of modeling land-use change in a framework that incorporates changes among all major land-use alternatives. With the notable exception of urban land, our findings highlight the importance of net returns to a variety of land-use categories as determinants of land area change for particular uses. The results indicate the margins that are most active for particular land-use categories and the variables that consequently have the greatest influence on the area allocated to these uses. These factors represent the most powerful market-based mechanisms for the government to influence land-use changes either unintentionally or through deliberate policy interventions.

[36] This conclusion could vary regionally. In contrast to our national-level results, Plantinga and Ahn (2002) estimate that in the U.S. South Central region policies to convert land to forests would be less costly than policies to retain existing forests.

DATA APPENDIX

Land Use

Our data on land use is from USDA National Resources Inventory (NRI). "Cropland" includes row- and close-grown crops, fallow, pasture, and haylands, in rotation with crops, permanent haylands, vineyards, orchards, and nurseries. "Pasture" includes land managed for introduced forage for livestock grazing. "Range" includes land under native or introduced forage suitable for grazing which receives only limited management. "Forests" are areas at least one acre in size and 100 feet in width that are least 10% stocked with trees with the potential to reach 13 feet at maturity. This translates to a canopy cover of at least 25%. "Urban lands" include areas in residential, industrial, commercial, or institutional uses. Parcels below ten acres, such as small parks and transport facilities, are also classed as urban if they are completely surrounded by urban lands. This definition excludes roads and other lands used for transportation in non-metropolitan areas, as these are separately identified by the NRI.

Crop Market Net Returns

Data on prices, yields, cash costs, and acres are used to compute weighted county-level market net returns for 21 major crops. State-level marketing-year-average prices and county-level yields are from the National Agricultural Statistics Service (NASS). Data on cash costs as a share of revenue at the state and regional level, respectively, are from the Census of Agriculture and the Economic Research Service's (ERS). County acreage from NASS and the Census of Agriculture provided weights for averaging across individual crops.

Government Payments

County-level estimates of total federal program payments per acre are from the Census of Agriculture and include receipts from deficiency payments, support price payments, indemnity programs, disaster payments, and payments for soil and water conservation projects. Conservation Reserve and Wetlands Reserve program payments are excluded.

Pasture Net Returns

Annual net returns per acre for pasture are estimated using pasture yields from the SOILS-5 database linked to the NRI, state prices for "other hay" from NASS, and per acre costs for hay and other field crops from the Census of Agriculture.

Range Net Returns

Annual rangeland net returns per acre are computed with forage yields from SOILS-5 and state-level per head grazing rates for private lands from ERS.

Forest Net Returns

We use a 5% interest rate to annualize the estimated net present value of a weighted average of sawtimber revenues from different forest types based on prices, yields, costs, and acres. State-level stumpage prices were gathered from state and federal agencies and private data services. Regional merchantable timber yield estimates for different forest types were obtained from Richard Birdsey of the U.S. Forest Service. Regional replanting and annual management costs were derived from Moulton and Richards (1990) and Dubois, McNabb and Straka (1999). The Faustmann formula was used to compute the optimal rotation age, assuming forests start newly planted at year zero. County acreage and timber output data from the Forest Inventory and Analysis (FIA) and Timber Product Output (TPO) surveys of the U.S. Forest Service provided weights for averaging across individual forest types and species, respectively.

Urban Net Returns

Annual urban net returns per acre are estimated as the median value of a recently-developed parcel, less the value of structures, annualized at a 5% interest rate. Median county-level prices for single family homes were constructed from the decennial Census of Population and Housing Public Use Microdata Samples and the Office of Federal Housing Enterprise Oversight (OFHEO) House Price Index. Regional data on lot sizes and the value of land relative to structures for single-family homes were from the Characteristics of New Housing Reports (C-25 series) and the Survey of Construction (SOC) micro data from the Census Bureau.

Additional data descriptions are provided in Lubowski (2002) and are available upon request.

References

Adams, Darius M., Ralph J. Alig, Bruce A. McCarl, John M. Callaway, and Steven M. Winnett. 1999. "Minimum Cost Strategies for Sequestering Carbon in Forests." *Land Economics* 75 (Aug.): 360–74.

Alig, Ralph J., and Robert G. Healy. 1987. "Urban and Built-Up Land Area Changes in the United States." *Land Economics* 63 (Aug.): 215–26.

Alig, Ralph J., and David N. Wear. 1992. "Changes in Private Timberland: Statistics and Projections for 1952 to 2040." *Journal of Forestry* 90 (5): 31–37.

Alig, Ralph J., Darius M. Adams, and Bruce McCarl. 1998. "Impacts of Incorporating Land Exchanges between Forestry and Agriculture in Sector Models." *Journal of Agricultural and Applied Economics* 30 (2): 389–401.

Alig, Ralph J., Michael R. Dicks, and Robert J. Moulton. 1998. "Land-Use Dynamics Involving Forestland: Trends in the U.S. South." In *Proceedings of the 1988 Southern Forest Economics Workers Meeting*, ed. R. A. Kluender, N. B. Smith, and M. M. Corrigan, 9–23. Moticello, Ark: University of Arkansas.

Alig, Ralph J., Darius M. Adams, Bruce McCarl, J. M. Callaway, and Steven Winnett. 1997. "Assessing Effects of Mitigation Strategies for Global Climate Change with an Intertemporal Model of the U.S. Forest and Agriculture Sectors." *Environmental and Resource Economics* 9 (3): 259–74.

Anderson, Margot, and Richard Magleby. 1997. "Agricultural and Environmental Indicators, 1996–97." Agricultural Handbook No. 712.

Washington, D.C.: U.S. Department of Agriculture, Economic Research Service.

Ben-Akiva, Moshe, and Steven R. Lerman. 2000. *Discrete Choice Analysis: Theory and Application to Travel Demand*. Cambridge, Mass.: The MIT Press.

Bockstael, Nancy E., and Elena G. Irwin. 2000. "Economics and the Land Use-Environment Link." In *The International Yearbook of Environmental and Resource Economics*, ed. H. Folmer and T. Tietenberg. Northampton, Mass.: Edward Elgar Pub.

Carrion-Flores, Carmen, and Elena G. Irwin. 2004. "Determinants of Residential Land-Use Conversion and Sprawl at the Rural-Urban Fringe." *American Journal of Agricultural Economics* 86 (4): 889–904.

Cropper, Maureen, Jyotsna Puri, and Charles Griffiths. 2001. "Predicting the Location of Deforestation: The Role of Roads and Protected Areas in North Thailand." *Land Economics* 77 (May): 172–86.

Dixit, Avinash K., and Robert S. Pindyck. 1994. *Investment under Uncertainty*. Princeton, N.J.: Princeton University Press.

Dubois, Mark R., Kenneth McNabb, and Thomas J. Straka. 1999. "Cost Trends for Forestry Practices in the South." *Forest Farmer* 32 (1): 3–8.

Farm Service Agency (FSA), U.S. Department of Agriculture. 2006. *Conservation Reserve Program, Monthly Summary—July 2006*. Washington, D.C.: Farm Service Agency.

Flather, C. H., S. J. Brady, and M. S. Knowles. 1999. "Wildlife Resource Trends in the United States: A Technical Document Supporting the 2000 USDA Forest Service RPA Assessment." General Technical Report No. 33. Ft. Collins, Col.: U.S. Department of Agriculture Forest Service, Rocky Mountain Research Station.

Food and Agricultural Policy Research Institute (FAPRI). 2001. "Analysis of the Grain, Oilseed, and Cotton Provisions of the Agricultural Act of 2001, H.R. 2646." FAPRI-UMC Report No. 09-01. Columbia, Mo.: Food and Agricultural Policy Research Institute, University of Missouri.

Gardner, Bruce. 2002. "North American Agricultural Policies and Effects on Western Hemisphere Markets since 1995, with a Focus on Grains and Oilseeds." Department of Agricultural and Resource Economics Working Paper No. 02-12. College Park, Md.: University of Maryland.

Greene, William H. 2000. *Econometric Analysis*, 4th ed. Upper Saddle River, N.J.: Prentice Hall.

Hausman, Jerry , and Daniel McFadden. 1984. "Specification Tests for the Multinomial Logit Model." *Econometrica* 52 (5): 1219–40.

Heimlich, Ralph E., and William D. Anderson. 2001. *Development at the Urban Fringe and Beyond: Impacts on Agriculture and Rural Land*. Agricultural Economic Report No. 803. Washington, D.C.: U.S. Department of Agriculture, Economic Research Service.

Heimlich, Ralph E., Keith D. Wiebe, Roger Claassen, Dwight Gadsby, and Robert M. House. 1998. *Wetlands and Agriculture: Private Interests and Public Benefits*. Agricultural Economics Report No. 765 (AER 765). Washington, D.C.: U.S. Department of Agriculture, Economic Research Service.

Helms, Douglas. 1992. *Readings in the History of the Soil Conservation Service*. Washington, D.C.: U.S. Department of Agriculture, Soil Conservation Service.

Hoekman, Bernard, Francis Ng, and Marcelo Olarreaga. 2003. "Reducing Agricultural Tariffs Versus Domestic Support: What's More Important for Developing Countries?" World Bank Policy Research Working Paper No. 2918. Washington, D.C.: World Bank.

Lubowski, Ruben N. 2002. "Determinants of Land-Use Transitions in the United States: Econometric Analysis of Changes among the Major Land-Use Categories." PhD diss., Harvard University.

Lubowski, Ruben N., Shawn Bucholtz, Roger Claassen, Michael J. Roberts, Joseph C. Cooper, Anna Gueorguieva, and Robert Johansson. 2006a. *Environmental Effects of Agricultural Land-Use Change: The Role of Economics and Policy*. Economic Research Report No. 25. Washington, D.C.: U.S. Department of Agriculture, Economic Research Service.

Lubowski, Ruben N., Marlow Vesterby, Shawn Bucholtz, Alba Baez, and Michael J. Roberts. 2006b. *Major Uses of Land in the United States, 2002*. Economic Information Bulletin No. 14. Washington, D.C.: U.S. Department of Agriculture Economic Research Service.

McCarl, Bruce A., and Uwe A. Schneider. 2001. "Greenhouse Gas Mitigation in U.S. Agriculture and Forestry." *Science* 294 (21 Dec.): 2481–82.

McMillen, Daniel P. 1992. "Probit with Spatial Autocorrelation." *Journal of Regional Science* 32 (3): 335–48.

Moulton, Robert J., and Kenneth R. Richards. 1990. "Costs of Sequestering Carbon through Tree Planting and Forest Management in the United States." GTR WO-58. Washington,

D.C.: U.S. Department of Agriculture, Forest Service.

Nelson, Gerald C., and Daniel Hellerstein. 1997. "Do Roads Cause Deforestation? Using Satellite Images in Econometric Analysis of Land Use." *American Journal of Agricultural Economics* 79 (1): 80–88.

Nelson, G. C., Virginia Harris, and Steven W. Stone. 2001. "Deforestation, Land Use, and Property Rights: Empirical Evidence from Darien, Panama." *Land Economics* 77 (May): 187–205.

Newell, Richard G., and Robert N. Stavins. 2000. "Climate Change and Forest Sinks: Factors Affecting the Costs of Carbon Sequestration." *Journal of Environmental Economics and Management* 40 (3): 211–35.

Nusser, S. M., and J. J. Goebel. 1997. "The National Resources Inventory: A Long-Term Multi-Resource Monitoring Programme." *Environmental and Ecological Statistics* 4 (3): 181–204.

Oxfam International (Oxfam). 2003. "Cultivating Poverty: The Impact of U.S. Cotton Subsides on Africa." Oxfam Briefing Paper 30. Boston, Mass.: Oxfam.

Parks, Peter J., and Randall A. Kramer. 1995. "A Policy Simulation of the Wetlands Reserve Program." *Journal of Environmental Economics and Management* 28 (2): 223–40.

Pew Center for Civic Journalism (Pew). 2000. "Research—Straight Talk from Americans—2000." www.pewcenter.org/doingcj/research/r_ST2000nat1.html.

Plantinga, Andrew J., and SoEun Ahn. 2002. "Efficient Policies for Environmental Protection: An Econometric Analysis of Incentives for Land Conversion and Retention." *Journal of Agricultural and Resource Economics* 27 (1): 128–45.

Plantinga, Andrew J., and Joseph Buongiorno. 1990. "Determinants of Changes in Non-Industrial Private Timberland Ownership in the United States." *Journal of World Forest Resource Management* 5 (1): 29–46.

Plantinga, Andrew J., Ralph Alig, and Hsiang-tai Cheng. 2001. "The Supply of Land for Conservation Uses: Evidence from the Conservation Reserve Program." *Resources, Conservation and Recycling* 31 (3): 199–215.

Roberts, Michael J., and Shawn Bucholtz. 2005. "Slippage in the Conservation Reserve Program or Spurious Correlation? A Comment." *American Journal of Agricultural Economics* 87 (1): 244–50.

Roberts, Michael J., and Ruben N. Lubowski. 2007. "Enduring Impacts of Land Retirement Policies: Evidence from the Conservation Reserve Program." *Land Economics* 83 (Nov.): 516–38.

Roberts, Michael J., Craig Osteen, and Meredith Soule. 2004. *Risk, Government Programs, and the Environment.* Technical Bulletin No. 1908. Washington, D.C.: U.S. Department of Agriculture, Economic Research Service.

Rust, John. 1987. "Optimal Replacement of GMC Bus Engines: An Empirical Model of Harold Zurcher." *Econometrica* 55 (5): 999–1033.

Schatzki, Todd. 2003. "Options, Uncertainty, and Sunk Costs: An Empirical Analysis of Land Use Change." *Journal of Environmental Economics and Management* 46 (1): 86–105.

Smith, Brad W., Patrick D. Miles, John S. Vissage, and Scott A. Pugh. 2004. *Forest Resources of the United States, 2002: A Technical Document Supporting the USDA Forest Service 2005 Update of the RPA Assessment.* General Technical Report NC-241. St. Paul, Minn.: Forest Service, North Central Research Station, U.S. Department of Agriculture.

Soil Survey Staff. 2000. *Soil Ratings for Plant Growth—A System for Arraying Soils According to their Inherent Productivity and Suitability for Crops*, ed. C. S. Holzhey and H. R. Sinclair. Lincoln, Neb.: U.S. Department of Agriculture, Natural Resources Conservation Service, National Soil Survey Center.

Stavins, Robert N. 1999. "The Costs of Carbon Sequestration: A Revealed-Preference Approach." *American Economic Review* 89 (4): 994–1009.

Stavins, Robert N., and Adam B. Jaffe. 1990. "Unintended Impacts of Public Investments on Private Decisions: The Depletion of Forested Wetlands." *American Economic Review* 80 (3): 337–52.

Thornes, Paul, and Daniel P. Miller. 1998. "Land Value and Parcel Size: A Semiparametric Analysis." *Journal of Real Estate Finance and Economics* 17 (3): 233–44.

U.S. Department of Agriculture (USDA). 1973. "Land Capability Classification." Soil Conservation Service Handbook, No. 210. Washington, D.C.: U.S. Department of Agriculture.

U.S. Department of Agriculture Economic Research Service (ERS). 2003. "U.S. and State Farm Income Data." http://www.ers.usda.gov/data/FarmIncome/finfidmu.html.

U.S. Environmental Protection Agency (EPA). 2005. *Greenhouse Gas Mitigation Potential in U.S. Forestry and Agriculture.* EAP 430-R-05-006. Washington, D.C.: U.S. Environmental Protection Agency.

Vesterby, Marlowe, and Kenneth S. Krupa. 1995. "Major Uses of Land in the United States: 1997." Statistical Bulletin No. 973. Washington, D.C.: U.S. Department of Agriculture, Economic Research Service.

Westcott, Paul C., and J. Michael Price. 2001. "Analysis of the U.S. Commodity Loan Program with Marketing Loan Provisions." Agricultural Economic Report No. 801. Washington, D.C.: U.S. Department of Agriculture, Economic Research Service.

Westcott, Paul C., C. Edwin Young, and J. Michael Price. 2002. "The 2002 Farm Act: Provisions and Implications for Commodity Markets." Agriculture Information Bulletin No. 778. U.S. Department of Agriculture, Economic Research Service: Washington, D.C.

World Trade Organization (WTO). 2006. "Talks Suspended. Today There Are Only Losers." WTO News, DDA June/July 2006 Modalities: Summary 24 July. http://www.wto.org/english/news_e/news06_e/mod06_summary_24july_e.htm.

Wu, Junjie. 2000. "Slippage Effects and the Conservation Reserve Program." *American Journal of Agricultural Economics* 82 (4): 979–92.

[17]

What Is the Value of Terroir?

By Robin Cross, Andrew J. Plantinga, and Robert N. Stavins*

Wine producers and enthusiasts use the term "terroir," from the French *terre* (meaning land), to refer to the special characteristics of a place that impart unique qualities to the wine produced. The Appellation d'Origine Contrôlée (AOC) system in France, and similar systems adopted in other wine-producing countries, are based upon the geographic location of grape production, predicated on this notion of terroir. Under the US system, production regions are designated as American Viticultural Areas (AVAs), with finer geographical designations known as sub-AVAs. Such designations allow wineries to identify the geographical origin of the grapes used in producing their wines, and—equally important—seek to prevent producers outside an AVA from making false claims about the nature and origin of their wines.

What is the value of terroir in the American context? Does the "reality of terroir"—the location-specific geology and geography—predominate in determining the quality of wine? Does the "concept of terroir"—the location within an officially named appellation – impart additional value to grapes and wine? Does location within such an appellation impart additional value to vineyards?

The central question we address is whether measurable site attributes—such as slope, aspect, elevation, and soil type—or appellation designations are more important determinants of vineyard prices.[1] We do this by conducting a hedonic price analysis to investigate sales of vineyards in Oregon's Willamette Valley, one of the most important wine-producing regions in the United States.[2]

How should site attributes and sub-AVA designations influence vineyard prices? If site attributes significantly affect wine quality and if consumers are able to discriminate such quality, then vineyard prices would depend on site attributes,[3] and AVA designations might be redundant. Alternatively, consumers might not be able to discriminate among wines perfectly and might use AVA designations as signals of average quality of wines from respective areas, and/or might derive utility directly from drinking wines which they know to be of particular pedigree. In this case, site attributes and AVA designations would influence vineyard prices, with parameters for site attributes indicating how producers value intra-AVA differences in vineyard characteristics. Presumably, producers attach premiums to site attributes that enhance wine quality, provided that consumers can perceive and are willing to pay for such quality differences.

What if, at the extreme, variation in vineyard prices is explained completely by AVA designations (that is, site attributes are irrelevant)? This would indicate that terroir matters economically—as a concept, though not as a fundamental reality. In other words, producers recognize the value of the AVA designation because they

*Cross: Department of Agricultural and Resource Economics, Oregon State University, Corvallis, OR 97331 (e-mail: robin.cross@oregonstate.edu); Plantinga: Department of Agricultural and Resource Economics, Oregon State University, Corvallis, OR 97331 (e-mail: plantinga@oregonstate.edu); Stavins: John F. Kennedy School of Government, Harvard University, 79 John F. Kennedy Street, Cambridge, MA 02138 (e-mail: robert_stavins@harvard.edu). We thank Charles Mason and participants in a session at the 2011 ASSA Meetings for helpful comments on a previous version of this paper.

[1] The notion of terroir is sometimes extended beyond natural endowments to encompass the history and culture of a place, but we employ a narrower and more common definition of terroir focused on physical attributes of the location (Olivier Gergaud and Victor Ginsburgh 2008).

[2] The Willamette Valley is designated as an AVA. Within the valley, there are six sub-AVAs: Chehalem Mountains, Yamhill-Carlton District, Ribbon Ridge, Dundee Hills, McMinnville, and Eola-Amity Hills.

[3] Wine quality is affected not only by site attributes, but also by the quality of growing stock, as well as vineyard management, and the skills and resources of the wine maker.

know that consumers will pay more for the experience of drinking wine from designated areas.[4] However, if site attributes known to affect wine quality have no impact on vineyard prices, this suggests that consumers cannot discern quality differences. Any appreciation they might express for an area's terroir would essentially be founded on reputation, not reality.

In the next part of the paper, we discuss some related research from the wine economics literature. Then, in Section II, we describe the data we employ, as well as our estimation strategy. In Section III, we present our results plus some robustness checks. Section IV concludes.

I. Previous Literature

Our analysis is related to and builds upon previous work by others. In one recent study, Gergaud and Ginsburgh (2008) find that site attributes of vineyards in the Haut-Médoc appellation in Bordeaux have no effect on wine prices or ratings. Our study builds on this work by examining—in addition to site characteristics—the value assigned to appellation designations. Further, we are able to measure site characteristics more precisely by using GIS-based information to develop detailed physiographic profiles of parcels.

In another recent study, Orley Ashenfelter and Karl Storchmann (2010) investigate the effects of climate on vineyards in the Mosel Valley. As in our study, the authors have fine-scale data on vineyard characteristics. They find that site characteristics—including slope, orientation, soil types, soil depth, and altitude—as well as solar radiation are significant determinants of vineyard quality. We do not include climate variables in our analysis, because of trivial variation in rainfall, humidity, and wind across the Willamette Valley, but our site attribute variables proxy for the amount of solar radiation received by vineyards.

An important similarity and difference with our study concerns respective dependent variables. Gergaud and Ginsburgh (2008) and Ashenfelter and Storchmann (2010) employ measures that are based on wine prices or ratings.

[4] Likewise, producers might bid up the value of vineyards located in designated appellations because there is prestige associated with owning vineyards in these areas.

Our dependent variable, vineyard sale prices, is preferable in the context of our investigation into the value of terroir. Provided that non-vineyard inputs into wine production (labor, wine-making techniques, etc.) are variable and reproducible, profit maximization implies that the optimal levels of these inputs are implicit functions of vineyard attributes and input and output prices that are constant across the vineyards in our sample. As such, we can estimate the implicit prices of vineyard attributes using the simple hedonic equation presented below.

As we suggest above, if consumers have limited information about specific vineyards or are unable to judge differences in quality among wines, they might use appellation designations as signals of quality. In fact, the results of many blind taste tests indicate that consumers have very limited ability to distinguish intrinsic qualities of wine (sweetness, acidity, tannins, etc.), and instead judge quality by relying on extrinsic signals, such as price, origin, and wine-maker reputation. Roberta Veale and Pascale Quester (2008) found that tasters' perceptions of quality were strongly correlated with price and country of origin, but not with intrinsic qualities related to taste. Similarly, Robin Goldstein et al. (2008) found that when price information is withheld, nonexpert tasters show no preference for more expensive wines and even show a slight preference for less expensive wines.

II. Data and Estimation Strategy

We employ a new dataset on vineyard sales provided by Northwest Farm Credit Services, a lending institution specializing in agriculture. This database includes the universe of 104 sales of properties in the Willamette Valley between 1995 and 2007 that included vineyards and vinelands (land that can be developed for vineyards). In addition to the total sale price, the size, and the location of the property, each sale record includes an appraiser's estimate of the value of non-vineyard assets, such as dwellings and other buildings, winery equipment, and non-vineyard land. We subtract the estimated values of these non-vineyard assets from the total sale price to obtain the value of vineyards in each sale. Because terroir relates exclusively to nontransferable attributes of vineyards, we also subtract from the vineyard value the appraiser's estimate of the value of all vineyard enhancements,

TABLE 1—ESTIMATION RESULTS

	Parameter estimate	Standard error
Constant	8.582	0.333
Eola-Amity Hills sub-AVA	0.438	0.138
McMinnville sub-AVA	0.154	0.230
Yamhill-Carlton sub-AVA	0.529	0.135
Dundee Hills sub-AVA	0.852	0.143
Chehalem Mountains sub-AVA	0.482	0.125
Vineyard acreage	−0.005	0.002
Vineyard acreage squared	0.000014	0.000006
Share of parcel with best elevation	0.157	0.154
Share of parcel with other elevation	0.130	0.164
Share of parcel with south aspect	0.202	0.268
Share of parcel with southeast/west aspect	−0.088	0.267
Share of parcel with east or west aspect	0.270	0.471
Share of parcel with best soil	−0.030	0.157
Share of parcel with good soil	0.048	0. 137
Share of parcel with best slope	0.075	0.286

Dependent variable = log of vineyard price per acre

Observations = 104

Adjusted $R^2 = 0.422$

leaving the estimated price of land for vineyards. Our dependent variable is the logarithm of this "vineyard price" per acre for each property (in 2007 dollars).

The sales records include information about average characteristics of vineyards included in each sale, but we develop more precise measures, using GIS-based information on slope, aspect, elevation, and soils, with location of each parcel determined from tax lot boundaries and matched to GIS maps of physiographic variables. Parcels are divided into ten-meter pixels, and each pixel is classified according to 14 slope, 16 aspect, 86 elevation, and 8 soil group categories. Because the number of categories exceeds the number of observed sales, we combine the categories based on accepted knowledge about which vineyard attributes are most favorable or unfavorable, compute the percentages of each parcel in each of the aggregated categories, and use these as independent variables in our hedonic regressions.

In addition to site attribute variables, we include the vineyard area, the square of area, and indicator variables for the location of a parcel within a sub-AVA.[5] Parcels outside of sub-AVAs are demarcated as being in the Willamette Valley AVA. The average price for vineyards is about $10,000 per acre, with prices ranging from $2,500 to $42,000 per acre. Given that our sample includes only parcels with vineyards, we must have within-sample variation in site attributes if we hope to measure effects of terroir. Fortunately, our sample reflects a significant range of physiographic conditions.

III. Results and Robustness

We regress the log of vineyard price on the variables discussed above, and a constant term. Variables are defined such that we expect the coefficients on the included site attributes and sub-AVA designations to be positive. The results reveal that most of the estimated coefficients on site attributes are positive, but none is significantly different from zero (Table 1).[6] However, four of the five estimated coefficients for sub-AVA are significantly different from zero at better than the 99 percent confidence level.

Pairwise F-tests indicate that the coefficient on Dundee Hills is different from those on Eola-Amity Hills, Yamhill-Carlton, and Chehalem

[5] Note that the sub-AVA designations were not officially adopted until the end of the period covered by our data. However, prior to that time, the areas that would later be designated as sub-AVAs were well recognized by wine producers, and it was common practice to label the origin of wines using the same geographical terms.

[6] The results do not change appreciably if we use robust standard error estimates.

TABLE 2—VINEYARD PRICE PREMIUMS AND BOTTLE PRICES

Vineyard prices		Average bottle prices	
Appellation designation	Vineyard price premium	Wine spectator	Wine advocate
McMinnville	$ 0	$ 40	$ 43
Willamette Valley	$ 0	$ 46	$ 47
Eola-Amity Hills	$ 2,933	$ 51	$ 50
Chehalem Mountains	$ 3,306	$ 52	$ 55
Yamhill-Carlton	$ 3,721	$ 49	$ 48
Dundee Hills	$ 7,163	$ 59	$ 54

Note: Vineyard premium is the increment in sale price (2007 dollars per acre) due to location inside a sub-AVA.

Mountains, but that the latter three coefficients are statistically equivalent. The coefficients on parcel size indicate that the per-acre price falls as the parcel size increases, but at a diminishing rate. The adjusted R^2, 0.422, is reasonable for cross-sectional data.

Why are the impacts of site attributes on sales price insignificant? One potential problem is that the site attributes are measured over the entire parcel, not just the vineyard portion. We therefore restrict the sample to parcels that are at least 50 percent vineyards, leaving 83 observations with which to estimate the model. The results show little change from the basic model.[7] Qualitatively, the findings remain the same for cut-off values of 68 percent and 75 percent.

Another possible explanation for the insignificance of site attributes is that their effects could be masked by the sub-AVA designations. When we drop the sub-AVA variables, the variable for best soils becomes significant. However, further investigation shows that this variable is highly correlated with the Dundee Hills sub-AVA variable and that it is the Dundee Hills location, not better soils, that raises vineyard prices. A further check is to see if there is variation in the site attributes within sub-AVAs. In fact, we find similar variation in site attributes within and across sub-AVAs.

Finally, if the fundamentals of terroir—as opposed to the reputation associated with an appellation designation—have important influences on wine quality, then parcels that are outside, but close to, sub-AVAs should be valued more than parcels that are farther away. We therefore reestimate the basic model with a variable for the distance of a vineyard from a sub-AVA included. We find that its influence is insignificant.

The significant premiums we find to be associated with sub-AVA designations are not related to observable site attributes. One possible explanation is that bottle prices for wines from sub-AVAs command higher prices (for reputational reasons), which in turns bids up the prices of sub-AVA vineyards. To investigate this possibility, we examined data on 2006 pinot noir bottle prices for the Willamette Valley AVA and for each sub-AVA. From the database of *Wine Spectator* magazine, we obtained 243 observations, and from the *Wine Advocate* database, 310 observations. Rated wines from the most highly valued appellation, Dundee Hills, command an average premium of some $13 per bottle in the *Wine Spectator* data and about $7 per bottle in the *Wine Advocate* data (Table 2). Likewise, from the results reported in Table 1, we can estimate the price premiums associated with vineyards in various appellations. The Dundee Hills sub-AVA commands a premium of more than $7,000 per acre.

The rankings of vineyard price premiums are roughly consistent with the bottle price premiums. Dundee Hills ranks first in vineyard price premiums and first in both of the estimates of bottle price premiums, whereas McMinnville and Willamette Valley fall at the bottom of all three sets of rankings. Of course, bottle prices are also affected by the skills and reputations of wine makers, and so these results are only suggestive of correlation.[8] That said, the results are consistent with the notion that consumer demand

[7] Econometric results of this and all other robustness checks are found in Cross, Plantinga, and Stavins (2011).

[8] Because there is considerable variation in bottle prices within sub-AVAs, we cannot reject the null hypothesis that average bottle prices are the same across sub-AVAs.

for wines of better pedigree drives higher valuation of respective vineyards.

IV. Conclusions

We have estimated a hedonic model of vineyard prices in Oregon to examine whether such prices vary systematically with designated appellation, after controlling for site attributes. Despite using precise measures of site attributes, we do not find evidence of significant effects on vineyard prices, and a series of robustness tests does not alter this finding. However, we do find that vineyard prices are strongly determined by location within specific sub-AVAs. These appellations are supposed to reflect the area's terroir, but our finding that the physical characteristics of vineyards are not priced implicitly in land markets raises questions about whether sub-AVA designations have a fundamental connection with terroir.

Nevertheless, our results make clear that the concept of terroir matters economically, both to consumers and to wine producers. Buyers and sellers of vineyard parcels in the Willamette Valley of Oregon attach a significant premium to sub-AVA designations. One possibility is that buyers are less informed than sellers about how the attributes of a vineyard will affect wine quality and, therefore, rely on sub-AVA designations as quality signals. Either way, consumers are evidently willing to pay more for the experience of drinking wines from these areas. While they may not discriminate among wines in terms of their intrinsic qualities, consumers are apparently responding to extrinsic qualities of wines, such as price and area of origin.

REFERENCES

Ashenfelter, Orley, and Karl Storchmann. 2010. "Using Hedonic Models of Solar Radiation and Weather to Assess the Economic Effect of Climate Change: The Case of Mosel Valley Vineyards." *Review of Economics and Statistics*, 92(2): 333-49.

Cross, Robin, Andrew J. Plantinga, and Robert N. Stavins. 2011. "The Value of Terroir: Hedonic Estimation of Vineyard Sale Prices." National Bureau of Economic Research Working Paper 16762.

Gergaud, Olivier, and Victor Ginsburgh. 2008. "Natural Endowments, Production Technologies and the Quality of Wines in Bordeaux: Does Terroir Matter?" *Economic Journal*, 118(529): F142-57.

Goldstein, Robin, Johan Almenberg, Anna Dreber, John W. Emerson, Alexis Herschkowitsch, and Jacob Katz. 2008. "Do More Expensive Wines Taste Better? Evidence from a Large Sample of Blind Tastings." *Journal of Wine Economics*, 3(1): 1–9.

Veale, Roberta, and Pascale Quester. 2008. "Consumer Sensory Evaluations of Wine Quality: The Respective Influence of Price and Country of Origin." *Journal of Wine Economics*, 3(1): 10–29.

[18]

Available online at www.sciencedirect.com

ScienceDirect

Journal of Environmental Economics and Management 54 (2007) 181–198

JOURNAL OF ENVIRONMENTAL ECONOMICS AND MANAGEMENT

www.elsevier.com/locate/jeem

Water demand under alternative price structures

Sheila M. Olmstead[a,*], W. Michael Hanemann[b], Robert N. Stavins[c,d]

[a]School of Forestry and Environmental Studies, Yale University, 230 Prospect Street, New Haven, CT 06511, USA
[b]Department of Agricultural and Resource Economics, University of California, Berkeley, CA 94720, USA
[c]John F. Kennedy School of Government, Harvard University, Cambridge, MA 02138, USA
[d]Resources for the Future, 1616 M Street, NW, Washington, DC 20036, USA

Received 24 May 2006
Available online 23 June 2007

Abstract

We estimate the price elasticity of water demand with household-level data, structurally modeling the piecewise-linear budget constraints imposed by increasing block pricing. We develop a mathematical expression for the unconditional price elasticity of demand under increasing block prices and compare conditional and unconditional elasticities analytically and empirically. We test the hypothesis that price elasticity may depend on price structure, beyond technical differences in elasticity concepts. Due to the possibility of endogenous utility price structure choice, observed differences in elasticity across price structures may be due either to a behavioral response to price structure, or to underlying heterogeneity among water utility service areas.

JEL classification: Q21; Q25; Q28; L95; D12

Keywords: Water demand; Price elasticity; Non-linear pricing

1. Introduction

Water managers traditionally have maintained that consumers do not respond to price signals, so demand management has occurred most frequently through restrictions on specific water uses and requirements for the adoption of specific technologies. In theory, raising prices to bring about water conservation is less costly than implementing a command-and-control approach, even if the prices in question are inefficient. Water utilities increasingly hear this message and respond in many cases by implementing increasing block prices (IBPs), under which marginal prices increase with the quantity consumed. The price elasticity of water demand is a key variable of interest in this context, because: (1) regulators may use price to reduce demand during periods of scarcity; and (2) utilities often face zero-profit constraints, so the impact of price changes on total revenues is a concern.

*Corresponding author. Fax: +1 203 432 0026.
E-mail address: sheila.olmstead@yale.edu (S.M. Olmstead).

0095-0696/$ - see front matter © 2007 Elsevier Inc. All rights reserved.
doi:10.1016/j.jeem.2007.03.002

In 2000, approximately one-third of US urban residential water customers faced IBPs, up from 4% in 1982 [31,33]. Consumer responses to IBPs in the market for water are unclear, at best. Estimating water demand functions under non-linear price regimes requires econometric techniques that can separate demand from supply. The problem is that marginal prices change as consumption increases, so that, while the price schedule is set by the supplier, the actual price paid by the consumer and the quantity consumed are simultaneously determined.

Our analysis offers three main contributions to the economic literature on water demand. First, using the most price-diverse, detailed, household-level water demand data yet available, we exploit substantial variation in the price incentives faced by households to estimate price elasticities under increasing block and uniform marginal prices. Using a structural model that accounts for non-linear prices, we estimate a full-sample price elasticity of water demand of approximately −0.33. We are not the first to estimate a structural water demand model for consumers facing piecewise-linear budget constraints. However, previous applications have analyzed very restrictive samples. Ours is more broadly representative of US urban residential water demand.

Second, we explore the ways in which the concept of price elasticity under non-linear prices differs from that under a typical single marginal price. We find that the price elasticity of demand under IBPs is a complex function of the price and income elasticities of demand along particular budget segments. These results may be of some general interest, as non-linear prices predominate in markets for many goods and services other than water, including electricity, local and wireless telephone services, and labor supply under progressive income taxation.

Third, we consider the interesting and policy-relevant possibility that, aside from innate mathematical differences in what price elasticity is measuring, household price elasticity may be a function of a utility's choice of price structure. Our examination of this question is prompted by: (1) some clues in meta-analyses in the water demand literature that suggest this could be the case; and (2) the fact that the three previous applications of structural estimation under non-linear prices have all resulted in high price elasticity estimates, relative to the mean in the literature. We perform two tests of this hypothesis. Results suggest that price elasticity may be a function of price structure. However, there are a number of complications in this determination which are described further below. What we observe may be a behavioral response to price structure, or it may be due to underlying heterogeneity among utilities adopting different price structures.

In large parts of the US, scarce water supplies are a serious environmental concern. In some cases, water is being used at rates that may exceed those dictated by the efficiency criterion, particularly when externalities are considered.[1] Cities in arid states such as Texas and California have struggled to manage scarcity in the face of population growth, as well as consumer demand for swimming pools, landscaping and other water-intensive amenities. The administratively determined rate schedule sets water markets apart from many natural resource markets in which we would expect prices to reflect scarcity (e.g., oil and coal). Nonetheless, public agencies typically set water prices, thus the potential exists for establishment of rate structures that provide social-welfare-compatible incentives. This paper explores two important aspects of water pricing—how sensitive US households are to changes in water prices, and whether price elasticity differs under alternative rate structures. Research in this area can provide critical information to water resource managers.

The paper is organized as follows. In Section 2, we review the literature on block pricing and the price elasticity of residential water demand. In Section 3 we describe our models, and in Section 4 we describe the data. In Section 5 we describe the results, and in Section 6 we conclude.

2. The literature and the economic context

2.1. Block pricing and efficiency

Urban residential water pricing typically takes one of three forms: (1) a uniform marginal price; (2) IBPs; or (3) decreasing block prices. Each of these price structures is typically accompanied by a fixed water service fee. Under a uniform price (UP), households pay a single volumetric marginal price at all levels of consumption. Increasing block structures charge higher marginal prices for higher quantities consumed, resulting in a water

[1] For example, in many parts of the US, annual water use is maintained by groundwater mining [8].

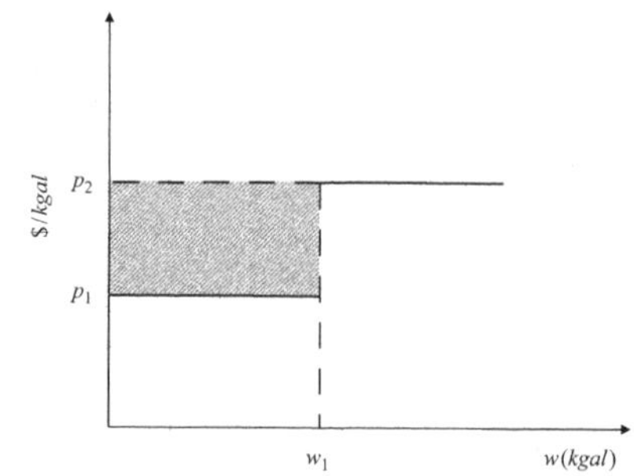

Fig. 1. Two-tier increasing block price structure, in which w is water (units); p_1 the marginal price of water in block 1; p_2 the marginal price of water in block 2; and w_1 the boundary quantity between blocks.

supply function that resembles a staircase ascending from left to right (Fig. 1); decreasing block structures are stacked in the opposite direction.

Water prices in North America generally lie below the long-run marginal cost (LRMC) of water supply, the efficient price [13,36]. Utilities' adoption of IBPs may expand the fraction of consumption priced at the LRMC. The LRMC can be greater than short-run average cost, because LRMC reflects the cost of new supply acquisition, and new supplies are typically more costly to develop than current supplies [13]. Thus, pricing all units at LRMC may cause utility revenues to exceed current expenses, sometimes substantially [25,10]. The efficient way to deal with this would be to rebate net revenues in some lump-sum fashion. IBPs instead charge something approaching LRMC for "marginal" uses (lawn-watering and the like), while meeting rate of return constraints through the manipulation of block cutoffs and infra-marginal prices. Increasing the fraction of consumption that faces efficient prices would result in welfare gains. But even if the highest-tier price in an IBP schedule does reflect LRMC, some welfare losses must result from marginal subsidies to consumers facing lower-tier prices. Thus, the relative efficiency of uniform and IBP structures depends on many things.

Water prices are usually set by elected or appointed entities, such as city councils or, for the relatively small number of investor-owned water utilities, by public utility commissions. Consumers may or may not influence the adoption of particular price structures, and even the magnitude of marginal price, through the political process.[2] We do not directly model the avenues through which water prices are set. Instead, we focus on the demand side—how estimates of elasticity capture different phenomena under different price structures, and whether consumers react differently to different price structures, establishing that endogenous price structure choice by utilities may be a serious obstacle to answering the second question.

2.2. *Consumer responses to water prices and price structure*

Previous analyses suggest that water demand is inelastic. In a meta-analysis of 124 estimates generated between 1963 and 1993, accounting for the precision of estimates, the mean price elasticity is −0.51, the short-run median is −0.38, and the long-run median is −0.64, with 90% of all estimates between 0 and −0.75 [7].[3]

Under block pricing, consumers face a piecewise-linear budget constraint. Fig. 2 depicts the budget constraint in a two-tier IBP system with a set of hypothetical indifference curves. The consumer has three

[2]For example, in the City of Corpus Christi, Texas, a local taxpayers' organization successfully lobbied for a regulation limiting water price increases by the public utility to the rate of inflation during the early 1990s, despite a series of severe regional droughts. In our experience, most of the citizen involvement in water rate setting has been aimed at preventing price increases, rather than promoting them.

[3]In a more recent meta-analysis of almost 300 price elasticity studies, 1963–1998, the mean price elasticity is −0.41 [4].

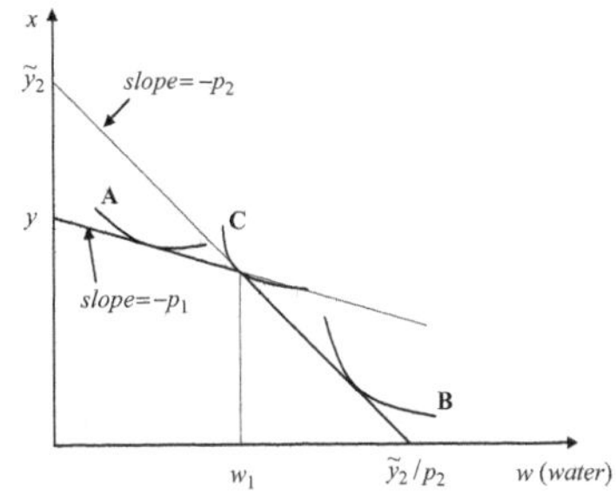

Fig. 2. Utility maximization under a two-tier increasing block price structure, where x is all goods other than water, and y is income.

reasonable consumption choices: consume on the interior of segment one, on the interior of segment two, or at the kink point—the quantity at which the marginal price increase occurs.[4]

Price elasticity under IBPs is typically modeled as the change in demand that results from a 1% change in all marginal prices at once.[5] Some of the largest price elasticity estimates, in the range of −1 to −2, have been obtained with structural analyses that model directly the piecewise-linear budget constraint created by block pricing [18,32], drawing upon an approach developed to estimate labor supply effects of progressive income taxation.[6] In addition, two meta-analyses have found that elasticity estimates where IBPs are implemented are higher than estimates of elasticity under linear marginal prices [7,4]. In comparing elasticities across different studies, the meta-analyses face many confounding factors that differentiate the studies in their samples. They control for some, but not all, of these factors. Another analysis at the utility, rather than the household level, suggests that demand is more elastic under IBPs than under decreasing block structures [26]. Results like these have caused some conjecture that previous elasticity estimates were biased, since they had not accounted for piecewise-linearity, or that IBPs themselves are responsible for greater sensitivity of demand to price [4,7,26].

Differences in elasticity estimates across price structures may arise from many sources. First, price elasticity as a mathematical measure captures something different under uniform and IBP structures. A price elasticity under IBPs includes two components absent from price elasticity under a constant marginal price—the elasticities of the probabilities that a household will locate at each possible kink or segment along its budget constraint, and the response of infra-marginal subsidies to price changes, essentially a secondary income effect.[7] In the Appendix, we derive a mathematical expression for price elasticity under IBPs and contrast this with elasticity under UPs. We show that structural IBP elasticity estimates that account for these extra components of the demand response to price changes may be either lower or higher than estimates of elasticity conditional on consumers remaining within their current blocks. Thus, while differences may arise from the correct estimation of an unconditional IBP elasticity, this cannot explain *a priori* why elasticity estimates under IBPs tend to be higher than those under UPs.

[4]Which outcome occurs depends on the shape of the indifference curves relative to the budget set; this is represented by the three alternative indifference curves in Fig. 2. Because, as modeled below, water is an essential good, we can rule out the possibility that consumption occurs at the corner solution of zero consumption. Under decreasing block structures, budget sets are non-convex, enabling multiple tangencies between indifference curves and the budget constraint. The theory and empirics of the non-convex case are straightforward extensions of the convex case; and they have been developed generally [24,23] for labor supply [3,15] and for water demand [17,18].

[5]The alternative is to consider the demand response to a change in just one of the block prices. However, considering a change in *all* of the block prices is more intuitive because then, no matter what marginal price a consumer faces, she experiences a 1% price increase on the margin. For consumers in the upper tiers of an IBP, the resulting changes in infra-marginal prices can be "rebated" by subtracting increased infra-marginal expenditures from household income. We discuss this further in Section 3.

[6]Prior to Hewitt and Hanemann [18], only three published studies had estimated price elasticities in the elastic range [5,6,19].

[7]This secondary income effect is due to "virtual income," a concept introduced in detail in Section 3.

Are there other potential explanations? Intuitively, IBPs introduce at least one threshold beyond which consumption causes an increase in the marginal water price, thus they may simply make prices more salient to consumers. Improvements in the presentation of water price information on water bills have been shown to increase consumers' price responsiveness, and IBPs may provide a similar signal [9]. Behavioral reactions to non-linear prices have been observed in the context of government tax and transfer programs [20].

3. Demand models

The structural discrete/continuous choice (DCC) model was developed in the study of labor supply by Hausman [3,15] and generalized by Moffitt [23,24]. The theoretically similar context implied by discrete/continuous consumption problems was developed by Hanemann [12], and first applied to water demand by Hewitt [17].

In the present analysis, we use a log–log demand function, described by (1)[8]

$$\ln w = Z\delta + \alpha \ln p + \gamma \ln y + \eta + \varepsilon. \tag{1}$$

The dependent variable is the log of daily household water demand. The marginal price of water is p, and income y. Water demand, with the exception of the small fraction used for drinking, is derived demand in which the primary demand is for water-consuming goods and services, such as clean laundry, indoor bathroom use, and green lawns. Thus, in addition to price and income, the demand models also include household characteristics. These variables, as well as daily weather observations and city fixed effects, are included in the matrix Z. The model has two error terms. The first source of error (η) represents heterogeneous water consumption preferences among households not explained by the household characteristics in Z. The second source of error (ε) reflects random error unobservable to both the household and the analyst. From the household's perspective, actual use may not coincide with optimal use due to leaks or other causes. From the analyst's perspective, ε captures these optimization errors, as well as the usual measurement error. We assume that the two errors are independent and normally distributed with mean zero and variances σ^2_η and σ^2_ε.

If all households face UPs for water (i.e., the price does not vary with quantity consumed), the demand function in (1) combined with normal error terms yields a conventional log-likelihood function for a log-normal regression (2), where w is observed consumption and $w^*(.)$ is optimal consumption. Maximizing (2) would produce parameter estimates equal to ordinary least squares (OLS) estimates

$$\ln L = \sum \ln\left(\frac{1}{\sqrt{2\pi}}\frac{\exp(-(s)^2/2)}{\sigma_v}\right), \tag{2}$$

where

$$v = \eta + \varepsilon,$$
$$s = (\ln w - \ln w^*(Z, p, y; \delta, \alpha, \gamma))/\sigma_v.$$

However, with IBPs, demand becomes more complicated. With K blocks, each with block price p_k, the blocks are separated by $K-1$ switching points or "kinks" denoted w_k. There is now a key distinction between conditional demand functions and the unconditional demand function. Conditional demand is defined as the quantity the individual consumes conditional on the fact that her consumption is in the kth price block. This is simply given by the demand function in (1) evaluated at the marginal price p_k and income equal to $\tilde{y}_k = y + d_k$, where

$$d_k = \begin{cases} 0 & \text{if } k = 1, \\ \sum_{j=1}^{k-1} (p_{j+1} - p_j) w_k & \text{if } k > 1. \end{cases} \tag{3}$$

For households consuming anywhere beyond the first block, there is a wedge between the marginal and average price of water. The implicit subsidy from the infra-marginal rates is accounted for by adding to

[8] This functional form is common in water demand analysis, due in part to the typically extreme right-skewness of demand. We use it here to be able to compare our estimates with other DCC models [18,34].

income the difference, d_k, between what a household would pay if all units were charged at the marginal price, and what they actually pay. This income supplement for a household consuming in the second block of a two-tier price structure would be equal to the cross-hatched region in Fig. 1.[9] The sum of income and d_k is commonly called "virtual income," here denoted $\tilde{y}_k$. There is a separate conditional demand function for each block. By contrast, there is but a single unconditional demand function, and this characterizes the consumer's overall choice—not just how much to consume conditional on being in a given price block, but also which price block to choose.[10]

Take, for example, a household's choice when $K = 2$, and x is consumption of all goods other than water (Fig. 2). If a household's demand for water, conditional on being in block 1, is less than the kink point w_1, then this is also its unconditional demand, and the household will consume in the first block (e.g., curve A). If conditional demand associated with block 2 exceeds w_1, then unconditional demand coincides with the conditional demand for block 2, and the household will consume in the second block (curve B). There is also the possibility that the household might consume exactly at the kink point, w_1 (curve C). This occurs if both the conditional demand function for block 1 yields a demand larger than w_1 and the conditional demand function for block 2 yields demand smaller than w_1. In that case, the unconditional demand is w_1, adjusting for random error (ε). Such households would like to locate on either of the dotted lines extending the two budget segments from the kink point in Fig. 2, but these choices are simply not available to them. All of this is captured by (4), unconditional demand, a function of conditional demands and kink points, where w is observed consumption, $\underline{w}_k^*(Z, p_k, \tilde{y}_k; \delta, \alpha, \gamma)$ is optimal consumption on block k, and w_k is consumption at kink point k

$$\ln w = \begin{cases} \ln \underline{w}_1^*(Z, p_1, \tilde{y}_1; \delta, \alpha, \gamma) + \eta + \varepsilon \\ \quad \text{if} \quad -\infty < \eta < \ln w_1 - \ln \underline{w}_1^*(Z, p_1, \tilde{y}_1; \delta, \alpha, \gamma), \\ \ln w_1 + \varepsilon \\ \quad \text{if} \quad \ln w_1 - \ln \underline{w}_1^*(Z, p_1, \tilde{y}_1; \delta, \alpha, \gamma) < \eta < \ln w_1 - \ln \underline{w}_2^*(Z, p_2, \tilde{y}_2; \delta, \alpha, \gamma), \\ \ln \underline{w}_2^*(Z, p_2, \tilde{y}_2; \delta, \alpha, \gamma) + \eta + \varepsilon \\ \quad \text{if} \quad \ln w_1 - \ln \underline{w}_2^*(Z, p_2, \tilde{y}_2; \delta, \alpha, \gamma) < \eta < \ln w_2 - \ln \underline{w}_2^*(Z, p_2, \tilde{y}_2; \delta, \alpha, \gamma) \\ \ldots \\ \ln w_{K-1} + \varepsilon \\ \quad \text{if} \quad \ln w_{K-1} - \ln \underline{w}_{K-1}^*(Z, p_{K-1}, \tilde{y}_{K-1}; \delta, \alpha, \gamma) < \eta < \ln w_{K-1} - \ln \underline{w}_K^*(Z, p_K, \tilde{y}_K; \delta, \alpha, \gamma), \\ \ln \underline{w}_K^*(Z, p_K, \tilde{y}_K; \delta, \alpha, \gamma) + \eta + \varepsilon \\ \quad \text{if} \quad \ln w_{K-1} - \ln \underline{w}_K^*(Z, p_K, \tilde{y}_K; \delta, \alpha, \gamma) < \eta < \infty. \end{cases} \tag{4}$$

Given the simultaneous determination of consumption and block price, if a typical single-error stochastic specification were employed, the size of the error term, marginal price, and virtual income would be systematically correlated. Thus, OLS estimates of the demand parameters will be biased and inconsistent. In the literature, this problem has been addressed with instrumental variables (IV) models, such as two-stage least squares, for labor supply and water and energy demand [1,6,16,27,28]. While such models can estimate downward-sloping demand curves, they exhibit two key limitations. First, simultaneous equations models can be used only to estimate elasticities conditional on remaining within the observed consumption block. Second, IV methods disregard the fact that some households consume within the neighborhood of a kink point, where it is unclear which value of marginal price should be assigned (in an econometric model that includes one or more error terms). Arbitrary treatment of these observations, such as assigning them to one block or another or dropping them from the sample, cannot be reconciled with utility theory. The DCC model, a maximum likelihood model, addresses these issues. The log-likelihood function for $K > 1$ is given by (5), in which Φ is the standard normal cumulative distribution function.[11] The first summation accounts for consumption along K

[9]The treatment of infra-marginal price changes as lump-sum changes in household income was developed in the labor supply literature [3,11], and the electricity demand literature [29,35]. The first application to water demand was by Billings and Agthe [2].

[10]With uniform prices, by contrast, the unconditional demand function is given by (1) directly.

[11]The derivation of the IBP portion of the likelihood function is described in [30].

linear budget segments, and the second summation for consumption at $K-1$ kinks

$$\ln L = \sum \ln \left[\begin{array}{l} \sum_{k=1}^{K} \left(\frac{1}{\sqrt{2\pi}} \frac{\exp(-(s_k)^2/2)}{\sigma_v} \right) (\Phi(r_k) - \Phi(n_k)) \\ + \sum_{k=1}^{K-1} \left(\frac{1}{\sqrt{2\pi}} \frac{\exp(-(u_k)^2/2)}{\sigma_\varepsilon} \right) (\Phi(m_k) - \Phi(t_k)) \end{array} \right], \quad (5)$$

where

$$v = \eta + \varepsilon, \quad t_k = (\ln w_k - \ln \underline{w}_k^*(.))/\sigma_\eta,$$

$$\rho = \text{corr}(v, \eta), \quad r_k = (t_k - \rho s_k) \Big/ \sqrt{1 - \rho^2},$$

$$s_k = (\ln w_i - \ln \underline{w}_k^*(.))/\sigma_v, \quad m_k = (\ln w_k - \ln \underline{w}_{k+1}^*(.))/\sigma_\eta,$$

$$u_k = (\ln w_i - \ln w_k)/\sigma_\varepsilon, \quad n_k = (m_{k-1} - \rho s_k) \Big/ \sqrt{1 - \rho^2}.$$

The estimated parameters can be used to calculate the expected unconditional demand and to measure the price elasticity of demand. In the Appendix, we provide analytical formulas for the expected unconditional demand and the price elasticity of demand. The overall price elasticity of demand is defined with respect to an equi-proportional change in all block prices.[12]

4. The data

The data comprise 1082 households in 11 urban areas in the United States and Canada, served by 16 public water utilities.[13] Table 1 provides descriptive statistics.

Daily household water demand and weather conditions are observed over two weeks in an arid season and two weeks in a wet season. Daily demand data were gathered by automated data loggers, attached to household water meters by utility staff. Households move seasonally across price blocks. During the wet season, for example, among households facing two-tier IBPs, 56% of household-days are observed in the first block, and 44% in the second. During the dry season, first-block observations are 25% of the total, and second-block 75%.[14] Given the econometric similarities between our analysis of water demand and labor supply analysis under progressive income taxation, it is interesting to discuss the degree to which consumers "cluster" at kink points. Among households facing IBPs in our data, approximately 40% of demand observations lie within 5% of a kink point (from above or below).

Each household faces one price structure throughout a season, but some price structures change between the two seasons. Thus, the data include 26 price structures: 8 two-tier IBP structures, 10 four-tier IBP structures, and 8 uniform structures. Marginal prices range from $0 to $4.96 per thousand gallons (kgal), with a mean of $1.71/kgal. Assuming that the four weeks of water demand we observe are representative of annual patterns, the total average annual household expenditure on water in this sample, including fixed charges, would be about $327.

Household demographic data were collected by survey.[15] These include lot size, square footage of homes, number of bathrooms, and family size. Home age is also included in the water demand equation, and we

[12]In the empirical analysis below, the elasticity is estimated by imposing a 1% change in each block price and numerically calculating the resulting percentage change in expected unconditional demand.

[13]The data were gathered by Mayer et al. [22] for a study sponsored by the American Waterworks Association Research Foundation. Sample utilities include Denver Water (CO); Eugene Water and Electric Board (OR); Seattle Public Utilities (WA); City of Bellevue Utilities Department (WA); Highline Water District (WA); Northshore Utility District (WA); City of San Diego Water Department (CA); Tampa Water Department (FL); Phoenix Water Department (AZ); City of Tempe Water Utilities (AZ); City of Scottsdale Water Department (AZ); Regional Water Services, Municipality of Waterloo (ONT); City of Cambridge (ONT); Walnut Valley Water District (CA); City of Lompoc Utilities Department (CA); and Las Virgenes Municipal Water District (CA).

[14]Similar movement can be seen for households facing four-tier IBPs. The share of household-days observed in the first block among four-tier households is 37% in the wet season, and 18% in the dry season; third-block observations comprise 24% of household-days in the wet season, and 62% in the dry season.

[15]Households chosen for the study were randomly sampled from a subset of utilities' customer databases: single-family households. Sampling procedures, response rates, and statistical tests for selection bias are described in [22].

Table 1
Descriptive statistics

Variable	Description	Units	Mean	Std. dev.	Min.	Max.
w	Daily household water demand	kgal/day	0.40	0.58	0.00	9.78
p_1	Marginal price in block 1	\$/kgal	1.45	0.54	0.50	3.70
p_2	Marginal price in block 2	\$/kgal	1.84	0.40	0.84	4.06
p_3	Marginal price in block 3	\$/kgal	2.43	0.87	0.93	4.96
p_4	Marginal price in block 4	\$/kgal	3.28	1.30	0.99	5.98
income	Gross annual household income	\$000/yr	69.81	67.67	5.00	388.64
w_1	Water quantity at kink 1	kgal/day	0.21	0.08	0.06	0.37
w_2	Water quantity at kink 2	kgal/day	0.36	0.09	0.30	0.50
w_3	Water quantity at kink 3	kgal/day	1.82	0.80	0.75	2.49
block p	Price structure: 1 if IBP; 0 if UP	0/1	0.61	0.49	0	1
seas	Season: 1 if irrigation season; 0 if not	0/1	0.51	0.50	0	1
weath	Evapotranspiration less effective rainfall	mm	5.06	8.42	−46.15	19.37
max t	Maximum daily temperature	°C	24.12	8.78	0	42.78
famsz	Number of residents in household	Number	2.79	1.34	1	9
bthrm	Number of bathrooms in household	Number	2.58	1.30	1	7
sqft	Approximate area of home	000 ft^2	2.02	0.82	0.40	4.37
lotsz	Approximate area of lot	000 ft^2	10.87	9.22	1.00	45.77
age	Approximate age of home	yrs/10	2.88	1.62	0.07	5
evap	Evaporative cooling: 1 if yes; 0 if no	0/1	0.09	0.28	0	1

Notes: City fixed effects not summarized here. Approximately 10% of observations drawn from each city (Las Virgenes Municipal Water District, Seattle, San Diego, Tampa, Phoenix, Tempe/Scottsdale, Waterloo/Cambridge, Walnut Valley Water District, Lompoc).

expect that both very old and new homes may use less water than "middle aged" homes.[16] We - also control for the presence or absence of evaporative cooling, which substitutes water for electricity in air conditioning.[17]

We include a set of daily weather variables. Maximum daily temperature is represented by max$t$, weath represents the moisture requirements of green lawns not met by precipitation, and seas is a dummy variable set equal to one during the arid (peak outdoor watering) season.[18] Finally, the models include a set of dummy variables that represent the 11 urban areas included in the study: Denver, Eugene, Seattle, San Diego, Tampa, Phoenix, Tempe/Scottsdale, Waterloo/Cambridge (Ontario), Walnut Valley, Las Virgenes, and Lompoc, accounting for variations in geography and other factors not addressed by the other covariates.

5. Empirical results

The full-sample DCC model, with the likelihood function the sum of (2) for uniform-price households and (5) for IBP households, is our baseline model. We explore the possibility of variation in price elasticity with price structure by estimating two further demand models: the baseline DCC model, but for IBP households

[16]Very old homes are likely to have smaller connections to their city water system, and also fewer water-using appliances, such as dishwashers and jacuzzis, than do newer homes. The newest homes were built after the passage of local ordinances in the 1980s and 1990s requiring water-conserving toilets and showerheads.

[17]In our sample, households with evaporative coolers use on average 40% more water per day. We include this variable to avoid biasing the other parameter estimates, particularly the income coefficient estimate, since evaporative cooling is relatively more common in low-income households.

[18]Lawn moisture needs are captured through estimated evapotranspiration (ET), less effective rainfall (equal to 0.6* measured precipitation). We use Hargreaves' approximation to the Penman–Monteith evapotranspiration equation, which requires mean, minimum, and maximum daily temperature, degrees latitude (to estimate a solar radiation parameter), and a readily available constant associated with the "crop" of interest—grass.

only; and a panel random-effects model for UP households only. We then perform some additional analysis regarding the possibility of endogenous price structures.

5.1. Full-sample discrete/continuous choice model

Table 2 reports results of the full-sample DCC model in column 1. The DCC parameter estimates amount to elasticities conditional on block location. Table 3 reports the means and standard deviations

Table 2
Water demand model estimates

	DCC full sample	Split sample models		DCC full sample block pricing test
		DCC IBP households only	Random effects UP households only	
Variable	(1)	(2)	(3)	(4)
ln price	−0.3407*	−0.6411*	−0.3258	−0.2330*
	(0.0298)	(0.0424)	(0.5644)	(0.0677)
block *p**ln price				−0.1106#
				(0.0626)
ln income	0.1306*	0.1959*	0.0432	0.1308*
	(0.0118)	(0.0163)	(0.0499)	(0.0118)
seas	0.3072*	0.3370*	0.2656*	0.3075*
	(0.0247)	(0.0330)	(0.0331)	(0.0247)
weath	0.0079*	0.0094*	0.0054*	0.0076*
	(0.0013)	(0.0017)	(0.0016)	(0.0013)
max *t*	0.0196*	0.0238*	0.0193*	0.0200*
	(0.0018)	(0.0024)	(0.0025)	(0.0018)
famsz	0.1960*	0.1959*	0.2095*	0.1957*
	(0.0056)	(0.0080)	(0.0225)	(0.0056)
bthrm	0.0585*	0.0553*	0.0498	0.0590*
	(0.0093)	(0.0127)	(0.0393)	(0.0093)
sqft	0.1257*	0.2001*	−0.0066	0.1253*
	(0.0140)	(0.0191)	(0.0600)	(0.0140)
lotsz	0.0065*	0.0097*	−0.0010	0.0064*
	(0.0009)	(0.0011)	(0.0046)	(0.0009)
age	0.0869*	0.1814*	−0.0870	0.0888*
	(0.0219)	(0.0304)	(0.0940)	(0.0219)
age2	−0.0137*	−0.0252*	0.0109	−0.0141*
	(0.0036)	(0.0050)	(0.0160)	(0.0036)
evap	0.2278*	0.2416*	−0.8139	0.2281*
	(0.0300)	(0.0334)	(0.6129)	(0.0300)
lasv	0.2590*	0.3391*		0.2570*
	(0.0408)	(0.0486)		(0.0409)
seat	−0.1232*	−0.0482		−0.1566*
	(0.0379)	(0.0491)		(0.0424)
sandg	0.0129	0.1361*		0.0114
	(0.0365)	(0.0416)		(0.0366)
tampa	−0.3879*	−0.3225*		−0.3919*
	(0.0373)	(0.0427)		(0.0374)
phx	−0.0044	−0.0271		−0.0086
	(0.0385)	(0.0444)		(0.0387)
tscot	−0.1033*	−0.1436*		−0.1064*
	(0.0359)	(0.0406)		(0.0360)
eug	0.0050		−0.0018	0.0340
	(0.0387)		(0.5846)	(0.0420)

Table 2 (*continued*)

	DCC full sample	Split sample models		DCC full sample block pricing test
		DCC IBP households only	Random effects UP households only	
Variable	(1)	(2)	(3)	(4)
wcamb	−0.1939*		−0.1646	−0.2545*
	(0.0362)		(0.1568)	(0.0499)
wvall	0.1784*		0.3272*	0.1021#
	(0.0387)		(0.1320)	(0.0580)
lomp	−0.0647		−0.0440	−0.1512*
	(0.0373)		(0.1132)	(0.0615)
constant	−3.6994*	−4.3288*	−3.1353*	−3.7068*
	(0.0652)	(0.0888)	(0.6854)	(0.0654)
σ_η	1.0769*	1.1493*		1.0767*
	(0.0103)	(0.0115)		(0.0103)
σ_ε	0.3552*	0.4220*		0.3562*
	(0.0277)	(0.0227)		(0.0277)

Notes: *Significant at $\alpha = 0.05$. #Significant at $\alpha = 0.10$. Standard errors in parentheses below estimates. Dependent variable is natural log of daily water demand. Models in columns 1, 2 and 4 are discrete/continuous choice (DCC), estimated by maximum likelihood. Model in column 3 is random effects GLS regression. City fixed effects are relative to Denver, CO, except in column 3, where they are relative to Seattle, WA. For full-sample models (columns 1 and 4), $N = 25{,}666$. For IBP split-sample model (column 2), $N = 15{,}621$; and for UP split-sample model (column 3), $N = 10{,}045$.

Table 3
Simulated price and income elasticities of demand

Discrete continuous choice model	Elasticity estimates			
	Parametric		Nonparametric	
	Price	Income	Price	Income
Full sample	−0.3309*	0.1261*	−0.3319*	0.1273*
	(0.0062)	(0.0031)	(0.0366)	(0.0133)
Increasing-block households only	−0.5893*	0.1786*	−0.6090*	0.1865*
	(0.0437)	(0.0138)	(0.0568)	(0.0159)
Random effects model				
Uniform price households only	−0.3258	0.0432		
	(0.5644)	(0.0499)		

Notes: *Significant at $\alpha = 0.05$. Reported estimates are the means of bootstrapped elasticities, and reported standard errors (in parentheses) are the standard deviations of bootstrapped elasticities. Simulated DCC elasticity estimates for variables other than price and income do not vary appreciably from coefficient estimates reported in Table 2. Elasticity estimates for random effects model are repeated from Table 2, for comparison purposes (no simulations necessary for this model).

of unconditional elasticity estimates bootstrapped parametrically and non-parametrically. These unconditional elasticities include the impacts of price changes on the probability of block and/or kink switching.[19]

[19]For the non-parametric simulations, we draw 100 random samples of size N, with replacement, from the data; take the expectation of conditional demand in exponential form; simulate a 1% change in price or income; calculate the resulting percent change in expected demand; and report the mean of bootstrapped estimates. For the parametric simulations, we generate 100 random draws from each error distribution and estimate elasticities for each pair of error draws, using the original data and parameter estimates and taking expectations of conditional demand.

The parameter of greatest interest, price elasticity, indicates that a 1% price increase will reduce demand by 0.33%, all else equal.[20] Our simulated price elasticity, produced by exploiting the widest variation yet available in price, price structure, and individual household characteristics, is comparable to the central tendency of estimates over the past 40 years, but much lower than those previously estimated using the DCC approach with smaller, more restrictive samples.[21] Hewitt and Hanemann, for example, obtain an elasticity almost six times the magnitude of ours [18]. Note that the simulated unconditional elasticity estimate is smaller in absolute value than the price parameter estimate, the conditional elasticity.[22] In this sample, block switching dampens the price elasticity, though it need not have done so in theory, as we show in the Appendix.

There are two intuitive reasons for this dampening effect. First, households may stick at kink points, where their price elasticity is zero. Second, if the price falls, and optimal consumption increases enough to move the household to a higher block, movement across the kink point immediately results in a marginal price increase (to the price of the new block), dampening the increase in demand we might expect from the original price decrease. The same would be true for a price increase, which might push the household into a lower block, at which it suddenly faces a lower marginal price. In our sample, these two effects seem to dominate any potential influences that, mathematically, could push this difference in the other direction.[23]

Weather, income and the other household-level characteristics affect demand in expected ways. Results regarding home age are weak, but as expected. The highest water demand occurs among homes aged 20–40 years; both newer homes and older homes use less water.[24] Our income elasticity estimate (0.13) is low compared with previous studies, usually between 0.2 and 0.6 [13]. Few previous studies have included household-level housing characteristics, which are strongly correlated with income. Their omission may have overestimated the influence of income on water demand.[25] Maximization of the likelihood function also provides estimates of σ_η and σ_ε. Consistent with the tax literature, household heterogeneity is responsible for more of the observed variation in water demand than random error.

5.2. *Exploring elasticity magnitude and price structure*

When we estimate the DCC model for households facing IBPs only, we obtain a simulated price elasticity of approximately −0.59. When we estimate a panel random-effects model for households facing uniform marginal prices only, we obtain an elasticity estimate of −0.33, and the estimate is statistically insignificant, even with about 400 households in this group. This pattern reflects what we see in the existing literature, even though the 95% confidence intervals for these two estimates overlap, as the UP estimate is noisy. In comparisons of studies using different samples, we observe more elastic demand among block-price samples than among uniform-price samples. The observed differences are much more significant than the small differences between our DCC price coefficients and elasticity estimates, which can be accounted for by block switching. Does price elasticity vary with price structure, all else equal, or is something else going on?

[20]Recall that price elasticity in our DCC models is a percent change in quantity demanded for a 1% increase in all prices in a household's price structure, including the impact on virtual income of increases in infra-marginal prices. Our elasticity estimates are a mixture of short-run and long-run. Much of the variation in price comes from the cross-section. But there is also substantial variation within individual households over time, due to IBPs. There is also a small amount of variation in price over time due to utility price changes.

[21]As a point of reference, we also estimated a random effects model for the full sample to demonstrate the price endogeneity problem. The estimated price elasticity is 1.40, reflecting upward-sloping price schedules, not upward-sloping demand.

[22]This is true of Hewitt and Hanemann, as well–simulated elasticities, allowing block switching, are smaller than price coefficient estimates [18].

[23]In general, these include both virtual income effects (which unambiguously raise the unconditional price elasticity, relative to the conditional elasticity), and the impact of price changes on the probabilities of the household's location on each segment and kink in its budget constraint (which might work in either direction). See the Appendix for a careful discussion of these effects.

[24]We have also estimated a model replacing the quadratic home age specification with the following home age categories: pre-1960; 1960–1969; 1970–1979; 1980–1989; and 1990–1996. Results are qualitatively similar. All else equal, water consumption peaks among homes built between 1970 and 1979. Both newer and older homes use less water. There were no significant changes to the other demand parameters in this alternative model.

[25]If we estimate our DCC model omitting all household characteristics, except income, we obtain an income elasticity of 0.32.

5.2.1. Tests of simple theoretical explanations

To examine the relationship between price elasticity and price structure, we investigate whether the observed difference can be explained by theory. First, we would expect price elasticities to be larger in the long run than in the short run. It is possible that this explains the differences seen in the previous literature. It does not explain the difference we see in our data, since all households (IBP and UP) were observed over approximately the same time period.

Second, if the average magnitude of prices were higher in IBP cities than in uniform-price cities, this could explain the observed difference in the magnitude of price elasticity. In our sample, the average magnitude of marginal price in uniform-price cities is \$1.72 per kgal, while in block-price cities it is \$1.78 (assuming that marginal price for IBP households is price in the block of observed consumption). With a difference this small, it is unlikely that heterogeneity in the average magnitude of prices is responsible for the difference in elasticities we observe.

Third, if the percentage of total income devoted to water consumption were lower among households in UP cities, on average, than among households facing IBPs, this might explain the observed difference in elasticities. Estimated annual expenditures on water consumption (including fixed charges) in our sample are approximately 0.5% of annual income; the budget share for water does not differ significantly across price structures.

We have eliminated some likely candidates from the list of potential explanations for variation in price elasticity with price structure. Two possibilities remain. First, it is conceivable that part of the observed difference in price elasticity of demand across price structures is driven by the shape of price structures. But correlation does not demonstrate causality. The second possibility is that price structures, themselves, are endogenous.

5.2.2. The possibility of endogenous price structures at the utility level

Our concern is that some unobserved characteristic of households within a community drives a utility's choice of price structure, and that this characteristic may be correlated with price elasticity. For example, environmentally conscious households may pay more attention to water prices, resulting in higher elasticity, and relatively "green" communities may also be more likely to adopt IBPs, seen as conservation tools. Conversely, communities facing long-term aridity and long growing seasons may lead utilities to adopt IBPs; these same communities may be places in which households place a high premium on outdoor water consumption and are thus less price sensitive.[26] The significant differences in coefficient estimates in columns 2 and 3 of Table 2 support the possibility of unobserved heterogeneity across UP and IBP households that may be correlated with water demand.

Ideally, to test for the possibility of endogenous price structures, we would estimate a two-stage model, first predicting price structure at the utility level, then estimating elasticity, given the predicted price structure. With the appropriate choice of instruments, this would presumably give us unbiased estimates of price elasticity, free of the effects of underlying heterogeneity in price structure choice among utilities. However, the data include only 16 water utilities, a sample too small to support the first stage of such a selection model.

We take a second-best approach. We test our full-sample models, allowing the price elasticity parameter to vary with price structure, but constraining the remaining demand parameters to be equal. It is likely that our model includes some endogenous variation in price structure choice. However, our data are the best known to us for this purpose, thus we perform what should be considered, at best, an informal test. In addition, the amount of price variation within UP cities, alone, is very small—the reason we were unable to estimate a statistically significant elasticity for those cities in our split-sample models.[27] Results, reported in the fourth column of Table 2, should be interpreted with these caveats in mind.

Like the split-sample models, the model with a block pricing/price interaction term suggests that demand among IBP households is more elastic than among UP households. The other demand parameters are

[26]For estimates of low price elasticity among households with strong preferences for outdoor consumption, see [21].

[27]Taking into account the variation over time and space, we have 13 uniform price observations. Variation in block prices (both across utilities and within the individual price structures, themselves) is much more extensive.

essentially unchanged, with the exception of some of the city fixed effects—not surprising, as the block-pricing interaction is now accounting for some of the variation absorbed by city fixed effects in prior models.

These results do not rule out the possibility of a behavioral reaction to the structure of water prices, due to increased attention to price or any other factor. However, given our remaining selection concerns, we cannot say definitively that IBPs tend to increase elasticity. We cannot dismiss the hypothesis that underlying city-level heterogeneity in characteristics such as aridity and conservation programs, which may be correlated with both consumers' price elasticity and utilities' price schedule choice, explains part of the observed pattern in the literature of higher price elasticity under IBPs.

6. Conclusion

This paper considers both aspects of the separation of demand and supply under non-linear pricing—simultaneous choice of price block and quantity consumed at the household level, and endogenous price structure at the utility level. We estimate the price elasticity of water demand among urban households facing increasing block and uniform marginal prices, using a structural DCC model that accommodates piecewise-linear budget constraints. Previous estimates of the price elasticity of water demand using DCC models have been very high relative to the central tendency of estimates in the literature using other models. Our estimates, which exploit a broadly representative and highly detailed sample of North American single-family households are much closer to the central tendency of prior estimates. While the DCC model is theoretically the appropriate model for estimating elasticities under IBPs, our results suggest that these models may not induce a significant break in how we think about consumer responses to non-linear water prices, as they did when introduced in the literature on the wage elasticity of labor supply under progressive income taxation.

That said, an additional lesson of the analysis is that elasticity under IBPs measures an inherently different phenomenon than elasticity under linear marginal prices. In the Appendix, we develop an expression for what is captured by an unconditional price elasticity under IBPs, and compare our conditional and unconditional estimates on this basis. While the differences between conditional and unconditional elasticities of water demand may not be economically significant in our case, non-linear prices are present in many other markets (e.g., cellular phone service, electricity), and an understanding of what an elasticity is measuring is critical, especially in regulated markets in which price is an important policy lever (for example, a tool to promote energy conservation).

Finally, in response to a literature that suggests that price elasticity varies with the shape of administratively determined supply curves, in ways unconnected to the mathematical difference between price elasticity under different price structures, we estimate separate price elasticities for two sub-samples of households—those facing IBPs and those facing uniform marginal prices. The elasticities do appear to differ (indeed, we fail to identify a price elasticity significantly different from zero for the uniform-price households, alone), and we are able to eliminate a number of simple explanations that would be consistent with economic theory. In a second full sample model, allowing price elasticity to vary with price structure and constraining all other demand parameters to be equal, we again observe a greater price elasticity under IBPs. Neither of these tests accounts for selection on price structure at the utility level. Thus, while the possibility of a behavioral response to price structure cannot be ruled out, neither can we definitively identify such an effect. The collection of data suitable for estimating a two-stage model of water demand under endogenous price structures would be a productive area for future research.

Regardless of their potential implications for price elasticity, an important efficiency justification for the implementation of IBPs remains. To the extent that IBPs increase the portion of consumers facing efficient prices for a scarce resource on the margin, they may well be welfare improving (and environmentally beneficial at the same time). Exploration of the efficiency advantages of IBPs is another area for further research.

Acknowledgments

This work was supported in part by Grant #SES-0095348 from the National Science Foundation, Economics Program. For comments and guidance, we are grateful to Li Gan, Tony Gómez-Ibáñez, Julie Hewitt, Jeffrey Liebman, Erin Mansur, Todd Olmstead, Ellen Pint, Bob Triest, Frank Wolak, Richard

Zeckhauser, and seminar participants at Yale, Harvard, Stanford, the University of California, Santa Barbara, and the National Bureau of Economic Research. Special thanks to Arik Levinson and two anonymous referees, whose suggestions greatly improved the paper. Graeme Auld provided excellent research assistance. The authors, alone, are responsible for any remaining errors.

Appendix A. How price affects demand with increasing block prices

In the presence of IBPs, the concept of how price affects demand is, itself, somewhat elusive: since there are multiple prices corresponding to the multiple rate blocks, which price is being considered? Does one consider a single-block price or all block prices? If all prices are considered, how is the response of demand to an across-the board change in all prices with an IBP different from the response of demand to a change in a conventional flat-rate price? In this Appendix we answer some of these questions. Since the results may be of some general interest, we start with a general model of demand with multiplicative random terms, deriving some general results regarding IBPs. We then specialize the results to the log–log functional form in (1).

We use the following notation for a general model of demand. The general conditional demand function with multiplicative random terms is written as

$$w = w(p_k, y + d_k)e^{\varepsilon}e^{\eta}, \tag{A.1}$$

where ε and η are independently normally distributed with zero mean and variances σ^2_{ε} and σ_{η}^2. Recall that w_k denotes the switching point between blocks k and $k+1$ and $w(p_k, y + d_k)e^{\varepsilon}e^{\eta}$ is consumption within the kth block. Define

$$\begin{aligned} a_k &\equiv w_{k-1}/w(p_k, y + d_k), \\ b_k &\equiv w_k/w(p_k, y + d_k), \\ c_k &\equiv w_k/w(p_{k+1}, y + d_{k+1}). \end{aligned} \tag{A.2}$$

In the presence of an IBP schedule with K blocks (and $K-1$ switching points), the utility-maximizing consumer ends up either somewhere along a block or at one of the switching points. Let π_k denote the probability that the consumer chooses some quantity within the kth block, and let λ_k denote the probability that the consumer chooses to locate at the switching point w_k. Then

$$\begin{aligned} \pi_k &= \Phi\left(\frac{\ln b_k}{\sigma_\eta}\right) - \Phi\left(\frac{\ln a_k}{\sigma_\eta}\right), \quad k = (1, \ldots, K), \\ \lambda_k &= \Phi\left(\frac{\ln c_k}{\sigma_\eta}\right) - \Phi\left(\frac{\ln b_k}{\sigma_\eta}\right), \quad k = (1, \ldots, K-1), \end{aligned} \tag{A.3}$$

where Φ is the standard normal cumulative distribution function. Observe that

$$\sum_{k=1}^{K} \pi_k + \sum_{k=1}^{K-1} \lambda_k = 1. \tag{A.4}$$

For a moment, leave the distribution of η unspecified. With the multiplicative random terms, the expected quantity demanded by the consumer in the face of IBPs—the mean of the unconditional demand function (denoted W)—takes the form:

$$W = \left[\sum_{k=1}^{K} w(p_k, y + d_k)e^{\sigma_\varepsilon^2/2} \int_{a_k}^{b_k} \xi f(\xi)\,\mathrm{d}\xi\right] + \sum_{k=1}^{K-1} w_k e^{\sigma_\varepsilon^2/2}\lambda_k, \tag{A.5}$$

where $\xi = e^{\eta}$; the term with the integral is the partial mean of ξ over the interval $[a_k, b_k]$. When η is normal, making ξ lognormal, the formula reduces to

$$W = \left[\sum_{k=1}^{K} w(p_k, y + d_k)e^{\sigma_\varepsilon^2/2}\pi_k^*\right] + \sum_{k=1}^{K-1} w_k e^{\sigma_\varepsilon^2/2}\lambda_k, \tag{A.6}$$

where

$$\pi_k^* = \Phi\left(\frac{\ln b_k}{\sigma_\eta} - \sigma_\eta\right) - \Phi\left(\frac{\ln a_k}{\sigma_\eta} - \sigma_\eta\right). \tag{A.7}$$

It follows from (A.6) that expected unconditional demand depends on the block prices in a complex manner. By way of illustration, suppose there is a reduction in the price of the second block, p_2. This affects the consumer's demand in several different ways. Not surprisingly, the change in p_2 has *no* effect on consumption in the first block, $dw(p_1, y)/dp_2 = 0$ or on the probability of being in that block, $d\pi_1/dp_2 = 0$. However, it does affect demand in *all* other blocks. If the consumer is located in the second block and stays there, his demand changes not only because of the price reduction but also because the change in p_2 causes the income adjustment term d_2 to change $\partial d_2/\partial p_2 = w_1$. The total impact on consumption within the second block is

$$dw(p_2, y + d_2)/dp_2 = \partial w(p_2, y + d_2)/\partial p_2 + w_1 \partial w(p_2, y + d_2)/\partial y. \tag{A.8}$$

If the commodity is a normal good with a non-negative income effect, this total derivative is negative[28]; hence, consumption in the second block unambiguously increases.

The price reduction unambiguously reduces the probability of being at the switching point at the *low* end of the second block, w_1, since

$$\frac{d\lambda_1}{dp_2} = -\frac{\phi(c_1)}{\sigma_\eta w(p_2, y + d_2)} \frac{dw(p_2, y + d_2)}{dp_2} > 0, \tag{A.9}$$

where $\phi(.)$ is the standard unit normal PDF. However, some of the other effects are ambiguous. Starting with the probability of being in the second block,

$$\frac{d\pi_2}{dp_2} = \frac{1}{\sigma_\eta w(p_2, y + d_2)} \frac{dw(p_2, y + d_2)}{dp_2} [\phi(\ln a_2/\sigma_\eta) - \phi(\ln b_2/\sigma_\eta)]. \tag{A.10}$$

Since $dw(p_2, y + d_2)/dp_2 < 0$, and since the standard unit normal PDF is symmetric and unimodal with a mode of zero, one obtains

$$\frac{d\pi_2}{dp_2} \geqslant (\leqslant) 0 \quad \text{as } w_1 w_2 \leqslant (\geqslant) w(p_2, y + d_2)^2. \tag{A.11}$$

The effect of the price change can go either way because, while the lower price in block 2 encourages the consumer to use more in that block, it may also push him into a higher consumption block. Similarly, with the probability of being at the switchpoint w_2,

$$\frac{d\lambda_2}{dp_2} = \frac{1}{\sigma_\eta w(p_2, y + d_2)} \left[\phi\left(\frac{\ln c_2}{\sigma_\eta}\right) \frac{dw(p_2, y + d_2)}{dp_2} + (w_2 - w_1)\phi\left(\frac{\ln b_2}{\sigma_\eta}\right) \frac{w_2}{w_3} \frac{\partial w(p_3, y + d_3)}{\partial y}\right]. \tag{A.12}$$

Since $dw(p_2, y + d_2)/dp_2 < 0$ while $(w_2 - w_1)\partial w(p_3, y + d_3)/dy > 0$ and $w_2/w_3 < 1$, a complex set of factors determines whether the sign of this derivative is positive (the likelihood of being at the switching point rises) or negative (it falls). For blocks $k > 2$,

$$\frac{d\lambda_k}{dp_2} = \frac{1}{\sigma_\eta w(p_{k+1}, y + d_{k+1})} \left[\phi\left(\frac{\ln c_k}{\sigma_\eta}\right) \frac{\partial w(p_{k+1}, y + d_{k+1})}{\partial y} + \phi\left(\frac{\ln b_k}{\sigma_\eta}\right) \frac{w_{k+1}}{w_k} \frac{\partial w(p_k, y + d_k)}{\partial y}\right] \tag{A.13}$$

and

$$\frac{d\pi_k}{dp_2} = \frac{(w_2 - w_1)}{\sigma_\eta w(p_k, y + d_k)} \frac{\partial w(p_k, y + d_k)}{\partial y} \left[\phi\left(\frac{\ln b_k}{\sigma_\eta}\right) - \phi\left(\frac{\ln a_k}{\sigma_\eta}\right)\right], \tag{A.14}$$

where again the sign can be positive or negative.

[28]This is because then:

$$0 \geqslant \partial w(p_2, y + d_2)/\partial p_2 + w(p_2, y + d_2)\partial w(p_2, y + d_2)/\partial y$$
$$\geqslant \partial w(p_2, y + d_2)/\partial p_2 + w_1 \partial w(p_2, y + d_2)/\partial y.$$

In some sense, it seems clear that, when the commodity is a normal good and p_2 decreases, overall consumption unambiguously increases. This is because, from (A.4) and the fact that $d\lambda_1/dp_2 > 0$, it follows that

$$\frac{d\Pr\{\text{consumption} \geqslant w_1\}}{dp_2} = \frac{d\left[\sum_{k=2}^{K}\pi_k + \sum_{k=2}^{K-1}\lambda_k\right]}{dp_2} < 0. \tag{A.15}$$

Moreover, for $k > 2$,

$$\frac{dw(p_k, y + d_k)}{dp_2} = (w_1 - w_2)\frac{\partial w(p_k, y + d_k)}{dy} < 0. \tag{A.16}$$

All of these shifts are the responses to a change in a single block price. Things become much more complicated if several block prices change. In order to characterize the overall price elasticity of demand, we proceed as follows. Let θ be a shift variable which affects all block prices proportionately; the conditional demand functions can now be written $w(\theta p_k, y + \theta d_k)$.

The unconditional expected demand can now be viewed as a function of θ, $W(\theta)$. The overall change in expected demand with respect to a change in θ, evaluated at $\theta = 1$, is

$$\begin{aligned}\frac{dW(1)}{d\theta} &= e^{\sigma_\varepsilon^2/2}\left[\sum_{k=1}^{K}\frac{dw(p_k, y + d_k)}{d\theta}\pi_k^* + \sum_{k=1}^{K} w(p_k, y + d_k)\frac{d\pi_k^*}{d\theta} + \sum_{k=1}^{K-1} w_k\frac{d\lambda_k}{d\theta}\right] \\ &= e^{\sigma_\varepsilon^2/2}\left[\sum_{k=1}^{K}\frac{dw(p_k, y + d_k)}{d\theta}\left\{\pi_k^* + \frac{\phi((\ln b_k/\sigma_\eta) - \sigma_\eta) - \phi((\ln a_k/\sigma_\eta) - \sigma_\eta)}{\sigma_\eta}\right\} + \sum_{k=1}^{K-1} w_k\frac{d\lambda_k}{d\theta}\right],\end{aligned} \tag{A.17}$$

where the total derivative of conditional demand along a segment with respect to θ is

$$\frac{dw(p_k, y + d_k)}{d\theta} = \frac{\partial w(p_k, y + d_k)}{\partial p_k} + d_k\frac{\partial w(p_k, y + d_k)}{\partial y} \tag{A.18}$$

and

$$\frac{d\lambda_k}{d\theta} = \frac{1}{\sigma_\eta w(p_k, y + d_k)}\left[\phi\left(\frac{\ln b_k}{\sigma_\eta}\right)\frac{dw(p_k, y + d_k)}{d\theta} - \frac{w(p_k, y + d_k)}{w(p_{k+1}, y + d_{k+1})}\phi\left(\frac{\ln a_k}{\sigma_\eta}\right)\frac{dw(p_{k+1}, y + d_{k+1})}{d\theta}\right]. \tag{A.19}$$

In the particular case where the conditional demand function takes the log–log form in (1), with some manipulation (A.19) can be written in elasticity form as

$$\begin{aligned}\frac{1}{W(1)}\frac{dW(1)}{d\theta} &= \alpha\left[\sum_{k=1}^{K}\frac{w(p_k, y + d_k)\psi_k}{p_k} + \sum_{k=1}^{K-1} w_k\chi_k\right]\Big/\Omega \\ &\quad + \gamma\left[\sum_{k=1}^{K}\frac{w(p_k, y + d_k)\psi_k d_k}{y + d_k} + \sum_{k=1}^{K-1} w_k\tau_k\right]\Big/\Omega,\end{aligned} \tag{A.20}$$

where α is the price elasticity of demand in the conditional log–log demand function (1) and γ is the income elasticity of demand in (1), and

$$\begin{aligned}\psi_k &\equiv \pi_k^* + \frac{1}{\sigma_\eta}\left[\phi\left(\frac{\ln b_k}{\sigma_\eta} - \sigma_\eta\right) - \phi\left(\frac{\ln a_k}{\sigma_\eta} - \sigma_\eta\right)\right], \\ \chi_k &\equiv \frac{1}{\sigma_\eta}\left[\phi\left(\frac{\ln b_k}{\sigma_\eta}\right)\frac{1}{p_k} - \phi\left(\frac{\ln a_k}{\sigma_\eta}\right)\frac{1}{p_{k+1}}\right], \\ \tau_k &\equiv \frac{1}{\sigma_\eta}\left[\phi\left(\frac{\ln b_k}{\sigma_\eta}\right)\frac{d_k}{y + d_k} - \phi\left(\frac{\ln a_k}{\sigma_\eta}\right)\frac{d_{k+1}}{y + d_{k+1}}\right], \\ \Omega &\equiv \sum_{k=1}^{K} w(p_k, y + d_k)\pi_k^* + \sum_{k=1}^{K-1} w_k\lambda_k.\end{aligned}$$

It follows that overall elasticity of expected demand with respect to θ, namely, $1/W(1)(dW(1))/d\theta$, which may be viewed as the analog of the price elasticity of demand for IBP, is a complex combination of both the price and income elasticities of demand associated with the conditional (segment) demand functions. In general, this overall elasticity is different from the simple price elasticity of the conditional demand function, α, but it does not appear that either the direction or the magnitude of the difference can be signed in the general case represented by (A.20). This is because several factors are involved in the formation of the overall elasticity.

The earlier results for a single block price change, involving (A.8)–(A.16), can shed some light on these factors. Assume, as before, that the good is a normal good. If p_2 falls and the consumer still stays in the second block, two things happen. Demand goes up because of the reduced price; but this is tempered by a decrease in demand caused by the decline in the virtual income for that block, $\tilde{y}_2$, which is triggered by the reduced differential between p_1 and p_2. Assuming there are more than two blocks, the consumer may want to move up out of the second block. Two possibilities need to be considered. (A) One possibility is that the consumer can become "trapped" at the kink point between the second and third blocks. (B) Even if the consumer does succeed in moving into a higher block, two things happen: he faces a higher (unchanged) price, p_3, but he also experiences a positive income effect from the increase in virtual income, $\tilde{y}_3$, caused by the decrease in p_2. While (A) dampens the increase in demand triggered by the reduction in p_2, (B) could in principle either dampen or amplify it depending on the relative strength of the income effect. Recall, also, that the effects on π_k and λ_k for $k>1$ are ambiguous.

In the analysis in this paper, we estimate an unconditional price elasticity that is smaller in absolute value than the conditional price elasticity. This suggests that, in our sample, the dampening effects of sticking at kink points and facing higher (lower) prices in a potential new block if prices rise (fall) must outweigh any amplifying effects of changes in virtual income or potential amplifying effects of changes in the probabilities of locating on any particular segment or switch point. This possibility is supported by the fact that the d_k are tiny fractions of household income in our case. Hence, the income effect associated with infra-marginal price changes is likely to be very small. This is not necessarily true of other markets characterized by piecewise linear budget constraints. In the case of labor supply, for example, the income effects created by infra-marginal tax rates appear to be very large [14]. With water and electricity, in contrast, the infra-marginal rate subsidies associated with IBPs are found in the literature to have almost no effect because of low prices, small differences between successive price blocks, and the fact that annual household water and energy expenditures are typically a small fraction of annual income [6,16,28]. Though in the general case the difference between conditional and unconditional elasticities under IBPs cannot be signed, what we observe may be typical in the context of water demand.

References

[1] D.E. Agthe, R. Billings, J.L. Dobra, K. Raffiee, A simultaneous equation demand model for block rates, Water Resources Res. 22 (1986) 1–4.

[2] R. Billings, D.E. Agthe, Price elasticities for water: a case of increasing block rates, Land Econ. 56 (1980) 73–84.

[3] G. Burtless, J.A. Hausman, The effect of taxation on labor supply: evaluating the Gary income maintenance experiment, J. Polit. Economy 86 (1978) 1101–1130.

[4] J.M. Dalhuisen, R.J.G.M. Florax, H.L.F. de Groot, P. Nijkamp, Price and income elasticities of residential water demand: a meta-analysis, Land Econ. 79 (2003) 292–308.

[5] L.E. Danielson, An analysis of residential demand for water using micro time-series data, Water Resources Res. 22 (1979) 763–767.

[6] S.C. Deller, D.L. Chicoine, G. Ramamurthy, Instrumental variables approach to rural water service demand, Southern Econ. J. 53 (1986) 333–346.

[7] M. Espey, J. Espey, W.D. Shaw, Price elasticity of residential demand for water: a meta-analysis, Water Resources Res. 33 (1997) 1369–1374.

[8] K.D. Frederick, G.E. Schwarz, Socioeconomic impacts of climate variability and change on US water resources, RFF Discussion Paper 00-21, Resources for the Future, Washington, DC, 2000.

[9] S. Gaudin, Effect of price information on residential water demand, Appl. Econ. 38 (2006) 383–393.

[10] D.C. Hall, Public choice and water rate design, in: A. Dinar (Ed.), The Political Economy of Water Pricing Implementation, Oxford, New York, 2000.

[11] R.E. Hall, Wages, income and hours of work in the US labor force, in: G. Cain, H. Watts (Eds.), Income Maintenance and Labor Supply, Rand McNally, Chicago, 1973.

[12] W.M. Hanemann, Discrete/continuous models of consumer demand, Econometrica 52 (1984) 541–561.
[13] W.M. Hanemann, Prices and rate structures, in: D. Baumann, J. Boland, W. Hanemann (Eds.), Urban Water Demand Management and Planning, McGraw-Hill, New York, 1997.
[14] J.A. Hausman, Labor supply, in: H. Aaron, J. Pechman (Eds.), How Taxes Affect Economic Behavior, Brookings Institution, Washington, DC, 1981.
[15] J.A. Hausman, Taxes and labor supply, in: A. Auerbach, M. Feldstein (Eds.), The Handbook of Public Economics, North-Holland, Amsterdam, 1985.
[16] J.A. Hausman, M. Kinnucan, D. McFadden, A two-level electricity demand model: evaluation of the Connecticut time-of-day pricing test, J. Econometrics 10 (1979) 263–289.
[17] J.A. Hewitt, Watering households: the two-error discrete-continuous choice model of residential water demand, Ph.D. Dissertation, University of California, Berkeley, CA, 1993.
[18] J.A. Hewitt, W.M. Hanemann, A discrete/continuous choice approach to residential water demand under block rate pricing, Land Econ. 71 (1995) 173–192.
[19] C.W. Howe, F.P. Lineweaver, The impact of price on residential water demand and its relation to system demand and price structure, Water Resources Res. 3 (1967) 13–32.
[20] J.B. Liebman, R.J. Zeckhauser, Schmeduling, Working Paper, Kennedy School of Government, Harvard University, 2004.
[21] E.T. Mansur, S.M. Olmstead, The value of scarce water: measuring the inefficiency of municipal regulations, Working Paper, 2006.
[22] P.W. Mayer, W.B. DeOreo, E.M. Opitz, J.C. Kiefer, W.Y. Davis, B. Dziegielewski, J.O. Nelson, Residential end uses of water, American Water Works Association Research Foundation, Denver, 1988.
[23] R. Moffitt, The econometrics of piecewise-linear budget constraints: a survey and exposition of the maximum likelihood method, J. Bus. Econ. Statist. 4 (1986) 317–328.
[24] R. Moffitt, The econometrics of kinked budget constraints, J. Econ. Perspect. 4 (1990) 119–139.
[25] J.E.T. Moncur, R.L. Pollock, Scarcity rents for water: a valuation and pricing model, Land Econ. 64 (1988) 62–72.
[26] M.L. Nieswiadomy, S.L. Cobb, Impact of pricing structure selectivity on urban water demand, Contemp. Pol. Issues 11 (1993) 101–113.
[27] M.L. Nieswiadomy, D.J. Molina, Urban water demand estimates under increasing block rates, Growth Change 19 (1988) 1–12.
[28] M.L. Nieswiadomy, D.J. Molina, Comparing residential water demand estimates under decreasing and increasing block rates using household data, Land Econ. 65 (1989) 280–289.
[29] J.A. Nordin, A proposed modification of Taylor's demand analysis: comment, Bell J. Econ. 7 (1976) 719–721.
[30] S.M. Olmstead, Reduced-form v. structural models of water demand, J. Bus. Econ. Statist., forthcoming.
[31] Organization for Economic Cooperation and Development, Household water pricing in OECD countries, OECD, Paris, 1999.
[32] E.M. Pint, Household responses to increased water rates during the California drought, Land Econ. 75 (1999) 246–266.
[33] Raftelis Environmental Consulting Group, Inc., Raftelis environmental consulting group 1998 water and wastewater rate survey, Raftelis Environmental Consulting, Charlotte, NC, 1999.
[34] P. Rietveld, J. Rouwendal, B. Zwart, Block rate pricing of water in Indonesia: an analysis of welfare effects, Bull. of Indonesian Econ. Stud. 36 (2000) 73–92.
[35] L.D. Taylor, The demand for electricity: a survey, Bell J. Econ. 6 (1975) 74–110.
[36] C. Timmins, Demand-side technology standards under inefficient pricing regimes: are they effective water conservation tools in the long run?, Environ. Resource Econ. 26 (2003) 107–124.

PART VI

DOMESTIC CLIMATE CHANGE POLICY

[19]

Oxford Review of Economic Policy, Volume 24, Number 2, 2008, pp.298–321

Addressing climate change with a comprehensive US cap-and-trade system

Robert N. Stavins*

Abstract There is growing impetus for a domestic US climate policy that can provide meaningful reductions in emissions of carbon dioxide (CO_2) and other greenhouse gases. I describe and analyse an up-stream, economy-wide CO_2 cap-and-trade system which implements a gradual trajectory of emissions reductions (with inclusion over time of non-CO_2 greenhouse gases), and includes mechanisms to reduce cost uncertainty. Initially, half of the allowances are allocated through auction and half through free distribution, with the share being auctioned gradually increasing to 100 per cent over 25 years. The system provides for linkage with emission-reduction credit projects in other countries, harmonization over time with effective cap-and-trade systems in other countries and regions, and appropriate linkage with actions taken in other countries, in order to establish a level playing field among domestically produced and imported products.

Key words: cap-and-trade system, carbon dioxide, greenhouse gas emissions, global climate change, carbon taxes

JEL classification: Q540, Q280, Q380, Q480, Q580

I. Introduction

The impetus for a meaningful US climate policy is growing. Scientific evidence has increased (Intergovernmental Panel on Climate Change, 2007*a*,*b*), public concern has been magnified, and many people perceive what they believe to be evidence of climate change in progress. Such concern is reinforced by the aggressive positions of key advocacy groups, which are no longer limited on this issue to the usual environmental interest groups; religious lobbies, for example, have also been vocal. This has been reflected in greatly heightened attention by the news media. The overall result is that a large and growing share of the US population now believes that government action is warranted (Bannon *et al.*, 2007).

In the absence of federal policy, regions, states, and even cities have moved forward with their own proposals for policies intended to reduce the emissions of carbon dioxide (CO_2) and other greenhouse gases. Partly in response to fears of a fractured set of regional policies, an increasing number of large corporations, sometimes acting individually, and at other times

*John F. Kennedy School of Government, Harvard University, National Bureau of Economic Research, and Resources for the Future, e-mail: robert_stavins@harvard.edu
doi: 10.1093/oxrep/grn017

in coalitions—together with environmental advocacy groups—have announced their support for serious national action. Building upon this is the April 2007 US Supreme Court decision that the Administration has the legislative authority to regulate CO_2 emissions,[1] as well as ongoing pressure from European and other nations for the United States to re-establish its international credibility in this realm by enacting a meaningful domestic climate policy.

Thus, momentum is clearly building towards the enaction of a domestic climate-change policy. But there should be no mistake about it—meaningful action to address global climate change will be costly. This is a key 'inconvenient truth' that must be recognized when policy-makers construct and evaluate proposals, because a policy's specific design will greatly affect its ability to achieve its environmental goals, its costs, and the distribution of those costs.

There is general consensus among economists and policy analysts that a market-based policy instrument targeting CO_2 emissions—and potentially some non-CO_2 greenhouse-gas (GHG) emissions—should be a central element of any domestic climate policy. This is reflected in international assessments of national policy instruments, as well (Intergovernmental Panel on Climate Change, 2007*c*). While there are trade-offs between two alternative market-based instruments—a cap-and-trade system and a carbon tax—the best and most likely approach for the short to medium term in the United States is a cap-and-trade system.

It is critical to identify the most effective, lowest-cost, and most equitable policy design at the outset, because any policy design once in place can be difficult to change (Repetto, 2007). The environmental integrity of a domestic cap-and-trade system for climate change can be maximized and its costs and risks minimized by: targeting all fossil-fuel-related CO_2 emissions through an upstream, economy-wide cap; setting a trajectory of caps over time that begins modestly and gradually becomes more stringent, establishing a long-run price signal to encourage investment; adopting mechanisms to protect against cost uncertainty; and including linkages with the climate-policy actions of other countries. Importantly, by providing the option to mitigate economic impacts through the distribution of emission allowances, this approach can establish consensus for a policy that achieves meaningful emission reductions. It is for these reasons and others that cap-and-trade systems have been used increasingly in the United States to address an array of environmental problems, including the phase-out of leaded gasoline in the 1980s, the reduction of sulphur dioxide (SO_2) and nitrogen oxide (NO_X) emissions from power plants beginning in 1995, and the phase-out of chlorofluorocarbons (CFCs) (Stavins, 2003).

A well-designed cap-and-trade system will minimize the costs of achieving any given emissions target. While firms have flexibility regarding precisely how much they emit, because they have to surrender an allowance for each ton of their emissions they will undertake all emission reductions that are less costly than the market price of an allowance. Through trading, this allowance price adjusts until emissions are brought down to the level of the cap. Firms' ability to trade emission allowances creates a market in which allowances migrate towards their highest-valued use, covering those emissions that are the most costly to reduce. Conversely, as a result of trading, the emission reductions undertaken to meet the cap are those that are least costly to achieve.

The cost of achieving significant emission reductions in future years will depend critically on the availability and cost of low- or non-emitting technologies. A cap-and-trade system that establishes caps extending decades into the future provides important price signals and

[1] Massachusetts *et al.* v. Environmental Protection Agency *et al.*, No. 05-1120, argued 29 November 2006, decided 2 April 2007.

hence incentives for firms to invest in the development and deployment of such technologies, thereby lowering the future costs of achieving emission reductions.

Even a credible long-run cap-and-trade system may provide insufficient incentives for investment in technology development because it would not address certain well-known factors (market failures) that discourage such investment, such as those associated with the public-good nature of the knowledge that comes from research and development efforts (Jaffe *et al.*, 2005; Newell, 2007). Thus, a cap-and-trade system alone will not encourage the socially desirable level of investment in research, development, and deployment of new technologies that could reduce future emission-reduction costs. To achieve this desired level of investment, additional policies may be necessary to provide additional government funding or to increase incentives for private funding of such research activities.

(i) Previous use of cap-and-trade systems

Over the past two decades tradable permit systems have been adopted for pollution control with increasing frequency in the United States (Tietenberg, 1997), as well as other parts of the world. The first important example of a trading programme in the United States was the leaded gasoline phase-down that occurred in the 1980s. Although not strictly a cap-and-trade system, the phase-down included features, such as trading and banking of environmental credits, that brought it closer than other credit programmes to the cap-and-trade model and resulted in significant cost savings. The lead programme was successful in meeting its environmental targets, and the system was cost-effective, with estimated cost savings of about $250 m per year (Nichols, 1997). Also, the programme provided measurable incentives for cost-saving technology diffusion (Kerr and Newell, 2000).

A cap-and-trade system was also used in the United States to help comply with the Montreal Protocol, an international agreement aimed at slowing the rate of stratospheric ozone depletion. The Protocol called for reductions in the use of CFCs and halons, the primary chemical groups thought to lead to depletion. The timetable for the phase-out of CFCs was accelerated, and the system appears to have been relatively cost-effective.

The most important application made in the United States of a market-based instrument for environmental protection is arguably the cap-and-trade system that regulates SO_2 emissions, the primary precursor of acid rain, established under the US Clean Air Act Amendments of 1990 (Ellerman *et al.*, 2000). The programme is intended to reduce SO_2 and nitrogen-oxide (NO_2) emissions by 10 m tons and 2 m tons, respectively, from 1980 levels (Burtraw *et al.*, 1998). A robust market of SO_2 allowance trading emerged from the programme, resulting in cost savings of the order of $1 billion annually, compared with the costs under some command-and-control regulatory alternatives (Carlson *et al.*, 2000). The programme has also had a significant environmental impact: SO_2 emissions from the power sector decreased from 15.7 m tons in 1990 to 10.2 m tons in 2005 (US Environmental Protection Agency, 2005).

In 1994, California's South Coast Air Quality Management District launched a cap-and-trade programme to reduce nitrogen-oxide and SO_2 emissions in the Los Angeles area (Harrison, 2003). This Regional Clean Air Incentives Market (RECLAIM) programme set an aggregate cap on NO_x and SO_2 emissions for all significant sources, with an ambitious goal of reducing aggregate emissions by 70 per cent by 2003. Trading under the RECLAIM programme was restricted in several ways, with positive and negative consequences. But despite problems, RECLAIM has generated environmental benefits, with NO_x emissions in the regulated area falling by 60 per cent and sulphurous oxide (SO_x) emissions by 50 per cent.

Furthermore, the programme has reduced compliance costs for regulated facilities, with the best available analysis suggesting 42 per cent cost savings, amounting to \$58 m annually (Anderson, 1997).

Finally, in 1999, under US Environmental Protection Agency guidance, 12 north-eastern states and the District of Columbia implemented a regional NO_x cap-and-trade system to reduce compliance costs associated with the Ozone Transport Commission (OTC) regulations of the 1990 Amendments to the Clean Air Act. Emissions caps for two zones from 1999 to 2003 were 35 and 45 per cent of 1990 emissions, respectively. Compliance cost savings of 40–47 per cent have been estimated for the period 1999–2003, compared to a base case of continued command-and-control regulation without trading or banking (Farrell *et al.*, 1999).

(ii) CO_2 and greenhouse-gas cap-and-trade systems

Although cap-and-trade has proven to be a cost-effective means to control conventional air pollutants, it has a very limited history as a method of reducing CO_2 emissions. Several ambitious programmes are in the planning stages or have been launched. First, the Kyoto Protocol, the international agreement that was signed in Japan in 1997, includes a provision for an international cap-and-trade system among countries, as well as two systems of project-level offsets. The Protocol's provisions have set the stage for the member states of the European Union to address their commitments using a regional cap-and-trade system.

By far the largest existing active cap-and-trade programme in the world is the European Union Emissions Trading Scheme (EU ETS) for CO_2 allowances, which has operated for the past 2 years with considerable success, despite some initial—and predictable—problems (Ellerman and Buchner, 2007). The 11,500 emitters regulated by the downstream programme include large sources such as oil refineries, combustion installations, coke ovens, cement factories, ferrous metal production, glass and ceramics production, and pulp and paper production, but the programme does not cover sources in the transportation, commercial, or residential sectors. Although the first phase, a pilot programme from 2005 to 2007, allows trading only in CO_2, the second phase, from 2008 to 2012, potentially broadens the programme to include other GHGs. In its first 2 years of operation, the EU ETS has produced a functioning CO_2 market, with weekly trading volumes ranging between 5 m and 15 m tons, with spikes in trading activity occurring along with major price changes. Apart from some problems with the programme's design and early implementation, it is much too soon to provide a definitive assessment of the system's performance.

A frequently discussed US CO_2 cap-and-trade system that has not yet been implemented is the Regional Greenhouse Gas Initiative (RGGI), a programme among 10 north-eastern states that will be implemented in 2009 and begin to cut emissions in 2015. RGGI is a downstream cap-and-trade programme intended to limit CO_2 emissions from power-sector sources. Beginning in 2015, the emissions cap will decrease by 2.5 per cent each year until it reaches an ultimate level 10 per cent below current emissions in 2019. This goal will require a reduction that is approximately 35 per cent below business-as-usual, or, equivalently, 13 per cent below 1990 emissions levels. RGGI only limits emissions from the power sector, and so incremental monitoring costs are low, because US power plants are already required to report their hourly CO_2 emissions to the federal government (under provisions for continuous emissions monitoring as part of the SO_2 allowance trading programme). The programme requires participating states to auction at least 25 per cent of their allowances; the remaining 75 per cent of allowances may be auctioned or distributed freely. Given that the system will not come into effect until 2009, at the earliest, it is obviously not possible to assess its performance.

Finally, California's Greenhouse Gas Solutions Act (Assembly Bill 32) was signed into law in 2006, is intended to begin in 2012 to reduce emissions to their 1990 levels by 2020, and may employ a cap-and-trade approach. Although the Global Warming Solutions Act does not require the use of market-based instruments, it does allow for their use, albeit with restrictions that they must not result in increased emissions of conventional, local air pollutants or toxics, that they must maximize environmental and economic benefits in California, and that they must account for localized economic and environmental justice concerns. This mixed set of objectives potentially interferes with the development of a sound policy mechanism. The Governor's Market Advisory Committee (2007) has recommended the implementation of a cap-and-trade programme, with a gradual phase-in of caps covering most sectors of the economy, and an allowance distribution system that uses both free distribution and auctions of allowances, with a shift towards more auctions in later years.

(iii) Organization of subsequent sections

Section II of the article describes a comprehensive US CO_2 cap-and-trade system, including a description of its key elements: a gradual trajectory of emissions reductions; tradable allowances; up-stream regulation with economy-wide effects; mechanisms to reduce cost uncertainty; allowance allocations that combine auctions with free distribution, with auctions becoming more important over time; availability of offsets for underground and biological carbon sequestration; supremacy over state and regional systems; and linkage with international emission-reduction credit and cap-and-trade systems and climate policies in other countries. Section III provides an economic assessment of the cap-and-trade system. Section IV compares the system with alternative approaches to the same policy goal. Finally, section V concludes.

II. The system

The United States can launch a scientifically sound, economically rational, and politically feasible approach to reducing its contributions to the increase in atmospheric concentrations of greenhouse gases by adopting an up-stream, economy-wide CO_2 cap-and-trade system which implements a gradual trajectory of emissions reductions over time, and includes mechanisms to reduce cost uncertainty, such as multi-year compliance periods, provisions for banking and borrowing, and possibly a cost-containment mechanism to protect against any extreme price volatility.

The permits in the system should be allocated through a combination of free distribution and open auction, in order to balance, on the one hand, legitimate concerns by some sectors and individuals who will be particularly burdened by this (or any) climate policy, with, on the other hand, the opportunity to achieve important public purposes with generated funds. The share of allowances freely allocated should decrease over time, as the private sector is able to adjust to the carbon constraints, with all allowances being auctioned after 25 years.[2]

[2] For a timely discussion of relevant auction design issues in carbon markets, see Burtraw *et al.* (2007).

In addition, it is important that offsets be made available both for underground and biological carbon sequestration, to provide for both short-term cost-effectiveness and long-term incentives for appropriate technological change. The federal cap-and-trade system can provide for supremacy over US regional, state, and local systems, to avoid duplication, double counting, and conflicting requirements. At the same time, it is also important to provide for harmonization over time with selective emission-reduction credit and cap-and-trade systems in other nations, as well as related international systems.

(i) Major though not exclusive focus on CO_2

Fossil-fuel-related CO_2 emissions, which accounted for nearly 85 per cent of the 7,147m metric tons of US GHG emissions in 2005, where tons are measured in CO_2-equivalent. Carbon-dioxide emissions arise from a broad range of activities involving the use of different fuels in many different economic sectors. In addition, biological sequestration and reductions in non-CO_2 GHG emissions can contribute substantially to minimizing the cost of limiting GHG concentrations (Reilly *et al.*, 2003; Stavins and Richards, 2005). Some non-CO_2 GHG emissions might be addressed under the same framework as CO_2 in a multi-gas cap-and-trade system.

(ii) A gradually increasing trajectory of emissions reductions over time

The long-term nature of the climate problem offers significant flexibility regarding when emission reductions actually occur. Policies taking advantage of this 'when flexibility' by setting annual emission targets that gradually increase in stringency can avoid many costs associated with taking stringent action too quickly, *without sacrificing environmental benefits* (Wigley *et al.*, 1996). Premature retirement of existing capital stock and production and siting bottlenecks that can arise in the context of rapid capital stock transitions can be avoided. In addition, gradually phased-in targets provide time to incorporate advanced technologies into long-lived investments (Jaffe *et al.*, 1999; Goulder, 2004). Thus, for any given cumulative emission target or associated atmospheric GHG concentration objective, a climate policy's cost can be reduced by gradually phasing in efforts to reduce emissions.

Because of the long-term nature of the climate problem and because of the need for technological change to bring about lower-cost emissions reductions, it is essential that the caps constitute a long-term trajectory. The development and eventual adoption of new low-carbon and other relevant technologies will depend on the predictability of future carbon prices, themselves brought about by the cap's constraints. Therefore, the cap-and-trade policy should incorporate medium-term to long-term targets, not just short-term targets.

For illustrative purposes in the cost assessment, I adopt and assess a pair of trajectories for the period 2012–50 to establish a reasonable range of possibilities. The less ambitious trajectory involves *stabilizing* CO_2 emissions at their 2008 level over the period 2012–50. This trajectory, in terms of its cumulative cap, lies within the range defined by the 2004 and 2007 recommendations of the National Commission on Energy Policy (2004, 2007). The more ambitious trajectory—again defined over the years 2012–50—involves reducing CO_2 emissions from their 2008 level to *50 per cent below their 1990 level by 2050*. This trajectory—defined by its cumulative cap—is consistent with the lower end of the range proposed by the US Climate Action Partnership (2007). This range of trajectories is consistent with the frequently cited global goal of stabilizing atmospheric concentrations of CO_2 at between 450 parts per million (ppm) and 550 ppm *if* all nations were to take commensurate action.[3]

(iii) Upstream point of regulation and economy-wide scope of coverage

In order to create *economy-wide coverage*, an *upstream point of regulation* should be employed, whereby allowances are surrendered based on the carbon content of fuels at the point of fossil-fuel extraction, import, processing, or distribution. This can be thought of as a system where regulation is at the mine-mouth, well-head, and point of import. First sellers of fossil fuels could be required to hold allowances: for coal, at the mine shipping terminus; for petroleum, at the refinery gate; for natural gas, at the first distribution point; and for imports, at the point of importation. Such a cap will effectively cover all sources of CO_2 emissions throughout the economy.

The upstream programme should include a credit mechanism to address the small portion of fossil fuels that are not combusted and to address the use of post-combustion emission-reduction technologies, such as carbon capture and sequestration (CCS). In addition, upstream regulation should include a credit-based programme for fossil-fuel exports so that they are not at a competitive disadvantage relative to supply from other countries that do not face any allowance requirements.

An economy-wide cap provides the greatest certainty that national emission targets will be achieved. Limiting the scope of coverage to a subset of emission sources leads to emissions uncertainty through two channels. First, changes in emissions from unregulated sources can cause national emissions to deviate from expected levels. Second, a limited scope of coverage can cause 'leakage', in which market adjustments resulting from a regulation lead to increased emissions from unregulated sources outside the cap that partially offset reductions under the cap.

An emission cap with broad coverage of emission sources reduces the cost of achieving a particular national emissions target. Three factors contribute to lower costs. First, a broader cap expands the pool of low-cost emission-reduction opportunities that can contribute to meeting a national target. Second, an economy-wide cap provides important flexibility to achieve emission targets given uncertainties in emission-reduction costs across sectors. Third, an economy-wide cap creates incentives for innovation in all sectors of the economy.

The point of regulation decision is a primary determinant of a cap-and-trade system's administrative costs through its effect on the number of sources that must be regulated. As the number of regulated sources increases, the administrative costs to regulators and firms rise. The upstream point of regulation makes an economy-wide cap-and-trade system administratively feasible, making it possible to cap nearly all US CO_2 emissions through regulation of just 2,000 upstream entities (Bluestein, 2005). A key advantage of an upstream programme is that it eliminates the regulatory need for facility-level GHG emissions inventories, which would be essential for monitoring and enforcing a cap-and-trade system that is implemented downstream at the point of emissions.

[3] 'Commensurate action' is defined as other countries taking action that is globally cost-effective, for example by employing cap-and-trade systems with the same allowance price or equivalent carbon taxes.

(iv) Elements that reduce cost uncertainty

Concern about cost uncertainty in the context of cap-and-trade systems derives from the possibility of unexpected, significant cost increases. The experience with the southern California RECLAIM cap-and-trade system for NO_x emissions is the frequently cited example. RECLAIM had *no* automatic mechanism to relax emission caps in the face of unexpectedly high costs and, in 2000, allowance prices spiked to more than 20 times their historical levels (Pizer, 2005). Cost uncertainty may increase the long-run cost of emission caps, because uncertainty about future allowance prices may deter firms from undertaking socially desirable, capital-intensive emission-reduction investments, forcing greater reliance on less capital-intensive, but more costly measures.

Allowance banking and borrowing can mitigate some of the undesirable consequences of cost uncertainty by giving firms the flexibility to shift the timing of emission reductions in the face of unexpectedly high or low costs. If the cost of achieving targets is unexpectedly and temporarily high, firms can use banked or borrowed allowances instead of undertaking costly reductions. Thus, banking and borrowing mitigate undesirable year-to-year variation in costs. Banking of allowances—undertaking extra emission reductions earlier, so that more allowances are available for use later—has added greatly to the cost effectiveness of previous cap-and-trade systems (Stavins, 2003), but banking provides little protection when costs remain high over extended periods, which could eventually lead to exhaustion of banked allowances. This problem may be particularly acute in a cap's early years, when relatively few allowances have been banked. Therefore, borrowing of allowances from future years' allocations can be a particularly useful form of cost protection in these early years. Of course, credible mechanisms need to be established to ensure that the use of borrowed allowances is offset through future emission reductions.

Banking and borrowing can be exceptionally important in reducing long-term cost uncertainty, but the possibility of dramatic short-term allowance-price volatility may call for the inclusion of a sensible cost-containment mechanism. Such a mechanism could allow capped sources to purchase additional allowances at a predetermined price, set sufficiently *high* to make it unlikely to have any effect unless allowance prices exhibited truly drastic spikes, *and* the revenues from the fee dedicated *exclusively* to financing emissions reductions by uncapped sources, such as of non-CO_2 greenhouse gases, or to buying back allowances in future years. This is very different from standard proposals for a 'safety valve,' both because environmental integrity (the cap) is maintained by using the fees exclusively to finance additional emissions reductions or buy back allowances in future years, and because the pre-determined price is set at a high level so that it has no effect unless there are drastic price spikes.

The pre-determined fee places a ceiling on allowance prices and hence on abatement costs, because no firms would undertake emission reductions more costly than the trigger price (Jacoby and Ellerman, 2002). To be used as an insurance mechanism, the fee should be set at the maximum incremental emission-reduction cost that society is willing to bear. At this level, the mechanism would be triggered only when costs are unexpectedly and unacceptably high.

(v) Allocation of allowances

While all allocation decisions have significant distributional consequences, whether allowances are auctioned or freely distributed can affect the programme's overall cost. Generally speaking, the choice between auctioning and freely allocating allowances does not influence firms' production and emission-reduction decisions. Firms face the same emissions cost regardless of the allocation method. Even when using an allowance that was received for free, a firm loses the opportunity to sell that allowance, and thereby recognizes this 'opportunity cost' in deciding whether to use an allowance. Consequently, in many respects, this allocation choice will not influence a cap's overall costs. But there are two ways that the choice to distribute allowances freely can affect a cap's cost.

First, auction revenue may be used in ways that reduce the costs of the existing tax system or fund other socially beneficial policies. Free allocations forgo such opportunities. Second, free allocations may affect electricity prices in regulated cost-of-service electricity markets, and thereby affect the extent to which reduced electricity demand contributes to limiting emissions cost-effectively.

In discussions about whether to auction or freely distribute allowances, much attention has been given to the opportunity to use auction revenue to reduce existing distortionary taxes on labour and capital. Use of auction revenue to reduce these taxes can stimulate economic activity, offsetting some of a cap's costs. Studies indicate that 'recycling' auction revenue by reducing personal income tax rates could offset 40–50 per cent of the economy-wide social costs that a cap would impose if allowances were freely distributed (Bovenberg and Goulder, 2003).

Achieving such gains may be difficult in practice, because climate policy would need to be tied to particular types of tax reform. The estimated cost-reductions are for policies in which auction revenue is used to reduce marginal tax rates that diminish incentives to work and invest. If, instead, auction revenue funded deductions or fixed tax credits, such tax reform would have a lesser effect (and perhaps no effect) on incentives to work and invest.

In general, auctioning generates revenue that can be put towards innumerable uses. Use of auction revenue to reduce tax rates is just one example. Other socially valuable uses of revenue could include reduction of the federal debt (including offsetting a cap's potentially adverse fiscal impacts), or funding desirable spending programmes (for example, research and development). On the other hand, some government uses of auction revenue may generate less economic value than could be realized by private-sector use of those funds. Thus, the opportunity to reduce the aggregate cost of a climate policy through auctioning, rather than freely distributing allowances, depends fundamentally on the use to which auction revenues are ultimately put.

While auctioning has the potential to reduce a climate policy's economy-wide costs, depending on how auction revenues are used, free distribution of allowances provides an opportunity to address the distribution of a climate policy's economic impacts. Free distribution of allowances can be used to redistribute a cap's economic burdens in ways that mitigate impacts on the most affected entities, and a sensible principle for allocation would be to try to compensate the most burdened sectors and individuals. Such redistribution of impacts may help establish consensus on a climate policy that achieves meaningful emission reductions.

Because free allocations may increase a cap's overall cost, it is important to consider what share of allowances need to be freely distributed to meet specific compensation objectives. A permanent allocation of all allowances to affected firms would, in aggregate, significantly overcompensate them for their financial losses (Goulder, 2000; Smith *et al.*, 2002; Bovenberg and Goulder, 2003). This is the case because much of the cost that a cap-and-trade system initially imposes on firms will be passed on to consumers in the form of higher prices. In effect, before any free allocation, firms are already partially compensated by changes in prices that result from the cap. Thus, freely allocating *all* allowances in perpetuity to affected firms would both overcompensate them in aggregate, and use up resources that could otherwise be put towards other uses.

Faced with important differences in the implications of free allocation and an auction, the best alternative is to *begin* with a hybrid approach, wherein half of the allowances are *initially* auctioned and half are freely distributed to entities that are burdened by the policy, including suppliers of primary fuels, electric power producers, energy-intensive manufacturers, and particularly trade-sensitive sectors. The share of allowances that are freely distributed should decline over time, until there is no free allocation 25 years into the programme. This is because over time the private sector will have an opportunity to adjust to the carbon constraints, including industries with long-lived capital assets. Thus, the justification for free distribution diminishes over time.[4] In the short term, however, free distribution provides flexibility to address distributional concerns that might otherwise impede initial agreement on a policy. The half that are initially auctioned will generate revenue that can be used for public purposes, including compensation for programme impacts on low-income consumers, public spending for related research and development, reduction of the federal deficit, and reduction of distortionary taxes.

Why this particular pattern of beginning with a 50–50 auction-free allocation, moving to 100 per cent auction over 25 years? This time-path of the numerical division between the share of allowances that is freely allocated and the share that is auctioned is consistent with analyses which have been carried out of the share of allowances that would need to be distributed freely to compensate firms for equity losses. In a series of analyses that considered the share of allowances that would be required *in perpetuity* for full compensation, Bovenberg and Goulder (2003) found that 13 per cent would be sufficient for compensation of the fossil-fuel-extraction sectors, and in a scenario consistent with the Bovenberg and Goulder study, Smith *et al.* (2002) found that 21 per cent would be needed to compensate primary energy producers and electricity generators.

The time-path recommended here for an economy-wide programme—50 per cent of allowances initially distributed freely, with this share declining steadily (linearly) to zero after 25 years—is equivalent in terms of present discounted value to perpetual allocations (as those previously analysed) of 15, 19, and 22 per cent, at real interest rates of 3, 4, and 5 per cent, respectively. Hence, the recommended allocation is consistent with the principle of targeting free allocations to burdened sectors in proportion to their relative burdens. It is also pragmatic to be more generous with the allocation in the early years of the programme.

(vi) Credits (offsets) for specified activities

The upstream programme should include selective use of the credit mechanism. First, credits should be issued for major non-combustion uses of fossil fuels, such as in some petrochemical feedstocks, as well as fuel exports. Second, credits should be issued for carbon capture and storage (CCS). Emission reductions from CCS technologies can be readily measured, and because there is no incentive to install CCS equipment absent a climate policy, emission reductions achieved by CCS are clearly additional. As CCS technologies may play a significant role in achieving long-run emission reduction goals (US Energy Information Administration, 2007; Deutch and Moniz, 2007), this credit mechanism is an essential component of the upstream cap.

Third, a programme of credits for selected cases of biological sequestration through land-use changes should be included. A cost-effective portfolio of climate technologies in the

[4] For discussion of the temporal dimension of climate policy, see Helm *et al.* (2005).

United States would include a substantial amount of biological carbon sequestration through afforestation and retarded deforestation (Stavins, 1999; Stavins and Richards, 2005; Lubowski *et al.*, 2006). Translating this into practical policy will be a considerable challenge, however, because of concerns about monitoring and enforcement, additionality, and permanence (Plantinga, 2007).

Fourth, provision should be made to provide coverage over time of non-CO_2 greenhouse gases. Although CO_2 is by far the most important anthropogenic greenhouse gas (84 per cent of radiative forcing linked with emissions in 2005), it is by no means the only greenhouse gas of concern. Carbon dioxide, methane (CH_4), nitrous oxide (N_2O), and three groups of fluorinated gases—sulphur hexafluoride (SF_6), hydrofluorocarbons (HFCs), and perfluorinated compounds PFCs—are the major greenhouse gases and the focus of the Kyoto Protocol. The non-CO_2 GHGs are significant in terms of their cumulative impact on climate change, representing about 16 per cent of radiative forcing in 2005. And because some emission reductions could be achieved at relatively low cost, their inclusion in a programme would be attractive in principle (Paltsev *et al.*, 2007*a*).

The sources of some of these gases are large in number and highly dispersed, making their inclusion in a cap-and-trade programme problematic. The answer may be to phase in regulation selectively over time with credit (offset) mechanisms, being careful to grant credits in CO_2-equivalent terms only for well-documented reductions. Over time, such approaches could be developed for industrial emissions of methane and NO_2 and for the manufacture of key industrial gases in the case of refrigerants (HFCs), circuits (PFCs), and transformers (SF_6). Thus, cap-and-trade of non-CO_2 GHGs would likely combine upstream and downstream points of regulation.

(vii) Linkage with other cap-and-trade systems and other nations' policies

Three distinct linkage issues are important. These are: the relationship of the national cap-and-trade system with existing state and regional systems in the United States; the linkage of the cap-and-trade system with other such systems in other parts of the world; and—more broadly—the relationship between the cap-and-trade system and other nations' climate policies.

First, there is the reality of various state and regional cap-and-trade systems for greenhouse gases in the United States. In the absence of a national climate policy, ten north-eastern states have developed a downstream cap-and-trade programme among electricity generators in their RGGI, and California is considering implementing a cap-and-trade programme at the state level. The economy-wide, national, upstream cap-and-trade system could take the place of any regional, state, and local systems to avoid duplication, double counting, and conflicting requirements (Stavins, 2007). It is likely that a decision will be reached on a national cap-and-trade system before any of the regional or state programmes have actually been implemented.

In the long run, linking of the US cap-and-trade system with such systems in other countries or regions, such as the EU ETS, will clearly be desirable to reduce the overall cost of reducing GHG emissions and achieving any global GHG concentration targets (Jaffe and Stavins, 2007). But there is a question of what level and type of linkage is desirable in the early years of the development of a US cap-and-trade system. In the short term, it may be best for the United States to focus on linkage with emission reduction credit (ERC) programmes, such as the Kyoto Protocol's Clean Development Mechanism (CDM), particularly if the CDM can be improved along the lines discussed at the 13th Conference of the Parties of the Framework Convention on Climate Change, in Bali, Indonesia, in December, 2007, namely to give greater emphasis to programme- and policy-based opportunities, as opposed to project-based opportunities, as a means to decrease the prevalence of additionality concerns.

First, by tapping low-cost emission-reduction opportunities in developing countries, linkage of the US system with CDM has a greater potential to achieve significant cost savings for the United States than does linkage with cap-and-trade systems in other industrialized countries (where abatement costs are more similar to those in the United States). Second, linkage with an ERC system such as CDM can only have the effect of decreasing domestic allowance prices, since transactions are uni-directional, i.e. US purchases of (low-cost) CDM credits. Third, the USA may have to choose between adopting a cost-containment mechanism and linking with cap-and-trade systems in other countries. It appears unlikely that the European Union would agree to linking its Emissions Trading Scheme with a US system that employed a safety valve or other such cost-containment measure. On the other hand, the USA could link with ERC systems, such as the CDM, even with a cost-containment measure in place.

Fourth, given that other cap-and-trade systems, such as that of the European Union, will likely be linked with CDM, linking the US system with CDM will have the effect of indirectly linking the US system with those other cap-and-trade systems, but in ways that avoid the short-term problems identified above. Fifth, such indirect linkage should reduce concerns about additionality normally associated with linking with CDM. If another country or region (for example, the European Union) has already linked with CDM, many of the credits that the US system would ultimately purchase would be used by other linked cap-and-trade systems if the United States did not link with CDM. Hence, for these credits, there is no incremental additionality concern regarding the US decision to link with CDM. Any US use of these credits would result in emission reductions in the other linked cap-and-trade systems that would otherwise have used the credits. Sixth, the indirect linkage created by a US link with CDM can achieve some and perhaps many of the cost savings that would arise from direct linkage with other cap-and-trade systems. This is because CDM credits can be sold on the secondary market, and so will ultimately go to the linked cap-and-trade system with the highest allowance price, pushing the allowance prices of the various cap-and-trade systems towards the convergence that would be achieved by direct linkage among cap-and-trade systems.

The fact that climate change is a global-commons phenomenon means that it can be sensible to condition the goals and operations of the US cap-and-trade programme on the GHG emissions reductions efforts that other countries are employing. One approach is to include a provision for the overall US emissions cap to be tightened when and if the President or the Congress determine that other major CO_2-emitting nations have taken specific climate-policy actions. Such 'issue linkage'—making the cap contingent upon the actions of other key countries—can make sense, particularly absent US participation in a binding international agreement. This links the *goals* of the US system with other countries' actions.

In addition, the *operation* of the cap-and-trade system should be linked with the actions of other key nations. As part of the cap-and-trade programme, imports of specific highly carbon-intensive goods (in terms of their emissions generated during manufacture) from countries which have not taken climate-policy actions comparable to those in the United States should be required to hold appropriate quantities of allowances (mirroring the allowance requirements on US sources). These allowances can be purchased from any participants in the domestic cap-and-trade system. This mechanism, if properly designed and implemented, can help establish a level playing-field in the market for domestically produced and imported products, and thereby can serve to reduce emissions leakage and induce key developing countries to join an international agreement (Morris and Hill, 2007).

There are some understandable concerns with such a mechanism. First, there is the economist's natural resistance to tampering with free international trade in order to achieve other ends. Second, there is the difficulty of making the needed calculations of appropriate quantities of allowances on imports of manufactured goods. Third, there is the inescapable irony that the United States might adopt a mechanism for use with other countries, which had recently been proposed by Europeans for use against the United States (although with a border tax) because of US non-ratification of the Kyoto Protocol. More broadly, there is the risk that this mechanism would be abused and inappropriately applied as a protectionist measure.

These concerns can be addressed by properly constraining the mechanism to apply only to primary highly energy-intensive commodities—such as iron and steel, aluminium, cement, bulk glass, paper, and, for that matter, fossil fuels. The requirement would not apply to countries that are taking comparable actions to reduce their GHG emissions, and exemptions could be provided for countries with very low levels of GHG emissions and the lowest levels of economic development.

In order to be compatible with World Trade Organization rules, it is key that the burden imposed on imported and domestic goods be roughly comparable, and that there not be discrimination among nations with similar conditions (Frankel, 2005; Pauwelyn, 2007). Also, this requirement should become binding only after 5–10 years, to allow time for an international climate agreement to be negotiated that includes all key countries in meaningful ways and thereby obviates the need for the mechanism (Aldy and Stavins, 2007).

(viii) Associated climate policies

The price signals generated by a well-functioning upstream cap-and-trade system will be insufficient for their purpose if there are remaining market failures that render those price signals ineffective. For example, there may be market failures other than the environmental externality of global climate change associated with energy-efficiency investments. If the magnitude of these non-environmental market failures is large enough and the cost of correcting them small enough to warrant policy intervention, then an argument can be made to attack these other market failures directly (Jaffe *et al.*, 1999).

Examples of such relevant market failures include information problems that lead consumers to under-value expected energy-cost savings when purchasing energy-consuming durable goods. Likewise, there is the principal–agent problem of landlords who may under-invest in energy-efficient appliances, because electricity costs are paid by tenants. Perhaps most important is the public-good nature of research and development, which leads to under-investment in R&D because knowledge generated may not be exclusive and so economic returns cannot be fully captured. To achieve the desired levels of investment, additional public policies—of various kinds, beyond the price signals generated by the cap-and-trade system—may be necessary (National Commission on Energy Policy, 2004, 2007).

III. Economic assessment

A considerable number of analytical models have been employed over the past several years to estimate the aggregate costs (and in some cases, the distributional impacts) of a cost-effective

Table 1: Anticipated CO_2 emissions reductions under two illustrative caps (million metric tons)

Scenario[a]		2005	2010	2015	2020	2025	2030	2035	2040	2045	2050
BAU	Emissions	5,984	6,517	6,995	7,357	7,915	8,518	9,283	10,013	10,871	11,656
Stabilize	Emissions	5,984	6,517	6,328	6,287	6,132	6,290	7,265	7,605	7,126	7,175
	Reduction[b]	0	0	−667	−1,070	−1,783	−2,228	−2,018	−2,408	−3,745	−4,481
	% reduction[c]	0	0	−10	−15	−23	−26	−22	−24	−34	−38
50% b/1990	Emissions	5,984	6,517	5,740	5,443	4,914	4,085	5,169	4,650	3,588	2,945
	Reduction	0	0	−1,255	−1,914	−3,001	−4,433	−4,114	−5,363	−7,283	−8,711
	% reduction	0	0	−18	−26	−38	−52	−44	−54	−67	−75

Notes: [a] 'BAU' (business as usual) is the reference case from Paltsev *et al.* (2007*a*,*b*); 'Stabilize' is based on the 287 cumulative carbon-dioxide-equivalent billion metric ton (CO_2-e bmt) case from Paltsev *et al.* (2007*a*,*b*); and '50% b/1990' refers to 2050 emissions capped at 50 per cent below the 1990 level, and is based on the 203 cumulative CO_2-e bmt case from Paltsev *et al.* (2007*a*,*b*). [b] Compared with BAU emissions in the same year. [c] Compared with BAU emissions in the same year.
Source: Paltsev *et al.* (2007*b*, pp. 1, 2, 3).

set of emissions-reduction actions to achieve various national CO_2 and GHG targets. Two models have had a distinctly US focus, and have been used to give particular attention to the costs associated with domestic cap-and-trade systems: the National Energy Modeling System (NEMS) of the US Department of Energy (US Energy Information Administration, 2007), and the Emissions Prediction and Policy Analysis (EPPA) model of the Massachusetts Institute of Technology's Joint Program on the Science and Policy of Global Change (Paltsev *et al.*, 2007*a*,*b*).

To provide illustrative empirical cost estimates, I draw upon recent results from MIT's EPPA model, both because of the recent vintage of the analysis and because the model was applied by its authors (Paltsev *et al.*, 2007*a*,*b*) to examining an upstream cap-and-trade system that is—in its stylized form—close to what is described here.

The first illustrative trajectory involves stabilizing CO_2 emissions at their 2008 level over the period from 2012 to 2050. This trajectory, in terms of its cumulative cap, lies within the range defined by the 2004 and 2007 recommendations of the National Commission on Energy Policy (2004, 2007). The second illustrative trajectory—also defined over the years 2012–50—involves reducing CO_2 emissions from their 2008 level to 50 per cent below their 1990 level by 2050. This trajectory—defined by its cumulative cap—is consistent with the lower end of the range proposed by the US Climate Action Partnership (2007).

The tradable CO_2 allowances have value because of their scarcity, and it is their market-determined price that provides incentives for cost-effective emissions reductions and investments that bring down abatement costs over time. As the required emissions reductions (relative to 'business as usual' (BAU)) increase over time under both cap trajectories (Table 1), the market prices of the allowances also increase, rising from \$18/ton of CO_2 in 2015 to \$70/ton of CO_2 in 2050 for the less aggressive policy, and rising from \$41/ton of CO_2 in 2015 to \$161/ton of CO_2 in 2050 for the more aggressive policy (Table 2).

Fossil-fuel prices are also predicted to change as a result of the cap-and-trade system, because of effects on the supply and demand for those fuels in various markets. As Table 2 indicates, the net effect of both caps on coal and petroleum prices is to depress those prices *relative* to what they would be in the absence of climate policy, because of reduced fuel demand. It is important to note, however, that although these prices include the effects of allowance prices on fossil-fuel supply and demand, they do not include the cost of allowances *per se*.

Table 2: Predicted CO_2 and fossil fuel prices[a] under two illustrative caps

	Scenario[b]	2005	2010	2015	2020	2025	2030	2035	2040	2045	2050
CO_2 price[c]	BAU	0	0	0	0	0	0	0	0	0	0
	Stabilize	0	0	18	22	26	32	39	47	57	70
	50% b/1990	0	0	41	50	61	74	90	109	133	161
Petroleum product	BAU	1.0	1.2	1.3	1.5	1.7	1.9	2.0	2.1	2.2	2.3
	Stabilize	1.0	1.2	1.3	1.5	1.6	1.7	1.4	1.4	1.5	1.5
	50% b/1990	1.0	1.2	1.3	1.5	1.5	1.6	1.3	1.4	1.3	1.2
Natural gas	BAU	1.0	1.1	1.3	1.5	1.7	2.0	2.3	2.7	3.1	3.6
	Stabilize	1.0	1.1	1.2	1.5	1.9	2.4	2.5	2.8	2.8	2.8
	50% b/1990	1.0	1.1	1.2	1.4	1.8	2.1	2.1	2.2	2.2	2.0
Coal	BAU	1,0	1.0	1.1	1.1	1.1	1.2	1.2	1.2	1.3	1.3
	Stabilize	1.0	1.0	1.0	1.0	1.0	1.0	1.0	1.1	1.1	1.2
	50% b/1990	1.0	1.0	1.0	1.0	1.0	1.0	1.0	1.1	1.1	1.2

Notes: [a] All fossil fuel prices are price indices, with 2005 set equal to 1. Note that the price indices do not include the cost of allowances, but do include the effects of changes in fossil-fuel supply and demand (induced by impacts of allowance prices on downstream users of respective fossil fuels). [b] 'BAU' (business as usual) is the reference case from Paltsev *et al.* (2007*a*,*b*); 'Stabilize' is based on the 287 cumulative CO_2-e bmt case from Paltsev *et al.* (2007*a*,*b*); and '50% b/1990' refers to 2050 emissions capped at 50 per cent below the 1990 level, and is based on the 203 cumulative CO_2-e bmt case from Paltsev *et al.* (2007*a*,*b*). [c] Year 2005 dollars per ton of CO_2-equivalent.
Source: Paltsev *et al.* (2007*b*, pp. 1, 2, 3).

Table 3: Relationship between CO_2 allowance prices and recent fuel prices

Fuel	Average base price[a] 2002–6	Added fuel cost for various allowance prices[b]		
		$25	$50	$100
Crude oil ($/bbl)	$40.00	$11.30 28%	$22.60 57%	$45.20 113%
Gasoline ($/gallon)	$1.82	$0.24 13%	$0.48 26%	$0.96 53%
Heating oil ($/gallon)	$1.35	$0.27 20%	$0.54 40%	$1.08 80%
Wellhead natural gas ($/mcf)	$5.40	$1.38 26%	$2.76 51%	$5.52 102%
Residential natural gas ($/mcf)	$11.05	$1.39 13%	$2.78 25%	$5.56 50%
Utility coal ($/short ton)	$26.70	$51.20 192%	$102.40 384%	$204.80 767%

Notes: [a] 2005 dollars. [b] Added cost does not include adjustment for the effects of respective cap-and-trade policies on producer prices.
Source: For base prices, Paltsev *et al.* (2007*a*); added fuel costs are from author's calculations, drawing upon Table 5, p. 53, in the same source.

As indicated above, the cap-and-trade system has the effect of reducing demand for fossil fuels relative to BAU conditions and hence reducing fossil-fuel prices *relative* to what those prices would be in the absence of policy. There is an important distinction, however, between the price of fuels themselves (Table 2) and the cost of using those fuels, which is illustrated in Table 3. For sample allowance prices of $25, $50, and $100/ton of CO_2, the added cost is estimated for major fuels, including crude oil, gasoline, heating oil, wellhead natural gas,

Table 4: Predicted aggregate costs: GDP impacts under two illustrative caps

	Scenario[a]	2005	2010	2015	2020	2025	2030	2035	2040	2045	2050
BAU[b]	GDP	11,981	14,339	16,921	19,773	22,846	26,459	30,534	34,929	39,530	44,210
% change GDP from BAU	Stabilize	0	0	−0.22	−0.38	−0.55	−0.68	−0.33	−0.29	−0.36	−0.28
	50% b/1990	0	0	−0.51	−0.79	−0.67	−0.56	−1.18	−1.00	−0.61	−0.48

Notes: [a] 'BAU' (business as usual) is the reference case from Paltsev *et al.* (2007*a*,*b*); 'Stabilize' is based on the 287 cumulative CO_2-e bmt case from Paltsev *et al.* (2007*a*,*b*); and '50% b/1990' refers to 2050 emissions capped at 50 per cent below the 1990 level, and is based on the 203 cumulative CO_2-e bmt case from Paltsev *et al.* (2007*a*,*b*). [b] Billions of year 2005 dollars.
Source: Paltsev *et al.* (2007*b*, pp. 1, 2, 3).

residential natural gas, and utility coal. These added costs of allowances to fuel users (which do not include the adjustment for the effects of the cap-and-trade policies on producer prices from Table 2) are compared with the average price of the respective fuels over a recent period of time.

The cap-and-trade system, like any regulatory initiative, affects the behaviour of individuals and firms, causing reallocation of resources, and thereby causing economic output to grow more slowly than it would in the absence of the policy. Impacts on gross domestic product (GDP) are measured relative to no policy (BAU), and so reductions in GDP do not indicate that output would be lower than current levels, but rather that output would be lower than it would otherwise be expected to be (Table 4).

Consistent with findings from other studies, the analysis indicates significant but affordable impacts on GDP, generally reductions below BAU of less than one-half of 1 per cent in each year of the programme for the less aggressive cap trajectory and ranging up to 1 per cent below BAU each year for the more aggressive policy (Table 4). These impacts on GDP by 2050 are equivalent to average annual GDP growth in the BAU case of 2.901 per cent, and average annual GDP growth of 2.895 and 2.891 per cent, respectively, under the two cap trajectories.

Despite the fact that aggregate impacts on economic output (GDP) are relatively small, there can be very substantial impacts on particular sectors or groups of people. Regardless of how allowances are distributed, most of the cost of the programme will be borne by consumers, facing higher prices of products, including electricity and gasoline—impacts that will continue as long as the programme is in place. Also, workers and investors in the energy sectors and energy-intensive industries will experience losses in the form of lower wages, job losses, or reduced stock values. Such impacts are temporary, and workers or investors who enter an industry after the policy takes effect typically do not experience such losses (Dinan, 2007). The fact that the policy is phased in gradually provides more time for firms and people to adapt.

The cost impacts can be regressive, because lower-income households spend a larger share of their income than wealthier households, and energy products account for a larger share of spending by low-income households than wealthier households. But the distributional impacts will depend greatly on the specifics of policy design, including how allowances are allocated and how auction revenues are used.

Certain sectors and firms will be particularly affected, including fossil-fuel producers, the electricity sector, and energy-intensive industries. Coal production will be the most affected because coal is the most carbon-intensive fuel and opportunities exist for electricity generators and some industrial consumers to switch to less carbon-intensive fuels. Petroleum sector output will be less affected, partly because demand for gasoline and other petroleum products is fairly insensitive to increased prices, at least in the short term. And it is uncertain whether a cap would benefit or adversely affect output and profitability of natural gas producers (US Energy Information Administration, 2003, 2006).

Among firms that consume fossil fuels and electricity, impacts will likely be most pronounced in energy- and emissions-intensive industries (Jorgenson *et al.*, 2000; Smith *et al.*, 2002; Bovenberg and Goulder, 2003; US Energy Information Administration, 2003). For example, some of the most affected industries will be petroleum refiners and manufacturers of chemicals, primary metals, and paper. Refiners experience both increased production costs for their production-related emissions and reduced demand as consumers seek to limit emissions from the use of petroleum products. Among industries experiencing similar increases in their costs, impacts will be greatest in globally competitive industries that are least able to pass through higher costs without experiencing reduced demand for their output.

Industry-level impacts may obscure significant variation in firm-level impacts within an industry. The electricity sector offers an important example of this point. Regional variation in electricity-sector impacts will be greater than in many other sectors because of regional differences in the composition of power plants (including fuel type), physical limits on interregional electricity trading, and state regulation of electricity markets. Increases in the cost of electricity generation depend on the carbon-intensity of a region's generation, which varies widely across the country.

While attention often focuses on a cap's impacts on particular industries, the ultimate burden will be borne by households, primarily in the form of increased expenditures on energy and other goods and services, but also through changes in labour income (including job losses) and investment income (i.e. stock and mutual fund returns) that arise from impacts on firms. Higher fuel prices will likely have a regressive effect on households, although the degree of regressivity may not be very great (Dinan, 2007). Further, this regressivity may be counterbalanced by the fact that adverse impacts on investment returns resulting from a cap's effect on the profitability of firms will fall most heavily on high-income households.

There are also distributional implications of the allowance allocation, and the aggregate value of allowances will be much greater than the total cost burden to the economy. The value of allowances will be two to four times greater than the total cost of the programme in most years under either of the cap trajectories. Therefore, even a partial free distribution of allowances provides an opportunity to address the distributional cost burdens of the policy by using allowances to compensate the most burdened sectors and individuals.

IV. Comparison with alternative approaches

The alternatives to the cap-and-trade approach most frequently considered by policy-makers for the purpose of reducing CO_2 and other GHG emissions are standards-based policies. In addition, among economists and some policy analysts, there has been discussion about the possible use of carbon taxes.

(i) Standards-based policies

Technology or performance standards are a commonly proposed means of achieving emission reductions. Examples include efficiency standards for appliances, vehicle fuel-economy standards, best-available control technology standards, and renewable portfolio standards for electricity generators.

Because of practical limitations, most standards to address CO_2 emissions would target energy use or emission rates from new capital equipment, such as appliances, cars, or electricity generators. The fact that standards would affect new, but not existing equipment limits the opportunity for near-term emission reductions. It also makes the level and timing of those reductions dependent on the rate of capital stock turnover, and thereby difficult to predict. Moreover, by increasing the cost of new capital stock without affecting the cost of using the existing capital stock, standards on new sources have the perverse effect of creating incentives to delay replacement of existing capital stock, which can significantly delay the achievement of emission reductions (Stavins, 2006).

When considered as an alternative to a well-designed cap-and-trade system, standards-based approaches are less cost-effective. Administrative limitations constrain the scope of sources that can be covered by a standards-based approach, compared with an upstream, broad-based cap-and-trade system, and standards may not target all determinants of emissions from covered sources. Consequently, they may not bring about many types of potentially cost-effective emission reductions. Also, standards often impose uniform requirements, even though the cost of emission reductions achieved by such standards may vary widely across regulated entities (Newell and Stavins, 2003).

Standards have also been proposed as complements to market-based policies. On the one hand, standards may needlessly restrict the flexibility that allows market-based policies to minimize the cost of achieving emission targets. If standards are applied within the umbrella of an economy-wide CO_2 cap-and-trade system, the standards will offer no additional CO_2 benefits, as long as the cap-and-trade system is binding, but depending upon the nature of the standard and its associated costs, its placement can drive up aggregate costs. On the other hand, as emphasized above, some market failures affecting the development and adoption of less emissions-intensive technologies may not be addressed by a cap-and-trade (or carbon tax) policy. Simply increasing the cost of emitting GHGs will not address the core sources of such market failures.

(ii) Carbon taxes

Both a carbon tax and a cap-and-trade system create a carbon price signal, but there is a fundamental difference in the way in which the level of that carbon price signal is determined. A carbon tax fixes the price of CO_2 emissions, and allows the quantity of emissions to adjust, whereas a cap-and-trade system fixes the quantity of aggregate emissions, and allows the price of CO_2 emissions to adjust.

A carbon tax (if implemented upstream and economy-wide) would appear to have some advantages over an equivalent upstream cap-and-trade system. First, is the simplicity of the carbon tax system, in which firms would not need to manage and trade allowances, and the government would not need to track allowance transactions and ownership. Experience with previous cap-and-trade systems, however, indicates that the costs of trading institutions are not significant. Whether a policy as significant as a meaningful national carbon tax would turn out to be simple in its implementation is an open question. Second, the tax approach avoids the political difficulties related to making allowance allocations among economic sectors, but would—on the other hand—create pressures for tax exemptions.

Third, a carbon tax would raise revenues that can be used for beneficial public purposes. Of course, an auction mechanism under a cap-and-trade system can do the same. Fourth, a tax approach eliminates the potential for price volatility that can exist under a cap-and-trade system. Some emissions-trading markets have exhibited significant volatility in their early years, including: the US NO_x Budget program (where prices increased in the presence of uncertainty about whether Maryland, a net supplier, would enter the programme on time); the RECLAIM programme in southern California (where price spikes were linked with flawed design and problems with electricity deregulation); and the EU ETS (where a dramatic price crash occurred when data revealed that the overall allocation had been *above* the BAU level). From an economic perspective, it makes sense to allow emissions to vary from year to year with economic conditions that affect aggregate abatement costs; and this happens automatically with a carbon tax. With a cap-and-trade system, this temporal flexibility needs to be built in through provisions for banking and borrowing, as described above.

There is also a set of apparent disadvantages of carbon taxes, relative to a cap-and-trade regime, that merits consideration. First among these is the over-riding resistance to new taxes in the current political climate. Second, in their simplest respective forms (a carbon tax *without* revenue recycling, and a cap-and-trade system *without* auctions), a carbon tax is more costly than a cap-and-trade system to the regulated sector, because with the former firms incur both abatement costs and the cost of tax payments to the government. In the case of the simplest cap-and-trade system, the regulated sector experiences only abatement costs, since the transfers associated with allowance purchase and sale remain within the private sector.

Third, cap-and-trade approaches leave distributional issues up to politicians, and provide a straightforward means to compensate burdened sectors. Of course, the compensation associated with free distribution of allowances based on historical activities can be mimicked under a tax regime, but it is legislatively more complex. The cap-and-trade approach avoids likely battles over tax exemptions among vulnerable industries and sectors that would drive up the costs of the programme, as more and more sources (emission-reduction opportunities) are exempted from the programme, thereby simultaneously compromising environmental integrity. Instead, a cap-and-trade system leads to battles over the allowance allocation, but these do not raise the overall cost of the programme nor affect its climate impacts. Some observers seem to worry about the political process's propensity under a cap-and-trade system to compensate sectors that effectively claim burdens (through free allowance allocations). A carbon tax is sensitive to the same pressures, and may be expected to succumb to them in ways that are ultimately much more harmful. This is the crucial political-economy distinction between the two approaches.

Fourth, a carbon tax provides much less certainty over emissions levels (in exchange for greater certainty over costs). Most climate policy proposals are for progressively greater cuts in emissions over time. Cap-and-trade is fundamentally well suited to this because it is a quantity-based approach. Progress under a carbon tax will be uncertain, mainly owing to variations in economic conditions.

Fifth and finally, a cap-and-trade system is much easier to harmonize with other countries' carbon mitigation programmes, which are more likely to employ cap-and-trade than tax approaches. Cap-and-trade systems generate a natural unit of exchange for harmonization: allowances denominated in units of carbon content of fossil fuels (or CO_2 emissions). In addition, cap-and-trade provides a convenient means—allowances traded between firms—to transfer resources for emissions reductions in developing countries. A carbon tax raises funds for the government that could likewise be used for this purpose, but such transfers would need to be between governments, and such transfers would be larger in magnitude than individual trades between sources under a cap-and-trade system, thereby reducing greatly the political feasibility of such arrangements.

Despite the differences between carbon taxes and cap-and-trade systems in specific implementations, the two approaches have much in common. Differences between the two approaches can begin to fade when various specific implementations of either programme are carried out. Hybrid schemes that include features of taxes and cap-and-trade systems blur the distinctions between the two (Parry and Pizer, 2007). In terms of the allocation mechanism, the government can auction allowances in a cap-and-trade system, thereby reproducing many of the properties of a tax approach. Mechanisms that deal with uncertainty in a cap-and-trade system also bring it close to a tax approach, including a cost-containment mechanism that places a cap on allowance prices, banking that creates a floor under prices, and borrowing that provides flexibility similar to a tax. To some degree, the dichotomous choice between taxes and permits can turn out to be a choice of design elements along a policy continuum.

In the meantime, debate continues among economists regarding cap-and-trade and carbon taxes. In a recent comparison of these two approaches, the Hamilton Project staff at the Brookings Institution concluded that a well-designed carbon tax and a well-designed cap-and-trade system would have similar economic effects (Furman *et al.*, 2007). Hence, they concluded, the two primary questions that should be used to decide between these two policy approaches are: (i) which is more politically feasible; and (ii) which is more likely to be well designed? In the context of the United States (and many other countries, for that matter), the answer to the first question is obvious. For the political-economy reasons I described above, the answer to the second question also favours cap-and-trade. In other words, it is important to identify and design policies that will be 'optimal in Washington', not just from the perspective of Cambridge, New Haven, or Berkeley.

V. Summary and conclusions

The need for a domestic US policy that seriously addresses climate change is increasingly apparent. A cap-and-trade system is the best and most likely approach for the United States in the short to medium term. Besides providing greater certainty about emissions levels, cap-and-trade offers an easy means (partial free distribution of allowances) of compensating for the inevitably unequal burdens imposed by climate policy; it is straightforward to harmonize with other countries' climate policies; it avoids the current political aversion in the United States to taxes; and it has a history of successful adoption.

The system outline in this article has several key features. It imposes an upstream cap on CO_2 emissions (carbon content measured at the point of fuel extraction, refining, distribution, or importation), with gradual inclusion of other greenhouse gases, to ensure economy-wide coverage while limiting the number of entities to be monitored. It sets a gradual downward trajectory of emissions ceilings over time, to minimize disruption and allow firms and households time to adapt. It also includes mechanisms to reduce cost uncertainty; these include provisions for banking and borrowing of allowances, and possibly a cost-containment mechanism (such as the sale of additional allowances during severe price spikes, with the revenues dedicated to bringing about additional emissions reductions) to protect against price volatility.

Initially, half of the programme's allowances would be allocated through auctioning and half through free distribution, primarily to those entities most burdened by the policy. This arrangement should help limit potential inequities while bolstering political support. The share distributed for free would phase out gradually over 25 years. The auctioned allowances would generate revenue that could be used for a variety of worthwhile public purposes. To increase the programme's short-term cost-effectiveness and create long-term incentives for technological development, entities that successfully implement carbon sequestration (biological or underground) would be eligible for offsets.

The system would operate at the federal level, eventually asserting supremacy over all regional, state, and local systems, while building on any institutions already developed at those levels. The system would also provide for linkage with international emissions-reduction credit arrangements, harmonization over time with effective cap-and-trade systems in other countries, and appropriate linkage with other actions taken abroad to maintain a level playing field between imports and import-competing domestic products. To address potential market failures that might render the system's price signals ineffective, certain complementary policies should be implemented, for example in the areas of consumer information and research and development.

Like other market-based emissions-reduction schemes, the one described here reduces compliance costs by offering regulated entities: rather than mandate specific measures on all sources, it allows emissions to be reduced however, wherever, and, to a great extent, whenever they are least costly. To illustrate the potential cost savings, this article has reported empirical cost estimates for two hypothetical time trajectories for emissions caps. The first stabilizes CO_2 emissions at their 2008 level by 2050, whereas the second reduces emissions from their 2008 level to 50 per cent below the 1990 level by 2050. Both are consistent with the often-cited global goal of stabilizing CO_2 atmospheric concentrations at between 450 and 550 ppm, provided all countries take commensurate action. The analysis found significant but affordable impacts on GDP under both trajectories: generally below 0.5 per cent a year for the less aggressive trajectory, and ranging up to 1 per cent a year for the more aggressive one.

We also explored the distributional implications of the programme. Illustrative estimates—which do not account for the offsetting effects of possible free allocation of allowances or redistribution of auction revenues—indicate a relatively small burden on fossil-fuel producers (about 4 per cent of the total), because most of the costs would be passed on to customers. Fossil-fuel-fired electricity generators also would bear a relatively small share, about 7 per cent, for analogous reasons. Business and industry would bear nearly 30 per cent of the total cost burden through their primary energy use, and about 25 per cent through their electricity use, for a total of about 55 per cent. The remaining roughly 35 per cent of costs would be borne by households.

The impact of any US policy will ultimately depend on the actions of other nations around the world. Without an effective global climate agreement, each country's optimal strategy is to free-ride on the actions of others. But if all countries do this, nothing will be accomplished, and the result will be the infamous tragedy of the commons. A cooperative solution—one that is scientifically sound, economically rational, and politically pragmatic—must remain the ultimate goal. Given these realities, a major strategic consideration in initiating a US climate policy should be to establish international credibility. The cap-and-trade system described and assessed in this article offers a way for the United States to demonstrate its commitment to an international solution while making its own real contribution to addressing climate change.

References

Aldy, J. E., and Stavins, R. N. (eds) (2007), *Architectures for Agreement: Addressing Global Climate Change in the Post-Kyoto World*. Cambridge, Cambridge University Press.

Anderson, R. (1997), *The US Experience with Economic Incentives in Environmental Pollution Control Policy*. Washington, DC, Environmental Law Institute.

Bannon, B., DeBell, M., Krosnick, J., Kopp, R., and Aldhous, P. (2007), 'Americans' Evaluations of Policies to Reduce Greenhouse Gas Emissions', Working Paper, June.

Bluestein, J. (2005), 'Upstream Regulation of CO_2', Presentation to the National Commission on Energy Policy Workshop, Washington, DC, 16 September.

Bovenberg, L. A., and Goulder, L. H. (2003), 'Confronting Industry-distributional Concerns in US Climate-change Policy', Discussion Paper (Les Séminaires de l'IDDRI no. 6), Institute on the Economics on the Environment and Sustainability.

Burtraw, D., Goeree, J., Holt, C., Palmer, K., and Shobe, W. (2007), 'Auction Design for Selling CO_2 Emission Allowances Under the Regional Greenhouse Gas Initiative', Final Report produced for RGGI, October.

— Krupnick, A., Mansur, E., Austin, D., and Farrell, D. (1998), 'The Costs and Benefits of Reducing Air Pollution Related to Acid Rain', *Contemporary Economic Policy*, **16**, 379–400.

Carlson, C., Burtraw, D., Cropper, M., and Palmer, K. (2000), 'Sulfur Dioxide Control by Electric Utilities: What Are the Gains from Trade?', Discussion Paper No. 98-44-REV, Washington, DC, Resources for the Future.

Deutch, J., and Moniz, E. J. (co-chairs) (2007), *The Future of Coal: Options for a Carbon-constrained World*. Cambridge, MA, Massachusetts Institute of Technology.

Dinan, T. (2007), *Trade-offs in Allocating Allowances for CO_2 Emissions*, Economic and Budget Issue Brief, Washington, DC, US Congressional Budget Office, 25 April.

Ellerman, D., and Buchner, B. (2007), 'The European Union Emissions Trading Scheme: Origins, Allocation, and Early Results', *Review of Environmental Economics and Policy*, **1**, 66–87.

— Joskow, P., Schmalensee, R., Montero, J.-P., and Bailey, E. (2000), *Markets for Clean Air: The US Acid Rain Program*, New York, Cambridge University Press.

Farrell, A., Carter, R., and Raufer, R. (1999), 'The NOx Budget: Market-based Control of Tropospheric Ozone in the Northeastern United States', *Resource and Energy Economics*, **21**,103–24.

Frankel, J. (2005), 'Climate and Trade: Links Between the Kyoto Protocol and WTO', *Environment*, **47**(7), 8–19, September.

Furman, J., Bordoff, J. E., Deshpande, M., and Noel, P. J. (2007), 'An Economic Strategy to Address Climate Change and Promote Energy Security', The Hamilton Project, Strategy Paper, Washington, DC, The Brookings Institution, October.

Goulder, L. H. (2000), 'Confronting the Adverse Industry Impacts of CO_2 Abatement Policies: What Does it Cost?', Climate Issues Brief No. 23, Washington, DC, Resources for the Future.

— (2004), 'Induced Technological Change and Climate Policy', Arlington, VA, Pew Center on Global Climate Change.

Harrison, D., Jr (2003), 'Ex Post Evaluation of the RECLAIM Emission Trading Program for the Los Angeles Air Basin', presented at Workshop on Ex Post Evaluation of Tradeable Permits: Methodological and Policy Issues, Paris, OECD Environment Directorate.

Helm, D., Hepburn, C., and Mash, R. (2005), 'Credible Carbon Policy', in D. R. Helm (ed.), *Climate-change Policy*, Oxford, Oxford University Press, 305–21.

Intergovernmental Panel on Climate Change (2007*a*), *Climate Change 2007—The Physical Science Basis: Working Group I Contribution to the Fourth Assessment Report of the IPCC*, Cambridge, Cambridge University Press.

— (2007*b*), *Climate Change 2007—Impacts, Adaptation and Vulnerability: Working Group II Contribution to the Fourth Assessment Report of the IPCC*. Cambridge, Cambridge University Press.

— (2007*c*), *Climate Change 2007—Mitigation of Climate Change: Working Group III Contribution to the Fourth Assessment Report of the IPCC*, Cambridge, Cambridge University Press.

Jacoby, H. D., and Ellerman, A. D. (2002), 'The Safety Valve and Climate Policy', MIT Joint Program on the Science and Policy of Global Change, Report No. 83, Cambridge, MA.

Jaffe, A. B., Newell, R. G., and Stavins, R. N. (1999), 'Energy–efficient Technologies and Climate Change Policies: Issues and Evidence', Climate Issues Brief No. 19, Washington, DC, Resources for the Future.
— (2005), 'A Tale of Two Market Failures: Technology and Environmental Policy', *Ecological Economics*, **54**, 164–74.
Jaffe, J., and Stavins, R. N. (2007), 'Linking Emissions Trading Systems', paper prepared for the International Emissions Trading Association, forthcoming.
Jorgenson, D. W., Goettle, R. J., Wilcoxen, P. J., and Ho, M. S. (2000), *The Role of Substitution in Understanding the Costs of Climate Change Policy,* Arlington, VA, Pew Center on Global Climate Change.
Kerr, S., and Newell, R. (2000), 'Policy-induced Technology Adoption: Evidence from the US Lead Phase-down', Washington, DC, Resources for the Future, draft manuscript.
Lubowski, R. N., Plantinga, A. J., and Stavins, R. N. (2006), 'Land-use Change and Carbon Sinks: Econometric Estimation of the Carbon Sequestration Supply Function', *Journal of Environmental Economics and Management*, **51**, 135–52.
Market Advisory Committee to the California Air Resources Board (2007), 'Recommendations for Designing a Greenhouse Gas Cap-and-trade System for California', 30 June.
Morris, M. G., and Hill, E. D. (2007), 'Trade is the Key to Climate Change', *The Energy Daily*, **35**(33), 20 February.
National Commission on Energy Policy (2004), *Ending the Energy Stalemate: A Bipartisan Strategy to Meet America's Energy Challenges*, Washington, DC, December.
— (2007), *Energy Policy Recommendations to the President and the 110th Congress*, Washington, DC, April.
Newell, R. G. (2007), 'Climate Technology Policy', Resources for the Future Climate Backgrounder, Washington, DC.
— Stavins, R. N. (2003), 'Cost Heterogeneity and the Potential Savings from Market-based Policies', *Journal of Regulatory Economics*, **23**, 43–59.
Nichols, A. (1997), 'Lead in Gasoline', in R. Morgenstern (ed.), *Economic Analyses at EPA: Assessing Regulatory Impact*, Washington, DC, Resources for the Future, 49–86.
Paltsev, S., Reilly, J. M., Jacoby, H. D., Gurgel, A. C., Metcalf, G. E., Sokolov, A. P., and Holak, J. F. (2007*a*), 'Assessment of US Cap-and-trade Proposals', Working Paper 13176, Cambridge, MA, National Bureau of Economic Research, June.
— Gurgel, A. C., Metcalf, G. E., Sokolov, A. P., and Holak, J. F. (2007*b*), 'Appendix C—Assessment of US Cap-and-trade Proposals: Details of Simulation Results,' Cambridge, MA, National Bureau of Economic Research, June.
Parry, I. W. H., and Pizer, W. A. (2007), 'Emissions Trading Versus CO_2 Taxes', mimeo.
Pauwelyn, J. (2007), 'US Federal Climate Policy and Competitiveness Concerns: The Limits and Options of International Trade Law', Working Paper 07-02, Nicholas Institute, Duke University, NI.
Pizer, W. (2005), 'Climate Policy Design Under Uncertainty', Discussion Paper No. 05–44, Resources for the Future, Washington, DC.
Plantinga, A. J. (2007), 'Land-use Change and Biological Carbon Sequestration', presentation at Workshop on Carbon Sequestration in Agriculture and Forestry, Thessaloniki, 27 June.
Reilly, J., Jacoby, H., and Prinn, R. (2003), 'Multi-gas Contributors to Global Climate Change: Climate Impacts and Mitigation Costs of Non-CO_2 Gases', Arlington, VA, Pew Center on Global Climate Change.
Repetto, R. (2007), 'National Climate Policy: Choosing the Right Architecture', Yale School of Forestry and Environmental Studies, June.
Smith, A., Ross, M., and Montgomery, D. (2002), 'Implications of Trading Implementation Design for Equity–Efficiency Trade-offs in Carbon Permit Allocations', Working Paper, Charles River Associates, Washington, DC.
Stavins, R. N. (1999), 'The Costs of Carbon Sequestration: A Revealed-preference Approach', *American Economic Review*, **89**(4), 994–1009.
— (2003), 'Experience with Market-based Environmental Policy Instruments', ch. 9 in K.-G. Mäler and J. Vincent (eds), *Handbook of Environmental Economics*, vol. I, Amsterdam, Elsevier Science, 355–435.
— (2006), 'Vintage-differentiated Environmental Regulation', *Stanford Environmental Law Journal*, **25**(1), 29–63.

Stavins, R. N. (2007), 'A US Cap-and-trade System to Address Global Climate Change', The Hamilton Project Discussion Paper 2007–13. Washington, DC, The Brookings Institution, October.

— Richards, K. R. (2005), *The Cost of US Forest-based Carbon Sequestration*, Arlington, VA, Pew Center on Global Climate Change, January.

Tietenberg, T. (1997), 'Tradeable Permits and the Control of Air Pollution in the United States', paper prepared for the 10th Anniversary Jubilee edition of *Zeitschrift Fürangewandte Umweltforschung*.

US Climate Action Partnership (2007), *A Call for Action—Consensus Principles and Recommendations from the US Climate Action Partnership: A Business and NGO Partnership*, Washington, DC, 22 January.

US Energy Information Administration (2003), *Analysis of S.139, the Climate Stewardship Act of 2003*, Washington, DC, US Department of Energy.

— (2006), *Energy Market Impacts of Alternative Greenhouse Gas Intensity Reduction Goals*, Washington, DC, US Department of Energy, March.

— (2007), *Energy Market and Economic Impacts of a Proposal to Reduce Greenhouse Gas Intensity with a Cap and Trade System*, Washington, DC, US Department of Energy, January.

US Environmental Protection Agency (2005), *Acid Rain Program: 2005 Progress Report*, Washington, DC, US EPA, Office of Air and Radiation, Clean Air Markets Division.

Wigley, T., Richels, R., and Edmonds, J. (1996), 'Economic and Environmental Choices in the Stabilization of Atmospheric CO_2 Concentrations', *Nature*, **379**(18), 240–3.

[20]

American Economic Review: Papers & Proceedings 2011, 101:3, 253–257
http://www.aeaweb.org/articles.php?doi=10.1257/aer.101.3.253

CRITICAL ISSUES IN NATIONAL CLIMATE POLICY DESIGN†

Challenges from State-Federal Interactions in US Climate Change Policy

By LAWRENCE H. GOULDER AND ROBERT N. STAVINS*

Throughout most of US history, state and local governments have had the primary responsibility for environmental protection (Richard L. Revesz 2001). However, since the passage of the National Environmental Policy Act in 1969, the federal role has increased significantly. But while federal environmental laws establish various standards, they typically leave states free to adopt more stringent requirements. Some states have adopted tighter thresholds for automobile emissions; others have created their own "Superfund" programs; and others have implemented their own state-based environmental protection acts (Revesz and Robert N. Stavins 2007).

The coexistence of state and federal programs is likely to continue in the context of US climate change policy. Although Congressional proposals for cap-and-trade or other forms of carbon pricing have stalled, federal climate policy moves forward, in part through greenhouse gas (GHG) regulation by the US EPA under the Clean Air Act, called for by the 2006 US Supreme Court decision in Massachusetts versus EPA, the Obama administration's subsequent "endangerment finding" that carbon dioxide (CO_2) and other greenhouse gases endanger public health and welfare, and the consequent designation in 2010 of CO_2 as a pollutant under the Clean Air Act.

Outside of the Clean Air Act, there is support in Congress and the administration for increasing the stringency of federal motor vehicle fuel efficiency standards (so-called Corporate Average Fuel Economy or CAFE standards). And there is ongoing interest in a "national renewable electricity standard" (RES), which would mandate that a given share of an electric company's production come from renewable sources (most likely wind power) or, in the case of a "clean energy standard" (CES), from an expanded list including nuclear and hydroelectric power.

At the state level, there is considerable climate policy activity as well. As of 2010, climate policies were being contemplated, developed, or implemented by more than half of the 50 states. At latest count, 30 states have implemented RES or CES programs.

The coexistence of federal and state policies raises important questions about how these policies will interact. In the presence of federal policies, to what extent will state efforts be cost effective? And how does the coexistence of state- and federal-level policies affect the ability of state efforts to achieve emissions reductions beyond those implied by federal policy?

I. Problematic Interactions from Overlapping Regulations

Some important interactions stem from the overlap of federal and state regulations. We examine such interactions as applied to two pairs of federal and state policies: state-level RES/CES and federal RES/CES programs; and California's Pavley II fuel-efficiency standards and federal CAFE standards. The first pairing is likely to occur; the second pairing will almost definitely occur.

†*Discussants:* Brian Murray, Duke University; Meredith Fowlie, University of California, Berkeley; Dallas Burtraw, Resources for the Future; Matthew Kahn, University of California, Los Angeles.

*Goulder: Department of Economics, Stanford University, Landau Economics Building, Stanford University, Stanford, CA 94305 (e-mail: goulder@stanford.edu); Stavins: John F. Kennedy School of Government, Harvard University, 79 John F. Kennedy St., Cambridge, MA 02138 (e-mail: robert_stavins@harvard.edu).

A. *Renewable Electricity or Clean Energy Standards*

Background.—Under the typical design of an RES or CES, generators earn tradable certificates or credits for each unit of renewable or clean energy they produce. At the end of the accounting period, each firm must surrender RES/CES certificates equivalent to its required level of renewable/clean energy production, defined as a specified share of its total production. Among the 30 states with such programs, most have targets of 15 to 20 percent of total electricity production by 2020. Some, such as California, have targets as high as 33 percent. Interest in Washington, DC in federal renewable electricity standards has been increasing. Key proposals in the Congress for a federal RES have targets ranging from 3 to 15 percent.

A renewable electricity standard with trading is closely related to the cap-and-trade approach to pollution control. Many state RES programs are intended as climate policies. Hence these can be thought of as CO_2 cap-and-trade systems where the difference in carbon intensity among the three fossil fuels—coal, petroleum, and natural gas—is ignored and (depending upon the treatment of other fuel sources) the zero-carbon properties of hydro and nuclear are ignored. The disregard for differences in carbon content limits the cost effectiveness of the RES. Cost effectiveness is also compromised because the RES does not directly put a price on the externalities associated with fossil-based electricity generation; instead it focuses on the ratio of renewable- to fossil-based generation.[1]

In assessing the implications of overlapping regulations, the simplest case to think about is where the federal RES (or CES) embraces all entities or activities covered by the states that are also adopting an RES.[2] Suppose the federal policy is already in place, and that a given state now introduces a policy that is more stringent that the federal one—in the sense that it requires reductions from sources within the state that are greater than would be achieved under the national program alone. How does this affect emissions reductions *nationwide*, as well as cost effectiveness?

Outcomes When Trading Is Allowed.—The answer depends on whether the federal program includes provisions for trading federal renewable energy credits. Consider the case where the federal program includes such provisions—which is the case with all existing proposals. And suppose that, prior to the imposition of the "green" state's tighter requirement, a firm was just meeting the federal requirement. In this case, by increasing its commitment to renewable electricity to comply with the tighter green state requirement, this firm will have more than met the federal requirement. It now has excess federal renewable credits and will wish to sell these credits to firms operating in states with a less stringent state-level RES (or none at all). The price of the federal credits will fall to coax out demand for the excess credits, and by purchasing these excess credits, firms in other states will not need to rely as much on renewable-sourced electricity.

Under these circumstances, "emissions leakage" could approach 100 percent: the increase in renewable electricity and the reduction in emissions achieved in the green state could be largely or fully offset by reduced renewable electricity and increased emissions in other states. Moreover, the green state's efforts would cause a worsening of the cost effectiveness of the nation's overall emissions-reduction efforts. In the absence of the green state's RES, the federal program would cause marginal abatement costs to be equated across states and sources. The green state's action now eliminates this equality, leading to higher marginal abatement costs in that state than in other states. Thus, overall costs rise with very little or no accompanying reduction in nationwide emissions. The same problems would arise if the federal program allowed a firm with operations in more than one state to meet the federal standard by averaging

[1] Because of its focus on a ratio or input intensity, the RES/CES is equivalent to the combination of a subsidy to electricity production and tax on emissions. As shown by Stephen Holland, Jonathan Hughes, and Christopher Knittel (2009), the subsidy component impedes cost effectiveness. For a broad examination of the interaction of overlapping policies intended to encourage the use of renewable sources of energy, see Carolyn Fischer and Louis Preonas (2010).

[2] Where national and state policies involve imperfectly overlapping coverage, the interactions depend upon which policy is more comprehensive in its scope of coverage, and which is more stringent for the sources covered by both policies. On this, see Goulder and Stavins (2010).

its performance across states with tighter and more lax state-level requirements.[3]

In principle, there are two ways in which the leakage described above can be avoided. One is for the federal government to allow a state (or group of states) a "carve out" from the federal program if it implemented or maintained a state program at least as stringent. There is abundant precedent for this in federal policy. The result would be two separate and distinct fuel-efficiency or RES/CES programs, with the federal system applying only to states that do not carve themselves out. The shadow price on fuel efficiency and the price of renewable credits will be different in some states than in the federal system, and so cost effectiveness will be compromised.

Another way to avoid these problematic interactions is through federal rules that preempt (that is, essentially, prohibit) state-level programs in the presence of a federal program. There is precedent for this approach as well. In this way, leakage is prevented, as well as the attendant loss of cost effectiveness. Also, this is a way of assuring that private industry does not face multiple standards. However, to the extent that a greener state's actions raise costs, those costs are borne by that state alone; therefore such states might oppose preemption on the grounds that states should have the freedom to decide whether to impose higher costs on themselves.

Outcomes When Trading Is Not Allowed.— An alternative federal design would involve no provision for renewable energy credit trading. A main virtue of trading is to encourage increased use of renewable energy sources where it is cheapest to do so. The alternative design would not have this virtue and thus would sacrifice cost effectiveness. In this case there is no longer a channel through which a firm facing a new and tighter state-level standard can reduce pressure on the federal constraint by selling excess federal credits. Here the more stringent green state standard does not cause emissions leakage to other states; hence it yields a reduction in nationwide emissions. However, a tightening of the federal standard instead of a tighter requirement in an individual state could accomplish the same reduction in emissions at lower cost.

B. *Motor-Vehicle Fuel Efficiency Standards*

Nested federal and state regulations lead to similar problems in the context of automobile fuel-efficiency standards or limits on greenhouse gas emissions per mile. By 2009, 14 states had moved to limit GHG emissions per mile. These so-called "Pavley standards"— named after Assemblywoman Fran Pavley, sponsor of the California legislation—require manufacturers to reduce per-mile GHG emissions by about 30 percent by 2016 and 45 percent by 2020 (California Air Resources Board 2008).

Since CO_2 emissions and gasoline use are nearly proportional, the Pavley standards effectively raise the fuel economy requirements for manufacturers in the states adopting the limits. These state-level actions are quantity-based regulations that can interact significantly with the existing Corporate Average Fuel Economy (CAFE) standards at the federal level. Consider an auto manufacturer that prior to the imposition of the Pavley limits was just meeting the CAFE standard. Now it must meet the (tougher) Pavley requirement through its sales of cars registered in the adopting states. In meeting the tougher Pavley requirements, its overall US average fuel economy exceeds the national requirement: the national constraint no longer binds. Hence the manufacturer is able to change the composition of its sales outside of the Pavley states; specifically, it can shift its sales toward larger cars with lower fuel economy.

Indeed, if all manufacturers were initially constrained by the national CAFE standard, the introduction of the Pavley requirements would lead to emissions leakage of 100 percent at the margin, because the reductions within the Pavley states would be completely offset by emissions increases outside of those states. Empirical estimates indicate that from 2009 through 2020, about 65 percent of the emissions reductions achieved in the new car market in the Pavley states would be offset by increased emissions in

[3] This result of near-100 percent leakage and the related cost-effectiveness outcomes are analyzed in more detail by Goulder and Stavins (2010). Similar outcomes apply in the context of a federal cap-and-trade system (Meghan McGuinness and A. Denny Ellerman 2008; Dallas Burtraw and Bill Shobe 2009; Goulder and Stavins 2010).

new car markets elsewhere (Goulder, Mark R. Jacobsen, and Arthur van Benthem 2009).[4]

In May 2009, the Obama administration reached an agreement with the 14 "Pavley states," according to which the United States would tighten the federal fuel economy requirements in such a way as to achieve effective reductions in GHGs per mile consistent with the Pavley initiative. In return, the 14 states agreed to abandon this first phase of the Pavley effort, which was no longer necessary, given the tightening of the federal standards. The Pavley states intend to introduce further tightening of the greenhouse-gas-per-mile standards after 2016. This would imply fuel economy standards more stringent than those applying at the federal level. Hence the leakage issue remains very much alive.

Despite this potential for leakage, the tougher state-level standards could conceivably accelerate the development of new technologies that auto manufacturers will eventually adopt throughout the nation, thereby leading to lower emissions and reduced fuel consumption. However, Goulder, Jacobsen, and van Benthem (2009) found that in the presence of the national CAFE standard, faster technological progress exacerbates the adverse fleet compositional impacts of state programs. As a result, in this case, greater technological progress yields relatively little benefit in terms of reduced fuel consumption.

C. *Price-Based Policies*

The problematic interactions identified above are specific to quantity-based regulations—whether market-based, such as cap-and-trade or renewable energy standards, or command-and-control, such as technology or performance standards. The situation is very different with price-based regulations. Suppose, for example, that a carbon tax were imposed at the federal level. If a state decided to impose new regulations requiring in-state reductions beyond what the federal tax would yield, the additional state-level reductions would not lead to offsetting increases elsewhere (apart from the usual "economic leakage"): the reductions in other states would remain governed by the federal carbon tax. Thus, price-based regulation at the federal level can avoid the types of problems discussed earlier, despite the regulatory overlap. However, it may be noted that the greener state's more aggressive action described here does not achieve nationwide reductions in the most cost-effective way: the same aggregate reductions could be achieved at lower cost with an increased federal carbon tax.

II. Potentially Positive Interactions

State and federal policies can interact along other dimensions, which may lead to positive outcomes. First, strategic interactions can arise between states and the federal government. In particular, state efforts can create pressure for more stringent federal policy. There is, in fact, a considerable history of California air standards having precisely this effect on federal policy developments, because industry is reluctant to face different standards in different parts of the country. For example, the California-led state-level tightening of greenhouse-gas-per mile standards helped bring about the subsequent tightening of federal CAFE standards.[5] Of course, such triggering of stronger federal policy is desirable only if the previous federal policy was insufficiently stringent.

Second, states can serve as laboratories for experimenting with innovative policy approaches. Approaches that prove successful on cost effectiveness or other dimensions could later be adopted at the federal level. The interaction here is one of information transfer. The case for state-level experimentation needs to be considered carefully: why the laboratories should be at the state, rather than national, level is not clear, and—in any event—there is some question regarding whether state authorities will allow their "laboratory" to be closed after the experiment has been completed and the information delivered.

[4] The same study found that another 5 percent of the emissions reduction would be offset by increased emissions from used cars, as the Pavley effort leads to lower scrap rates of older, less fuel-efficient automobiles.

[5] Similarly, there is broad agreement that the California-led state-level tightening of greenhouse-gas-per mile standards brought about the subsequent tightening of federal CAFE standards. Automakers did not wish to face different standards at the federal and state level. Hence they were willing to support tighter federal standards so long as the state standards were removed.

III. Rationales for State Action when Policies Do Not Overlap

When state and federal efforts do not overlap, the problems from nested regulation do not apply. In these circumstances, states can fill the regulatory gap. For example, states may address market failures not confronted by federal policy. One important such failure is the principal-agent problem of inadequate incentives—even in the face of efficient energy prices—for either owners or occupiers of rental properties to invest in energy-efficiency technologies, such as thermal insulation. This market failure is best addressed through building codes or zoning, both of which are probably better implemented at the state or local level, because of geographic differences in climate.

More broadly, the case can be made for state-level action when action that arguably is best taken at the federal level is not politically feasible. Since the externality from greenhouse gas emissions transcends national boundaries, climate change ideally should be addressed at the global level. However, just as political and institutional obstacles to a "world-government" initiative make national policies the best available option, so too can political obstacles to national efforts leave room for subnational efforts. In the United States, subnational efforts will remain critical in the absence of meaningful federal policy.

IV. Conclusions

The coexistence of state and federal policies raises questions about their interactions. Problems arise when state and federal policies overlap. Two regulatory contexts stand out: renewable electricity and clean energy standards; and automobile fuel-economy standards. Because of problematic interactions, state-level efforts may fail to reduce greenhouse gas emissions nationally and may reduce the cost effectiveness of the overall national effort. The difficulties from overlapping regulations can be avoided through price-based (as opposed to quantity-based) federal policy.

At the same time, the possibility exists of some specific positive interactions between state and federal climate policies. And more broadly, rationales exist for subnational actions when national policy is politically infeasible.

REFERENCES

Burtraw, Dallas, and Bill Shobe. 2009. "State and Local Climate Policy under a National Emissions Floor." Unpublished.

California Air Resources Board. 2008. *Comparison of Greenhouse Gas Reductions for the United States and Canada under US CAFE Standards and California Air Resources Board Greenhouse Gas Regulations: An Enhanced Technical Assessment.* California Environmental Protection Agency. Sacramento, February.

Fischer, Carolyn, and Louis Preonas. 2010. "Combining Policies for Renewable Energy: Is the Whole Less than the Sum of Its Parts?" *International Review of Environmental and Resource Economics*, 4(1): 51–92.

Goulder, Lawrence H., and Robert N. Stavins. 2010. "Interactions between State and Federal Climate Change Policies." National Bureau of Economic Research Working Paper 16123.

Goulder, Lawrence H., Mark R. Jacobsen, and Arthur A. van Benthem. 2009. "Unintended Consequences from Nested State and Federal Regulations: The Case of the Pavley Greenhouse-Gas-Per-Mile Limits." National Bureau of Economic Research Working Paper 15337.

Holland, Stephen P., Jonathan E. Hughes, and Christopher R. Knittel. 2009. "Greenhouse Gas Reductions under Low Carbon Fuel Standards?" *American Economic Journal: Economic Policy*, 1(1): 106–46.

McGuinness, Meghan, and A. Denny Ellerman. 2008. "The Effects of Interactions between Federal and State Climate Policies." Massachusetts Institute of Technology Center for Energy and Environmental Policy Research Working Paper WP-2008-004.

Revesz, Richard L. 2001. "Federalism and Environmental Regulation: A Public-Choice Analysis." *Harvard Law Review*, 115(2): 553–641.

Revesz, Richard L., and Robert N. Stavins. 2007. "Environmental Law." In *Handbook of Law and Economics*, Vol. 1. No. 27, Handbooks in Economics, ed. A. Mitchell Polinsky and Steven Shavell, 499–589. New York: Elsevier.

[21]

Available online at www.sciencedirect.com

SCIENCE @ DIRECT®

ELSEVIER

Journal of Environmental Economics and Management 51 (2006) 135–152

JOURNAL OF ENVIRONMENTAL ECONOMICS AND MANAGEMENT

www.elsevier.com/locate/jeem

Land-use change and carbon sinks: Econometric estimation of the carbon sequestration supply function

Ruben N. Lubowski[a,*], Andrew J. Plantinga[b], Robert N. Stavins[c,d]

[a]Economic Research Service, US Department of Agriculture, 1800 M Street, NW, Washington, DC 20036, USA
[b]Department of Agricultural and Resource Economics, Oregon State University, Corvallis, OR 97331, USA
[c]John F. Kennedy School of Government, Harvard University, 79 John F. Kennedy Street, Cambridge, MA 02138, USA
[d]Resources for the Future, 1616 M Street NW, Washington, DC 20036, USA

Received 18 January 2005
Available online 5 December 2005

Abstract

If the United States chooses to implement a greenhouse gas reduction program, it would be necessary to decide whether to include carbon sequestration policies—such as those that promote forestation and discourage deforestation—as part of the domestic portfolio of compliance activities. We investigate the cost of forest-based carbon sequestration by analyzing econometrically micro-data on revealed landowner preferences, modeling six major private land uses in a comprehensive analysis of the contiguous United States. The econometric estimates are used to simulate landowner responses to sequestration policies. We treat key commodity prices as endogenous and predict carbon storage changes with a carbon sink model. Our estimated sequestration costs exceed those from previous engineering cost analyses and sectoral optimization models. Our estimated sequestration supply function is similar to the carbon abatement supply function from energy-based analyses, suggesting that forest-based carbon sequestration merits consideration in a cost-effective portfolio of domestic US climate change strategies.

Keywords: Abatement; Carbon; Climate change; Costs; Forestry; Greenhouse gases; Land use; Land-use change; Sequestration

1. Introduction

Increased concern about the threat of global climate change has brought with it greater attention to the possibility of encouraging the growth of forests as a means of removing carbon dioxide (CO_2) from the atmosphere. The Kyoto Protocol to the Framework Convention on Climate Change [41] states that the participating nations can employ carbon sequestration as part of their portfolios of strategies to achieve their domestic CO_2 targets. Hence, if the United States chooses to implement a domestic greenhouse gas reduction program, it will be necessary to decide whether to include carbon sequestration policies—such as those that promote forestation and discourage deforestation—as part of its portfolio of compliance activities.

*Corresponding author. Fax: +1 202 694 5775.
E-mail addresses: rlubowski@ers.udsa.gov (R.N. Lubowski), plantinga@oregonstate.edu (A.J. Plantinga), robert_stavins@harvard.edu (R.N. Stavins).

0095-0696/$ - see front matter
doi:10.1016/j.jeem.2005.08.001

The costs of carbon sequestration would presumably be one major criterion in that decision. Since the late 1980s, it has been suggested that sufficient lands are available to use the approach to mitigate significant amounts of CO_2 emissions [14,17], and claims have been made that such forestry-based carbon sequestration is a relatively inexpensive means of addressing climate change [8,36]. We investigate the cost of supplying domestic forest-based carbon sequestration using an econometric model of the revealed preferences of landowners who can use their land for alternative purposes.

1.1. Previous studies of the costs of US forest-based carbon sequestration

Three general approaches have been used to estimate the costs of sequestering carbon in the United States: bottom-up engineering cost studies; optimization models that allow for behavioral response in the forest and agricultural sectors; and econometric analyses of the revealed preferences of landowners.[1]

Most previous studies of the costs of sequestering carbon in the United States have employed bottom-up engineering cost analyses or optimization models. In "bottom-up" or "engineering cost" methods, marginal cost schedules are constructed by adopting information on revenues and costs of production of alternative land uses on representative types or locations of land, and sorting these in ascending order of cost. In the earliest and simplest of these [8,21,23], the analysts estimated available land area, forest carbon accumulation rates, and land and planting costs for hypothetical carbon sequestration programs, and thereby derived the total amount of carbon that could be captured and the cost per ton of sequestration.[2] In two other cases, the engineering approach was modified to include an anticipated increase in agricultural land prices as a hypothetical carbon sequestration program expanded, and crop and pastureland were removed from agricultural production. This was done by drawing upon previous (exogenous) estimates of the price elasticity of demand for agricultural land [32,33].

A carbon sequestration program may increase agricultural land prices, leading landowners to convert unregulated forest lands to agricultural land, thereby offsetting some of the effects of a carbon sequestration program. Alig et al. [4] addressed this issue through the use of a two-sector, multi-period simulation model, in which the forest and agricultural sectors were linked, and the welfare of producers and consumers in the two sectors was maximized.[3]

A relatively small number of studies have used a revealed-preference approach, in which actual land-use changes have been analyzed to estimate relationships between land-use choices and relative returns in the forest and agricultural sectors, thereby leading to the simulation of carbon-sequestration cost functions [24,31,39].[4] These studies examined the relationship between observable historic events (changes in timber and agricultural product prices) and landowner responses (conversion of land into and out of forest), and statistically estimated a response function. The models then posited a hypothetical economic stimulus—for example, government subsidies for carbon sequestration—and estimated how landowners would respond.

In theory, there are a number of reasons why landowners' actual behavior regarding the disposition of their lands might not be well predicted by "engineering" or "least-cost" analyses [39]: land-use changes can involve irreversible investments in the face of uncertainty [27,35], and landowners may want to retain options for future land-use decisions [29]; there may be non-pecuniary returns (for example, esthetics, and recreation) to landowners from forest uses of land [30], as well as from agricultural uses; liquidity constraints or simple "decision-making inertia" may mean that economic incentives affect landowners only with some delay; and

[1]For surveys of carbon sequestration cost analyses, see: Sedjo et al. [37]; Richards and Stokes [34]; Stavins and Richards [40]; and van Kooten et al. [47].

[2]In a variation on this theme, Parks and Hardie [28] substituted the estimated foregone net revenues from agricultural production for observed sale and rental prices of agricultural land.

[3]Alig et al. [4] use the Forest and Agricultural Sector Optimization Model (FASOM). Adams et al. [1] use a precursor to this model to estimate marginal costs of carbon sequestration in the US. Later applications of FASOM are in Adams et al. [2], McCarl and Schneider [18], and Murray et al. [22]. Lewandrowski et al. [15] use the US Agricultural Sector Model (USMP), to estimate carbon sequestration costs from afforesting crop and pasture lands and other agricultural-sector activities. Our literature review focuses solely on analyses in the United States.

[4]For econometric applications focused on sequestration from changes in agricultural practices, see Antle et al. [5] and Kurkalova et al. [13].

there may be private, market benefits or costs of alternative land uses of which an analyst is unaware. The econometric cost analyses have sought to address some of these problems.

1.2. Distinguishing features of this analysis

Compared with previous econometric analyses of the costs of carbon sequestration, this study is distinguished by three principal features. First, rather than considering only the movement of land between forestry and agriculture, six major land uses are modeled: forest, crop, pasture, range, urban, and a Federally designated private use, the Conservation Reserve Program (CRP).[5] Accounting for future urbanization is particularly important if it reduces the land base available for conversion to forest.

Second, detailed micro-data are employed that are comprehensive of the contiguous United States. Drawing upon repeated observations of land uses and land characteristics on 844,000 sample points from the National Resources Inventory (NRI), we observe three land-use transitions between 1982 and 1997 for lands encompassing 91% of non-Federal lands in the contiguous United States. Previous econometric analyses have considered relatively small regions of the country and have not incorporated micro-data on land-use changes or land quality, a critical determinant of land-use decisions.

Third, key commodity prices are treated as endogenous in the simulations of the carbon sequestration supply function. A national-scale carbon sequestration policy will likely affect prices for forest and agricultural commodities, in turn changing the incentives of private landowners to convert land to or retain land in forest.

1.3. Preview of the paper

We model probabilities of transitions among land uses as functions of the anticipated economic returns to alternative uses. A nested logit specification is used for the transition probabilities to relax restrictions implicit with simpler logit models. The estimated parameters and respective elasticities are found to be consistent with expectations of economic theory. We find that the probability of a land parcel transitioning to a particular use increases as the county-average net returns to that land use increase, after accounting for the quality of the parcel.

We build upon the econometric results by simulating landowner responses to carbon sequestration policies, modeled as a combination of a tax on undesirable land-use changes and a subsidy for desirable ones. The simulations are iterative in nature, including feedback effects on commodity prices resulting from induced land-use changes. A carbon sink model is then used to derive changes in carbon storage associated with the sequence of land-use transitions estimated in the land-use simulation model, including forest carbon stored in biomass, litter, and soils, emissions associated with harvesting of merchantable wood, and agricultural carbon linked with soils and harvested biomass. Merging the results of these simulations with marginal costs numbers associated with the various levels of the subsidy/tax leads to the estimated carbon sequestration supply function.

We compare our estimated supply function with those from previous studies, and find that over the range of costs considered in previous studies our marginal cost estimates are considerably greater than those from engineering cost analyses and sectoral optimization models. Likewise, at low levels of sequestration, our marginal cost estimates exceed those from previous econometric analyses; this relationship is reversed, however, at higher levels of sequestration.

We also compare our estimated marginal costs of carbon sequestration with estimates of costs of energy-based carbon abatement. We find that the carbon sequestration supply function is roughly similar to the central tendency of the carbon abatement supply functions, which suggests that forest-based carbon sequestration merits consideration as part of a cost-effective portfolio of domestic US climate change strategies. For example, if emission reductions in the United States on the scale proposed under the Kyoto

[5]The CRP, established by the Food Security Act of 1985, provides annual rental payments to landowners voluntarily retiring environmentally sensitive land from crop production under 10- to 15-year contracts and currently enrolls 34.7 million acres (an area about the size of Iowa) [42].

Protocol were to be achieved entirely through domestic actions (forest-based sequestration and/or energy-based abatement activities) and with the type of policy incentive considered in this paper, our analysis implies that 33–44% of the reductions could be met cost-effectively through forest-based sequestration.

2. Econometric analysis of national land use

2.1. Econometric model

Consider a risk-neutral landowner facing the choice of allocating a parcel of land of uniform quality among a set of alternative uses.[6] We posit that landowners choose uses to maximize the present discounted value of the stream of expected net benefits from the land, and that landowners base their expectations of future land-use profits on current and historic values of relevant variables. Given these simplifying assumptions, the decision rule that emerges from the related dynamic optimization problem [16] is to choose the use with the highest expected one-period return at time t minus the current one-period expected opportunity cost of undertaking conversion. With K potential land uses ($j, k = 1, \ldots, K$), the owner of a unit of land in use j will choose the use k at time t that satisfies:

$$\arg\max_{k}(R_{kt} - rC_{jkt}) \geqslant R_{jt}, \tag{1}$$

where R_{jt} and R_{kt} represent the instantaneous expected net benefits at time t from a unit of land in use j and k, respectively; C_{jkt} is the expected marginal cost of converting one unit of land from use j to use k at time t ($C_{jjt} = 0$); and r is the discount rate. If the use k satisfying Eq. (1) equals j, then the land unit will remain in its current use at time t; otherwise, the landowner will reallocate the land to the use $k \neq j$ that maximizes expected net returns after conversion costs.

The landowner's profit function may be thought of as including both observed and unobserved components. Specific restrictions on the structure of the unobserved components yield alternative specifications of probabilistic models. An ordinary logit model is one obvious possibility, but the assumption of independent disturbances in the simple logit model implies that the ratio of the probabilities of any two choices is independent of the other alternatives.

We allow for differences in substitutability among alternatives using a *nested* logit specification, which imposes the property of "independence of irrelevant alternatives" within but not across specified subgroups ("nests") of choices. Dividing the choice set into mutually exclusive subgroups K_s ($s = 1, \ldots, S$), each containing elements l ($l = 1, \ldots, J_s$), we specify U_{ijkt}, the landowner's utility from allocating land parcel i in use j to use k at time t, as the sum of a component, V_{ijkt}, that is unique to the alternative k and another component, V_{ijst}, that is common to all the alternatives in K_s including k. Each of these components, in turn, includes an observed component plus an unobserved component characterized as a random error.

Under assumptions analogous to the standard logit model, the probability of choosing alternative k that is grouped in K_s can be expressed as the product of two terms: the probability, P_{ijst}, of choosing any of the alternatives within K_s; and the conditional probability, $P_{ijkt|s}$, of choosing k given the choice of K_s [6]. For land parcel i starting in use j, the probability of choosing land use $k \in K_s$ between time t and $t+1$ is thus:

$$P_{ijkt} = P_{ijst} \cdot P_{ijkt|s} = \frac{\exp(V_{ijst} + \tau_{st} I_{ijst})}{\sum_{s=1}^{S} \exp(V_{ijst} + \tau_{st} I_{ijst})} \frac{\exp(V_{ijkt})}{\sum_{l=1}^{J_s} \exp(V_{ijlt})}, \tag{2}$$

where τ_{st} are parameters, and $I_s = \ln \sum_{l=1}^{J_s} \exp(V_{ijlt})$. This "inclusive value" for nest K_s equals the expected utility for the choice of alternatives within a nest. The expression in Eq. (2) embodies the first-order Markov property since the probability of the parcel changing use depends only on exogenous covariates at time t.

Our chosen nesting structure is based on the premise that land uses with more similar land quality requirements are closer substitutes. We expect land quality to affect land-use net returns principally in terms of agricultural yields, and we measure land quality using the Land Capability Class (LCC), a summary measure

[6]If net returns and land conversion costs are linear in land quantity, the size of parcels will not affect the relative profitability of land-use options, so land-use choices for a heterogeneous parcel can be treated as the sum of land-use choices on constituent uniform-quality parcels.

of the suitability of land for producing crops [43]. Land in crops has the highest average quality as measured by the LCC, while pasture and CRP uses tend to be adopted on higher quality lands relative to forest and range uses. Crops, pasture, and CRP uses appear more similar in terms of the land quality required to generate a competitive level of returns, compared with forests and range. We incorporate these differences in land quality requirements by specifying our nested logit model with three nests: K_1 (crops, CRP, and pasture); K_2 (forest and range); and K_3 (urban). We model urban land use as a unique nest, due to its greater degree of irreversibility, and because land quality, as measured by the LCC, is likely to be a much less important determinant of urban development returns.

Landowners presumably compare net returns to alternative uses on particular parcels. Although we have land-use data at the parcel level, we lack parcel-level observations of net returns. Instead, we observe county-level average returns, and so to allow for parcel-level variation, we interact the average-return variables for each land use with parcel-level indicators of land quality. The magnitude of this land quality adjustment could be different for different land uses depending, for example, on the sensitivity of the different net returns to land quality as well as on the land quality distribution for the land in each use. We thus specify the component of utility that is unique to each alternative k as

$$V_{ijkt} = \alpha^0_{jkt} + \alpha^q_{jkt} LCC^q_{it} + \beta^0_{jkt} R_{kc} + \beta^q_{jkt} LCC^q_{it} R_{kc} + \varepsilon_{ijkt}, \tag{3}$$

where α^0_{jkt} is an alternative-specific intercept, α^q_{jkt}, β^0_{jkt}, and β^q_{jkt} are parameters,[7] R_{kc} is the level of net returns to use k in county c,[8] LCC^q_{it} is a dummy variable indicating whether parcel i is of quality q at time t, and ε_{ijkt} is the error term.[9] We lack data on the costs of changing land use, but we expect these costs to be closely related to land quality. Accordingly, we model conversion costs with the terms $\alpha^0_{jkt} + \alpha^q_{jkt} LCC^q_{it}$ in (3), which provide an intercept varying with land quality and initial use.

CRP participation depends on a different set of decisions than other land-use choices, because enrollment depends on both the landowner's bid, which includes a proposed rental rate, and the government's choice of whether to accept the bid, which depends on the environmental characteristics of a parcel as well as the cost. Because the program targets cropland, CRP rental rates are highly correlated with the profitability of cropping in a given locality. We account for the effect of crop net returns on the incentive to remain in cropland. Incentives to enroll in CRP are specified as a function of LCC, as lower land quality as measured by LCC has always been strongly associated with program eligibility. We would thus expect greater enrollment on lower quality lands. Eq. (3) thus becomes $V_{ijkt} = \alpha^0_{jkt} + \alpha^q_{jkt} LCC^q_{it} + \varepsilon_{ijkt}$ for $k =$ CRP.

For the component of utility that is constant across the alternatives within each nest, we include constant terms for the nest and interactions with the land quality indicator variables. For land parcel i in use j, the component of utility that is constant within each nest is thus:

$$V_{ijst} = \gamma^0_{jst} + \gamma^q_{jst} LCC^q_{it} + \tau_{st} I_{ijst}, \tag{4}$$

where γ^0_{jst} is a constant specific to nest s and γ^q_{jst} is a coefficient on the land quality indicators. Substituting Eqs. (3) and (4) into (2) yields a complete nested logit model for estimation.

---

[7]If the net return variables perfectly captured the net returns for each land use at the level of the individual parcel, then the coefficients on net returns should simply reflect the marginal utility of income and would not be expected to vary across the alternatives. We do not expect the coefficients to be equal in our specification, as the net return variables are county averages and the LCC interaction terms scale these averages based on parcel-level land quality differences. The magnitude of this adjustment is expected to vary across uses based on the sensitivity of the net returns to land quality and the quality distribution of the land initially in each use.

[8]Due to the logit structure of the model, this specification based on net return levels is identical to one where the net return variables are all rescaled as differences from the net returns to one base land-use category. Any component of utility that is constant across alternatives will drop out of the probabilities in (2), so only relative differences will determine the land-use decisions.

[9]The eight LCCs were merged into four groups (I–II, III–IV, V–VI, VII–VIII) to ensure enough observations in each land use and quality to estimate the nested logit model. An analysis based on a non-nested (conditional logit) version of the model indicated that our land-use simulations (see the next section) were not sensitive to this aggregation of the land quality classes.

2.2. Econometric estimation and results

We estimate the model using repeated observations of non-Federal land use from the NRI of the US Department of Agriculture's Natural Resources Conservation Service (NRCS). The NRI is a panel survey of land use and land characteristics on non-Federal lands conducted at 5-year intervals from 1982 to 1997 over the 48 contiguous United States. Data include approximately 844,000 point-level observations, each representing a land area given by a sampling weight [26]. We observe land use at each sample point in four survey years, providing information on land-use changes over three transitions: 1982–1987, 1987–1992, and 1992–1997. We focus on six land uses: crops, pasture, forest, urban, range, and CRP.[10] The land base in our analysis comprises 1.4 billion acres, representing about 74% of the total land area and about 91% of non-Federal land in the contiguous United States. Further details on the NRI data are provided in Appendix A.[11]

The NRI data reveal the disposition of land units across major uses over time. Land units generally remain in the same use. For example, of land parcels cropped in 1982, 84.2% remained in crops in 1997, while 7.3% had been converted to CRP, 4.6% to pasture, 1.6% to urban use, 1.3% to forest, and 0.9% to range. Of land parcels forested in 1982, 95.4% remained in forests in 1997, while 2.5% had been converted to urban, 1% to pasture, 0.5% to crops, another 0.5% to range, and 0% to CRP. The urban land-use category appears to be an absorbing state, with lands almost never converted from urban to non-urban uses.

Our dependent variable is the choice of land use in year, $t+5$, at each NRI point, and our covariates are the land use in year t, the land quality rating of the point, and proxies for the expected net returns from the land-use alternatives as of year t. To smooth temporary shocks from weather and other factors affecting net returns in particular years, we assume that landowners use an average of the annual net returns per acre to each land use over the most recent 5-year period as the basis for their expectations of future net returns. Denoting each year as t, we specify land-use choices observed at time $t+5$ as a function of the average land-use net returns between years $t-4$ and t, inclusive.

By assembling data from a variety of private and public sources, we constructed county-level estimates of annual net returns (per acre) for crops, pasture, forest, range, and urban uses for all 3014 counties in the 48 contiguous states (Appendix A). Net return estimates are thus constructed for each of our land-use categories, except for CRP, which is treated differently, as discussed above. The estimates for cropland include net returns from market sales as well as direct farm program payments. We report summary statistics on land-use net returns in Table 1.

We employ maximum likelihood methods to estimate the parameters of the model for the transition probabilities from each of four starting land uses (crops, pasture, forest, and range) to each of our six ending uses (crops, pasture, forest, range, urban, and CRP).[12] We estimate versions of the econometric model using data from two different time periods and then compare the results. First, we estimate the model using all observations from the 1992–1997 period, the latest transition for which NRI data are available. These estimates draw solely on cross-sectional variation in the independent variables. If there are variations over time in landowners' responses to economic incentives, possibly the result of permanent changes in factors not modeled by our analysis, then parameter estimates based on the most recent data will provide the best foundation for predicting future behavior.

Second, we estimate the model using a sample of pooled data from each of the three available transitions (1982–1987, 1987–1992, and 1992–1997). For each starting use, we stratify the pooled dataset by NRI point and select a random sample that includes only a single observation of each land unit. In this way, we include observations from each time period, but purge the sample of serial correlation due to parcel-specific unobserved effects, which would violate the assumption of independent disturbances in the logit model and potentially bias the estimates. The estimated coefficients reflect the average behavior of parcels in each land

[10] Public lands and transportation infrastructure are excluded from the analysis, as changes in these uses are not affected directly by private landowner decisions. We omit water bodies and barren lands as these uses are unlikely to vary over time. Finally, we exclude lands classified as marshlands and "miscellaneous," as data to measure net returns to these uses are not available.

[11] All the appendices are available as an online supplement to this article at http://www.aere.org/journal/index.html

[12] Because land virtually never transitions out of urban uses, we do not model these potential land-use changes and, for our simulations, assume land parcels remain urbanized with 100% probability. We estimated the model for lands starting in CRP over 1992–1997 only. Some parcels became eligible to exit the program in 1996, and changed to other uses by 1997.

Table 1
Summary statistics

Variables	Level	Mean and standard deviation by year[a]			
		1982	1987	1992	1997
Land use is crops (1 = yes, 0 = no)	NRI point	0.296 (0.456)	0.286 (0.452)	0.269 (0.443)	0.266 (0.441)
Land use is pasture (1 = yes, 0 = no)	NRI point	0.092 (0.289)	0.089 (0.285)	0.089 (0.284)	0.084 (0.278)
Land use is forest (1 = yes, 0 = no)	NRI point	0.282 (0.450)	0.284 (0.451)	0.284 (0.451)	0.286 (0.451)
Land in urban use (1 = yes, 0 = no)	NRI point	0.036 (0.187)	0.041 (0.197)	0.046 (0.209)	0.054 (0.225)
Land use is range (1 = yes, 0 = no)	NRI point	0.293 (0.455)	0.289 (0.453)	0.287 (0.452)	0.286 (0.452)
Land use is CRP (1 = yes, 0 = no)	NRI point	0 (0)	0.010 (0.098)	0.024 (0.153)	0.023 (0.024)
Land Capability Class 1–2 (1 = yes, 0 = no)	NRI point	0.232 (0.422)	0.232 (0.422)	0.231 (0.422)	0.231 (0.421)
Land Capability Class 3–4 (1 = yes, 0 = no)	NRI point	0.347 (0.476)	0.347 (0.476)	0.347 (0.476)	0.347 (0.476)
Land Capability Class 5–6 (1 = yes, 0 = no)	NRI point	0.215 (0.410)	0.214 (0.410)	0.215 (0.410)	0.215 (0.410)
Land Capability Class 7–8 (1 = yes, 0 = no)	NRI point	0.206 (0.404)	0.207 (0.404)	0.207 (0.405)	0.207 (0.405)
Crop net returns ($/acre/year)[b]	County	58.4 (38.4)	73.8 (45.0)	82.8 (48.2)	n/a
Pasture net returns ($/acre/year)	County	16.1 (11.3)	7.9 (8.0)	12.7 (9.0)	n/a
Forest net returns ($/acre/year)	County	6.0 (5.9)	9.0 (9.0)	17.2 (17.1)	n/a
Urban net returns ($/acre/year)	County	3302 (3120)	4651 (5591)	4475 (4418)	n/a
Range net returns ($/acre/year)	County	11.2 (10.1)	10.3 (8.9)	10.4 (9.3)	n/a

[a]Point-level variables are indicator variables weighted by NRI-point acreage weights. Net returns are lagged 5-year averages in 1990 dollars (deflated by the producer price index for all commodities) weighted by the county acreage in each land use given by the NRI.
[b]Equals the sum of the market-component of crop net returns plus direct government payments (excluding the Conservation and Wetlands Reserve Programs).

Table 2
Own-return land-use choice elasticities[a]

Initial land use	Final land use and time period[b]									
	Crops		Forest		Pasture		Range		Urban	
	1982–97	1992–97	1982–97	1992–97	1982–97	1992–97	1982–97	1992–97	1982–97	1992–97
Crops	0.192**	0.011**	0.332**	0.310**	0.090**	0.183**	0.477**	0.376**	0.156**	0.342**
	(0.005)	(0.001)	(0.024)	(0.043)	(0.017)	(0.031)	(0.036)	(0.048)	(0.005)	(0.016)
Forest	0.178**	0.295**	−0.000	0.001	0.091	−0.000	0.235**	0.232	0.511**	0.792**
	(0.039)	(0.064)	(0.000)	(0.055)	(0.179)	(0.000)	(0.033)	(0.330)	(0.010)	(0.058)
Pasture	0.306**	0.341**	0.023*	0.005	−0.005	−0.012	1.373**	1.042**	0.314**	0.331**
	(0.012)	(0.022)	(0.011)	(0.027)	(0.003)	(0.008)	(0.033)	(0.050)	(0.014)	(0.026)
Range	0.072	0.065	0.064**	0.127	0.159	0.399	−0.002	−0.001	0.385**	0.419**
	(0.069)	(0.229)	(0.023)	(0.906)	(0.700)	(0.417)	(0.001)	(0.971)	(0.036)	(0.031)

[a]Elasticities are evaluated at the means of the data and are the percentage change in the probability of choosing the final land use, conditional on being in the initial use, for a 1% change in the net returns to the final use. Standard errors, in parentheses, are estimated using the Delta Method [11]. * and ** denote significance at the 5% and 1% levels, respectively.

[b]There are no own-return elasticities for CRP as net returns from CRP are not directly specified in the econometric model.

use from 1982 to 1997, weighted by the acreage in each period. If changes in estimated parameters over time are due to transient shocks, then estimates based on a longer time period are potentially superior for predictive purposes.

The results for both the 1992–1997 and 1982–1997 periods indicate good fit of the model, and are consistent with the expected economic relationships.[13] Estimated parameters for the eight equations are reported in Appendix D. For each starting use, Table 2 reports the estimated elasticity of the probability of choosing each land-use alternative with respect to the net returns to that alternative ("own return elasticities"). These elasticities indicate the percentage change in the probability of a particular land-use change for a 1% change in the corresponding net returns. In 11 out of 20 cases for 1992–1997 and 14 out of 20 cases for 1982–1997, the own-return elasticities are positive and significantly different from zero at the 0.01 level. In the three cases for each set of estimates where the own-return elasticities are negative, they are never significantly different from zero, even at the 0.05 level. The cross-elasticities (not reported) are always opposite in sign to the own-return elasticities.[14]

The elasticities indicate that landowners starting with lands in either crops or pasture responded as anticipated to net returns from alternative land uses. For land starting in range, the own-return elasticities with respect to urban net returns are positive and significant, as well as the forest net return elasticity for 1982–1997. None of the other own-return elasticities (for land starting in range use) are significantly different from zero, suggesting that rangeland owners are relatively insensitive to the profitability of alternative uses, with the exception of urban development. This is reasonable, given that range lands tend to be of the lowest quality and thus unsuitable for other agricultural uses.

[13]For the transition-specific estimates, pseudo R^2 values (McFadden's likelihood ratio index) range from 0.68 to 0.95 and 0.71 to 0.95 for the 1992–1997 and 1982–1997 estimates, respectively. The positive and generally significant estimates of the inclusive value parameters together with tests [12] of a non-nested logit model using the same variables are consistent with violations of the "independence of irrelevant alternatives" hypothesis, supporting the use of the less restrictive nested specification. For land initially in pasture, one inclusive value parameter in both the 1982–97 and 1992–97 estimates lies outside the unit interval with 95% confidence. This suggests these estimates may not be globally consistent with random utility maximization [19]. Possible explanations are the crude data on pasture returns, as well as difficulties in distinguishing permanent pasture versus pasture in rotation with crops. Any misspecification in the pastureland equation, however, is expected to have small impacts on the overall results given that pasture accounted for a relatively small share of the total land base in 1997 (see Fig. 1).

[14]The cross-elasticities are the elasticities of the probability of choosing a particular use j with respect to the profits of a different use k. In the nested logit model, these can be of the same sign as the own-return elasticities when the inclusive value parameters are negative.

The elasticities with respect to forest net returns are especially important for our simulation model of carbon sequestration. These elasticities are positive, as expected, in seven out of eight cases. While the elasticity is negative for forestlands over 1982–1997, it is not significantly different from zero. The elasticity with respect to forest net returns is positive and significantly different from zero for lands starting in crops for both models. For lands starting in pasture and range, the forest net return elasticities are positive for both periods and significantly different from zero over 1982–1997.

A final econometric consideration is the fact that land parcels located near one another may have unobserved characteristics that are correlated across space. If such characteristics influence land-use decisions or if local land-use choices are interdependent, error terms will be correlated across space, leading to inconsistent and inefficient estimates in a logit model, due to induced heteroskedasticity [20]. We explored the potential importance of spatial dependence by eliminating observations near one another. Estimates with samples that included only one point within each of the NRI's primary sampling units produced results similar to estimates including all points, suggesting that spatial dependence is not a critical concern.

3. Simulation model of carbon sequestration

In order to estimate the carbon sequestration supply function, we conduct policy simulations and compute corresponding flows of carbon in terrestrial sinks. We examine a two-part policy involving a subsidy for the conversion of land to forest and a tax on the conversion of land out of forest. As noted by Stavins [39], a policy that only subsidizes forestation creates incentives for landowners to convert land out of forest and then back into forest in order to receive the subsidy. This entails inefficient expenditures on land conversion, which are discouraged by the tax on deforestation in the two-part policy. A second feature of the policy is a requirement that afforested lands remain in forest for a specified period of time.[15]

For each policy scenario, we specify the level of the subsidy and tax, denoted Z. For land moving into forest, the subsidy increases the annual net return to forest, and, for land moving out of forest, it reduces the annual net return to non-forest uses. Suppressing the parcel and time subscripts, indexing land uses by j and k, forest use by 1, and the relevant set of non-forest uses by -1, Z modifies the estimated transition probabilities P_{jk} in the following manner:

$$P_{j1} = f(\hat{\boldsymbol{\beta}}_{j1}, NR_1 + Z, \mathbf{NR}_{-1}), \tag{5a}$$

$$P_{11} = f(\hat{\boldsymbol{\beta}}_{11}, NR_1, \mathbf{NR}_{-1} - \mathbf{Z}), \tag{5b}$$

$$P_{1j} = f(\hat{\boldsymbol{\beta}}_{1j}, NR_1, \mathbf{NR}_{-1} - \mathbf{Z}), \tag{5c}$$

$$P_{jk} = f(\hat{\boldsymbol{\beta}}_{jk}, NR_1 + Z, \mathbf{NR}_{-1}) \tag{5d}$$

for all j and k not equal to 1, where $\hat{\boldsymbol{\beta}}$ is a vector of estimated parameters, NR_1 is the annual net return to forest land, $\mathbf{NR}_{-1}$ is a vector of annual net returns to the relevant non-forest uses, and $\mathbf{Z}$ is a conformable vector with each element equal to Z. In general, Z has the effect of increasing the probability that land shifts into forest and reducing the probability that it shifts out of forest.

Four issues merit comment. First, the policy mechanism implicit in our cost simulation model provides incentives for land-use changes, rather than for carbon sequestration directly. If the costs of administering these two types of policies are the same, carbon sequestration costs will be lower with a policy directly targeting carbon, because it will tend to convert land with the lowest costs per ton of carbon, rather than the lowest cost per acre of land. It is likely, however, that the administrative costs of the two approaches would differ significantly. In general, performance measures based on inputs to carbon sequestration—such as quantities of particular types of land—would be relatively easy to monitor, compared with the task of monitoring the quantity of carbon sequestered, relative to an assumed reference case [40].

[15] This is a common contractual arrangement under voluntary land conservation programs. For example, CRP contracts stipulate that land parcels must be kept in a specified conservation use for a period of 10–15 years.

Second, the subsidy and tax rates are identical to each other, and remain constant over time. If rates of carbon sequestration differ between land that is afforested and land on which deforestation is deterred, then equal subsidy and tax rates may not be efficient. We assume such equal rates, because it simplifies the policy simulations considerably. Moreover, analyses of efficient and cost-effective climate change mitigation have found that carbon prices should rise over time to delay the costs of stabilizing or reducing greenhouse gas concentrations [10,25]. If incentives for carbon sequestration were to increase over time, landowners with forward-looking price expectations would, in some instances, have incentives to delay carbon sequestration, even if it is profitable to undertake immediately [48]. To be consistent with the econometric model, we assume landowners have static price expectations, which implies that significant incentives do not exist to delay profitable carbon sequestration; and to simplify the simulations, we treat the subsidy/tax rate as constant over time.

Third, the probabilities in Eq. (5) are expressed as functions of R_1, the annualized net returns from timber harvests. Thus, we assume initially that all forested land is eligible for timber harvesting, including those lands converted to forest in response to the subsidy. We also consider policies designed to establish permanent forest stands. In this case, timber harvesting is prohibited on afforested lands, but still allowed on land that is forested at the beginning of the simulation. To model this policy, we set R_1 equal to zero for all transitions of land into forest (5a) and all transitions between non-forest uses (5d). In this case, landowners receive only the subsidy Z for afforesting their land. For land that begins in forest, and hence is ineligible for the forestation subsidy, R_1 is unmodified (5b and 5c).

Fourth and finally, the modified transition probabilities in (5) are used in the policy simulation model to generate estimates of acreage transitions between pairs of land-use categories for each level of Z. Land afforested in response to the subsidy is assumed to remain in forest throughout the policy simulation. Thus, for these forest lands, but not for land originally in forest, we set the probability in (5c) to zero. We also compute a baseline sequence of land-use transitions by setting Z in (5) equal to zero. Naturally, in computing this baseline, we do not require afforested land to remain in forest. The impact of the policy on carbon sequestration is measured relative to the baseline (that is, we net out the change in carbon storage that would have occurred in the absence of the policy). As a practical matter, baseline actions are not observable once the policy is in effect, implying that some landowners are subsidized for forestation they would have undertaken even without subsidies. In the policy simulations, all landowners are subsidized for forestation, and hence are required to retain their land in forest. But to measure the cost of the policy, we consider foregone net returns net of the baseline. In this way, our estimates measure the economic costs to society, rather than the financial costs to the government.

A policy simulation model, described in Appendix B, is used to simulate the response to the incentives for carbon sequestration and to estimate feedback effects on timber and crop prices resulting from induced land-use changes. Aggregating simulated land-use changes at the level of each NRI point, the simulation model generates national estimates of acreage transitions among land-use categories over long time horizons. A carbon sink model, described in Appendix C, is used to translate the estimated land-use transitions into projected carbon flows. This model accounts for changes in carbon stocks in the relevant biomass, soil, and product categories for each of the land uses. It is important to acknowledge that, while we use the best available information on carbon storage, considerable uncertainty surrounds these estimates. In addition, we have ignored some of the practical issues that would arise with a carbon sequestration policy, including monitoring and enforcement.[16]

The carbon sequestration supply function is derived using the procedure discussed by Stavins [39]. For the baseline and each level of the subsidy/tax, we record the areas of land afforested and deforested during each period of the simulation. Net forestation for a given subsidy/tax level is defined as the area afforested minus the area deforested, net of this quantity in the baseline. The product of net forestation and the subsidy/tax gives the opportunity cost of the policy. For example, if a payment of $100 is required to induce an extra acre of forestation, then this represents the foregone profits associated with the original use of the land. The total cost of the policy is computed as the present value of cumulative opportunity costs discounted from the

[16]For more discussion, see the special issue of Choices devoted to climate change [38].

midpoint of each 5-year interval. Likewise, we compute cumulative carbon flows relative to the baseline and the present value of carbon sequestered.[17]

Total costs and total carbon are computed for each subsidy/tax rate and expressed as annualized equivalents. Marginal costs of carbon sequestration equal the change in costs divided by the change in carbon, computed as the ratio of the increment in annualized total costs (from one subsidy/tax level to the next) to the corresponding increment in annualized total carbon. The supply schedule is constructed by arraying marginal costs against annualized total carbon.

4. Empirical results

Simulations are conducted for subsidy/tax rates (values of Z) ranging in \$50 increments from \$0 to \$350 per acre. The two versions of the econometric model—one estimated for the period, 1982–1997, and the other for the period, 1992–1997—produce very similar carbon sequestration supply functions for carbon prices below \$200 per ton. Our discussion below focuses on results generated with the 1992–1997 model.

4.1. Primary results

We simulate national total quantities of land in six categories of use over time for the baseline and policy scenarios. In the initial year (1997), there are approximately equal areas of cropland, rangeland, and forest (about 400 million acres each), while the areas of land in pasture, urban, and CRP vary from 33 to 119 million acres (Fig. 1). Under the baseline scenario, the largest change is in the area of urban land, which increases from 76 to 495 million acres during the 250-year simulation period (Fig. 1). The areas of land in agricultural uses (crops, range, pasture, and CRP) decline throughout the simulation period, with cropland experiencing the greatest absolute loss. Forest land increases initially but begins to decline after about 50 years. After 250 years, there is a relatively small net decline in forest area in the baseline.

When a \$100 per acre subsidy/tax is introduced, forest area almost doubles during the simulation period, from 405 to 754 million acres (Fig. 2). Most of the increase is due to increased forestation and, correspondingly, there are larger declines by the end of the simulation in the areas of cropland, rangeland, pasture, and CRP. The increase in urban land area is also smaller than in the baseline, as more land is converted to or retained in forest. The area of cropland increases for several decades before declining. By encouraging conversion of cropland to forests, the subsidy/tax has positive effects on net returns to cropland, raising the probability that land moves into crops. Initially, there is a net increase in cropland as more land moves into crops from other agricultural uses (pasture, range, and the CRP) than is converted from crops to forest and urban. But as the remaining acres of other agricultural lands decline, flows of land into crops fall, and there is a net decline in cropland.

While the general pattern of land-use changes is similar at other levels of the subsidy/tax, in a given year the increment in forest area declines as the rate of the subsidy/tax increases.[18] In addition, at the highest subsidy/tax rates, forest area increases rapidly in early years of the simulation before quickly leveling off. For example, at subsidy/tax rates above \$250 per acre, the area of forest more than doubles by 25 years into the simulation, but after 25 years, relatively little forest is added, because a large share of the land base has already been afforested.

There is an increase in carbon storage when agricultural lands are converted to forests or land is retained in forest.[19] Fig. 3 portrays carbon flows relative to the baseline for selected subsidy/tax rates. There is a small negative carbon flow in the first period of the simulation for all levels of the subsidy/tax, because cropland area increases initially under the subsidy/tax scenarios due to higher net returns induced by the policy, whereas cropland area declines initially in the baseline. The net increase in cropland results from the conversion of

[17] See Richards and Stokes [34] for an assessment of alternative approaches to intertemporal carbon accounting in carbon sequestration cost studies.

[18] This is true except at low levels of the subsidy/tax, where the increments increase with the subsidy/tax. As discussed below, this results in our supply function having a convex portion.

[19] Appendix C provides details on estimated carbon storage levels under different land uses.

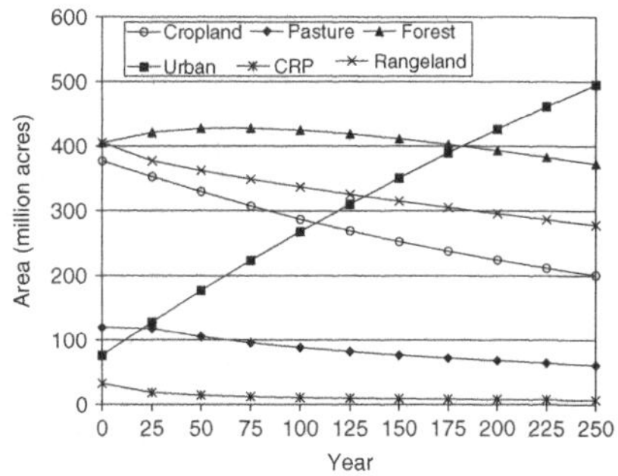

Fig. 1. The area of land by use in the baseline scenario.

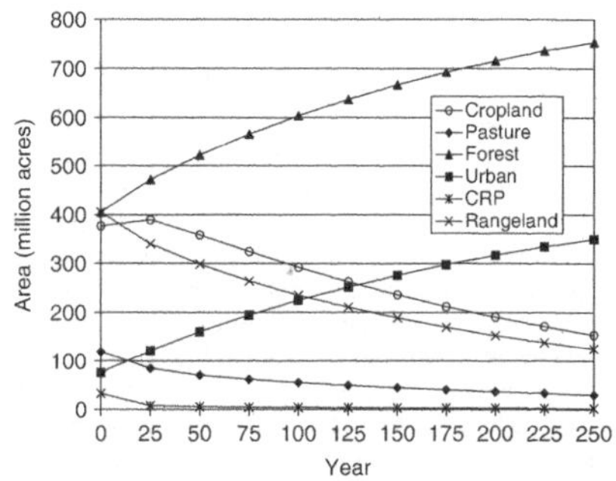

Fig. 2. The area of land by use with a $100 per-acre subsidy/tax.

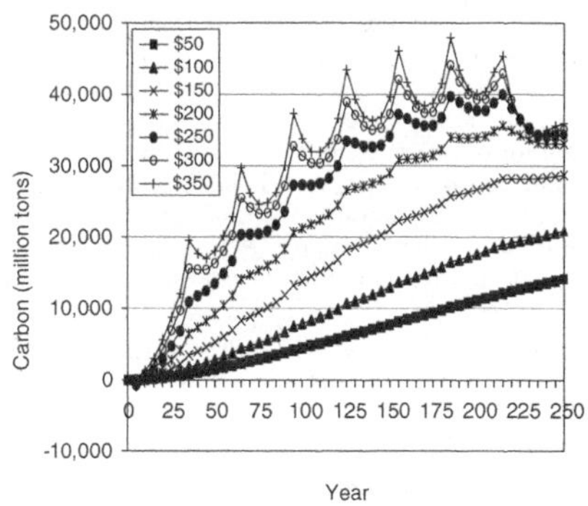

Fig. 3. The flow of carbon relative to the baseline with different subsidy/tax rates.

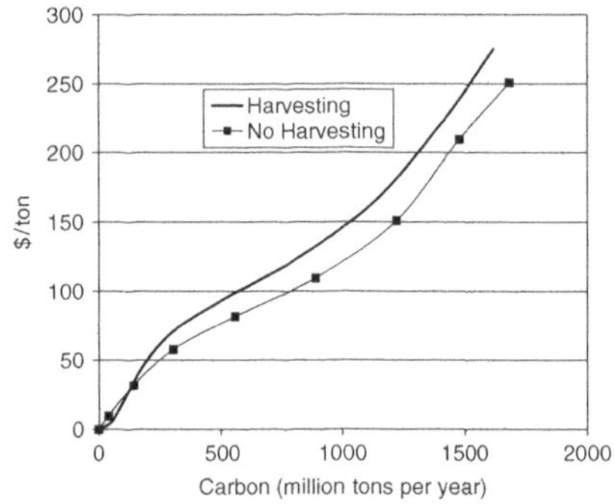

Fig. 4. The marginal costs of forest-based carbon sequestration in the United States.

pasture and rangeland, producing an immediate and negative flow of carbon (see Appendix C). Once new forests become established, transitions between non-forest lands contribute relatively little to the aggregate carbon supply. The $100 per acre subsidy/tax captures an additional 1.2 billion tons of carbon on non-forest land over the 250-year horizon by, for example, deterring conversion of pasture to cropland. However, transitions involving non-forest lands account for less than 1% of the net carbon sequestration induced by the policy.

Moving from lower to higher subsidy/tax rates, carbon flows increase, though eventually at a diminishing rate, due to the declining induced increment in forest area. At the highest subsidy rates, a scalloped pattern emerges, reflecting the periodicity of harvest and regrowth cycles. On afforested lands, there is a positive flow while trees grow to maturity. The flow is negative at the time of harvest and, then, the cycle repeats itself. The pattern is most pronounced at the highest subsidy/tax rates, because much of the land that will be afforested during the simulation is converted during the first years of the simulation. At lower subsidy/tax levels, the increase in forest area is more gradual, creating a more even pattern of harvest and regrowth.[20]

The land-use change and carbon flow results are combined to produce a marginal cost function for forest-based carbon sequestration. Econometric estimation and respective simulations were carried out both with the full set of three transitions over the period 1982–1997 and with the final transition of 1992–1997. As highlighted above, the share of land in urban uses has increased monotonically and dramatically, suggesting significant structural change. Because the simulations are forward-looking and span very long time horizons, we employ the more recent 1992–1997 econometric estimates and related land-use simulations in our calculations of the carbon sequestration marginal cost function.[21] This is portrayed in Fig. 4, where the solid line indicates the supply function for our basic scenario (allowing harvesting of wood products).

Although the marginal cost function is convex throughout most of its range, it is concave at low levels (below $70/ton). This is a consequence of the logistic specification used in the underlying econometric model. Recall that the vast majority of plots in the NRI sample remains in the same use over time; in other words, there is a low probability that land transitions out of its current use. At low marginal costs (low subsidy/tax),

[20]For most of the simulation, the flows trend upward, due largely to the addition of new forests over time, but after about 200 years, the flows drop to lower levels for the highest subsidy/tax rates. There are two reasons for this. First, throughout the simulation a large amount of carbon is stored in solid wood products and gradually released. Once forest area stabilizes, these negative carbon flows are no longer offset by the addition of new forest lands, which generate positive carbon flows. Second, carbon in forests that are never harvested reaches a maximum level after 12 decades (see Appendix C). The zero flows in unharvested forests result in a diminished aggregate carbon flow. At higher subsidy/tax rates, the negative flows from the product pools become apparent sooner, and zero flows in unharvested forests occur earlier.

[21]It should be noted, however, that the results from both sets of econometric estimations and related simulations are similar.

the logistic is a convex function of its arguments. As forest net returns increase in the policy scenarios, there is a range of subsidy/tax rates for which the land moving into forest increases at an increasing rate, implying that marginal costs of carbon sequestration increase at a decreasing rate at low levels of the subsidy/tax.[22]

4.2. *Sensitivity analysis*

In deriving the main results, we assume all forest lands (including lands converted to or retained in forest in response to policy) are periodically harvested. Another realistic policy scenario involves a contractual stipulation that prohibits timber harvests on lands enrolled in a carbon sequestration program. A prohibition on harvesting may increase the discounted value of carbon sequestered, because negative carbon flows resulting from harvesting are avoided. But harvesting restrictions also reduce or even eliminate revenues received by landowners, increasing the financial incentives needed to increase forest area. Thus, a priori, the net effect on the marginal costs of carbon sequestration of allowing timber harvesting is ambiguous.

Therefore, we also conduct simulations for a no-harvesting scenario. We find that prohibiting harvesting lowers the marginal costs of sequestration (Fig. 4). The key reason is that at each level of the subsidy/tax, total forest area is greater under the no-harvesting scenario. Although harvesting restrictions reduce the amount of land enrolled in the carbon sequestration program, they also result in higher timber prices relative to the corresponding harvesting case, which helps to retain existing (that is, non-program) lands in forest. In addition, at a 5% discount rate, the per-acre discounted carbon flows are greater under the no-harvesting scenario. In sum, more forest land outside the program and more carbon sequestration on program lands combine to produce lower marginal costs when harvesting is prohibited.

We also examined the sensitivity of the carbon sequestration supply function to alternative discount rates. In the econometric and simulation analysis reported above, the private net returns to forest and urban uses are annualized using a 5% rate (the returns to other land uses are measured in annual terms). We compare three alternatives (1%, 3%, and 7%) to the benchmark rate of 5%. Higher discount rates lower the present value of both costs and carbon flows. Hence, in theory, changes in the discount rate have an ambiguous effect on unit costs of the policy (the ratio of annualized dollars to annualized tons of carbon). Empirically, we find that higher discount rates increase unit costs, and decrease annualized carbon flows. Thus, the marginal cost curve for carbon sequestration shifts up as we move from lower to higher discount rates. For sequestration of 500 million tons of carbon per year, the marginal (average) cost per ton rises from $8 ($4) to $41 ($20), $93 ($53), and $164 ($102) at discount rates of 1%, 3%, 5%, and 7%, respectively.

4.3. *Comparison with results from other studies*

We compare our marginal cost estimates with those from previous studies of carbon sequestration in forests (Fig. 5) and carbon abatement (Fig. 6).[23] Over most of the range of carbon prices considered in previous forest-based carbon sequestration studies, our cost estimates are higher than those obtained using optimization models [1,7] and bottom-up engineering cost methods [33]. Comparing our estimates with those from the earlier econometric study by Stavins [39], we find similar costs at low carbon sequestration levels, but lower costs at higher carbon sequestration levels. In particular, the Stavins curve becomes close to vertical by about 600 million tons of carbon per year, whereas our curve is still relatively flat at this point. Part of the reason for this difference is the fact that Stavins extrapolates his econometric results from the Mississippi Delta region to US farm acreage totaling 551 million acres, including cropland and pasture. In the present study, crop and pasture acreage are somewhat lower, but rangeland and CRP are included, yielding a total of 933 million acres of privately owned agricultural land available for conversion to forest. As Figs. 1 and 2 indicate, a substantial amount of the increase in forest area in our simulations is due to conversion of

[22]The convexity of the marginal cost curve at higher marginal costs is due to the fixed size of the land base, rather than the passing the inflection point in the logistic probability curve. Net forestation in the first period increases at an increasing rate for all of the subsidy/tax levels considered. However, at higher subsidy/tax levels, the constraint on available land leads to decreasing incremental increases in total cumulative net forestation (and carbon sequestration).

[23]Stavins [39] reports and discusses the marginal cost estimates from previous studies.

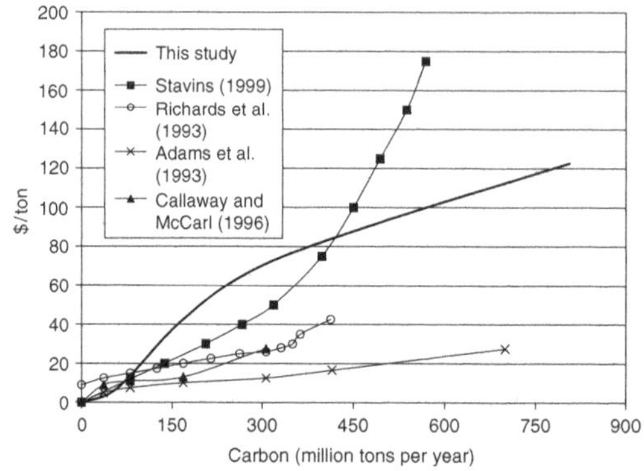

Fig. 5. Alternative estimates of the marginal costs of forest-based carbon sequestration in the United States.

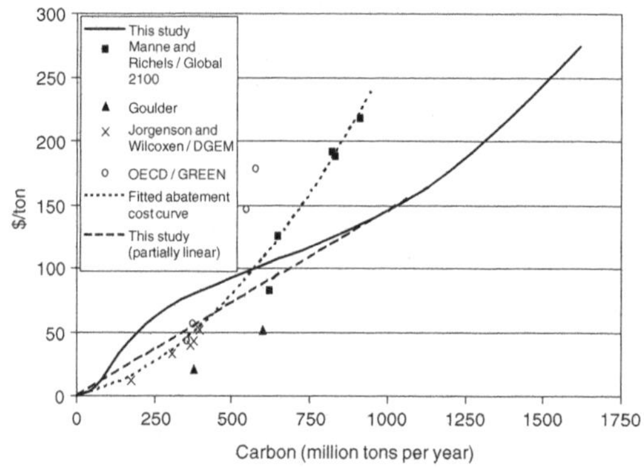

Fig. 6. Estimates of the marginal costs of forest-based carbon sequestration and carbon abatement in the United States. The abatement cost estimates are derived from Energy Modeling Forum [9]. See Stavins [39] for details.

rangeland to forest. Stavins' econometric model was fitted to data exclusively of transitions between cropland and forest.[24]

Finally, we compare the marginal costs of forest-based carbon sequestration with estimates of energy-based carbon abatement.[25] For estimates below $100 per ton, our estimates are comparable to those of Jorgensen and Wilcoxen, Manne and Richels, and OECD/Green—particularly if the initial part of our curve is treated as linear—and higher than those in Goulder. For costs above $100 per ton, our estimates are lower than the abatement studies considering costs in this range, namely OECD/Green and Manne and Richels.

[24]An additional explanation for the difference is that the 36 counties examined in the Stavins [39] study may simply not be representative of the United States along relevant dimensions.

[25]The abatement cost estimates are derived from estimates in the Energy Modeling Forum [9]. See Stavins [39] for details.

Given the relative positions of the supply functions, the results suggest that forest-based carbon sequestration merits consideration as part of a cost-effective portfolio of domestic US climate change strategies. Consider, for example, a domestic program that would bring US annual greenhouse gas emissions 7% below 1990 levels over the period, 2008–2012, as would be required by the Kyoto Protocol. This would entail a reduction of about 573 million short tons of carbon in the year 2010.[26] To compare the abatement and sequestration cost estimates explicitly, we fit a function to the point estimates from the abatement studies (the dotted line in Fig. 6).[27] Our analysis indicates that if cost-effective emission reductions[28] in the United States on the scale proposed under the Kyoto Protocol were to be achieved entirely through domestic actions (forest-based sequestration and/or energy-based abatement activities), 33% of the reduction would be achieved through forest-based sequestration (or up to 44% if the lower portion of our curve were linearized to join the convex section, the dashed line in Fig. 6).

5. Conclusions

In this econometric analysis of carbon sequestration costs, six major land uses were modeled, detailed micro-data of land use and land quality were employed that are comprehensive of the contiguous United States, and key commodity prices were treated as endogenous in the simulations of the carbon sequestration supply function. We compared our estimated carbon sequestration supply function with ones from previous studies, and found that over the range of carbon prices considered in most studies, our marginal cost estimates are greater than those from engineering cost analyses and sectoral optimization models. Because our cost estimates are derived from landowners' actual behavior regarding disposition of their lands, they may reflect factors such as option values associated with delaying irreversible land conversion, liquidity constraints, and unobserved benefits and costs of alternative land uses.

We find lower marginal costs of carbon sequestration when timber harvesting is prohibited on lands enrolled in the carbon sequestration program. Marginal costs fall because the additional present value costs of enrolling lands on which harvesting is prohibited are more than outweighed by the additional present value carbon sequestered. This result is reinforced by endogenous price effects. Restrictions on timber harvesting on enrolled lands raise timber prices, creating incentives for landowners to retain existing (non-program) lands in forest.

The national scope of our study allows us to compare directly our estimates of the marginal costs of carbon sequestration with estimates of costs from energy-based carbon abatement analyses. We find that the estimated carbon sequestration supply function is roughly similar to the central tendency of the carbon abatement supply function, indicating that about a third of the US target under the Kyoto Protocol would be cost-effectively achieved by employing forest-based sequestration policies, in addition to energy-based carbon abatement strategies. This suggests that forest-based carbon sequestration merits consideration as part of a cost-effective portfolio of domestic US climate change policies.

Acknowledgments

This paper is based, in part, on Lubowski's Ph.D. dissertation at Harvard University. The authors acknowledge financial support from the US Department of Energy and research and programming assistance from David Lewis and Shuzhen Nong. The authors thank Michael Roberts, the editor, and two anonymous referees for helpful comments, and take responsibility for all remaining errors. The opinions expressed are the authors' alone and do not necessarily correspond to those of any of the institutions with which they are affiliated.

[26]Estimated US net greenhouse gas emissions in 1990 were 1555 million short tons of carbon equivalent [46]. Projected 2010 energy-based carbon emissions are 1972 million short tons [44] and projected other greenhouse emissions minus sequestration removals are 47 million short tons of carbon equivalent [45].

[27]A quadratic form was chosen so as to minimize the Akaike [3] information criterion (AIC).

[28]The cost-effective portfolio equalizes the marginal costs of sequestration and abatement at the level that achieves the desired total reduction.

Appendix. Supplementary materials

Supplementary information (Appendices A to D) associated with this article can be found in the online version at doi:10.1016/j.jeem.2005.08.001 or in the online archive for supplementary material at http://www.aere.org/journal/index.html

References

[1] R.M. Adams, D.M. Adams, J.M. Callaway, C.C. Chang, B.A. McCarl, Sequestering carbon on agricultural land: social cost and impacts on timber markets, Contemporary Pol. Issues 11 (1) (1993) 76–87.

[2] D.M. Adams, R.J. Alig, B.A. McCarl, J.M. Callaway, S.M. Winnett, Minimum cost strategies for sequestering carbon in forests, Land Econ. 75 (3) (1999) 360–374.

[3] H. Akaike, A new look at statistical model identification, IEEE Trans. Automatic Control 19 (6) (1974) 719–722.

[4] R.J. Alig, D.M. Adams, B.A. McCarl, J.M. Callaway, S. Winnett, Assessing effects of mitigation strategies for global climate change with an intertemporal model of the US forest and agriculture sectors, Environ. Resource Econ. 9 (1997) 259–274.

[5] J. Antle, S. Capalbo, S. Mooney, E. Elliott, K. Paustian, Spatial heterogeneity, contract design, and the efficiency of carbon sequestration policies for agriculture, J. Environ. Econ. Manage. 42 (2) (2003) 231–250.

[6] M. Ben-Akiva, R. Lerman, Discrete Choice Analysis: Theory and Application to Travel Demand, The Massachusetts Institute of Technology Press, Cambridge, MA, 2000.

[7] J.M. Callaway, B.A. McCarl, The economic consequence of substituting carbon payments for crop subsidies in US agriculture, Environ. Resource Econ. 7 (1) (1996) 15–43.

[8] D. Dudek, A. LeBlanc, Offsetting new CO_2 emissions: a rational first greenhouse policy step, Contemporary Pol. Issues 8 (1990) 29–42.

[9] Energy Modeling Forum, Reducing Global Carbon Emissions: Cost and Policy Options, Stanford University, Stanford, CA, 1995.

[10] L.H. Goulder, K. Mathai, Optimal CO_2 abatement in the presence of induced technological change, J. Environ. Econ. Manage. 39 (1) (2000) 1–38.

[11] W.H. Greene, Econometric Analysis, fourth ed., Prentice-Hall, Upper Saddle River, NJ, 2000.

[12] J. Hausman, D. McFadden, Specification tests for the multinomial logit model, Econometrica 52 (5) (1984) 1219–1240.

[13] L. Kurkalova, C.L. Kling, J. Zhao, Multiple benefits of carbon-friendly agricultural practices: empirical assessment of conservation tillage in Iowa, Environ. Manage. 33 (4) (2004) 519–527.

[14] D. Lashof, D. Tirpak, Policy options for stabilizing global climate, Report to Congress, vol. II, US Environmental Protection Agency, Washington, DC, 1989.

[15] J. Lewandrowski, M. Peters, C. Jones, R. House, M. Sperow, M. Eve, K. Paustian, Economics of sequestering carbon in the us agricultural sector, US Department of Agriculture, Economic Research Service, Technical Bulletin 1909, Washington, DC, 2004.

[16] R.N. Lubowski, Determinants of land-use transitions in the United States: econometric analysis of changes among the major land-use categories, Ph.D. Dissertation, Harvard University, Cambridge, MA, 2002.

[17] G. Marland, The prospects of solving the CO_2 problem through global reforestation, US Department of Energy, DOE/NBB-0082, Washington, DC, 1988.

[18] B.A. McCarl, U.A. Schneider, Greenhouse gas mitigation in US agriculture and forestry, Science 294 (2001) 2481–2482.

[19] D. McFadden, Modeling the choice of residential location, in: A. Karlqvist, et al. (Eds.), Spatial Interaction Theory and Planning Models, North-Holland, Amsterdam, 1978.

[20] D.P. McMillen, Probit with spatial autocorrelation, J. Regional Sci. 32 (3) (1992) 335–348.

[21] R.J. Moulton, K.R. Richards, Costs of sequestering carbon through tree planting and forest management in the United States, US Department of Agriculture, Forest Service, General Technical Report WO-58, Washington, DC, 1990.

[22] B.C. Murray, B.A. McCarl, H.C. Lee, Estimating leakage from forest carbon sequestration programs, Land Econ. 80 (1) (2004) 109–124.

[23] New York State Energy Office, Analysis of carbon reduction in New York State, Report of the New York State Energy Office, in consultation with NYS Department of Environmental Conservation and NYS Department of Public Service, New York, NY, 1991.

[24] R.G. Newell, R.N. Stavins, Climate change and forest sinks: factors affecting the costs of carbon sequestration, J. Environ. Econ. Manage. 40 (3) (2000) 211–235.

[25] W.D. Nordhaus, J. Boyer, Warming the World: Economic Models of Global Warming, The Massachusetts Institute of Technology Press, Cambridge, MA, 2000.

[26] S.M. Nusser, J.J. Goebel, The national resources inventory: a long-term multi-resource monitoring programme, Environ. Ecol. Statist. 4 (3) (1997) 181–204.

[27] P. Parks, Explaining 'irrational' land use: risk aversion and marginal agricultural land, J. Environ. Econ. Manage. 28 (1) (1995) 34–47.

[28] P. Parks, I. Hardie, Least-cost forest carbon reserves: cost-effective subsidies to convert marginal agricultural land to forests, Land Econ. 71 (1) (1995) 122–136.

[29] R.S. Pindyck, Irreversibility, uncertainty, and investment, J. Econ. Lit. 29 (3) (1995) 1110–1152.

[30] A.J. Plantinga, The costs of carbon sequestration in forests: a positive analysis, in: R.A. Sedjo, N. Sampson, J. Wisniewski (Eds.), Economics of Carbon Sequestration in Forestry, CRC Press, New York, NY, 1997, pp. 269–277.
[31] A.J. Plantinga, T. Mauldin, D. Miller, An econometric analysis of the costs of sequestering carbon in forests, Amer. J. Agr. Econ. 81 (4) (1999) 812–824.
[32] K.R. Richards, Estimating costs of carbon sequestration for a United States greenhouse gas policy, Report prepared for Charles River Associates, Inc., Cambridge, MA, 1997.
[33] K.R. Richards, R.J. Moulton, R.A. Birdsey, Costs of creating carbon sinks in the US, Energy Conserv. Manage. 34 (9–11) (1993) 905–912.
[34] K.R. Richards, C. Stokes, A review of forest carbon sequestration cost studies: a dozen years of research, Climatic Change 63 (1–2) (2004) 1–48.
[35] T. Schatzki, Options, uncertainty, and sunk costs: an empirical analysis of land use change, J. Environ. Econ. Manage. 46 (1) (2003) 86–105.
[36] R.A. Sedjo, A. Solomon, in: N. Rosenberg, W. Easterling, P. Crosson, J. Darmstadter (Eds.), Greenhouse Warming: Abatement and Adaptation, RFF Press, Washington, DC, 1989, pp. 110–119.
[37] R.A. Sedjo, J. Wisniewski, A. Sample, J. Kinsman, The economics of managing carbon via forestry: assessment of existing studies, Environ. Resource Econ. 6 (1995) 139–165.
[38] B. Sohngen (Ed.), Climate change and agriculture: impacts and mitigation, Choices, 19(3), 2004.
[39] R.N. Stavins, The costs of carbon sequestration: a revealed-preference approach, Amer. Econ. Rev. 89 (4) (1999) 994–1009.
[40] R.N. Stavins, K.R. Richards, The Cost of Supplying Forest-Based Carbon Sequestration in the United States, The Pew Center on Global Climate Change, Arlington, VA, 2005.
[41] United Nations General Assembly, United Nations Framework Convention on Climate Change, A/AC.237/18, New York, NY, 1997.
[42] US Department of Agriculture, Farm service agency, conservation reserve program monthly summary: October 2004, Washington, DC, 2004. Accessed June 8, 2005. http://www.fsa.usda.gov/dafp/cepd/crp_statistics.htm
[43] US Department of Agriculture, Land capability classification, Soil Conservation Service Handbook No. 210, Washington, DC, 1973.
[44] US Department of Energy, Energy information administration, Annual Energy Outlook, DOE/EIA-0383, Washington, DC, 2004.
[45] US Department of State, US Climate Action Report, Washington, DC, 2002.
[46] US Environmental Protection Agency, Inventory of US greenhouse gas emissions and sinks: 1990–2002, EPA 430-R-04-003, Washington, DC, 2004.
[47] G.C. van Kooten, A.J. Eagle, J. Manley, T. Smolak, How costly are carbon offsets? A meta-analysis of carbon forest sinks, Environ. Sci. Pol. 7 (4) (2004) 239–251.
[48] K. van 't Veldt, A.J. Plantinga, Carbon sequestration or abatement? The effect of rising carbon prices on the optimal portfolio of greenhouse gas abatement strategies, J. Environ. Econ. Manage. 50 (1) (2005) 59–81.

[22]

Climate Change and Forest Sinks: Factors Affecting the Costs of Carbon Sequestration[1]

Richard G. Newell

Resources for the Future, Washington, DC 20036

and

Robert N. Stavins[2]

John F. Kennedy School of Government, Harvard University, Cambridge, Massachusetts 02138; and Resources for the Future, Washington, DC 20036

Received September 11, 1998; revised May 4, 1999; published online August 10, 2000

The possibility of encouraging the growth of forests as a means of sequestering carbon dioxide has received considerable attention, partly because of evidence that this can be a relatively inexpensive means of combating climate change. But how sensitive are such estimates to specific conditions? We examine the sensitivity of carbon sequestration costs to changes in critical factors, including the nature of management and deforestation regimes, silvicultural species, relative prices, and discount rates. © 2000 Academic Press

1. INTRODUCTION

The Kyoto Protocol to the United Nations Framework Convention on Climate Change [37] establishes the principle that carbon sequestration can be used by participating nations to help meet their respective net emission reduction targets for carbon dioxide (CO_2) and other greenhouse gases.[3] Several studies have found that growing trees to sequester carbon could provide relatively low-cost net

[1] Valuable comments on previous versions of this paper were provided by Lawrence Goulder, William Nordhaus, Andrew Plantinga, Kenneth Richards, Roger Sedjo, two anonymous referees, an associate editor, and participants in seminars at the Universities of California at Los Angeles and Santa Barbara, Maryland, Michigan, and Texas, Harvard, Stanford, and Yale Universities, Resources for the Future, and the National Bureau of Economic Research. The authors alone are responsible for any errors.

[2] Address correspondence to: Professor Robert N. Stavins, John F. Kennedy School of Government, Harvard University, 79 John F. Kennedy St., Cambridge, MA 02138, phone: 617-495-1820, Fax: 617-496-3783. E-mail: robert_stavins@harvard.edu

[3] After fossil-fuel combustion, deforestation is the second largest source of carbon dioxide emissions to the atmosphere. Estimates of annual global emissions from deforestation range from 0.6 to 2.8 billion tons, compared with slightly less than 6.0 billion tons annually from fossil-fuel combustion, cement manufacturing, and natural gas flaring, combined [10, 31]. There are three pathways along which carbon sequestration is of relevance for atmospheric concentrations of carbon dioxide: carbon storage in biological ecosystems, carbon storage in durable wood products, and substitution of biomass fuels for fossil fuels [24]. The analysis in this paper considers the first two pathways. For further discussion, see Parks et al. [18].

0095-0696/00 $35.00

emission reductions for a number of countries [3], including the United States [1, 4, 19, 20, 23, 33].[4]

When and if the United States chooses to ratify the Kyoto Protocol and/or subsequent international agreements, it will be necessary to decide whether carbon sequestration policies—such as those that promote forestation[5] and discourage deforestation—should be part of the domestic portfolio of compliance activities. The potential cost-effectiveness of carbon sequestration activities will presumably be a major criterion, and so it is important to ask what factors affect the costs of such programs. We examine the sensitivity of sequestration costs to changes in key factors, including the nature of the management regimes, silvicultural species, relative prices, and discount rates.

Our analytical model takes account of current silvicultural understanding of the intertemporal linkages between deforestation and carbon emissions, on the one hand, and between forestation and carbon sequestration, on the other. Furthermore, our analysis uses a methodology whereby econometric estimates of the costs of carbon sequestration are derived from observations of landowners' actual behavior when confronted with the opportunity costs of alternative land uses [33]. This is in contrast with "engineering" or "least cost" approaches used to estimate the costs of carbon sequestration, of which even the best are unlikely to capture important elements of landowner behavior, such as the effects of irreversible investment under uncertainty, non-pecuniary returns from land use, liquidity constraints, decision making inertia, and other costs and benefits of land use of which the analyst is unaware.[6]

In summary, we find, first, that the costs of carbon sequestration can be greater if trees are periodically harvested, rather than permanently established. Second, higher discount rates imply higher marginal costs and non-monotonic changes in the amount of carbon sequestered. Third, higher agricultural prices lead to higher marginal costs or reduced sequestration. Fourth, retarded deforestation can sequester carbon at substantially lower costs than increased forestation. These results depend in part on the time profile of sequestration and the amount of carbon

[4] There is a range of estimates of the relevant marginal cost function. These various estimates are compared by Stavins [33], whose own estimates are significantly greater than the others for more ambitious sequestration programs.

[5] Distinctions are sometimes made in the forestry literature between "afforestation" and "reforestation," where the former refers to changes from non-forest to forest production on lands that have not been forested during the preceding 50 years or more, and the latter refers to changes to forest production on lands that have more recently been deforested [11]. In our analysis, there is no reason to make this distinction, and so we simply refer to any change *to* forest use as "forestation." This is in contrast to a change *from* forest use of land—"deforestation."

[6] The simplest of previous analyses derived single point estimates of average costs associated with particular sequestration levels [8, 13, 14, 27, 29], sometimes assuming that the opportunity costs of land are zero [7, 16, 38, 39]. "Engineering/costing models" have constructed marginal cost schedules by adopting land rental rates or purchase costs derived from surveys for representative types or locations of land, and then sorting these in ascending order of cost [15, 23]. Simulation models include a model of lost profits due to removing land from agricultural production [19], a mathematical programming model of the agricultural sector and the timber market [1, 2], a related model incorporating the effects of agricultural price support programs [4], and a dynamic simulation model of forestry [35]. An analysis by Plantinga [20] adopts land-use elasticities from an econometric study to estimate sequestration costs, an approach similar in some respects to the methodology used here. For surveys of the literature, see Richards and Stokes [24] and Sedjo et al. [28, 30].

released upon harvest, both of which may vary by species, geographic location, and management regime, and are subject to scientific uncertainty.

In Section 2 of the paper, we describe the analytical model; in Section 3, we carry out simulations for various scenarios and thereby examine the sensitivity of the marginal cost of carbon sequestration; and in Section 4, we offer some conclusions.

2. ANALYTICAL MODEL

We draw upon econometrically estimated parameters of a structural model of land use, layer upon it a model of the relationships that link changes in alternative land uses with changes in the time paths of CO_2 emission and sequestration, and examine the sensitivity of carbon sequestration costs to key underlying factors. Our analysis focuses on the empirically relevant land-use options of forest and farm.[7]

2.1. *A Structural, Empirical Model of Land Use*

In previous work with a different policy motivation, Stavins and Jaffe [34] developed a dynamic optimization model of a landowner's decision of whether to keep land in its status quo use or convert it to serve another purpose.[8] Landowners are assumed to observe current and past values of economic, hydrologic, and climatic factors relevant to decisions regarding the use of their lands for forestry or agricultural production and on this basis form expectations of future values of respective variables. Given this information, landowners attempt to maximize the expected long-term economic return to the set of productive activities that can be carried out on their land. They face ongoing decisions of whether to keep land in its current state—either forested or agricultural use—or to convert the land to the other state. Relevant factors a landowner would be expected to consider include: typical agricultural and forestry revenues for the area, the quality of a specific land parcel for agricultural production, agricultural costs of production, and the cost of converting land from a forested state to use as cropland. Thus, we anticipate that a risk-neutral landowner will seek to maximize the present discounted value of the stream of expected future returns.

We summarize the formal statement of the landowner's problem in the Appendix, where the application of control theoretic methods yields a pair of necessary conditions for changes in land use. The first necessary condition implies that a parcel of cropland should be converted to forestry use if the present value of expected net forest revenue exceeds the present value of expected net agricultural revenue. Stated formally, forestation (conversion of agricultural cropland to forest) occurs if a parcel is cropland and if

$$\left(F_{it} - D_{it} - A_{it} \cdot q_{ijt} + M_{it}\right) > 0, \tag{1}$$

[7] In both industrialized nations and in developing countries, nearly all deforestation is associated with conversion to agricultural use [11].

[8] A detailed description of the dynamic optimization model and the derivation of the econometrically estimatable model is found in Stavins and Jaffe [34], while Stavins [32] provides an illustration of the use of the model for environmental simulation.

where i indexes counties, j indexes individual land parcels, and t indexes time; upper case letters are stocks or present values; lower case letters are flows; F is forest net revenue, equal to the expected present value of annual net income from forestry per acre (i.e., stumpage value); D is the expected present value of the income loss (when converting to forest) due to delay of first harvest for one rotation period; A is the expected present value of the future stream of typical agricultural revenues per acre; q is a parcel-specific index of feasibility of agricultural production, including effects of soil quality and soil moisture; and M is the expected cost of agricultural production per acre, expressed as the present value of an infinite future stream.

On the other hand, a forested parcel should be converted to cropland if the present value of expected net agricultural revenue exceeds the present value of expected net forest revenue plus the cost of conversion. That is, deforestation occurs if a parcel is forested and if

$$\left(A_{it} \cdot q_{ijt} - M_{it} - C_{it}^{\alpha P_{it}} - \left(F_{it} - W_{it}\right)\right) > 0, \qquad (2)$$

where C is the average cost of conversion per acre, P is the Palmer hydrological drought index, and W_{it} is the windfall of net revenue per acre from a one-time clear cut of forest (prior to conversion to agricultural use).

Inequalities (1) and (2) imply that all land in a county (of given quality) will be in the same use in the steady state. In reality, counties are observed to be a mix of forest and farmland. Although this may partly reflect deviations from the steady state, it is due largely to the *heterogeneity* of land, particularly regarding its suitability for agriculture. As shown in Stavins and Jaffe [34], such unobserved heterogeneity can be parameterized within an econometrically estimatable model so that the individual necessary conditions for land-use changes aggregate into a single-equation model, in which the parameters of the basic benefit–cost relationships and of the underlying, unobserved heterogeneity can be estimated simultaneously.

The complete model yields a set of econometrically estimatable equations, as shown in the Appendix. Using panel data for 36 counties, comprising approximately 13 million acres of land, in Arkansas, Louisiana, and Mississippi, during the period 1935–1984, the parameters of the complete model were estimated with nonlinear least squares procedures [34]. Table I provides descriptive statistics of the major variables used in the simulation analysis.

2.2. *A Dynamic Simulation Model of Future Land Use*

Our initial step in moving from an estimated model of historical land use to a model of carbon sequestration involves introducing relevant silvicultural elements into the necessary conditions previously derived. There are three principal silvicultural dimensions to be considered: symmetries and asymmetries between forestation and deforestation, alternative species for forestation, and alternative manage-

TABLE I
Descriptive Statistics[a]

Variable	Mean	Standard deviation	Minimum	Maximum
Gross agricultural revenue ($/acre/year)	259.04	44.58	184.77	376.03
Agricultural production cost ($/acre/year)	220.39	52.03	143.61	359.81
Forest revenue[b] ($/acre/year)				
Mixed stand	19.29	7.45	6.71	38.36
Pine stand	58.96	23.38	19.92	118.24
Tree-farm establishment cost ($/acre)	92.00	0.00	92.00	92.00
Conversion cost ($/acre)[c]	27.71	0.00	27.71	27.71
Carbon sequestration due to forestation[d] (tons/acre)				
Natural regrowth of mixed stand, periodically harvested	43.36	0.00	43.36	43.36
Natural regrowth of mixed stand, no harvest	50.59	0.00	50.59	50.59
Pine plantation, periodically harvested	41.05	0.00	41.05	41.05
Pine plantation, no harvest	49.99	0.00	49.99	49.99
Carbon emissions due to deforestation[e] (tons/acre)	51.83	0.00	51.83	51.83
Interest rate[f]	5%	0.00	5%	5%

[a] The sample is of 36 counties in Arkansas, Louisiana, and Mississippi, located within the Lower Mississippi Alluvial Plain. All monetary amounts are in 1990 dollars; means are unweighted county averages.

[b] Gross forest revenue minus harvesting costs; an annuity of stumpage values.

[c] The historical analysis uses actual conversion costs, varying by year.

[d] Present-value equivalent of life-cycle sequestration.

[e] Present-value equivalent of life-cycle emissions.

[f] The historical analysis uses actual, real interest rates; simulations of future scenarios use the 5% real rate.

ment regimes. Two of the equations from the land use model need to be adjusted for this purpose,

$$q_{it}^{y} = \left[\frac{F_{it} - D_{it} + M_{it}}{A_{it}}\right] \tag{3}$$

$$q_{it}^{x} = \left[\frac{F_{it} - W_{it} + M_{it}}{A_{it} - C_{it}^{\alpha P_{it}}}\right], \tag{4}$$

where, for each county i at time t, q^{y} is the threshold value of land quality (i.e., suitability for agriculture) below which the incentive for forestation manifests itself, and q^{x} is the threshold value of land quality above which the incentive for deforestation manifests itself.

First, we note that Eqs. (3) and (4) already exhibit two significant asymmetries between forestation and deforestation. Forestation produces a supply of timber (and an associated forest-revenue stream) only with some delay, since the first harvest subsequent to establishment occurs at the completion of the first rotation, while deforestation involves an immediate, one-time revenue windfall from cutting of the stand, net of a loss of future revenues from continued forest production. Additionally, under actual management practices during the sample period of historical analysis, costs were associated with converting forestland to agricultural cropland, but no costs were involved with essentially abandoning cropland and allowing it to return to a forested state. For the simulations associated with carbon sequestration policies, however, we need to allow for the possibility of "tree farming," that is, intensive management of the forest, which brings with it significant costs of establishment.

Second, there is the choice of species. In the econometric analysis, only mixed stands[9] were considered to reflect historical reality, but in the carbon-sequestration context it is important to consider the possibility of both mixed stands and tree farms (plantations of pure pine). We develop revenue streams for both, based upon observed practice in the region.[10]

The third silvicultural dimension is the choice of management regime. The historical analysis assumed that all forests were periodically harvested for their timber. For purposes of carbon sequestration, however, we should consider not only such conventional management regimes, but also the possibility of establishing "permanent stands" that are never harvested. These three silvicultural considerations lead to the respecification of Eq. (3),

$$q_{its}^{y} = \left[\frac{F_{its} - D_{its} + M_{it} - K_{it}}{A_{it}} \right], \tag{5}$$

where subscript s indicates species and K is the cost associated with establishing a pine-based tree farm.[11] For the case of permanent (unharvested) stands, F and D are set equal to zero. Combining variable values associated with these silvicultural dimensions into logical sets yields four scenarios to be investigated: natural regrowth of a mixed stand, with and without periodic harvesting, and establishment of a pine plantation, with and without periodic harvesting.

2.3. *Generating a Forest Supply Function*

Next, we introduce some policy-inspired modifications to develop a forest supply function. First, note that dynamic simulations of fitted values of the model, employing current/expected values of all variables (including prices), will generate

[9] Mixed stands of appropriate shares of various species of hardwoods and softwoods, specific to each county and time period, were included in the data used for econometric estimation. The calculated revenue streams draw upon price data for both sawlogs and pulpwood in proportion to use, based upon 55-year rotations.

[10] The tree-farm revenue streams represent a mix of 80% loblolly pine and 20% slash pine, based upon practice in the area [5]. We use a rotation length of 45 years for loblolly and 30 years for slash pine, also reflecting standard practice [15].

[11] Forest establishment costs include the costs of planting (purchase of seedlings, site preparation, and transplanting), post-planting treatments, and care required to ensure establishment [15]. We adopt a value of $92/acre ($1990), based upon estimates by Richards et al. [23] for converted cropland in the Delta (three-state) region.

baseline predictions of future forestation and/or deforestation [32].[12] These results constitute our baseline for policy analysis. Second, we can simulate what land-use changes would be forthcoming with changed values of specific variables. In general, we can examine the consequences of public policies that affect the economic incentives faced by landowners. The difference in forestation/deforestation between the first (baseline) and the second (counterfactual) simulation is the predicted impact of a given policy.

In order to generate a representation of the forest supply function, several types of policies can be considered. A payment (subsidy) could be offered for every acre of (agricultural) land that is newly forested. But this would provide an incentive for landowners to cut down existing forests simply to replant in a later year in exchange for the government payment. On the other hand, a tax could be levied on each acre of land that is deforested. But such an approach would provide no added incentive for forestation of land that is not currently in that state. One solution is to think of a two-part policy that combines a subsidy on the flow of newly forested land with a tax on the flow of (new) deforestation. As a first approximation, the two price instruments can be set equal, although this is not necessarily most efficient.

We simulate this policy by treating the subsidy as an increment to forest revenues in the forestation part of the model (Eq. (4)) and treating the tax payment as an increment to conversion or production costs in the deforestation part of the model (Eq. (5)). Letting Z represent the subsidy and tax, the threshold equations ((3) and (4)) for forestation and deforestation, respectively, become

$$q_{its}^{y} = \left[\frac{(F_{its} - D_{its} + Z_{it}) + M_{it} - K_{it}}{A_{it}} \right] \tag{6}$$

$$q_{its}^{x} = \left[\frac{F_{its} - W_{its} + (M_{it} + Z_{it})}{A_{it} - C_{it}^{\alpha P_{it}}} \right]. \tag{7}$$

Thus, a dynamic simulation based upon Eqs. (6) and (7) in conjunction with the other equations of the model (see the Appendix), in which the variable Z is set equal to zero, will generate a baseline quantity of forestation/deforestation over a given time period. By carrying out simulations for various values of Z over the period and subtracting the results of each from the baseline results, we can trace out a forest acreage supply function, with marginal cost per acre (Z) arrayed in a schedule with total change in acreage over the time period, relative to the baseline.[13]

[12] Statistical tests, reported in Stavins and Jaffe [34], indicate a high degree of structural (and parametric) stability of the model over the 50-year time period of estimation. It is therefore possible to carry out future factual and counter-factual simulations. Extrapolations of historical trends would imply future increases in the relative price of timber to agricultural crops, but extrapolations of historical trends of relative yields would favor agriculture. Not knowing what the future will bring, the baseline simulations employ constant values of all variables, including real prices and yields. Nevertheless, the baseline simulations exhibit changes in land use over time, both because of the partial-adjustment nature of the model and because modifications of silvicultural practices are assumed for both baseline and policy simulations, as is explained later.

[13] This is a partial-equilibrium analysis of a 36-county region. If a national analysis were being carried out, it would be necessary to allow for price endogeneity, i.e., allow for land-use changes induced by changes in Z to affect agricultural and forest product prices. On this, see Stavins [33].

It might be argued that since the policy intervention we model is a tax/subsidy on land use, not on carbon emissions and sequestration, it does not lead to the true minimum carbon-sequestration marginal cost function. This may seem to be a valid criticism in the narrowest analytic sense, but it is not valid in a realistic policy context. It would be virtually impossible to levy a tax on carbon emissions or a subsidy on sequestration, because the costs of administering such policy interventions would be prohibitive. Looked at this way, such an instrument would likely be *more* costly per unit of carbon sequestered than would the deforestation tax/forestation subsidy policy considered here.[14]

2.4. *Computing the Marginal Cost of Carbon Sequestration*

For any parcel of land, there are several types of comparisons that could be made between the time-paths of carbon emissions/sequestration in a baseline and a policy simulation. First, we can consider a parcel that is continually in cropland in both simulations, in which case it exhibits zero net carbon sequestration/emission over the long run in both, and so the policy impact is also zero.[15] Second, a parcel may continually be in a forested state in both simulations, in which case it sequesters carbon in both simulations, but net sequestration due to the policy intervention is again zero. Third, a parcel may be in agricultural use in the baseline, but forestation takes place in the policy simulation in year t; here, net carbon sequestration due to the policy intervention will be the time-path of annual sequestration that commences in year t. Fourth, a parcel may be in a forested state in the baseline, but deforestation takes place in the policy simulation in year t; then the net carbon emissions due to the policy intervention will be the time-path of annual *emissions* that commence in year t, assuming durable wood products are produced from merchantable timber.

Carbon-Sequestration Time Profiles

The next step, conceptually, is to link specific time paths of carbon sequestration (and emissions) with forestation and deforestation. Scientific understanding of these linkages is evolving; we draw upon recent biological models and employ a set of temporal carbon yield curves based on Moulton and Richards [15] and Richards et al. [23].[16] Figure 1 provides a pictorial representation of one example of the time-path of carbon sequestration and emission linked with a specific forest management regime. In the example, the time profile of cumulative carbon

[14] This is not to suggest that a uniform tax/subsidy would be the first-best policy. A more efficient but still practical policy instrument might well involve a non-uniform tax/subsidy, set in accordance with regional and other factors.

[15] With constant relative prices in the baseline, the time-path of policy-induced changes in land use in the model is always such that individual counties are characterized by increases or decreases in forested acreage, *relative to the baseline*, but never both.

[16] Nordhaus [17] and Richards et al. [23] also use carbon yield curves, while many other sequestration cost studies have used point estimates of average flows.

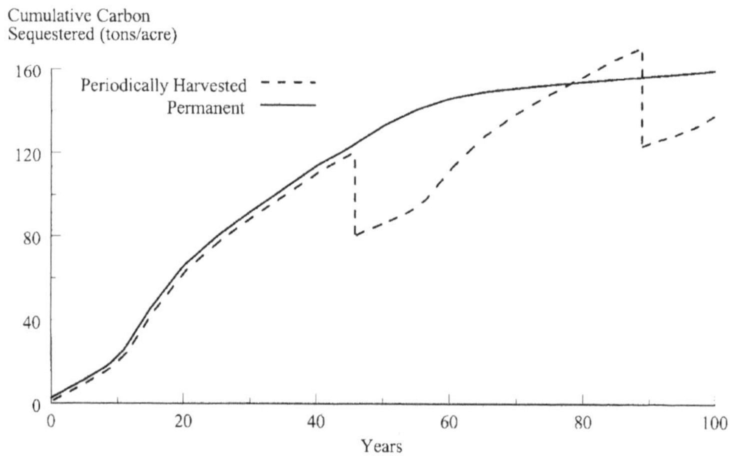

Source: Based on data from Moulton and Richards (1990) and Richards (1994).

FIG. 1. Time profile of carbon sequestration (Loblolly pine in delta states region).

sequestration is for establishing a new loblolly pine plantation. Carbon sequestration occurs in four components of the forest: trees, understory vegetation, forest floor, and soil.[17] When a plantation is managed as a permanent stand, cumulative sequestration increases monotonically, with the magnitude of annual increments declining so that an equilibrium quantity of sequestration is essentially reached within 100 years, as material decay comes into balance with natural growth.

The figure also shows the sequestration path for a stand that is periodically harvested. In this case, carbon accrues at the same rate as in a permanent stand until the first harvest, when carbon is released as a result of harvesting, processing, and manufacturing of derivative products.[18] Much of the carbon sequestered in wood products is also released to the atmosphere, although this occurs with

[17] Although shares vary greatly among forest types, reference points are: tree carbon contains about 80% of ecosystem carbon, soil carbon about 15%, forest litter 3%, and the understory 2% [23]. Variation in these shares is significant; for some species, soil carbon accounts for nearly 50% of total forest carbon.

[18] Our calculations of releases from the understory, forest floor, soil, and non-merchantable timber are based upon Moulton and Richards [15] and Richards et al. [23]. The share of total forest carbon that actually ends up in merchantable wood varies considerably by species. A reasonable reference point is about 40%. Much of the remaining 60% is released at the time of harvest and in the manufacturing process (in both cases through combustion), the major exception being soil carbon, which exhibits much slower decay.

considerable delay as wood products decay.[19] In this scenario, the forest is replanted and the process begins again.[20]

Although the carbon yield curve with harvesting in Fig. 1 eventually moves above the yield curve for a "permanent" stand, this need not be case. It depends upon the share of carbon that is initially sequestered in wood products and upon those products' decay rates (plus the decay rate of soil carbon). With zero decay rates, the peaks in the harvesting yield curve would increase monotonically, but with positive decay rates the locus of the peaks approaches a steady-state quantity of sequestration, because eventually decay in the stock of carbon stored in existing wood products offsets the amount of new carbon sequestered through tree growth. That steady-state quantity can, in theory, lie above or below the level associated with the equilibrium level of the "permanent" yield curve.[21]

Discounting Carbon Costs and Benefits

Recognizing the intertemporal nature of net carbon sequestration raises a question: how can we associate a number—the marginal cost of carbon sequestration—with diverse units of carbon that are sequestered in different years over long time horizons? Previous sequestration studies have used a variety of methods to calculate costs in terms of dollars per ton, the desired units for a cost-effectiveness comparison. These approaches have been classified as "flow summation," "mean carbon storage," and "levelization" [24].

The "flow summation" approach is the simplest: the present value of costs is divided by the total tons of carbon sequestered, regardless of when sequestration occurs. This summary statistic fails to take into account the time profile of sequestration, and second, the measure is very sensitive to the length of the time horizon selected for calculation (in the case of periodic-harvesting scenarios). Furthermore, assuming that not only costs but also benefits of sequestration are to be discounted over time, this approach implies that marginal benefits of sequestra-

[19] As Sedjo et al. [30] point out, examinations of the long-term effects of timber growth on carbon sequestration are "highly dependent upon the assumptions of the life-cycle of the wood products" (p. 23). Harmon et al. [9] found this to be the case in their scientific review. The two critical parameters are the assumed length of the life-cycle of wood products and the assumed share of timber biomass that goes into long-lived wood products. Drawing upon the work of Row [25], Row and Phelps [26], and Turner et al. [36], we develop a time-path of gradual decay of wood products over time, based upon an appropriately weighted average of pulpwood, sawlog, hardwood, and softwood estimates from Plantinga and Birdsey [21]. The final profile is such that one year following harvest, 83% of the carbon in wood products remains sequestered; this percentage falls to 76% after 10 years, and 25% after 100 years (and is assumed to be constant thereafter). At an interest rate of 5%, the present-value equivalent sequestration is approximately 75%, identical to that assumed by Nordhaus [17].

[20] Another potential scenario, which we do not consider, is that harvested wood is used for fuel. If this is to produce electricity or liquid fuels such as methanol, thereby substituting for fossil-fuel use, then there would be two additional effects to consider: (1) the net impact on atmospheric CO_2 emissions of each unit of forestation would be significantly enhanced, and (2) the demand for wood would be increased, which would matter in a general-equilibrium setting. On the other hand, the general-equilibrium effects of bringing a new source of wood to the market would also need to be considered.

[21] There has been a significant amount of debate within the scientific community about the relative superiority of these two regimes in terms of their carbon-sequestration potential. Harmon et al. [9] find that old growth forests are superior to periodic harvesting approaches in their ability to sequester carbon, but Kershaw et al. [12] demonstrate that this is dependent upon specific circumstances.

tion are increasing exponentially over time at the discount rate. A similar summary statistic is based upon "mean carbon storage." In this case, the present value of costs is divided by the numerical average of annual carbon storage. This statistic suffers from the same problems as the first.

The third alternative—"levelization"—seems most reasonable: the discounted present value of costs is divided by the discounted present value of tons sequestered. Alternatively (and equivalently), an annuity of present value costs is divided by an annuity of present value tons. This is the approach we use. It may be thought of as assuming that the marginal damages associated with additional units of atmospheric carbon are constant and that benefits (avoided damages) and costs are to be discounted at the same rate. Note that such an assumption of constant marginal benefits is approximately correct if damages are essentially proportional to the rate of climate change, which many studies have asserted.[22]

Specifically, we define the present values (in year t) of the time-paths of carbon sequestration and carbon emissions associated with forestation or deforestation occurring in year t as Ω_t^S and Ω_t^E, respectively. Thus, the total, present-value equivalent net carbon sequestration/emissions associated with any baseline or policy simulation are calculated as

$$PV(SEQ) = \sum_{i=1}^{36}\left[\sum_{t=0}^{90}\left(FORCH_{it}^a \cdot D_{it}^a \cdot \Omega_t^S - FORCH_{it}^c \cdot D_{it}^c \cdot \Omega_t^E\right)\cdot(1+r)^{-t}\right], \tag{8}$$

where

$$\Omega_t^S = \sum_{h=t}^{90} CS_h \cdot (1+r)^{t-h} \tag{9}$$

$$\Omega_t^E = \sum_{h=t}^{90} CE_h \cdot (1+r)^{t-h}, \tag{10}$$

and where $FORCH^a$ and $FORCH^c$ are forestation and deforestation, respectively, as a share of total county area (see the Appendix for formulae), D^a and D^c are dummy variables for forestation and deforestation, respectively, and CS_h and CE_h are, respectively, annual incremental carbon sequestration and carbon emissions per acre under individual scenarios.

We develop the constituent carbon yield curves for various forest species, location, and management conditions, and initially use a 5% discount rate. The present-value equivalent carbon-sequestration measure associated with natural

[22] If the marginal damages of carbon emissions were expected to change at some rate g over time, an appropriate modification of the levelization procedure could entail reducing the discount rate for carbon by the rate g. For monotonically increasing sequestration time profiles this modification would raise the present-value tons of carbon and lower the marginal cost of carbon sequestration if marginal damages were growing over time (i.e., $g > 0$); it would do the opposite if damages were expected to fall. For non-monotonic sequestration paths, such as those involving periodic harvesting, the effect depends on the specific shape of the path; $g > 0$ could in principle raise or lower present-value carbon. For the scenarios we investigate, such a modification—which is equivalent to lowering the discount rate (for $g > 0$)—also raises the present-value carbon for the harvesting scenarios, but not by as much as for the non-harvesting scenarios.

regrowth of a mixed stand is 43.36 tons if periodically harvested and 50.59 tons if permanent; for a pine plantation the values are 41.05 if periodically harvested and 49.99 tons if permanent.[23] Additionally, we calculate present-value carbon emission measures for deforestation with sale of merchantable timber (51.83 tons). These values are also reported in Tables I and IV. As described above, these values depend on the time profile of sequestration and the amount of carbon released upon harvest, both of which may vary by species, geographic location, and management regime, and are subject to scientific uncertainty. Silvicultural scenarios with more rapid carbon accumulation and less emissions upon harvest will exhibit higher carbon present values and thus lower costs of carbon sequestration per ton.

Since we derive marginal costs on an annual per acre basis using the tax/subsidy scheme, Z, we first convert present value tons of carbon to an equivalent annuity, $AV(SEQ)$, as

$$AV(SEQ) = \frac{PV(SEQ)}{PVFAC(r)}, \tag{11}$$

where $PVFAC$ is a present-value factor used to annualize the present value at rate r. We then divide the carbon-sequestration annuity by the total acreage of forestation, $TFORCH$, relative to the baseline in order to place it on a per acre basis. Lastly, we compute the marginal cost per ton of carbon sequestration MC for each scenario by dividing marginal cost per acre per year by the per acre carbon-sequestration annuity:

$$MC = \frac{Z}{\left(\dfrac{AV(SEQ)}{TFORCH}\right)}. \tag{12}$$

As discussed below, Table II illustrates this computation for a periodically harvested pine plantation.

3. THE COSTS OF CARBON SEQUESTRATION

The results of dynamic land-use simulations for the 90-year period from 1990 to 2080 constitute the fundamental inputs into the final carbon simulation model consisting of Eqs. (8), (9), and (10).[24] A 90-year period was used to allow at least

[23] The yield curves provided in Fig. 1 are simply examples for one species, loblolly pine. The growth curves that underlie respective yield curves are themselves a function, partly, of precipitation and temperature, both of which are presumably affected in the long run by atmospheric concentrations of CO_2 and induced climate change [6]. We ignore this endogeneity to climate change in estimating sequestration costs, as have all previous studies. Likewise, all studies have ignored potential economic endogeneity of relevant variables to climate change. The mixed-stand carbon paths are weighted averages from hardwood and pine constituents, assuming 55% hardwoods and 45% southern pine [5]. The assumed density of carbon in merchantable hardwoods is from Moulton and Richards [15] for Delta state hardwoods. In the case of softwoods (pines), density and assumed rotation length are for loblolly pine and slash pine [15], weighted as 80% and 20%, respectively, of total softwoods. Carbon-sequestration patterns and merchantable wood volumes for pine are based on Richards et al. [23] for cropland in the Delta region.

[24] In a prior step, the econometrically estimated parameters were used with newly available data for 1989 to simulate total forested acreage per county in 1989, the base year for the simulations.

TABLE II
Land Change and Carbon-Sequestration Costs and Quantities, Periodically Harvested Pine Plantation

Marginal cost per acre ($/acre/year) *Z*	Forestation relative to baseline (1,000's acres) *TFORCH*	Average cost per acre ($/acre/year)	Annualized carbon sequestration relative to baseline (1,000's tons/year) *AV(SEQ)*	Marginal cost of carbon sequestration ($/ton) *MC* = *Z*/[*AV(SEQ)*/*TFORCH*]	Average cost of carbon sequestration ($/ton)
0	0	0.00	0	0.00	0.00
10	518	10.00	784	6.61	6.61
20	1,057	15.10	1,600	13.21	9.97
30	1,615	20.25	2,445	19.82	13.38
40	2,192	25.45	3,319	26.42	16.81
50	2,787	30.69	4,219	33.03	20.27
60	3,398	35.96	5,145	39.63	23.76
70	3,893	41.27	5,895	46.24	27.26
80	4,224	46.60	6,395	52.84	30.78
90	4,455	51.95	6,745	59.45	34.31
100	4,653	57.32	7,045	66.05	37.86
200	6,579	105.63	9,961	135.97	69.77
300	7,484	129.15	11,332	202.03	85.31
400	7,897	142.25	11,957	268.05	93.96
500	8,212	155.98	12,434	334.11	103.03
600	8,470	169.22	12,825	400.18	111.77
700	8,689	182.74	13,156	466.22	120.71
800	8,874	195.72	13,437	532.20	129.28
900	9.038	208.21	13,685	598.31	137.53
1,000	9,178	219.53	13,897	664.35	145.01

Notes: Variable symbols are given at the bottom of certain headings to illustrate how figures were computed (see Section 2.3). Discount rate is 5%; baseline forestation is 52,000 acres; baseline carbon sequestration is 4.6 million tons.

one rotation of each forest species; given the consequences of discounting, the results are not fundamentally affected by the length of the period of analysis once that period exceeds 50 years or so. Different time-paths of annual carbon increments, CS_h and CE_h, and different cost and revenue streams of forestation and deforestation are associated with each of the four scenarios to be examined.

As previously described, simulations are employed to trace out the supply curve of net carbon sequestration, in which the marginal costs of carbon sequestration, measured in dollars per ton, are arrayed in a schedule with net annualized[25] carbon sequestration (relative to the baseline). Table II provides the results for one scenario, a periodically harvested pine plantation, with the sale of merchantable timber when/if deforestation occurs. We focus initially on this scenario and provide detailed results for it, by way of example. The relatively attractive forest revenues associated with this management regime result in a small amount of net forestation taking place in the baseline simulation, a gain of about 52 thousand

[25] As explained above, both dollars of costs and tons of sequestration (and emission) are discounted. Hence, annual sequestration refers to an annuity that is equivalent to a respective present value for a given discount rate.

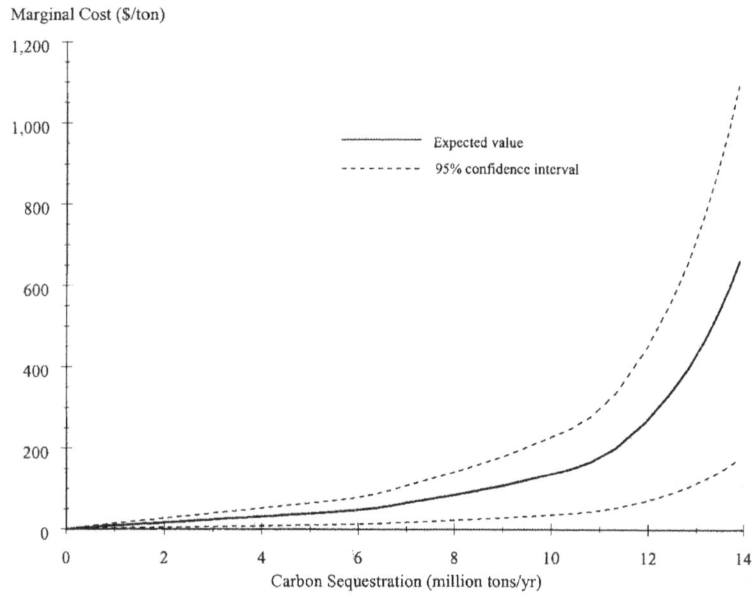

FIG. 2. Marginal cost of carbon sequestration (scenario #3—periodically harvested pine plantation).

acres (over the 90-year study period). Baseline net carbon sequestration is approximately 4.6 million tons annually. As can be seen in Table II and Fig. 2, the marginal costs of carbon sequestration increase approximately linearly until these costs are about $66 per ton, where annual sequestration relative to the baseline has reached about 7 million tons. This level of sequestration is associated with a land-use tax/subsidy of $100 per acre and net forestation relative to baseline of 4.7 million acres.

Beyond this point, marginal costs increasingly depart from a linear trend. Beyond about $200 per ton, they turn steeply upward. Indeed, the marginal cost function appears to be nearly asymptotic to a sequestration level of about 15 to 16 million tons annually (Figure 2).[26] This is not surprising. Such an implicit limit would be associated in the model with net forestation of about 10.5 million acres,

[26] Although the assumption of exogenous prices becomes less tenable as land-use impacts become more severe, it is nevertheless true that the relevant agricultural prices (and to a lesser degree, stumpage values) are determined on national and international markets of which the study region represents only a trivial share. In any event, however, the reliability of the model's predictions decreases as we move further outside the range of the data on which the underlying econometric parameters were estimated.

TABLE III
Costs of Carbon Sequestration for Alternative Silvicultural Scenarios for 5 Million Tons of Sequestration above Baseline

	Alternative silvicultural scenarios			
Species regime	Natural regrowth of mixed stand		Pine plantation	
Management regime	Periodic harvest	No harvest	Periodic harvest	No harvest
Scenario	#1	#2	#3	#4
Baseline change in forestation (1,000 acres)	−259	−297	52	−69
Baseline carbon sequestration (1,000 tons)	4,005	3,931	4,578	4,368
Marginal cost per acre ($/acre/year)	55.80	49.20	58.40	49.10
Forestation relative to baseline (1,000 acres)	3,074	2,662	3,301	2,710
Average cost per acre ($/acre/year)	33.80	30.31	35.12	30.23
Forestation carbon sequestration (tons/acre)	43.36	50.59	41.05	49.99
Deforestation carbon emissions (tons/acre)	51.83	51.83	51.83	51.83
Annualized carbon sequestration (1,000 tons/year)	5,000	5,000	5,000	5,000
Marginal cost of carbon sequestration ($/ton)	34.33	26.30	38.57	26.61
Average cost of carbon sequestration ($/ton)	20.79	16.20	23.20	16.38

Note: Discount rate is 5%.

for a total forested area of 13 million acres, just shy of the total area of the 36 counties of the study region.[27]

3.1. *Alternative Silvicultural Scenarios*

Simulated costs of carbon sequestration are summarized in Table III for four scenarios. In scenario #1, all forestation is assumed to be through natural regrowth of mixed stands that are periodically harvested. The more modest forest revenues associated with this management regime (relative to the pine plantation) result in net deforestation taking place in the baseline simulation, a loss of about 260 thousand acres. The marginal cost of carbon sequestration is about $34 when 5 million tons are sequestered annually.

[27] An advantage of our revealed-preference approach, compared with the usual engineering approaches, is that because the simulation model's parameters are econometrically estimated, those parameters have associated with them not only estimated values (coefficients), but also estimated standard errors. Hence, we can provide a richer description of the marginal cost function through the use of stochastic (Monte Carlo) simulations, drawing upon the relevant variance–covariance matrix. Based upon these simulations, Fig. 2 provides not only a set of point estimates of the marginal cost function, but also the 95% confidence interval around that function. There is also uncertainty associated with a number of the variables employed in the analysis. Hence, the figure probably presents an under-estimate of the true error bounds.

If we modify the previous scenario to eliminate periodic harvesting (thus setting the forest revenue stream for *new* forests equal to zero), deforestation increases somewhat in the baseline (scenario #2, Table III).[28] The timber revenue stream in scenario #1 was forestalling some conversion of forest to agriculture; with the elimination of this revenue stream in scenario #2, deforestation increases. On its own, preventing periodic harvesting of timber would tend to increase the marginal costs of carbon sequestration, since the net opportunity costs associated with an agriculture/forestry change increase. Indeed, this modest loss of expected revenue (about 13%) does cause a modest decrease in the total amount of induced forestation that occurs relative to the case with harvesting (scenario #1). But the time-path of carbon sequestration without harvesting is sufficiently favorable to overcome this effect, so that the marginal costs of sequestration are actually *less* in the no-harvest cases than in those cases where periodic harvesting is permitted. For example, the marginal cost of carbon sequestration is now only \$26 (compared with \$34 in the presence of periodic harvesting) when 5 million tons are sequestered annually.

The picture changes somewhat when we allow for tree farms of pure pine to be established as the regime of forestation. Now the economic incentives that exist in the baseline actually cause little or no deforestation to occur. Potential annual revenues from forestry are significantly greater than in the case of mixed stands, but up-front plantation establishment costs partially mitigate this effect. Overall, a given land-use tax/subsidy brings about greater net forestation in the pure pine case, but this effect is overwhelmed by the differences in carbon-sequestration potential, and so the periodic pine scenario (#3) exhibits *greater* marginal sequestration costs than the periodic mixed-stand case (scenario #1). The difference in carbon sequestration is being driven by the fact that retarded deforestation is responsible for a considerable part of the net carbon sequestration (relative to baseline) for the mixed stands, but in the pine plantation case, we find that all of the carbon sequestration in scenario #3 is due to forestation (which in present-value equivalent terms provides substantially less carbon saved per acre). Scenario #4, the pine plantation without periodic harvesting, provides an intermediate case, which yields results quite similar to the related mixed-stand scenario (#2), because the absence of periodic harvesting eliminates one of the major economic differences and the carbon yield curves themselves are similar.

3.2. *Discount Rates*

Because of the long time horizons employed in the analysis, it is natural to ask about the sensitivity of the results to the assumed discount rate (5%). Changing the discount rate has two types of effects on the simulations. First, many of the economic variables take on new values. One example is the trade-off between foregone future forest revenues F and the immediate windfall of revenue from

[28] Note that the alternative scenarios imply alternative parameter values for each pair of baseline and counter-factual simulations. What is critical for our marginal cost calculations is that any pair of baseline and counter-factual simulations employs identical assumptions (parameter values), with the exception, of course, of Z_{it}, the tax/subsidy that generates the counter-factuals and leads to our marginal cost estimates.

TABLE IV
Present-Value Equivalent Carbon Sequestration and Emissions with Alternative Discount Rates

	Alternative discount rates			
Carbon sequestration and emissions	2.5%	5.0%	7.5%	10.0%
Present-value equivalent carbon sequestration (tons per acre)				
Natural regrowth of mixed stand				
Periodic harvest (scenario #1)	61.90	43.36	30.63	22.72
No periodic harvest (scenario #2)	91.48	50.59	32.85	23.52
Pine plantation				
Periodic harvest (scenario #3)	54.66	41.05	30.76	23.75
No periodic harvest (scenario #4)	80.68	49.99	34.33	25.25
Present-value equivalent carbon emissions (tons per acre)				
Deforestation	54.28	51.83	50.99	50.55

carrying, W. Second, the present-value equivalent tons per acre of sequestration are affected by changing discount rates (Table IV).[29]

In Table V, we examine the impact of changing discount rates on three output variables: marginal sequestration costs, induced forestation, and induced carbon sequestration. The sensitivity analysis is carried out for two pine-plantation scenarios—periodically harvested (#3) and no periodic harvests (#4). First, we find that as the discount rate increases (from 2.5% to 10%), marginal sequestration costs increase monotonically, as expected. The simplest explanation of this effect is that the present-value equivalent sequestration decreases with increased interest rates. The magnitude of the impact is similar to that reported by [23], who found that raising the discount rate in their analysis from 3 to 7% nearly doubled marginal costs.

Next, we find that as the discount rate increases, the forestation caused by a given ($50/acre) subsidy/tax increases. This is also as anticipated, since the up-front subsidy/tax becomes more important, relative to discounted future flows of net revenue, with the increased discount rate. Finally, and most interesting, as the discount rate increases, the impact on induced carbon sequestration is not monotonic: at first increasing interest rates increase induced sequestration, but then they have the opposite effect, decreasing carbon sequestration. The explanation is that there are two factors at work here: land-use changes and the present-value equivalent of carbon sequestration per acre. At first, the land-use effect is dominant, and so with higher interest rates, we find more induced forestation and so more sequestration, but then the effect of smaller present values of carbon sequestration per acre becomes dominant, and so carbon sequestration begins to decrease with higher discount rates. The effect is particularly dramatic in scenario

[29] The rotation period may also be responsive to changes in the discount rate. The extent of the response will depend on the range of discount rates analyzed and the sensitivity of stumpage values to changes in rotation period. While the effect can, in principle, be substantial, it is not for the species and range of discount rates we analyze, and the ultimate effect on annualized carbon yields and sequestration costs is very small.

TABLE V
Discount Rate Sensitivity of the Cost and Quantity
of Carbon Sequestration, Pine Plantation

Carbon sequestration and forestation costs and quantities	Alternative discount rates			
	2.5%	5.0%	7.5%	10.0%
Marginal cost of sequestration ($/ton)				
(Sequestration = 5 million tons/year)				
Periodic harvest (scenario #3)	33	39	58	92
No periodic harvest (scenario #4)	18	27	46	81
Forestation relative to baseline (1,000 acres)				
(Subsidy/tax = $50/acre)				
Periodic harvest (scenario #3)	1,467	2,787	4,368	6,131
No periodic harvest (scenario #4)	1,453	2,763	4,336	6.092
Carbon sequestration relative to baseline				
(1,000 tons/year) (subsidy/tax = $50/acre)				
Periodic harvest (scenario #3)	3,271	4,219	4,302	3,928
No periodic harvest (scenario #4)	4,460	5,099	4,832	4,242

#4, where there is no periodic harvesting, since the fall in present-value carbon equivalents is greatest in that case.

3.3. *The Economic Environment*

It is of particular interest to ask what would happen to the estimated quantities of carbon sequestration and marginal costs if there were significant changes in the economic environment. The baseline simulation with recent price data reflects the reality currently being experienced in the study area—minimal, although not trivial, deforestation. In contrast to this, other parts of the United States—such as New England and the Middle Atlantic states—began to experience positive net rates of forestation as early as the middle of the 19th century. Such background patterns of land-use changes are potentially important. By modifying the assumed level of agricultural product prices in the analysis, we can produce baseline simulations with significant amounts of forestation or deforestation occurring (in the absence of policy intervention), and then investigate the consequences of policy interventions in these new dynamic contexts. We focus here on sensitivity analysis for the periodically harvested pine plantation scenario.

Thus, we change agricultural product prices (in both the baseline and policy simulations) and observe what happens to net forestation and sequestration. As can be seen in Table VI, increasing agricultural prices produces baseline simulations with significant deforestation. What are the impacts of such price changes on carbon sequestration *relative to baseline* at a given level of policy intervention, such as a land-use subsidy/tax of $50 per acre? Not surprisingly, we find that induced sequestration decreases monotonically as the background agricultural product price level increases. The change, however, is by no means linear. The context of low agricultural prices (30% below the base case) increases induced sequestration by 80%, whereas the high price context (30% above the base case) decreases induced sequestration by only 25%.

TABLE VI
Sensitivity of Results to Agricultural Prices, Periodically Harvested Pine Plantation

Carbon sequestration and forestation costs and quantities	Departures from base case agricultural product prices						
	−30%	−20%	−10%	Base case	+10%	+20%	+30%
Baseline forestation/deforestation (1,000 acres)	5,968	3,317	1,430	52	−977	−1,758	−2,362
Marginal cost of carbon sequestration ($/ton) (Sequestration = 5 million tons/yr)	21.93	26.88	32.44	37.91	38.87	39.60	40.94
Carbon sequestration relative to baseline (1,000 tons/year) (subsidy/tax = $50/acre)	7,656	6,212	5,094	4,219	3,914	3,669	3,183

Note: Discount rate is 5%.

The same non-linear impact is seen when we observe the effect of agricultural price changes on the marginal costs of sequestration, again in Table VI. Marginal sequestration costs increase monotonically as we increase the background context of agricultural prices. This is as expected, since the opportunity cost of the land increases. Once again, the change is far from linear; decreases in agricultural prices have a much greater impact than do increases. This happens because higher agricultural product prices result in a substantial amount of deforestation in the baseline. As a result, the effect of a given tax/subsidy—in the context of high agricultural prices—is not only to increase forestation, but also to *retard deforestation*. And the carbon consequences of a unit of retarded deforestation (51.83 tons per acre from Table II) are significantly greater than those associated with a unit of forestation (41.05 tons per acre from Table II), in terms of present-value equivalents. The increased "carbon efficiency" of the policy intervention in the context of a high level of background deforestation thus reduces the marginal costs of sequestration below what they otherwise would be in the context of high agricultural prices.

4. CONCLUSIONS

When and if the United States chooses to ratify the Kyoto Protocol or subsequent international agreements, it will be necessary to decide whether carbon sequestration policies should be part of the domestic portfolio of U.S. compliance activities. For this reason, we have examined the sensitivity of sequestration costs to changes in key factors, including the nature of the management and deforestation regimes, silvicultural species, relative prices, and discount rates.

What conclusions can be drawn from these quantitative results? First, there is the somewhat surprising finding that marginal sequestration costs can be greater for cases with periodic harvesting of timber. Despite the fact that opportunity costs for landowners are less, the more favorable sequestration pattern provided by permanent stands can counteract and overwhelm this effect.[30]

[30] A consistent set of assumptions is employed in the baseline and policy simulations underlying each scenario. This means that comparisons across scenarios typically involve different amounts of deforestation (or forestation) in respective baselines.

Second, changing the discount rate has two types of effects: many of the economic variables take on new values, and the present-value equivalent tons per acre of sequestration are affected. As the discount rate increases, the *marginal costs* of sequestration increase monotonically, because the present-value equivalent sequestration decreases. But as the discount rate increases, the impact on the *quantity* of induced carbon sequestration is not monotonic, because two factors work in opposite directions: forestation increases, but the present-value equivalent of carbon sequestration per acre decreases.

Third, background patterns of land-use changes are potentially important, a reality that we investigated by varying the baseline level of agricultural product prices. We found that induced sequestration decreases monotonically and non-linearly as the background agricultural product price level increases. Likewise, marginal sequestration costs increase monotonically and non-linearly as agricultural prices increase because the opportunity cost of the land increases.

Fourth and finally, there is the striking asymmetry between the marginal costs of carbon sequestration through forestation and those through retarded deforestation. This provides another argument for focusing carbon-sequestration efforts in areas of relatively high rates of deforestation, such as in tropical forests. In addition to the fact that these areas are more efficient engines of carbon storage than temperate forests and in addition to the lower opportunity costs of land that we would ordinarily anticipate to be associated with such areas, there is the additional reality that in an intertemporal economic context, retarded deforestation provides carbon conservation at much lower marginal costs than does forestation of the same area.[31] Of course this would have to be considered alongside other conditions present in any particular context, such as institutional concerns pertaining to administrative feasibility and the strength of property rights.

For many countries, carbon sequestration through forestation or retarded deforestation may be a cost-effective approach to contributing to reduced global atmospheric concentrations of CO_2. This seems most likely to be true for developing nations, although even for highly industrialized countries such as the United States, carbon sequestration through land-use changes could arguably be part of a cost-effective portfolio of short-term strategies [33]. Whether and to what degree "forestry instruments" belong in individual nations' global climate policy portfolios will depend upon geographic, institutional, and economic characteristics of countries and key local characteristics of forestry and land-use practices [22]. The investigation reported in this paper represents one step along the way to such comprehensive analysis.

APPENDIX: THE DYNAMIC OPTIMIZATION PROBLEM

A risk-neutral landowner will seek to maximize the present discounted value of the stream of expected future returns,

$$\max_{\{g_{ijt}, v_{ijt}\}} \int_0^\infty \Big[(A_{it} q_{ijt} - M_{it})(g_{ijt} - v_{ijt}) - C_{it}^{\alpha P_{it}} g_{ijt} + f_{it} S_{ijt} + W_{it} g_{ijt} - D_{it} v_{ijt} \Big] e^{-r_i t}\, dt$$

[31] Additionally, many would argue that the non-climate change benefits of retarding tropical deforestation typically exceed those of increased forestation in temperate zones, because of the preservation of biological diversity in these exceptionally rich ecologies.

$$\textit{subject to}: \quad \dot{S}_{ijt} = v_{ijt} - g_{ijt}$$
$$0 \leq g_{ijt} \leq \bar{g}_{ijt}$$
$$0 \leq v_{ijt} \leq \bar{v}_{ijt},$$

where I indexes counties, j indexes individual land parcels, and t indexes time; upper case letters are stocks or present values and lower case letters are flows.[32] The variables are:

A_{it} discounted present value of the future stream of typical expected agricultural revenues per acre in county I and time t;
q_{ijt} parcel-specific index of feasibility of agricultural production, including effects of soil quality and soil moisture;
g_{ijt} acres of land converted from forested to agricultural use (deforestation);
v_{ijt} acres of cropland returned to a forested condition (forestation);
M_{it} expected cost of agricultural production per acre, expressed as the discounted present value of an infinite future stream;
C_{it} average cost of conversion per acre;
P_{it} the Palmer hydrological drought index and α is a parameter to be estimated, to allow precipitation and soil moisture to influence conversion costs;
f_{it} expected annual net income from forestry per acre (annuity of stumpage value);
S_{ijt} stock (acres) of forest;
r_t real interest rate;
W_{it} windfall of net revenue per acre from clear cut of forest, prior to conversion to agriculture;
D_{it} expected present discounted value of loss of income (when converting to forest) due to gradual regrowth of forest (first harvest of forest does not occur until the year $t + R$, where R is the exogenously determined rotation length);
$\bar{g}_{ijt}$ maximum feasible rate of deforestation, defined such that

$$\int_t^{t+\Delta} \left[\bar{g}_{ij\tau}\right] d\tau = S_{ijt}$$

for arbitrarily small interval, Δ, over which $\bar{g}_{ij\tau}$ is constant; and
$\bar{v}_{ijt}$ maximum feasible rate of forestation, defined such that

$$\int_t^{t+\Delta} \left[\bar{v}_{ij\tau}\right] d\tau = T_{ijt} - S_{ijt}$$

for arbitrarily small interval, Δ, over which $\bar{v}_{ij\tau}$ is constant.

The application of control theoretic methods yields a pair of necessary conditions for changes in land use [34]. Forestation (conversion of agricultural cropland to forest) occurs if a parcel is cropland and if

$$\left(F_{it} - D_{it} - A_{it} \cdot q_{ijt} + M_{it}\right) > 0, \tag{A1}$$

[32] This specification implies that all prices and costs are exogenously determined in broader national or international markets, a reasonable assumption in the present application.

where F is forest net revenue, equal to f_{it}/r_t. On the other hand, deforestation occurs if a parcel is forested and if

$$\left(A_{it} \cdot q_{ijt} - M_{it} - C_{it}^{\alpha P_{it}} - (F_{it} - W_{it})\right) > 0. \tag{A2}$$

These inequalities imply that all land in a county will be in the same use in the steady state. In reality, counties are observed to be a mix of forest and farmland, due largely to the *heterogeneity* of land. If conversion costs are allowed to be heterogeneous across land parcels (within counties) and flood-control projects affect conversion costs as well as agricultural feasibility (yields), then the conversion cost term in the first equation in the Appendix (i.e., the objective function) is multiplied by q_{ijt}. As shown in [34], such unobserved heterogeneity can be parameterized within an econometrically estimatable model so that the *individual* necessary conditions for land-use changes aggregate into a single-equation model, in which the parameters of the basic benefit–cost relationships and of the underlying, unobserved heterogeneity can be estimated simultaneously. The complete model yields the following set of econometrically estimatable equations:

$$FORCH_{it} = FORCH_{it}^{a} \cdot D_{it}^{a} - FORCH_{it}^{c} \cdot D_{it}^{c} + \lambda_i + \phi_{it}$$

$$FORCH_{it}^{a} = \gamma_a \cdot \left[d_{it} \cdot \left[F\left[\frac{\log(q_{it}^{y}) - \mu(1 + \beta_2 E_{it})}{\sigma(1 + \beta_3 E_{it})} \right] \right] + (1 - d_{it}) - \left[\frac{S}{T} \right]_{i,t-1} \right]$$

$$FORCH_{it}^{c} = \gamma_c \cdot \left[d_{it} \cdot \left[1 - F\left[\frac{\log(q_{it}^{x}) - \mu(1 + \beta_2 E_{it})}{\sigma(1 + \beta_3 E_{it})} \right] \right] + \left[\frac{S}{T} \right]_{i,t-1} - 1 \right]$$

$$d_{it} = \left[\frac{1}{1 + e^{-(N_i + \beta_1 E_{it})}} \right]$$

$$q_{it}^{y} = \left[\frac{F_{it} - D_{it} + M_{it}}{A_{it}} \right] \tag{A3}$$

$$q_{it}^{x} = \left[\frac{F_{it} - W_{it} + M_{it}}{A_{it} - C_{it}^{\alpha P_{it}}} \right], \tag{A4}$$

where all Greek letters are parameters that can be estimated econometrically;[33]

FORCH	change in forest land as a share of total county area;
$FORCH^a$	forestation (abandonment of cropland) as a share of total county area;
$FORCH^c$	deforestation (conversion of forest) as a share of total county area;
D^a and D^c	dummy variables for forestation and deforestation, respectively;
ϕ	an independent (but not necessarily homoscedastic) error term;

[33] The econometrically estimatable coefficients have the following interpretations: λ_i is a county-level fixed-effect parameter; γ_a and γ_c are partial adjustment coefficients for forestation and deforestation; μ is the mean of the unobserved land-quality distribution; σ is the standard deviation of that distribution; α is the effect of weather on conversion costs; β_1 is the effect of government flood-control programs on agricultural feasibility; β_2 is the effect of these programs on the heterogeneity mean; and β_3 is the effect of programs on the standard deviation.

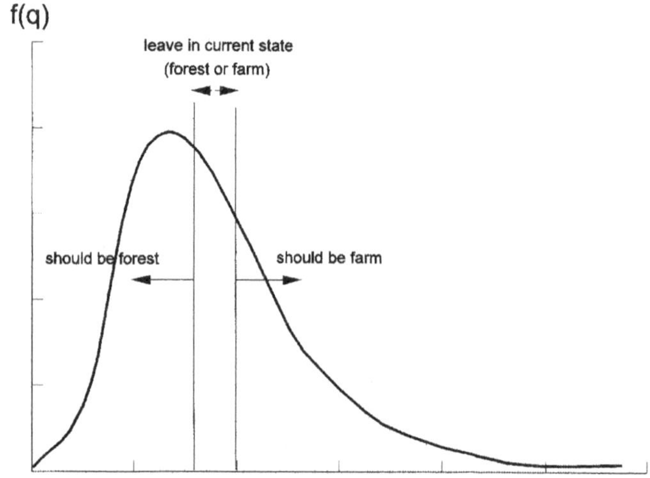

FIG. 3. The distribution of land quality and economic thresholds of forestation and deforestation.

d	probability that agricultural production is feasible;
q^y	threshold value of (unobserved) land quality (suitability for agriculture) below which the incentive for forestation manifests itself;
q^x	threshold value of land quality above which the incentive for deforestation manifests itself; E is an index of the share of a county that has been artificially protected from flooding by Federal programs (by time t);
E	index of share of county artificially protected from periodic flooding;
S	stock (acres) of forest;
F	cumulative, standard normal distribution function;
T	total county area; and
N	share of a county that is naturally protected from periodic flooding.

A simplified, pictorial representation of the model is provided in Fig. 3. The skewed distribution in the figure represents the parameterized lognormal distribution of unobserved land quality; and q^y_{it} and q^x_{it} are the forestation and deforestation thresholds, respectively. Note that each is a (different) function of the benefits and costs of forest production relative to agricultural production. The asymmetries between Eqs. (3) and (4) cause the separation between the two thresholds (where economic signals suggest to leave land in its existing state, whether that be forest or farm). Thus, if expected forest revenues increase, both thresholds shift to the right and we would anticipate that some quantity of farmland would be converted to forest uses. Likewise, an increase in expected agricultural prices means a shift of the two thresholds to the left, and consequent deforestation.

REFERENCES

1. R. M. Adams, D. M. Adams, J. M. Callaway, C. Chang, and B. A. McCarl, Sequestering carbon on agricultural land: Social cost and impacts on timber markets, *Contemp. Policy Issues* **11**, 76–87 (1993).
2. R. Alig, D. Adams, B. McCarl, J. M. Callaway, and S. Winnett, Assessing effects of mitigation strategies for global climate change with an intertemporal model of the U.S. forest and agricultural sectors, *Environ. Resour. Econom.* **9**, 259–274 (1997).
3. J. P. Bruce, H. Lee, and E. F. Haites, "Climate Change 1995: Economic and Social Dimensions of Climate Change," contribution of Working Group III to the Second Assessment Report of the Intergovernmental Panel on Climate Change, Cambridge Univ. Press, New York (1996).
4. J. M. Callaway and B. McCarl, The economic consequences of substituting carbon payments for crop subsidies in U.S. agriculture, *Environ. Resour. Econom.* **7**, 15–43 (1996).
5. R. Daniels, personal communication, Mississippi State University, Agricultural Extension Service, Starkville, MS (1994).
6. R. K. Dixon, S. Brown, R. A. Houghton, A. M. Solomon, M. C. Trexler, and J. Wisniewski, Carbon pools and flux of global forest ecosystems, *Science* **263**, 185–190 (1994).
7. R. K. Dixon, J. K. Winjum, K. J. Andrasko, J. J. Lee, and P. E. Schroeder, Integrated land-use systems: Assessment of promising agroforest and alternative land-use practices to enhance carbon conservation and sequestration, *Climatic Change* **27**, 71–92 (1994).
8. D. J. Dudek and A. LeBlanc, Offsetting new CO_2 emissions: A rational first greenhouse policy step, *Contemp. Policy Issues* **8**, 29–42 (1990).
9. M. E. Harmon, W. K. Farrell, and J. F. Franklin, Effects on carbon storage of conversion of old-growth forest to young forests, *Science* **247**, 699–702 (1990).
10. R. A. Houghton, Tropical deforestation and atmospheric carbon dioxide, *Climatic Change* **19**, 99–118 (1991).
11. C. J. Jepma, M. Asaduzzaman, I. Mintzer, S. Maya, and M. Al-Moneef, "A Generic Assessment of Response Options," Draft, Intergovernmental Panel on Climate Change, Report of Working Group III. Geneva, Switzerland (January 20, 1995).
12. J. A. Kershaw, C. D. Oliver, and T. M. Hinckley, Effect of harvest of old growth douglas-fir stands and subsequent management on carbon dioxide levels in the atmosphere, *J. Sustainable Forestry* **1**, 61–77 (1993).
13. G. Marland, "The Prospect of Solving the CO_2 Problem through Global Reforestation," DOE/NBB-0082, U.S. Department of Energy, Washington, DC (1988).
14. O. Masera, M. R. Bellon, and G. Segura, Forestry options for sequestering carbon in Mexico: Comparative economic analysis of three case studies, *presented at* Terrestrial Carbon Sequestration: An Economic Synthesis Workshop, Bergendal, Sweden, May 15–19 (1995).
15. R. J. Moulton and K. R. Richards, "Costs of Sequestering Carbon Through Tree Planting and Forest Management in The United States," GTR WO-58, U.S. Department of Agriculture, Forest Service, Washington, DC (1990).
16. New York State Energy Office, "Analysis of Carbon Reduction in New York State," Albany (1993).
17. W. D. Nordhaus, The cost of slowing climate change: A survey, *Energy J.* **12**, No. 1, 37–65 (1991).
18. P. J. Parks, D. O. Hall, B. Kriström, O. R. Masera, R. J. Moulton, A. J. Plantinga, J. N. Swisher, and J. K. Winjum, An economic approach to planting trees for carbon storage, *in* "Economics of Carbon Sequestration in Forestry" (R. A. Sedjo, R. N. Sampson, and J. Wisniewski, Eds.), CRC Press, Boca Raton, FL (1997).
19. P. J. Parks and I. W. Hardie, Least-cost forest carbon reserves: Cost-effective subsidies to convert marginal agricultural land to forests, *Land Econom.* **71**, 122–136 (1995).
20. A. J. Plantinga, The costs of carbon sequestration in forests: A positive analysis, *presented at* Terrestrial Carbon Sequestration: An Economic Synthesis Workshop, Bergendal, Sweden, May 15–19 (1995).
21. A. J. Plantinga and R. A. Birdsey, Carbon fluxes resulting from U.S. private timberland management, *Climatic Change* **23**, 37–53 (1993).
22. K. R. Richards, R. Alig, J. D. Kinsman, M. Palo, and B. Sohngen, Consideration of country and forestry/land-use characteristics in choosing forestry instruments to achieve climate mitigation goals, *in* "Economics of Carbon Sequestration in Forestry" (R. A. Sedjo, R. N. Sampson, and J. Wisniewski, Eds.), CRC Press, Boca Raton, FL (1997).

23. K. R. Richards, R. J. Moulton, and R. A. Birdsey, Costs of creating carbon sinks in the U.S., *Energy Conservation Management* **34**, 905–912 (1993).
24. K. R. Richards and C. Stokes, "Regional Studies of Carbon Sequestration: A Review and Critique," Mimeo, Pacific Northwest Laboratory, Washington, DC (June 1995).
25. C. Row, HARVCARB: Modeling forest management and the carbon balance, *presented at* the Western Forest Economics Meeting, Welches, Oregon, May 4–6 (1992).
26. C. Row and R. B. Phelps, Tracing the flow of carbon through the U.S. forest products sector, *in* "Proceedings of the 19th World Congress of the International Union of Forest Research Organizations, Montreal, Canada, August 5" (1990).
27. E. S. Rubin, R. N. Cooper, R. A. Frosch, T. H. Lee, G. Marland, A. H. Rosenfeld, and D. D. Stine, Realistic mitigation options for global warming, *Science* **257**, 148–149, 261–266 (1992).
28. R. A. Sedjo, R. N. Sampson, and J. Wisniewski (Eds.), "Economics of Carbon Sequestration in Forestry," CRC Press, Boca Raton, FL (1997).
29. R. A. Sedjo and A. M. Solomon, Climate and forests, *in* "Greenhouse Warming: Abatement and Adaptation" (N. J. Rosenberg, W. E. Easterling III, P. R. Crosson, and J. Darmstadter, Eds.), Resources for the Future, Washington, DC (1989).
30. R. A. Sedjo, J. Wisniewski, A. Sample, and J. D. Kinsman, "Managing Carbon via Forestry: Assessment of Some Economic Studies," Discussion Paper 95-06, Resources for the Future, Washington, DC (November 1995).
31. T. M. Smith, W. P. Cramer, R. K. Dixon, R. Leemans, R. P. Nelson, and A. M. Solomon, The global terrestrial carbon cycle, *Water Air Soil Pollution* **70**, 19–37 (1993).
32. R. N. Stavins, Alternative renewable resource strategies: A simulation of optimal use, *J. Environ. Econom. Management* **19**, 143–159 (1990).
33. R. N. Stavins, The costs of carbon sequestration: A revealed-preference approach, *Amer. Econom. Rev.* **89**, 994–1009 (1999).
34. R. N. Stavins and A. B. Jaffe, Unintended impacts of public investments on private decisions: The depletion of forested wetlands, *Amer. Econom. Rev.* **80**, 337–352 (1990).
35. S. Swinehart, "Afforestation as a Method of Carbon Sequestration: A Cost–Benefit Analysis," Ph.D. dissertation, Department of Operations Research/Engineering-Economic Systems, Stanford University, Stanford, CA (1996).
36. D. P. Turner, J. Lee, G. J. Koperper, and J. R. Barker (Eds.), "The Forest Sector Carbon Budget of the United States: Carbon Pools and Flux Under Alternative Policy Assumptions," U.S. Environmental Protection Agency, Corvallis, OR (1993).
37. United Nations General Assembly, "United Nations Framework Convention on Climate Change," A/AC.237/18 (Part II)/Add.I and A/AC.237/18 (Part II/Add.1/Corr.1), New York (1997).
38. G. C. Van Kooten, L. Arthur, and W. Wilson, Potential to sequester carbon in Canadian forests: Some economic considerations, *Canad. Public Policy* **18**, 127–138 (1992).
39. J. K. Winjum, R. K. Dixon, and P. E. Schroeder, Estimation of the global potential of forest and agroforest management practices to sequester carbon, *Water Air Soil Pollution* **62**, 213–227 (1992).

PART VII

INTERNATIONAL CLIMATE CHANGE POLICY

[23]

ELSEVIER

Available online at www.sciencedirect.com

SCIENCE @ DIRECT®

Climate Policy 3 (2003) 373–397

CLIMATE POLICY

www.climatepolicy.com

Thirteen plus one: a comparison of global climate policy architectures

Joseph E. Aldy [a], Scott Barrett [b], Robert N. Stavins [c,*]

[a] *Department of Economics, Harvard University, Cambridge, MA, USA*

[b] *Paul H. Nitze School of Advanced International Studies, Johns Hopkins University, Baltimore, MD, USA*

[c] *John F. Kennedy School of Government, Harvard University and Resources for the Future, Cambridge, MA, USA*

Received 3 March 2003; received in revised form 20 July 2003; accepted 9 September 2003

Abstract

We critically review the Kyoto Protocol and thirteen alternative policy architectures for addressing the threat of global climate change. We employ six criteria to evaluate the policy proposals: environmental outcome, dynamic efficiency, cost-effectiveness, equity, flexibility in the presence of new information, and incentives for participation and compliance. The Kyoto Protocol does not fare well on a number of criteria, but none of the alternative proposals fare well along all six dimensions. We identify several major themes among the alternative proposals: Kyoto is "too little, too fast"; developing countries (DCs) should play a more substantial role and receive incentives to participate; implementation should focus on market-based approaches, especially those with price mechanisms; and participation and compliance incentives are inadequately addressed by most proposals. Our investigation reveals tensions among several of the evaluative criteria, such as between environmental outcome and efficiency, and between cost-effectiveness and incentives for participation and compliance.

Keywords: Policy architecture; Kyoto Protocol; Efficiency; Cost-effectiveness; Equity; Participation; Compliance

1. Introduction

There is a growing consensus among both natural scientists and economists that a credible approach is needed to address the threat of global climate change (Watson, 2001). The Kyoto Protocol to the UN Framework Convention on Climate Change (FCCC) may or may not come into force, but serious questions have been raised regarding the Protocol's anticipated performance. A variety of alternative policy architectures[1] have been proposed, including thirteen that have been developed by economists

* Corresponding author.

E-mail addresses: sbarrett@jhu.edu (S. Barrett), aldy@fas.harvard.edu, robert_stavins@harvard.edu (R.N. Stavins).

[1] The general importance of focusing on policy "architecture" and institutions in the global climate domain was first noted by Schmalensee (1996, 1998). See also Victor and Salt (1995), Stavins (1997), Sandalow and Bowles (2001).

1469-3062/$ – see front matter
doi:10.1016/j.clipol.2003.09.004

and other scholars. In this paper, we review these thirteen proposals plus the Kyoto Protocol in the light of key performance criteria. Our review emphasizes economic and design aspects of climate regimes. We give less attention to the politics of regime formation. The latter is important, particularly in comparing Kyoto with the alternatives. While Kyoto has been shaped and molded by a political process, the alternatives have been designed only "on paper". Nonetheless, the proposals have been developed in response to perceived design flaws in the Kyoto approach, and hence it is reasonable and, we believe, useful to compare Kyoto with the alternatives on the basis of their fundamental architectures.

In Section 2 of the paper, we explore the six criteria for assessing global climate policy regimes. In Section 3, we describe the fundamental characteristics of the Kyoto Protocol and alternative policy architectures; and in Section 4, we synthesize the key architectural elements of the proposed climate regimes. In Section 5, we conclude.

2. Alternative criteria for assessing global climate policy regimes

Six potential criteria can guide an assessment of proposed global climate policy regimes: (1) the environmental outcome; (2) dynamic efficiency; (3) dynamic cost-effectiveness; (4) distributional equity (both cross-sectional and intertemporal); (5) flexibility in the presence of new information; and (6) participation and compliance.[2] The last criterion has frequently been ignored, but is essential for achievement of the first four criteria. While climate negotiations have focused on cost-effectiveness, attainment of this objective requires more than flexible mechanisms; it requires full participation. Similarly, the negotiations leading up to Kyoto emphasized the need for stringent targets, but these are of little or no environmental benefit if participation is low or if parties fail to comply.

2.1. The environmental outcome

For many participants in the global climate policy process, particularly non-economists, the most important criterion for assessing alternative policy regimes is the likely magnitude of environmental outcomes.[3] The stock nature of climate change complicates any ranking of policies, because policies may differ in their respective time paths of changes in net emissions. Policies can take a wide variety of emissions paths to achieve the same steady-state atmospheric concentrations (Wigley et al., 1996). While an economic assessment would indicate similar if not identical impacts from climate change at any given steady-state concentration, the rate of climatic change would vary across policies and result in different transitional impacts (on the cost side of the ledger).

---

[2] By "participation" we mean whether a country is a party to a treaty intended to mitigate emissions. By "compliance" we mean the degree to which a country that is a party to such a treaty implements the treaty's obligations.

[3] From an economic perspective, the environmental outcome criterion is redundant, given the dynamic efficiency criterion. Any policy that satisfies dynamic efficiency delivers the socially optimal environmental outcome. But policy debates in international climate negotiations reveal much more interest in environmental outcomes (or proxies, such as emissions reductions) than in efficiency. To some observers, "more emission reductions are better," whatever a comparison of benefits and costs may suggest, but fundamental trade-offs cannot be avoided, even from a strictly environmental perspective. Under some plans, for example, emission reductions would be achieved by greater use of nuclear power, an outcome that would not be welcomed by most champions of the environmental outcome criterion.

Environmental effectiveness is difficult to measure in this domain. Uncertainties abound in predicting global climate change, particularly in terms of economically-relevant geographic scales; quantifying biophysical impacts; and monetizing impacts, especially for non-market goods and services. In order to estimate the impacts of particular policies, it is necessary to estimate what countries would have done in the absence of policy, but such counterfactual baselines cannot be observed.[4]

The need to assess the environmental effectiveness of climate policies raises another difficulty if the policies induce "emissions leakage". If an international (but not fully global) climate policy results in differences in marginal compliance costs among countries, then emissions may "leak" from participating high-cost countries to non-participating low- or zero-cost countries through one of two economic channels.

First, a policy may foster comparative advantage for low-cost countries (for example, countries without emissions commitments) in the production of greenhouse-gas-intensive goods and services. There may be a shift in production of greenhouse-gas-intensive output. Some firms may relocate manufacturing plants from countries with emissions commitments (and higher energy costs) to countries without emissions commitments. In this case, countries with commitments may comply with their obligations, but some of their emissions reductions would be offset by increases in emissions in countries without commitments.

Second, the higher energy costs associated with compliance would reduce world energy demand, depressing oil and coal prices. Countries without emissions commitments would consume more fossil fuels, offsetting some of the emissions reductions by countries with commitments. Therefore, a "narrow-but-deep" agreement may not significantly reduce net emissions, but largely redistribute emissions.[5]

2.2. *Dynamic efficiency*

It seems reasonable to ask whether the gains outweigh the losses of a public policy, and thus determine whether society as a whole is made better-off as a result of the policy. A global climate policy that achieves maximum aggregate *net benefits* is said to be *efficient*.[6] Of course, global climate change policy must address actions, impacts, benefits, and costs that occur over very long time horizons. Greenhouse gases accumulate in the atmosphere because of their slow natural decay rates. Moreover, thermal ocean mixing can delay climatic response. In addition, private sector responses to climate change policies can require significant changes in long-lived capital stock: 50–70 years for electricity generators, and 60–100 years for residential buildings, for example (Jaffe et al., 1999). Exogenous and

[4] The Helsinki Protocol, for example, required that parties reduce their sulfur dioxide emissions by 30%. Actual emissions fell by more than 30%. This might suggest that the agreement succeeded environmentally, but the agreement did not significantly affect behavior (Levy, 1993; Barrett, 2003). In the absence of the treaty, most of the emission reductions would likely have occurred anyway.

[5] Estimates of the magnitude of leakage vary widely. In the case of a unilateral reduction in emissions by the European Union, estimated leakage rates range from 2 to 80% (see Fisher et al., 1996). That is, for every 100 tonnes of carbon abated by the EU, global emissions could fall from 20 to 98 tonnes.

[6] Note that Pareto efficiency requires not only that aggregate net benefits be maximized but that no one be made worse-off by the policy change. The practical question for policy is whether it is essential that no one *actually* be made worse-off or that no one *potentially* be made worse-off. This weaker condition is the frequently employed Kaldor-Hicks criterion. See Stavins (2001a) for a primer on global climate economics. Although efficiency is an important criterion for sound policy analysis, most economists think of benefit–cost analysis as no more than a tool to assist in decision making. Virtually all would agree, however, that the information in a well-done benefit–cost analysis can be of great value in helping to make decisions about risk reduction policies (Arrow et al., 1996).

endogenous technological change can have great bearing on global climate change and policies to address it, especially over the long-term. For all three reasons, *dynamic efficiency* is an important criterion to employ.[7]

To assess dynamic efficiency, economic analysis must account for people's preferences over consumption across time. Individuals' rates of time preference and expectations about future income growth (which affects future consumption opportunities) influence how individuals discount future consumption (Goulder and Stavins, 2002). Individual behavior reflects these preferences in market activities, such as buying and selling government bonds, home mortgages, and credit card balances. At the societal level, concerns for intergenerational equity loom large, in part because explicit markets for intergenerational trade-offs do not exist. As a general matter, future net benefits should be discounted, but the appropriate value of the discount rate for very long-time horizons is unclear.[8]

Uncertainty in estimating benefits and costs also characterizes the global climate change problem. These uncertainties, the intertemporal nature of the problem, and the *irreversible* characteristics of investment in climate protection (economic, if not physical irreversibility), create the conditions for decision-makers to value delaying investment decisions (known as *quasi-option value*) until more information is available (Arrow and Fisher, 1974; Hanemann, 1989; Dixit and Pindyck, 1994; Kolstad and Toman, 2000). This value should, in principle, be included in the calculation of benefits and costs. In the global climate context, the irreversibilities include both the accumulation of greenhouse gases in the atmosphere and the accumulation of capital investments that cannot easily be reversed.[9]

2.3. *Cost-effectiveness*

Although assessing the costs of global climate change mitigation is by no means a trivial task, it is vastly easier than estimating the benefits of policy action. The uncertainty in quantifying regionally disaggregated biophysical impacts of climate change and the difficulty of monetizing some categories of biophysical impacts substantially hinders such benefit estimation. A less demanding criterion circumvents the benefits estimation problem by focusing on cost-effectiveness, in which the least costly means of achieving some given target or goal (which may or may not be efficient) is identified.[10]

Cost-effectiveness analysis cannot be used to compare policies with different benefit streams. Moreover, relying on cost-effectiveness as an assessment criterion can lead to the identification of a low-cost way of doing something that is fundamentally not sensible in economic terms. That is, it can lead to the identification of "fast trains to the wrong station". On the other hand, this kind of analysis can reveal that some policy measures are simply inferior to others.

[7] The application of this criterion to climate change policies is challenging in those cases where the policy proposals do not specify long-term emissions paths. This also complicates the assessment of the environmental effectiveness of climate policies, such as the Kyoto Protocol, which sets an emission constraint only for a 5-year period.

[8] Comprehensive summaries are by Lind (1982) and Portney and Weyant (1999). Recent research has suggested a theoretical basis for employing lower discount rates for longer run analyses (Weitzman, 2001), and empirical research has illustrated this in the presence of uncertainty (Newell and Pizer, 2000). One study has suggested that for climate change a negative discount rate is conceivable (Dasgupta et al., 1999).

[9] These two effects push a stochastic benefit–cost analysis of global climate policy in opposite directions. Which is dominant? Although it has been argued that the second effect is more important (Kolstad, 1996), it is ultimately an empirical question (Ulph and Ulph, 1997; Narain and Fisher, 2000).

[10] Efficiency subsumes cost-effectiveness. A policy that satisfies the efficiency criterion provides the optimal path of net emissions, and requires the cost-effective implementation of this path.

2.4. Equity

An exceptionally important issue in global climate change policy is the distribution of the benefits and costs of policy action, both cross-sectionally and over time. Although economists have given considerable thought over the years to the possibility of using weights to incorporate distributional considerations into determinations of efficiency, there is no consensus, nor likely to be one, on what those weights ought to be. It is reasonable, instead, to estimate benefits and costs, and separately provide as much information as possible to decision makers about winners and losers.

Assessments of international, intra-national, and intergenerational distributions of the benefits and costs of alternative policy regimes are necessary for the identification of equitable climate strategies. A number of criteria merit consideration (Goulder, 2000). First, the criterion of *responsibility* would suggest that—other things equal—those nations that are most responsible for the accumulation of greenhouse gases in the atmosphere should take on the greatest burden for containing the problem. Second, the criterion of *ability to pay* implies that the wealthier nations that possess greater capacity should lead the response to the global climate change problem. Third, the criterion of the *distribution of benefits* suggests that those nations which stand to benefit most from action taken ought to take on greater shares of the cost burden. The first two considerations suggest that industrialized nations should bear the principal burdens for dealing with the prospect of climate change. The third criterion suggests that developing countries (DCs) should shoulder a relatively higher burden, as compared with the other equity criteria.[11]

Though the focus of our attention is with mitigation, it is important to note that climate policy may also involve adaptation. Since future climate change is unavoidable, given the accumulation of greenhouse gases in the atmosphere that has already occurred and their relatively gradual decay rates, individuals, institutions, and ecosystems will need to adapt. Hence, an efficient response to climate change would include a substantial adaptation component. Just as the burden for emissions mitigation raises questions for equity, so does adaptation. For example, the notions of responsibility and ability to pay could translate into technical and financial assistance from industrialized countries to developing countries for adaptation, as well as for mitigation.

The long time horizon of the global climate change problem and potential policy responses raises important issues of *intergenerational equity*. While some have called for the use of low discount rates for long time horizons (such as in the climate context), the use of discounting in benefit–cost analysis has ambiguous effects. A zero or low discount rate might not favor future generations, because it would distort investment decisions among climate change policies and other policies (in which a standard, positive discount rate would guide the decision-making). As a result, society may pass up opportunities to employ other, non-climate policies that could benefit future generations. A world with "too much" climate change investment and "too little" non-climate policy investment may make future generations worse-off, not better-off. More broadly, Schelling (1998) has highlighted the trade-off that may exist between policies to address intergenerational equity and those that address (current) distributional equity: acting to protect future generations (who presumably will be better-off materially than current ones) means that fewer resources will be available to help today's poor in developing countries.

[11] Note, however, that while developing countries may be relatively the most vulnerable to climate change, the absolute benefits of mitigation are likely to be greater for industrialized countries. This is because industrialized countries have much higher levels of national income, and, thus more to lose economically for a given percentage loss in gross domestic product.

2.5. Policy flexibility in the presence of new information

Because uncertainty in benefits and costs characterizes global climate change, the risks of premature or unnecessary actions need to be compared with the risks of failing to take actions that subsequently prove to be warranted (Goulder, 2000). Hence, many economic analyses have advocated a sequential decision-making approach to climate change policy that would facilitate the modification and adaptation of policies as new information reduces uncertainties. Because such new information is potentially of great value, flexible policies (that adapt to new information) have significant advantages over more rigid policy mechanisms (Arrow et al., 1996).

2.6. Participation and compliance

When economists consider domestic environmental problems, they ordinarily put aside participation and compliance issues by assuming (quite reasonably in some cases) the existence of a sovereign government vested with effective, coercive powers. In the international domain, however, full national sovereignty for individual nations means that free rider problems would likely undermine adequate participation and compliance. Countries can engage in free riding behavior through either non-participation or non-compliance, and so at a fundamental level this criterion asks whether a given climate change policy architecture is likely to deter free riding.

A truly efficient (and, hence, cost-effective) climate change agreement would secure full participation by all countries, with each and every country mitigating its emissions to the point where its own marginal abatement costs equaled the sum of marginal benefits globally. But taking the behavior of other countries as given, each country can do better by mitigating only up to the point where its own marginal benefit equals its marginal cost. As long as global marginal benefits exceed every nation's own marginal benefits, countries will either want to avoid participating or avoid complying fully, if they do participate. Successful international cooperation must change these incentives.

Full participation and compliance are necessary but not sufficient conditions for a dynamically efficient and cost-effective climate change policy. For example, an international agreement may be watered down so much that every country participates and complies, and yet the treaty achieves next to nothing. Ideally, a treaty would sustain full participation and compliance while at the same time implementing the dynamically efficient level of climate mitigation. The constraint of sovereignty, however, may make this ideal unattainable (Barrett, 2003).[12]

One alternative is a "narrow-but-deep" agreement—one that achieves substantial per-party mitigation, but attracts relatively little participation. Another alternative is a "broad-but-shallow" agreement—one that achieves relatively little per-country mitigation, but attracts nearly full participation among nations. Current understanding of the benefit and cost functions characterizing climate change suggests that the latter type of policy is more likely to satisfy the dynamic efficiency criterion. Since marginal emissions control costs increase steeply, a broad-but-shallow policy would result in lower overall costs (Barrett, 2002). Moreover, a broad-but-shallow policy could mitigate emissions leakage.

[12] A review of the theory and practice of international cooperation suggests there is not a trade-off between full participation and full compliance (Barrett, 1999, 2003). While participation in international environmental agreements varies widely, compliance appears to be full, irrespective of participation level (Chayes and Chayes, 1995). One reason is that customary international law requires that countries comply with agreements to which they are parties. Another reason is that countries can easily—and legally—avoid the need to comply simply by failing to participate.

Table 1
Alternative international policy architectures for global climate change

Author	Relatively modest short-term goals	Provision for increased developing country participation over time	Use of market-based instruments	Cost constraints through hybrid instruments	Provisions of incentives for participation and compliance
Aldy et al. (2001)	✓	✓	✓	✓	
Barrett (2001, 2003)	✓	✓			✓
Benedick (2001)	✓	✓	✓		
Bradford (2002)		✓	✓		
Cooper (1998, 2001)			✓	✓	
Hahn (1998)	✓	✓	✓		
McKibbin and Wilcoxen (1997, 2000)	✓	✓	✓	✓	
Nordhaus (1998, 2002)	✓	✓	✓		
Schelling (1997, 1998)	✓				
Schmalensee (1996, 1998)	✓	✓	✓		
Stavins (2001b)	✓	✓	✓	✓	
Stewart and Wiener (2001)	✓	✓	✓		
Victor (2001)		✓	✓	✓	✓

3. Alternative global climate treaty architectures

We now turn to a review of the fundamental architecture of the Framework Convention on Climate Change (FCCC) and the Kyoto Protocol, and the fundamental characteristics of thirteen alternative international approaches to the climate change problem (Table 1).

3.1. The Kyoto Protocol of the Framework Convention on Climate Change

At the United Nations Conference on Environment and Development, held in 1992 in Rio de Janeiro, Brazil, agreement was reached on the Framework Convention on Climate Change, which established as its ultimate objective the "stabilization of greenhouse-gas concentrations in the atmosphere at a level that would prevent dangerous anthropogenic interference with the climate system." As an interim step, the FCCC imposed a non-binding goal of reducing greenhouse-gas emissions by industrialized countries (the so-called Annex I countries[13]) to their 1990 levels by the year 2000. The FCCC allowed countries flexibility to develop and implement their own domestic policies to achieve their goals, and provided additional flexibility by allowing Annex I countries to sponsor emissions abatement and sequestration projects in other countries through "joint implementation" (a form of project-based emissions trading). Signed initially by 161 nations, the FCCC entered into force in January, 1994 after being ratified by 50 countries (including the United States). Today, the FCCC has 187 parties, more than any other international environmental agreement (Barrett, 2003).

In December 1997, some 160 countries negotiated the Kyoto Protocol to the Framework Convention. Subsequent negotiations filled in many of the details of the Protocol, and the treaty was substantially

[13] These include the developed nations plus economies in transition.

completed by November 2001. While the Protocol maintains the principle of differentiated responsibilities by the industrialized and developing worlds, it imposes ambitious targets and timetables for emissions reductions by industrialized nations, and it expands significantly the opportunities for countries to achieve their commitments cost-effectively through emissions trading and other "flexible mechanisms". The agreement stipulates "binding" commitments, although, as we explain below, the nature of the requirement that commitments be binding remains unsettled.

The Kyoto Protocol provides specific greenhouse-gas emissions commitments for 38 industrialized (Annex B) countries for the 2008–2012 "commitment period".[14] These emissions targets are expressed relative to countries' emissions in the year 1990.[15] The relative commitments range from 8% below 1990 levels (for the European Union) to 10% *above* 1990 levels (in the case of Australia).[16] Considering the growth of some economies subsequent to 1990, and the essential collapse of others, the range of implicit targets is much greater, with the United States facing a target of about 30% below business-as-usual (BAU) levels in 2012, and Russia and other economies in transition facing targets that would allow substantial *increases* in emissions *above* anticipated BAU levels in 2012. Likewise, Germany's apparently ambitious Kyoto target of an 8% reduction translates into a targeted increase in emissions, due to the post-1990 reunification of the two German nations, and the United Kingdom's target of an 8% reduction likewise translates into a targeted emissions increase, due to the privatization of British coal mining and the opening up of North Sea natural gas sources.[17] These targets apply to six classes of greenhouse gases: carbon dioxide (CO_2), methane, nitrous oxide, hydroflourocarbons (HFCs), perfluorocarbons (PFCs), and sulfur hexafluoride (SF_6).[18]

The Kyoto Protocol acknowledged and reinforced the FCCC's principle of "common but differentiated responsibilities". Unlike industrialized nations, developing countries have no specific obligations to abate greenhouse-gas emissions under the Protocol, and the Protocol provides no mechanism for developing countries to adopt emissions commitments voluntarily.[19]

The Kyoto Protocol includes three flexibility mechanisms that can help countries achieve their commitments at lower costs: international emissions trading, joint implementation, and the Clean Development Mechanism (CDM). The international emissions trading mechanism allows Annex B countries to trade portions of their assigned amounts (targeted emissions) with one another, whereas the joint implementation mechanism allows these same countries to cooperate on projects and transfer emissions allowances

[14] The Kyoto Protocol designates those countries with emissions commitments as Annex B countries. With only a few exceptions, the set of countries with Annex B commitments is identical to the set of Annex I countries in the FCCC.

[15] Transition economy countries were allowed to use a base year other than 1990 if their economic transition from central planning began prior to that date. Countries may also choose to employ 1995 as the base year for measuring changes in emissions of the synthetic greenhouse gases (hydrofluorocarbons, perfluorocarbons, and sulfur hexafluoride).

[16] Note, however, that the European Union can reallocate its emissions allocations among the 15 EU members through the "bubble" provision of the Kyoto Protocol (Article 4). Under this reallocation, the emissions commitments of EU members would range between 28 and below to 27% above 1990 levels.

[17] It is this "hot air" in the German and British targets that render European Union compliance under its Kyoto bubble possible at low cost. According to one recent estimate, fully 80% of the EU targets will be achieved by Germany and the United Kingdom (Andersen, 2002).

[18] The agreement does not specify any obligations beyond 2012. These would have to be negotiated in a new agreement (possibly, an amendment), which would be binding only on the countries that ratified it, provided that the new agreement entered into legal force.

[19] This has confounded the attempts of two countries—Argentina and Kazakhstan—to adopt emissions commitments and participate within the Kyoto framework.

on the basis of such projects. The CDM allows Annex B countries to finance projects in non-Annex B countries in exchange for credits towards meeting their own emission reduction commitments.

The Protocol implicitly allows for trading across different types of gases and some limited trading across time. Since emissions commitments represent the weighted sum of a country's net emissions of greenhouse gases,[20] the Protocol implicitly allows inter-gas trading. By focusing on *net* emissions, the Protocol allows for (potentially cost-effective) substitution of carbon sequestration for greenhouse-gas abatement. Emissions quotas refer to 5-year averages, and countries are allowed to bank and borrow emissions allowances within this 5-year window. Countries may also bank (but not borrow) emissions allowances for use in future, as-yet-undefined commitment periods.

Unlike the underlying FCCC, the Kyoto Protocol stipulates that emissions commitments are legally binding. At the same time, however, Article 18 of the Kyoto Protocol prohibits use of a compliance mechanism entailing "binding consequences" unless adopted by means of an amendment. A compliance mechanism was agreed to in Bonn in July 2001: any industrialized country that fails to comply with its first commitment period obligation must make up for this shortfall in the second commitment period *with a 30% penalty*. This mechanism will fail to alter behavior for several reasons. First, by Article 18, the penalty cannot apply to any country without that country consenting to be bound by it, assuming that the penalty were even included in a future amendment. Second, the penalty applies to a future control period, and a country would have to agree to the emission limit applying in such a period. A country could insist on a generous future limit, taking any sting out of the penalty. Finally, the mechanism relies entirely on self-punishment. Other countries do not take any actions to enforce compliance with the compliance mechanism itself.

The Kyoto architecture can be summarized as including four elements: ambitious, short-term reduction targets for industrialized countries; no emissions obligations for developing countries; flexibility for countries to achieve their commitments through market-based mechanisms; and non-compliance sanctioned with a penalty (not yet binding) linked to commitments in subsequent periods.

As of July 2003, the Kyoto Protocol had been ratified by 111 parties (nations) to the FCCC. To enter into force, Kyoto must be ratified by at least 55 countries, accounting for at least 55% of 1990 Annex I CO_2 emissions. Only the latter trigger for entry into force remained to be fulfilled. As of July 2003, 29 Annex I countries had ratified the Kyoto Protocol, accounting for 44.2% of 1990 Annex I emissions.[21]

Entry into force only awaits ratification by Russia. To secure the participation of Russia (and other Annex I countries, including Japan), negotiating parties made concessions in Bonn and Marrakech. Giving these countries more (sink) allowances effectively relaxed the emissions constraints negotiated previously in Kyoto. These changes reduced the environmental effectiveness of the protocol and illustrate the potential trade-offs between participation incentives and the environmental outcome of an international agreement. This may hint at a key consequence of the Kyoto agreement: it may not achieve high participation and compliance while reducing emissions substantially. For example, while Canada has ratified the agreement, it has signaled its intention to count exports of "clean energy" to the United States towards its

[20] Greenhouse-gas emissions are aggregated based on their 100-year global warming potentials, and are reported in terms of "carbon dioxide equivalent". Refer to Schmalensee (1993), Reilly et al. (1999) and Manne and Richels (2000) for a discussion of the economics of global warming potentials.

[21] These included the member states of the European Union, Canada, Japan, the Czech Republic, Estonia, Latvia, Norway, Romania, and Slovakia.

emission reduction obligations.[22] This accounting, if implemented, would violate the treaty, suggesting that compliance may prove to be a significant problem for the Kyoto Protocol, even if the agreement enters into force.

Environmentalists have supported the Kyoto Protocol partly because it has been "the only game in town" and partly because of the expectation that, with time, the emission limitations achieved by this agreement can be strengthened. The agreement, however, is unlikely to achieve substantial mitigation, in the short term or in the long term, partly because it is ill-equipped to promote participation and compliance. This further undermines dynamic efficiency, although analyses of the Kyoto agreement with US participation show that this policy's global emissions path would be severely sub-optimal (Nordhaus, 2001). Kyoto does incorporate mechanisms aimed at promoting cost-effectiveness, but the success of these ultimately depends on the ability of the agreement to achieve the agreed emission limits.

The agreement champions "distributional equity" by placing the burden of mitigation upon industrialized countries, but in so doing it sacrifices efficiency and cost-effectiveness. The Clean Development Mechanism, likely to bear substantial transaction costs, cannot correct for this failure.

By establishing the precedent of negotiating one 5-year commitment period at a time, the Kyoto agreement promotes a flexible approach that can account for new information in subsequent negotiations and emissions commitments. This approach also highlights a tension between a flexible, adaptive regime that incorporates new information, and a policy that provides more certainty by setting longer-term emissions commitments. The additional certainty of long-term commitments may provide sufficient incentive for investments in long-lived capital that may not occur under a system of periodically negotiated 5-year periods. Some proposed long-term emissions paths may fail to provide such an incentive, however, if they are not dynamically consistent (Aldy et al., 2001). The Kyoto agreement also fails to promote participation and compliance, evident by the US withdrawal from the agreement, the lack of a requirement that developing countries adopt emissions commitments, and the weak self-enforcement regime under Article 18.

3.2. Alternatives to the Kyoto Protocol

Thus, the Kyoto Protocol's architecture has been criticized on a variety of grounds, including: it imposes high costs and unfair burdens on some industrialized countries; it does not require that developing countries take on emissions commitments;[23] it provides ineffective incentives for participation and compliance; and it generates modest short-term climate benefits while failing to provide a long-term solution. In

[22] See, for example, Steven Chase, "Liberal MPs Threaten to Withdraw Kyoto Votes", *Globe and Mail*, 13 September 13 2002, p. A4. The Government of Canada's press release on the ratification says that "Canada will ... continue to work with the international community for improvements to the Kyoto Protocol regime, whether in the recognition of cleaner energy exports or in identifying how more countries can be engaged in setting and achieving targets." See http://www.climatechange.gc.ca/english/whats_new/20021217.htm.

[23] It has been suggested that it is important for all countries—developing, as well as industrialized—to take on emission targets, because: (1) developing countries will account for more than half of global emissions by 2020, if not before; (2) developing countries provide the greatest opportunities now for relatively low-cost emissions reductions; and (3) if developing countries are not included, Annex B abatement will shift comparative advantage in the production of carbon-intensive goods and services outside of that coalition of countries, and render developing economies *more* carbon-intensive than they otherwise would be (see our discussion of emissions leakage in Section 2.1, above).

response to these and other perceived flaws[24] in the agreement and in response to uncertainty regarding the agreement's future given the declared non-participation by the United States, a variety of alternatives have been proposed.[25] Perhaps unsurprisingly, all of these proposed alternatives have been developed by researchers working in the United States. Indeed, we are not aware of any comprehensive proposals developed elsewhere—at least not since the Kyoto Protocol was negotiated. The proposals reviewed here have been advanced in venues ranging from one-page editorials to book-length manuscripts. In the remainder of this section, we briefly describe the key architectural elements of each proposal, taking the proposals in alphabetical order according to their authors.

3.2.1. A hybrid international trading program (Aldy et al., 2001)

This first proposal is for a hybrid international trading instrument that combines an international trading mechanism, not unlike that found in the Kyoto Protocol, with a safety-valve or price ceiling, to be implemented by an international agency making available additional permits at a fixed price.[26] Proceeds from the sale of additional permits would finance climate change research and aid developing countries' efforts to abate greenhouse-gas emissions. Developing countries would be included in the short term via voluntary measures and in the longer term via mandatory commitments (Table 2).

The near-term environmental effectiveness of this proposal would depend upon the breadth of actual participation by developing countries.[27] The use of a safety-valve may limit emissions reductions in contrast with a pure quantity system, but could deliver more environmental benefits than a pure quantity system if the proceeds from the safety-valve provide sufficient incentive for developing countries to adopt emissions commitments. The hybrid price–quantity approach under this policy could approach efficiency in light of uncertainty in both costs and benefits. Support for full, unrestricted emissions trading could also promote cost-effective attainment of emissions commitments, especially once developing country participation becomes mandatory.

The authors of this proposal recommend a sequential process for determining emissions commitments and safety-valve prices in lieu of fixing such commitments over the long term, allowing for policy adaptation to new information over time. Reflecting the *ability to pay* notion of equity, they suggest that developing countries should participate to the extent possible in the near term, with those adopting emissions commitments participating and enjoying economic gains from international emissions trading. Funds accruing from the sale of extra emissions permits through the safety-valve could support emissions abatement efforts in developing countries as well.

To promote compliance, Aldy, Orszag, and Stiglitz recommend experimenting with social and economic (trade) sanctions. Some critics believe that trade restrictions would be difficult to calculate and prone to

[24] Most other critiques of the Protocol may be thought of as referring more to the details than to the general structure (architecture) of the agreement. For example, Hahn and Stavins (1999) note that the international emissions trading program outlined in Article 17 is unlikely to be truly cost-effective if it is implemented through a heterogeneous set of domestic policy instruments.

[25] Obviously, one clear distinction between the Kyoto Protocol and these alternatives lies in the fact that the Kyoto agreement represents the outcome of years of international negotiations and, with the ratification of Russia, will serve as the basis for international climate change policy through 2012. The alternative proposals have not undergone the same real-world evaluation as the Kyoto Protocol, which could be inferred as ex ante political acceptance of the Kyoto approach by ratifying countries.

[26] The hybrid approach combining a tradable permit system with a "tax" (elastic supply of additional permits) has been examined by McKibbin and Wilcoxen (1997), Kopp et al. (1997) and Kopp et al. (1999), building on earlier work by Weitzman (1974) and Roberts and Spence (1976). A recent assessment is provided by Jacoby and Ellerman (2002).

[27] Our assessments of the alternative policy architectures are summarized in Table 2.

Table 2
Alternative international policy architectures for global climate change

Alternative	Environmental outcome	Dynamic efficiency	Cost-effectiveness	Distributional equity	Flexibility	Incentives for participation and compliance
Kyoto Protocol	Probably low, given short-term nature of commitments, and poor incentives for participation and compliance.	Requires reductions that are too large in short run, and silent on reductions required for long run.	Flexible mechanisms help cost-effectiveness, but non-participation by key countries reduces cost-effectiveness; CDM burdened by transactions costs.	Only industrial countries (ICs) face targets, but developing countries (DCs) help shape rules. DCs receive some adaptation assistance.	Emission ceilings are locked in, but only for 5-year periods.	Incentives for participation and compliance are very weak.
Aldy et al. (2001)	Depends on safety-valve price and extent of developing country participation.	Allows for policies that could be consistent with dynamic efficiency.	International emissions trading with a safety-valve would likely result in common price for all participants.	Delays mandatory emissions commitments by DCs. Safety-valve funds to DCs for abatement efforts.	Commitments and safety-valve price adjusted over time in response to new information.	Use of sanctions, especially on trade, to promote compliance. Incentives for developing country participation.
Barrett (2001, 2003)	Depends on the agreed standards.	Technology lock-in may impair efficiency, but increased R&D may also lower costs.	Would not equalize marginal costs across all sectors.	R&D funded according to UN scale. ICs pay for technology adoption by DCs; adaptation funded by ICs.	R&D protocol provides information about technologies to lower costs, but standards may create lock-in.	R&D investment, economies of scale, network externalities, and trade restrictions create incentives for participation. No need to enforce compliance.
Benedick (2001)	Depends on levels for R&D, technology standards, etc.	Technology lock-in may be a problem, but public sector R&D may lower costs.	Would not be a global agreement, and would not equalize marginal costs across all sectors.	ICs to transfer new technologies to DCs. US to show leadership in reducing emissions unilaterally.	R&D would provide more information about new technologies.	Participation deliberately restricted, at least initially and in some areas. No explicit mention of compliance.
Bradford (2002)	Would depend on the magnitude of financial contributions to the central authority.	Could potentially support a dynamically efficient outcome.	Common offer bid for emissions allowances to all countries would insure cost-effectiveness.	Financing obligations would reflect ability to pay and expected benefits from mitigating climate change.	Central authority could adjust emissions allowances purchases with new information over time.	Does not explicitly address enforcement of financing obligations.
Cooper (1998, 2001)	Would depend on the level of the carbon tax.	Could potentially support a dynamically efficient outcome.	Common carbon tax would be cost-effective.	Tax would be uniform, but part of revenue could be redistributed to DCs.	Tax level can be changed, to adjust to new information.	Does not incorporate explicit mechanisms. Relies on a "commitment" to treaty objectives plus transparency.
Hahn (1998)	Depends upon levels at which instruments are set.	Depends upon levels and time paths of instruments.	Could be cost-effective, due to reliance on market-based and related instruments.	Depends upon allocations.	Very flexible; instruments that perform best are continued.	No attention is given to participation and compliance.
McKibbin and Wilcoxen (1997, 2000, 2002)	Relatively low carbon emissions price implies modest near-term emissions reductions.	Could potentially support a dynamically efficient outcome.	Common carbon price across all countries supports cost-effective implementation.	DCs would receive emissions endowments in excess of current emissions.	Decadal negotiations to select carbon price allows for accounting of new information.	Does not substantially address participation or compliance issues.
Nordhaus (1998, 2002)	Relatively low carbon tax implies modest near-term emissions reductions.	Could potentially support a dynamically efficient outcome.	Harmonized carbon tax insures cost-effective implementation among participating countries.	Participation conditional on per capita income. DCs would also likely receive financial transfers.	Periodic international votes allows for adjusting carbon tax to new information.	Promotes compliance through trade measures. Developing country participation supported through financial transfers.

Schelling (1997, 1998)	Would probably have little effect on emissions.	Does not front-load mitigation. Promotes R&D to reduce future mitigation costs.	Would aim to reduce emissions globally.	Financial transfers to DCs.	Emphasizes the need to act, rather than to meet a particular target.	Enforcement of compliance not needed by design.
Schmalensee (1996, 1998)	Little effect in short run, but significant effects in long term.	If targets are sufficient, could be dynamically efficient.	Could be cost-effective, due to reliance on market-based and related instruments.	Little attention given to distributional equity in the cross-section, but could provide intertemporal equity.	Quite flexible, due to focus on beginning with modest targets.	No attention given to participation and compliance issues.
Stavins (2001b)	Abatement would be very modest in the short term, but much more ambitious in the long term.	If targets are sufficient, could be dynamically efficient.	Could be cost-effective, due to reliance on tradable permits, carbon taxes, and hybrid systems.	Addresses cross-sectional distributional equity through allocation of permits and use of growth targets.	Long-term targets are flexible, to allow for effects of learning.	Little attention to participation and compliance, except for incentives for DCs.
Stewart and Wiener (2001)	Would depend on the magnitude of the "headroom" allowances given to DCs.	Dynamic efficiency weakened by participation and compliance problems.	Reliance on an expanded CDM, and participation and compliance problems undermine cost-effectiveness.	Headroom allowances to DCs plus emissions trading provide potential economic gains to poor countries.	Emission commitments would need to be periodically negotiated.	Similar to Kyoto Protocol, with exception of incentives from "headroom" allowances.
Victor (2001)	Similar in targets to KP, but with safety-valve sales of additional permits.	Better than KP in its emission path, but not defined.	Includes flexible mechanisms of Kyoto Protocol; hence, can be cost-effective.	By bringing DCs into set of nations facing binding constraints only as they become more wealthy, equity is addressed.	Subsequent periods would need to be renegotiated.	Compliance is considered through buyer liability scheme, but participation is not addressed.

political manipulation. They could also allow for a sequence of reprisals and counter-reprisals that would damage trade relations.

3.2.2. A research and development protocol (Barrett, 2001, 2003)

This proposal is for a fundamentally different approach, emphasizing common incentives for climate-friendly technology research and development (R&D), rather than targets and time tables. Barrett's approach includes a research and development protocol that would support collaborative research, and protocols that would require common standards for technologies identified through collaborative research efforts. Barrett maintains that the departure from emissions commitments and market-based instruments is the necessary cost of designing a participation- and compliance-compatible regime. His proposal also includes a protocol aimed at making some short-term progress, but without reliance on international enforcement.

This proposal could potentially support a high degree of environmental effectiveness, depending on the payoffs to the cooperative R&D efforts,[28] but the system would neither be efficient nor cost-effective, not least because the technology standards would not apply to every sector of the global economy, and may entail some technological lock-in.[29] Equity is considered explicitly, and flexibility would be aided by the public investment in basic research, although it may be hampered by lock-in. The real strength in this proposal lies in the incentives it would create for participation and compliance. Basic R&D would lower the cost of developing new technologies. Economies of scale, network externalities, and automatic trade restrictions would also make diffusion of the new technologies more attractive to each country, the greater the number of countries adopting the technology protocols. Minimum participation would need to be set high enough to ensure that participation was tipped.

3.2.3. A portfolio approach (Benedick, 2001)

This proposal is similar to Barrett's approach, and emphasizes (long term) international standards and incentives for technology innovation and diffusion, but also includes a renegotiation of the Kyoto targets and a process in which participation in negotiations expands over time. The approach is to adopt a portfolio of policies, including a small carbon tax to fund new technology research, to move the international community toward a desirable technology strategy.

The proposal can reasonably be viewed as a blend of several other proposals. Like Barrett, Benedick proposes the adoption of standards, but unlike Barrett's, this proposal is not strategic in focus. The proposal is aimed at achieving environmental effectiveness, but it would be less successful in supporting efficiency or cost-effectiveness, partly because it focuses on the advantage of limiting negotiations to a relatively small subset of countries. Enforcement is not considered.

3.2.4. International emissions trading without a cap (Bradford, 2002)

This proposal is the equivalent of an international emissions trading program without a fixed cap on emissions. All nations, including developing countries, are allocated permits equivalent to their anticipated business-as-usual time path of emissions. Periodically, an international authority offers to purchase (and retire) emissions allowances. Distributional issues are handled through the financing of the international

[28] For other commentaries on the potential for employing technological cooperation as a central architectural element, see Buchner et al. (2002), Edmonds et al. (2001), Flannery (2001) and Jacoby (1998).

[29] This may also be a problem with other proposals.

authority, with differential funding responsibilities being established on the basis of per capita income levels and other criteria.

The environmental effectiveness of Bradford's proposal would depend on the magnitude of countries' contributions to the central authority responsible for purchasing emissions allowances. While countries have incentives to sell emissions reductions (so long as the bidding price exceeds marginal costs), the proposal may not adequately induce participation in the financing scheme. This approach reveals the costs of participation in a much more transparent manner than other policies, such as straight quantity-based systems that allow for devolution to the private sector. It has the advantage of letting countries know how much they would be spending—in total—on climate change mitigation. However, it also has the disadvantage, because of its transparency, of possibly becoming a lightning rod for political opponents.

Subject to this financing participation constraint, the central authority could purchase emissions allowances from countries over time consistent with a dynamically efficient emissions path. The process of soliciting bids for emissions allowances would result in cost-effective emissions abatement. This policy reflects several equity principles. Bradford recommends that a country's financial contribution to the central authority depend on its per capita income (ability to pay) and the benefits it will incur from climate change mitigation (distribution of benefits). As new information becomes available, the central authority could adjust its plans to purchase emissions allowances accordingly, allowing for substantial policy flexibility, so long as countries adequately finance this authority. There is no suggestion how the agreement would enforce either contributions to the international authority or the emissions limits associated with the purchase scheme.

3.2.5. Harmonized domestic carbon taxes (Cooper, 1998, 2001)

This proposal is conceptually distant from the Kyoto framework: instead of multilateral negotiations over national emissions quotas, countries would negotiate a set of common actions aimed at achieving global emissions targets. In particular, a harmonized carbon tax would be used by all participating nations—industrialized and developing alike—to tax their domestic carbon usage at a common rate, thereby achieving cost-effectiveness.

This proposal by Cooper (1998, 2001) for a uniform, harmonized carbon tax raises a number of problems. First, developing countries may argue that it is unfair that they should adopt the same tax as industrialized countries, given that the latter are largely responsible for the climate change problem in the first place. Second, developing countries may have little incentive to adopt such a tax (or, indeed, substantial mitigation effected by a different policy instrument), though transfers could be used to promote participation by developing countries. Finally, adoption of a harmonized tax may create incentives for gaming behavior. Countries may change their tax codes, for example, to neutralize the effect of a carbon tax.[30]

Overall, Cooper's proposal for a uniform carbon tax shares many of the strengths of an ideal quantity-based approach (with perfect trading). Indeed, under certain conditions, the fundamental difference between the two approaches depends only on the existence of uncertainty in mitigation costs (Weitzman, 1974). The weakness in Cooper's proposal is also similar to the weakness in quantity-based proposals, and in the Kyoto Protocol itself: the difficulty in enforcing participation and compliance. Cooper

[30] Iterative approaches may be vulnerable to similar gaming behavior. The Kyoto Protocol's CDM, for example, creates incentives for "paper trades", and the Kyoto emission limits for certain key countries were diluted in the Conference of the Parties (six) negotiations in Bonn and Marrakech. Bradford's bidding approach may also be vulnerable to strategic manipulation.

(2000) addresses the compliance issue, but drawing on Chayes and Chayes (1995), he argues that explicit mechanisms for compliance are not needed (refer to Section 4 for a discussion of this point).

3.2.6. A portfolio of case studies (Hahn, 1998)

Another significant departure from the "targets and time tables model" is provided by Hahn (1998), who proposes experimentation with multiple "case studies" of potential policy instruments to abate greenhouse-gas emissions in the short term, including: coordinated measures; an emissions tax; tradable emission permits among some set of industrialized nations; tradable emission permits among industrialized nations with joint implementation for developing countries; and a hybrid system.

It is difficult to assess this approach, because it depends upon the particular set of case studies that are examined, and upon their performance (and hence ultimate adoption). It can be said, however, that the approach could well be cost-effective, given the apparent focus on market-based instruments. Also, this is a highly flexible approach, given the reliance on learning by doing. Essentially no attention is given to issues of participation and compliance.

3.2.7. Domestic hybrid trading schemes (McKibbin and Wilcoxen, 1997, 2000, 2002)

This is largely a price-based approach, with two domestic markets for tradable permits—one for annual emissions and another for perpetuities ("endowments"). As in other proposals, governments would provide a safety-valve of permit sales at a fixed price (which can rise over time). Endowments would be made to both industrialized and developing countries, the latter well in excess of current or anticipated emissions to allow for economic growth. In this proposal, there is no international trading system.

The McKibbin and Wilcoxen proposal would likely deliver modest environmental effectiveness in the near term with its fixed emissions permit price of US$ 10 per tonne. As a relatively flexible policy approach, the authors advocate international negotiations to set the emissions permit price once per decade. To induce developing country participation, McKibbin and Wilcoxen recommend that developing countries receive emissions endowments in excess of current emissions. In contrast with proposals that allow developing countries to gain from selling unused emissions allowances to developed countries, this proposal implies only modest costs in the near term (from investments anticipating positive domestic permit prices in the future). Although it distinguishes between developed and developing countries in terms of their near-term emissions abatement, the policy may address equity concerns to a lesser degree than other proposals that use some form of implicit side-payments for developing countries. The lack of any explicit or implicit side-payments may result in non-participation by these countries.

Like several other proposals, this architecture does not address the broader participation and compliance incentives problem. If these problems were overcome, however, the approach could achieve cost-effective emissions abatement in the long term (once developing countries' emissions permit prices equaled the price in developed countries), and perhaps close to cost-effectiveness in the short term. This could support an emissions path broadly consistent with dynamic efficiency, depending on the emissions prices set at the decadal climate policy negotiations.

3.2.8. An efficient set of harmonized carbon taxes (Nordhaus, 1998)

This proposal combines the notion of harmonized carbon taxes with attention to the efficiency of the targets. This is done by setting the harmonized carbon tax at the efficient level through a dynamic benefit–cost analysis, where the benefits are determined through an international voting mechanism which is intended to reflect countries' true willingness-to-pay. As in other proposals, developing countries participate only when their per capita incomes reach particular threshold levels. Compliance is promoted through import duties, which are levied on goods from non-participant countries, based on carbon content, with goods from poor countries exempted.[31]

The proposed harmonized carbon taxes in the Nordhaus proposal would likely deliver modest environmental effectiveness in the near term. He recommends a low carbon tax applied only to countries above an income threshold. Consistent with an ability-to-pay notion of distributional equity, this proposal requires emissions abatement efforts only among wealthy countries. Nordhaus suggests that some form of financial side-payment will be necessary to induce developing country participation. While the policy explicitly incorporates a mechanism based on a dynamic benefit–cost analysis, the distinction between participating and non-participating countries undermines the dynamic efficiency and cost-effectiveness of the proposal.[32] By allowing countries to reveal periodically their willingness to pay for climate protection, this policy structure can adjust and adapt to new information.

3.2.9. A global climate Marshall Plan (Schelling, 1997, 1998, 2002)

This proposed climate policy architecture is inspired by the process of dividing up resources made available by a "Marshall Plan". The approach would be for the industrialized countries to accept mutually agreed actions. International mechanisms in pursuit of targets and time tables (such as international permit trading) are dismissed, although domestic market-based instruments are recommended.

Schelling's proposal starts from the premise that substantial enforcement, requiring effective penalties for non-compliance, is needed but cannot be supported by existing international institutions (which do not have the authority to levy credible threats of significant penalties for non-compliance). His approach is thus to look at measures that might be adopted without such penalties. Starting from this premise, Schelling's proposal would presumably not reduce emissions substantially, and so would not receive high marks for environmental effectiveness. He recognizes the desirability of efficiency and cost-effectiveness, but his proposal does not incorporate mechanisms aimed at satisfying these criteria. However, financial transfers to developing countries would help reduce emissions cost-effectively, while at the same time satisfying some notion of equity. Schelling recognizes the need for flexibility, but his approach has neither advantages nor disadvantages in this regard as compared with most of the alternatives. Participation and compliance are not directly enforced, nor does this approach require enforcement. Indeed, it is predicated on enforcement being inevitably weak.

3.2.10. A broad-but-shallow beginning (Schmalensee, 1996, 1998)

Schmalensee (1996, 1998) focuses on two dimensions of an international climate change agreement: the breadth of the coalition of countries that are meaningfully participating; and the depth of their commitment (stringency of targets). He argues that the most productive first step is to include as many countries as possible, but not require very severe reductions. He terms this a broad and shallow approach, in contrast with the Kyoto Protocol, which he characterizes as being narrow and deep. Schmalensee's argument is

[31] In a subsequent paper, Nordhaus (2002) provides a detailed comparison of price-based and quantity-based architectures, in which he strongly favors the former. The architecture which is recommended includes harmonized carbon taxes with relatively modest short-term tax levels that increase significantly over time, combined with a mechanism for developing countries to take on increasing degrees of responsibility over time.

[32] With full compliance, the policy is cost-effective among all participating countries (through the harmonized carbon tax), but implicit prices on carbon emissions would vary among participating and non-participating countries.

that getting a large number of nations to make commitments now is the best strategy for building the depth and breadth necessary in the long term to address the problem in meaningful ways.

In terms of environmental effectiveness, this approach would have relatively small effects in the short term, but significant effects in the long term. If the targets are sufficient, it could thus be dynamically efficient. Likewise, this approach could be cost-effective, due to its reliance on market-based instruments to achieve targets. Little attention is given to cross-sectional equity, but the time path may be consistent with intertemporal equity. The approach provides considerable flexibility by starting off with modest targets. No attention is given to participation and compliance.

3.2.11. A three-part policy architecture (Stavins, 2001b)

This policy architecture is consistent with the Framework Convention on Climate Change, but departs from the Kyoto Protocol: (1) all countries participate, with an explicit mechanism providing for voluntary accession by developing countries, and a trigger, linked with per capita income, which would require developing countries to take on "growth targets," commitments that are a function of per capita income and other negotiated factors;[33] (2) in aggregate, short-term targets that are moderate yet rigid, and long-term targets—put in place now—that are much more ambitious (in order to induce needed technological change), but flexible to respond to learning; and (3) market-based instruments, including international permit trading, possibly with a safety-valve.

In terms of environmental effectiveness, abatement would be very modest in the short term, but much more ambitious in the long term. Depending upon the specific time path of adopted targets, this could yield dynamic efficiency. In any event, there is considerable promise of cost-effectiveness, due to reliance on tradable permits, carbon taxes, and related hybrid systems. Cross-sectional distributional equity can be addressed through the initial allocation of permits to developing countries and the use of growth targets. Flexibility is maintained by using firm (but modest) targets in the short term, but flexible (but more ambitious) targets in the long term, to allow for response to learning. Like many of the architectural proposals, there is little direct attention given to participation and compliance, although the incentives for developing countries to participate could be considerable.

3.2.12. Using quotas to attract developing countries (Stewart and Wiener, 2001)

This approach focuses on increasing developing country participation by four instruments: (1) a streamlined CDM; (2) voluntary participation in emissions trading without emissions quotas;[34] (3) mechanisms for voluntary accession to the emissions quota system; and (4) automatic graduation to the quota system given particular per capita incomes having been reached. Stewart and Wiener (2001) specifically focus on the need to secure the participation of major developing countries such as China and India by giving these countries "headroom allowances"—allocations in excess of their likely BAU emissions.

The proposal is intended to increase participation in a Kyoto-like agreement by reducing emissions reduction burdens in the near term. The authors suggest that developing countries should have the option to adopt voluntary emissions commitments with headroom emissions allowances as the necessary, implicit side-payment. With international emissions trading, this could further reduce the environmental

[33] See, for example Frankel (1999). In 1999, the Argentine government offered to take on an emissions commitment indexed to its economic growth. Lutter (2000) provides an analysis.

[34] This would be the international version of an "emission reduction credit" program, as opposed to a "cap-and-trade" program (see Stavins, 2002 for definitions and examples), and can be thought of—in the Kyoto Protocol context—as a national-level CDM policy.

effectiveness from the current status of the Kyoto agreement. While this provides an economic incentive for developing countries to participate, it does not address the participation and compliance incentive problems for developed countries or for developing countries once they are no longer net exporters of emissions allowances in future commitment periods.

The participation and compliance concerns hinder dynamic efficiency and cost-effectiveness, although the policy does aim to improve on both dimensions by expanding opportunities to take advantage of low-cost emissions abatement in developing countries. The proposal addresses equity concerns by requiring developed countries to finance emissions abatement in developing countries, whether through the expanded CDM or trade in emissions allowances under developing country commitments. It maintains essentially the same structure and timing as the Kyoto agreement, and so is comparable in its flexibility with respect to new information.

3.2.13. Increasing compliance through buyer liability (Victor, 2001)

Finally, Victor (2001) proposes an approach that in the short term is similar to the Kyoto Protocol, except that individual countries can buy unlimited numbers of allowances at a specified price (safety-valve). In the short term, developing countries participate through the CDM, but in the long term, a graduation mechanism is proposed for developing countries as they reach particular incomes and then must adopt either quotas (as developed countries) or growth targets. Compliance, he argues, would be promoted through a buyer liability scheme. Under this arrangement, if the seller of a permit did not reduce its emissions as promised, the buyer could not claim the emission credit. Since buyers of emission credits are most likely to be private entities in the liberal democracies, Victor argues that the buyer liability rule could be reliably enforced by domestic institutions in these countries.

The environmental effectiveness of this approach is necessarily compromised by its reliance on a safety-valve approach in the short term, but the overall result could be a time path of emissions reductions much closer to the dynamically efficient path than that likely to be forthcoming with the Kyoto Protocol, for example. Reliance on market-based instruments facilitates cost-effectiveness, and distributional equity is addressed through the use of growth targets. Although compliance is considered through the buyer liability scheme, the proposal does not address participation.

4. Major themes among the policy architectures

Although there is considerable diversity among these thirteen alternatives to the Kyoto Protocol, a number of themes emerge (some of which are shared by *all* of the proposals): use of relatively moderate short-term goals; provision for increased developing country participation over time; use of market-based mechanisms; cost constraints through hybrid instruments; and provision of incentives for participation and compliance (Table 1).

First, many of the proposals reflect a general concern that the Kyoto commitments are "too little, too fast," that is, insufficient to do much about the climate change problem, but excessively ambitious (and hence costly) in the short term. Therefore, nearly all of the proposals feature commitments which are moderate in the short-term and become much more stringent in the long-term.

Second, many proposals maintain that developing countries must play a more significant role over time. Several proposals (Aldy et al., 2001; Schmalensee, 1996, 1998; Stavins, 2001b; Stewart and Wiener, 2001) would require developing countries to take on emission commitments in the near term. These proposals plus others (McKibbin and Wilcoxen, 1997, 2000, 2002; Nordhaus, 1998; Victor, 2001) recommend some form of graduation: an income threshold above which nations must take on emission commitments. Other proposals include developing country participation in forms that do not involve emission commitments (Bradford, 2002; Hahn, 1998). Finally, one proposal would have developing countries participate in the financing of research and development activities, but with contributions reflecting their differentiated responsibilities and capabilities (Barrett, 2001, 2002).

Third, a number of proposals provide positive incentives for developing country participation. Some would require that developing countries adopt emission ceilings but with "head room" so that these countries could become net exporters of emission allowances, providing the resources needed to finance their abatement (Aldy et al., 2001; Stavins, 2001b; Stewart and Wiener, 2001). The proposal by Barrett (2001, 2003) would have developing countries be bound by the technology standards incorporated in separate protocols, but the diffusion of these technologies in developing countries would be financed by industrialized countries, also an element of Benedick's (2001) proposal.

Fourth, nearly all of the proposals would allow, encourage, or require implementation through market-based instruments. While Cooper (1998, 2001) advocates harmonized carbon taxes, most proposals favor hybrid quota-tax schemes (Aldy et al., 2001; Hahn, 1998, McKibbin and Wilcoxen, 1997, 2000, 2002; Victor, 2001) or tradable permit systems (Hahn, 1998; Stavins, 2001b; Stewart and Wiener, 2001). Bradford's (2002) proposal could potentially achieve a similar, cost-effective outcome by implementing a permit purchase scheme. Efficient implementation of all such instruments would allow countries to achieve their emissions commitments at lower cost, which would presumably increase the likelihood that they would comply with their commitments. It is precisely because of concerns about compliance/participation, however, that Barrett (2001, 2003) departs from the conventional economists' prescription of market-based instruments, and advocates collaborative R&D combined with technology standards.[35] Schelling's (1997, 1998) proposal also drops the premise that targets and timetables can be enforced internationally, although his proposal is not far from unilateralism.

Fifth, in response to concerns about the costs of complying with emissions commitments, many proposals recommend a hybrid tax-quota or pure price regimes to set a cap on marginal costs (Aldy et al., 2001; Cooper, 1998, 2001; McKibbin and Wilcoxen, 1997, 2000, 2002; Stavins, 2001b; Victor, 2001). Other proposals take different approaches to limiting the costs of climate policy (Barrett, 2001, 2003; Bradford, 2002; Hahn, 1998).

Sixth, although there is widespread recognition that the Kyoto Protocol does not provide effective incentives for participation and compliance, most proposals give relatively little attention to this aspect of an international climate agreement, the chief exceptions being Barrett (2001, 2003), Victor (2001), and Wiener (1999, 2001). More broadly, advocates of policies to ensure low costs of attaining emissions commitments believe that such low costs will provide incentives for participation and compliance. Other authors (Aldy et al., 2001; Nordhaus, 1998) argue that a treaty such as the Kyoto Protocol can, in principle, be enforced by means of trade restrictions. Barrett (2003) shows how trade restrictions helped enforce other agreements, but he doubts the efficacy of trade restrictions in enforcing a climate agreement relying on either targets and timetables or taxes. Cooper (2000) rejects trade restrictions for a different

[35] This proposal has its own problems (lack of cost-effectiveness, for example), but the important point here is that the proposal arose from a concern about international enforcement. For most other proposals, enforcement is either assumed not to be a problem or, as in the Kyoto Protocol itself, an enforcement mechanism is added almost as an afterthought to an approach that is advocated for other virtues.

reason. Citing Chayes and Chayes (1995), he argues more broadly that sanctions are not needed and that transparency in governmental actions (monitoring) should provide sufficient incentive for compliance.[36]

Any pragmatic proposal addressing developing country participation must confront a difficult trade-off if the United States is to participate as well. If a proposal includes aggressive developing country commitments, the agreement may be expected to fail to elicit developing country participation, since nearly all developing countries—and many industrialized countries as well—believe that the industrialized world should take on binding emissions commitments first.[37] On the other hand, if a proposal recommends no near-term emissions commitments by developing countries, the agreement may be expected to fail to gain political acceptance in the United States, as evidenced by the unanimous (95–0) passage of the Byrd–Hagel Resolution in the US Senate (1997), which called for similar treatment of industrialized and developing countries in any international agreement on global climate change.[38] The notion of allowing short-term developing country participation *exclusively* through the CDM (Victor, 2001) would likely garner little support from the United States, since such participation would fall well short of the criteria specified by the Byrd–Hagel Resolution.

5. Conclusions

As the international community considers efforts to address the risks posed by global climate change, a number of issues have been raised about the design and potential impacts of climate change policy. These issues have served as the basis for our review of the Kyoto Protocol and thirteen alternative climate change policies. We have employed six criteria by which to evaluate the Protocol and these other proposals: environmental outcome, dynamic efficiency, dynamic cost-effectiveness, distributional equity, flexibility in the presence of new information, and participation and compliance. The Kyoto Protocol does not fare well along several of these dimensions, including environmental outcome, dynamic efficiency,

[36] The reasoning embraced by Chayes and Chayes (1995) is flawed in a number of respects. First, evidence that countries comply is not evidence that compliance is not a problem. Countries may only be complying with agreements that do not seek to change behavior or that only aim to coordinate. Second, the Chayes's do not consider the participation problem. Under the rules of international law, countries are expected to comply with treaties to which they become a party, but they are not required to participate. A country that worried about its ability to comply would thus choose not to participate. For critiques of the Chayes' reasoning, see Downs et al. (1996) and Barrett (1999, 2003).

[37] A coalition of developing countries rebuffed New Zealand's call during the 1997 Kyoto Conference of the Parties for developing country commitments. The coalition has succeeded in keeping the topic off the agendas of subsequent Conferences of the Parties. While the case has been made that developing countries with emissions commitments could become net exporters of emissions allowances and thus enjoy gains from trade (Yellen, 1998), this argument has generated little support in the developing world. Likewise, though Cooper (1998, 2001) suggests that developing country governments would be favorably disposed toward imposing (harmonized) carbon taxes for public finance (if not environmental) reasons, no developing country has adopted such a tax. We note as well that a number of major developing countries have substantial energy subsidies (International Energy Agency, 1999).

[38] Senate Resolution 98 (12 June 1997) states that the United States should not be a signatory to any agreement under the Framework Convention on Climate Change which would "mandate new commitments to limit or reduce greenhouse-gas emissions for the Annex I Parties, unless the protocol or other agreement also mandates new specific scheduled commitments to limit or reduce greenhouse-gas emissions for Developing Country Parties within the same compliance period." The resolution refers to five developing countries by name: China, Mexico, India, Brazil, and South Korea. Subsequently, the Clinton Administration employed the phrase, "meaningful participation by key developing countries" (Eizenstat, 1998) in a "strategic interpretation" of the Byrd–Hagel Resolution.

cost-effectiveness, and participation and compliance. And none of the thirteen alternative proposals we evaluated fared well along all six dimensions.

This reflects, in part, the fundamental tensions among the criteria. For example, the typical policy trade-off between efficiency and equity also characterizes the climate change problem, especially with respect to the role of developing countries. Fully efficient and cost-effective emissions abatement would require developing countries to actively limit their greenhouse-gas emissions, while ability-to-pay and responsibility notions of equity would likely exempt most developing countries from emissions abatement. Some proposals attempt to address this conflict by allocating "headroom" emissions allowances to developing countries and allowing them to participate in international emissions trading, but this weakens environmental outcomes. A number of proposals advocate cost-effective implementation, but this would only occur conditional on participation and compliance by all countries—a condition most proposals do not effectively address. In contrast, proposals that focus on creating incentives for participation and compliance do so at the expense of cost-effective implementation. And pursuing policies with a primary goal of maximizing the environmental outcome may be inconsistent with dynamic efficiency, and could undermine participation and compliance incentives.

In our review of the 13 policy proposals, we identified several common themes: (1) Kyoto is "too little, too fast"—high costs for a small set of participants with very modest environmental gains; (2) developing countries should play a more substantial role over time; (3) incentives should be provided to promote developing country participation; (4) implementation should focus on market-based approaches; (5) price mechanisms should be a key element of market-based approaches, either through hybrid tax-quota schemes or emissions taxes; and (6) participation and compliance incentives merit more consideration and thought than evident in most proposals.

These major themes and the tensions among the relevant criteria for evaluating climate change policy illustrate the challenge in developing and implementing an international policy regime that can effectively mitigate climate change risks in an efficient, fair, and inclusive manner. Common to most policy proposals, including the Kyoto Protocol, is an emphasis on cost-effective implementation policies and inadequate concern for participation and compliance incentives.

In light of the current international context in which countries representing a majority of the world's greenhouse-gas emissions will not take on emissions commitments under the Kyoto Protocol, future analytic work should focus on policies that can attempt to reconcile efficiency and cost-effectiveness with participation and compliance. This is certainly no easy task, given the domestic and international political constraints on climate change policy. Policy-makers would certainly benefit from quantitative assessments of these trade-offs. Various modeling teams have evaluated the costs and benefits of the Kyoto Protocol. Comparable assessments are needed of alternative policy strategies, especially those that are substantially different in form. Finally, to the extent that the Kyoto process goes forward, researchers and policy-makers alike should take advantage of the opportunity to gain information about climate change policy *in practice* in order to exploit a positive attribute of the Kyoto agreement—the flexibility to adapt to new information.

Acknowledgements

This paper draws, in part, upon Barrett and Stavins (2002), and benefitted from comments from three anonymous referees, as well as the editor of this journal. Financial support was provided by the U.S. Environmental Protection Agency. The authors are solely responsible for any errors.

References

Aldy, J.E., Orszag, P.R., Stiglitz, J.E., 2001. Climate change: an agenda for global collective action. In: Prepared for the Conference on the Timing of Climate Change Policies. Pew Center on Global Climate Change, Washington, DC, October 2001.

Andersen, M.S., 2002. Regulation or coordination: european climate policy between Scylla and Charybdis. In: Paper Presented at 10th Symposium of the Egon Sohmen Foundation, Climate Protection and Emissions Trading—US and European Views. Dresden, Germany, 25–26 October.

Arrow, K.J., Fisher, A.C., 1974. Environmental preservation, uncertainty, and irreversibility. Q. J. Econ. 88, 312–319.

Arrow, K.J., Parikh, J., Pillet, G., et al., 1996. Decision-making frameworks for addressing climate change. In: Bruce, J.P., Lee, H., Haites, E.F. (Eds.), Climate Change 1995: Economic and Social Dimensions of Climate Change. Cambridge University Press, Cambridge, pp. 53–77.

Barrett, S., 1999. A theory of full international cooperation. J. Theor. Polit. 11 (4), 519–541.

Barrett, S., 2001. Towards a Better Climate Treaty. Policy Matters, 01-29, AEI-Brookings Joint Center for Regulatory Studies Washington, DC, November; Reprinted in World Econ. 3 (2), 35–45.

Barrett, S., 2002. Consensus treaties. J. Inst. Theor. Econ. 158 (4), 529–547.

Barrett, S., 2003. Environment and Statecraft. Oxford University Press, Oxford.

Barrett, S., Stavins, R., 2002. Increasing Participation and Compliance in International Climate Change Agreements. Working Paper, John F. Kennedy School of Government, Harvard University, Cambridge, MA, 13 August.

Benedick, R.E., 2001. Striking a new deal on climate change. Issues Sci. Technol. Fall, 71–76.

Bradford, D.F., 2002. Improving on Kyoto: Greenhouse-Gas Control as the Purchase of a Global Public Good. Princeton University Working Paper, 30 April Draft.

Buchner, B., Carraro, C., Cersosimo, I., Marchiori, C., 2002. Back to Kyoto? US Participation and the Linkage between R&D and Climate Cooperation. FEEM Climate Modelling and Policy Unit, Venice, Italy.

Chayes, A., Chayes, A., 1995. The New Sovereignty. Harvard University Press, Cambridge, MA.

Cooper, R., 1998. Toward a real treaty on global warming. Foreign Affairs 77 (2), 66–79.

Cooper, R., 2000. International approaches to global climate change. World Bank Res. Observ. 15 (2), 145–172.

Cooper, R., 2001. The Kyoto Protocol: a flawed concept. Environ. Law Rep. 31, 11,484–11,492.

Dasgupta, P., Mäler, K.-G., Barrett, S., 1999. Intergenerational equity, social discount rates, and global warming. In: Portney, P.R., Weyant, J.P. (Eds.), Discounting and Intergenerational Equity. Johns Hopkins University Press, Baltimore, MD, pp. 51–77.

Dixit, A., Pindyck, R., 1994. Investment Under Uncertainty. Princeton University Press, New York.

Downs, G.W., Rocke, D.M., Barsoon, P.N., 1996. Is the good news about compliance good news about cooperation? Int. Org. 50, 379–406.

Edmonds, J., Roop, J.M., Scott, M., 2001. Technological change and its effects on mitigation costs. In: Claussen, E., Cochran, V.A., Davis, D.P. (Eds.), Climate Change Science, Strategies, and Solutions. Brill Publishing, Boston, pp. 209–226.

Eizenstat, S., 1998. Testimony before the Senate Foreign Relations Committee. United States Senate, Washington, DC, 11 February 1998.

Fisher, B.S., Barrett, S., Bohm, P., Kuroda, M., Mubazi, J.K.E., Shah, A., Stavins, R.N., 1996. An economic assessment of policy instruments for combating climate change. Intergovernmental Panel on Climate Change, Climate Change 1995: Economic and Social Dimensions of Climate Change. Cambridge University Press, Cambridge, pp. 397–439.

Flannery, B.P., 2001. Alternative climate proposals: an efforts-based approach. In: Presentation Prepared for the International Conference on Climate Policy—Do We Need a New Approach? Exxon Mobil Corporation, Venice, 6–8 September.

Frankel, J., 1999. Greenhouse-Gas Emissions. Policy Brief No. 52, Brookings Institution, Washington, DC.

Goulder, L.H., 2000. Central themes and research results in global climate change policy. In: Paper Prepared for the Intergovernmental Panel on Climate Change. Working Group III, Third Assessment Report, Mimeo, Stanford, CA, May.

Goulder, L.H., Stavins, R.N., 2002. An eye on the future: how economists' controversial practice of discounting really affects the evaluation of environmental policies. Nature 419, 673–674.

Hahn, R.W., 1998. The Economics and Politics of Climate Change. American Enterprise Institute Press, Washington, DC.

Hahn, R.W., Stavins, R.N., 1999. What Has the Kyoto Protocol Wrought? The Real Architecture of International Tradable Permit Markets. American Enterprise Institute Press, Washington, DC.

Hanemann, W.M., 1989. Information and the concept of option value. J. Environ. Econ. Manage. 16 (1), 23–37.

International Energy Agency, 1999. World Energy Outlook 1999 Insights: Looking at Energy Prices: Getting the Prices Right. IEA, Paris.

Jacoby, H.D., 1998. The Uses and Misuses of Technology Development as a Component of Climate Policy. MIT Report No. 43, MIT Joint Program on the Science and Policy of Climate Change, Cambridge, MA.
Jacoby, H.D., Ellerman, A.D., 2002. The 'Safety-valve' and Climate Policy. MIT Report No. 83, MIT Joint Program on the Science and Policy of Climate Change, Cambridge, MA.
Jaffe, A.B., Newell, R.N., Stavins, R.N., 1999. The induced innovation hypothesis and energy-saving technological change. Q. J. Econ. 114 (3), 941–975.
Kolstad, C.D., 1996. Learning and stock effects in environmental regulation: the case of greenhouse-gas emissions. J. Environ. Econ. Manage. 31, 1–18.
Kolstad, C.D., Toman, M., 2000. The Economics of Climate Policy. Resources for the Future Discussion Paper 00-45, Washington, DC.
Kopp, R., Morgenstern, R., Pizer, W., 1997. Something for Everyone: A Climate Policy That Both Environmentalists and Industry Can Live With. Policy Brief. Resources for the Future, Washington, DC.
Kopp, R., Morgenstern, R., Pizer, W., Toman, M., 1999. A Proposal for Credible Early Action in US Climate Policy. Weathervane Feature. Resources for the Future, Washington, DC.
Levy, M.A., 1993. European acid rain: the power of Tote-Board Diplomacy. In: Keohane, R.O., Haas, P.M., Levy, M.A. (Eds.), Institutions for the Earth: Sources of Effective Environmental Protection. MIT Press, Cambridge, MA, pp. 75–132.
Lind, R.C. (Ed.), 1982. Discounting for Time and Risk in Energy Policy. Resources for the Future, Washington, DC.
Lutter, R., 2000. Developing countries' greenhouse emissions: uncertainty and implications for participation in the Kyoto Protocol. Energy J. 21 (4), 93–120.
Manne, A.S., Richels, R.G., 2000. A Multi-Gas Approach to Climate Policy—With and Without GWPs. No. 44/2000, Fondazione Eni Enrico Mattei Note di Lavoro, June.
McKibbin, W.J., Wilcoxen, P.J., 1997. A Better Way to Slow Global Climate Change. Policy Brief No. 17, Brookings Institution, Washington, DC.
McKibbin, W.J., Wilcoxen, P.J., 2000. Moving Beyond Kyoto. Policy Brief No. 66, Brookings Institution, Washington, DC.
McKibbin, W.J., Wilcoxen, P.J., 2002. The role of economics in climate change policy. J. Econ. Perspect. 16 (2), 107–129.
Narain, U., Fisher, A., 2000. Irreversibility, Uncertainty, and Catastrophic Global Warming. Giannini Foundation Working Paper 843, University of California, Department of Agricultural and Resource Economics, Berkeley.
Newell, R., Pizer, W., 2000. Discounting the Distant Future: How Much Do Uncertain Rates Increase Valuations? Resources for the Future Discussion Paper 00-45.
Nordhaus, W.D., 1998. Is the Kyoto Protocol a Dead Duck? Are There Any Live Ducks Around? Comparison of Alternative Global Tradable Emissions Regimes. Working Paper, Department of Economics, Yale University, New Haven, CT, 31 July.
Nordhaus, W.D., 2001. Global warming economics. Science 294, 1283–1284.
Nordhaus, W.D., 2002. After Kyoto: alternative mechanisms to control global warming. In: Paper Prepared for Presentation at the Annual Meetings of the Allied Social Science Associations. Atlanta, GA, 4 January.
Portney, P.R., Weyant, J.P. (Eds.), 1999. Discounting and Intergenerational Equity. Resources for the Future, Washington, DC.
Reilly, J., Prinn, R., Harnisch, J., Fitzmaurice, J., Jacoby, H., Kicklighter, D., Melillo, J., Stone, P., Sokolov, A., Wang, C., 1999. Multi-gas assessment of the Kyoto Protocol. Nature 401, 549–555.
Roberts, M.J., Spence, M., 1976. Effluent charges and licenses under uncertainty. J. Publ. Econ. 5, 193–208.
Sandalow, D.B., Bowles, I.A., 2001. Fundamentals of treaty-making on climate change. Science 292, 1839–1840.
Schelling, T.C., 1997. The cost of combating global warming: facing the tradeoffs. Foreign Affairs 76 (6), 8–14.
Schelling, T.C., 1998. Costs and Benefits of Greenhouse-Gas Reduction. American Enterprise Institute Press, Washington, DC.
Schelling, T.C., 2002. What makes greenhouse sense? Time to rethink the Kyoto Protocol. Foreign Affairs 81 (3), 2–9.
Schmalensee, R., 1993. Comparing greenhouse gases for policy purposes. Energy J. 14 (1), 245–255.
Schmalensee, R., 1996. Greenhouse Policy Architecture and Institutions. MIT Joint Program on the Science and Policy of Climate Change, Cambridge, MA.
Schmalensee, R., 1998. In: Nordhaus, W.D. (Ed.), Greenhouse Policy Architecture and Institutions. Economics and Policy Issues in Climate Change. Resources for the Future, Washington, DC, pp. 137–158.
Stavins, R.N., 1997. Policy Instruments for Climate Change: How Can National Governments Address a Global Problem? University of Chicago Legal Forum 1997, pp. 293–329.
Stavins, R.N., 2001a. Economic analysis of global climate change policy: a primer. In: Claussen, E., Cochran, V.A., Davis, D.P. (Eds.), Climate Change: Science, Strategies, and Solutions. Brill Publishing, Boston.
Stavins, R.N., 2001b. Give Bush Time on Climate Issues. Boston Globe. 4 April.

Stavins, R.N., 2002. Experience with market-based environmental policy instruments. In: Mäler, K.-G., Vincent, J. (Eds.), Handbook of Environmental Economics. Elsevier, Amsterdam.
Stewart, R.B., Wiener, J.B., 2001. Reconstructing Climate Policy: The Paths Ahead. Policy Matters, 01-23, AEI-Brookings Joint Center for Regulatory Studies, Washington, DC, August.
Ulph, A., Ulph, D., 1997. Global warming, irreversibility, and learning. Econ. J. 107, 636–650.
Victor, D.G., 2001. International Agreements and the Struggle to Tame Carbon. 4 July.
Victor, D.G., Salt, J.E., 1995. Keeping the climate treaty relevant. Nature 373 (6512), 280–282.
Watson, R.T. (Ed.), 2001. Climate Change 2001: Synthesis Report. Contribution of Working Group I, II, and III to the Third Assessment Report of the Intergovernmental Panel on Climate Change. Cambridge University Press, Cambridge, UK.
Weitzman, M.L., 1974. Prices vs. quantities. Rev. Econ. Stud. 41 (4), 477–491.
Weitzman, M.L., 2001. Gamma discounting. Am. Econ. Rev. 91 (1), 260–271.
Wiener, J.B., 1999. Global environmental regulation: instrument choice in legal context. Yale Law J. 108, 677–800.
Wiener, J.B., 2001. Designing global climate regulation. In: Schneider, S., Rosencranz, A., Niles, J. (Eds.), The Stanford Reader in Global Climate Change Policy.
Wigley, T.M.L., Richels, R., Edmonds, J.A., 1996. Economic and environmental choices in the stabilization of atmospheric CO_2 concentrations. Nature 379, 240–243.
Yellen, J., 1998. Testimony before the House of Representatives Commerce Committee, Subcommittee on Energy and Power. United States House of Representatives, Washington, DC, 6 October.

[24]

Beyond Kyoto

Getting Serious About Climate Change

By Robert N. Stavins

The Day After Tomorrow, last summer's blockbuster film about the apocalyptic consequences of the greenhouse effect, had less scientific basis than *The Wizard of Oz*. Reality, however, is disturbing enough. There is now a near-consensus that man-made emissions of greenhouse gases are very likely to change the earth's climate in ways that most people will regret.

You've heard the story before, but it's worth repeating. Two trace constituents of the atmosphere, carbon dioxide and water vapor, create a thermal blanket for the planet, much the way glass on a greenhouse traps the sun's energy within. And a good thing, too: without greenhouse warming, the planet would be far too cold to be livable.

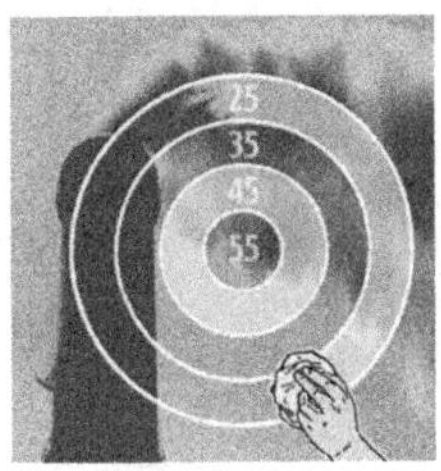

But the balance between too much and too little greenhouse effect is fairly delicate. Massive quantities of carbon dioxide are produced from the combustion of fossil fuels – coal, petroleum and natural gas. Meanwhile, the direct warming effects of carbon dioxide and other greenhouse gases – methane, nitrous oxide and halocarbons – are being indirectly amplified because the warming increases the evaporation of water, thereby increasing atmospheric concentrations of yet another greenhouse compound – water vapor.

Average global surface temperatures have risen by about one degree Fahrenheit since 1970. This fits the predictions of climate-change computer models that also take account of increases in atmospheric dust (which cools by reflecting sunlight) and variations of the sun's energy output. Changes in mid-continent temperatures and those at high latitudes have been two to four times greater than the global averages – also as predicted.

Balmier temperatures (which would be welcome in some places) are only a part of the story. The most important consequences of greenhouse

ALAN MAZETTI

CLIMATE CHANGE

gas concentrations are likely to be changes in patterns of precipitation and runoff, the melting of glaciers and icebergs, increases in sea level and changes in storm frequency and intensity. That's why scientists talk about global climate change rather than global warming.

But moving from predictions of average global temperature changes to regional climate impacts is problematic. The best computer models cannot yet produce reliable estimates of impacts on a localized basis.

What is obvious, though, is that emissions in one country affect the climate in every other. Hence the logic of the global pact on emissions hammered out in Kyoto, Japan in December 1997.

Four years later, the Bush Administration announced that it would not submit the Kyoto Protocol, which had been initialed by the Clinton White House, to the Senate for ratification. Of course, the Clinton Administration had chosen not to submit the agreement to the Senate either, and it is very unlikely that Al Gore, had he been elected in 2000, would have done so. For, even before the Kyoto conference, the Senate had resolved by a vote of 95-0 that it would not approve a climate-control pact along the lines of the Kyoto accord. More about the reasons for this Bronx cheer later.

The key requirement for implementing the Kyoto Protocol is that a minimum of 55 nations, representing 55 percent industrialized economies' emissions of carbon dioxide in 1990, must ratify the agreement. The United States accounted for more than a third of total carbon emissions in that year. But with Canada, Japan and Russia (which respectively produced 3 percent, 6 percent and 17 percent of the world's emissions) now on board, the Protocol takes effect this quarter.

The impact on total emissions, however, will be much less than anticipated in 1997. First, non-participation by the United States is critical because it is the largest emitter. Second, the targets were revised in 2001 in ways that lowered the costs for Canada, Japan and

©MARC RIBOUD/MAGNUM PHOTOS

ROBERT N. STAVINS is the Albert Pratt professor of business and government at Harvard's John F. Kennedy School of Government and director of the university's environmental economics program.

Russia in order to induce them to ratify the Protocol. The original Kyoto Protocol – if extended to the year 2050 – would have led to a 13 percent emissions reduction compared to 1990 emissions. Without the United States on board and with the 2001 revisions of the rules, the anticipated aggregate emission reduction for the year 2050 is only 1 to 2 percent.

What, then, can and should be done? Start with the fact that the cost of containing emissions varies enormously from country to country. And as long as global incremental benefits exceed every nation's benefits from its own effort to contain emissions, countries will either want to avoid participating or avoid complying if they do participate. Successful international cooperation must somehow change these incentives.

FUNDAMENTALS FIRST

To be worth the parchment it is written on,

©STEPHANIE MAZE/CORBIS

a post-Kyoto framework must be based on sound science and economics, as well as on realistic politics. Thus, it is hard to imagine a viable strategy that doesn't include (a) the means to ensure that all significant economies – both those industrialized and those on the way – are eventually involved, (b) a detailed schematic for getting to where good science and good economics say we ought to be in a reasonable amount of time, and (c) an emphasis on market-based (rather than centrally directed) policy instruments.

Near-Universal Participation

Broad participation is essential to address the "global commons" problem – the fact that greenhouse emissions from anywhere affect everywhere. The share of global emissions coming from relatively poor countries is significant and growing. In fact, developing countries may well account for more than half of global emissions by the year 2020. To some, the idea of making demands on poor countries seems unfair – the "haves" rather than the "have-nots" ought to do the heavy lifting, at least at first. But the simple reality is that developing countries provide the greatest opportunities for relatively low-cost emissions reductions. Hence, it would be unnecessarily costly to focus emissions-reductions activities exclusively in the developed world.

It's true that industrialized countries are responsible for the bulk of man-made concentrations of greenhouse gases in the atmosphere. But if the developing countries are not included from the beginning, producers of goods and services that generate a lot of emissions will have incentives to move plants outside of the coalition of participating countries. Thus, rather than helping developing countries shift to less-carbon-intensive paths of economic development, the industrialized world would actually be pushing them the other way.

Fear of this emissions-leakage phenomenon – and the associated movement of jobs – was the central reason behind the 1997 Byrd-Hagel Resolution in the United States Senate, which called for treating industrialized and developing countries similarly in any international agreement on global climate change. The resolution said that the United States should not be a signatory to any agreement under the Framework Convention on Climate Change that would "mandate new commitments to limit or reduce greenhouse gas emissions for [industrialized countries], unless the protocol or other agreement also mandates new specific scheduled commitments to limit or reduce greenhouse gas emissions for Developing Country Parties within the same compliance period." Indeed, the resolution refers to five developing countries by name: China, Mexico, India, Brazil and South Korea.

Still, on grounds of fairness, it is unreasonable to expect lower-income nations to pay the bill for emissions reduction in the short term. Hence, the policy dilemma. On the one hand, for purposes of environmental effectiveness and economic efficiency, developing countries need to participate. On the other, they cannot be expected to incur the consequent costs. But there is a way around this seeming anomaly.

First, a mechanism is needed for voluntary accession of developing countries into the group of nations pledged to meet specific emissions targets. Second, and much more important, an automated trigger is needed – for example, success in reaching some minimum standard of living – to bind new members to undertake needed emissions reduction.

An even better approach would be to set emissions targets for all countries, which would gradually become more stringent as individual counties become richer. Such targets could start at business-as-usual emissions levels. For that matter, the targets could be set above business-as-usual – although such headroom has been denigrated as "hot air" in the case of Russia's less-than onerous target created to draw the country into the Kyoto Protocol. If the program were combined with international emissions trading, in which one country could buy emissions-reduction credits from another, it could provide cash windfalls for poor countries even as they play a role in emissions control.

What's more, an emissions growth target need not be a one-size-fits-all number, but an equation that relates targeted emissions to per capita income and possibly to other factors as well. If a developing country were doing particularly well economically, its target would become proportionately more stringent. However, if it suffered an economic setback, the emissions target would be relaxed.

This is a natural extension of the division

of obligations built into the Kyoto Protocol. A Brookings Institution report by Jeffrey Frankel, a former Clinton White House economic adviser, calculated that for every 10 percent increase in per capita GDP, the Kyoto targets (on average) become about 1 percent more stringent. In 1999, the Argentine government offered to make the relationship explicit, tying its emissions commitment to economic growth.

In any event, two necessary characteristics of any formula for setting emissions targets are (a) that it not create perverse incentives to increase relatively low levels of emissions, and (b) that it is relatively simple, so as not to create impediments to negotiation. It also makes sense to combine emissions targets with a well-designed international tradable-permit program. That way, cost-effectiveness as well as the distribution of the burden could be addressed.

An Extended Time-Path for Emissions Reduction

Global climate change is a long-term problem because greenhouse gases remain in the atmosphere for decades or even centuries. The Kyoto Protocol does not sufficiently reflect this important reality – that is, the cumulative nature of the problem. The protocol has only short-term targets: an average 5 percent reduction from 1990 levels by 2008–2012. This may sound like a modest reduction, but it translates into a 25-30 percent reduction for the United States from its business-as-usual emissions path. That's because the U.S. economy grew exceptionally rapidly during the 1990s – a remarkable 37 percent growth for the decade. As a consequence, U.S. carbon emissions increased by more than 12 percent over the period, while Western European emissions increased by only 1 percent.

Thus, the contrast between Europe and elsewhere is dramatic. Emissions of carbon dioxide from the United Kingdom, Germany and Russia fell significantly after 1990 (the Kyoto Protocol's baseline year), for reasons having nothing to do with environmental policy. Emissions fell in Britain because of the downsizing of the domestic coal industry initiated by Prime Minister Margaret Thatcher's government (1979–1990), in Germany because reunification led to the closure of plants that gulped energy (and produced little of value) in the former East Germany, and in Russia because of its economic collapse.

Accordingly, some 80 percent of the European Union's carbon dioxide reductions under the Kyoto Protocol will be achieved by two countries – Germany and the UK – and is thus really only possible because the EU was counted as one emitting nation. This goes a long way to explaining why the resistance to Kyoto was so much greater in the United States.

The Kyoto Protocol's targets thus amount to too little, too fast: they will have very modest overall impact, yet still manage to be unreasonably stringent for countries that enjoyed significant economic growth after 1990. Two elements are needed for a workable fix: (a) moderate targets in the short term to avoid rendering trillions of dollars worth of machinery and energy infrastructure prematurely obsolete, and (b) more-stringent targets for the long term to motivate technological change, which in turn is needed to bring down the costs of reducing emissions. The longer-term targets should be flexible, because there is still considerable uncertainty about the consequences of emissions on climate and the cost-effectiveness of various abatement measures.

Specifically, emissions targets ought to start at business-as-usual levels, then gradually depart from these, so that the absolute level of emissions would increase in the near term – but at slower rates. In the long term, though, countries would have to reduce the total amount of gases emitted. For a conservative goal of stabilizing atmospheric concentrations of carbon dioxide at twice pre-industrial levels, it has been estimated that the cost-effective time path would lead to peak global emissions in 2030. Major emission reductions would take place in the second half of the century.

Such a time path, put in place now, would be consistent with what is often dismissed as politics as usual. In representative democracies, there are strong incentives to place costs on future voters and, if possible, future generations. But in the case of climate policy, gradual containment amounts to good science and good economics as well as pragmatic politics.

Market-Based Reduction Mechanisms

In principle, the Kyoto Protocol calls for working through the market rather than against it. There is widespread agreement that conventional regulatory approaches – so-called "command-and-control" policies – cannot do the job at acceptable costs. To keep costs down in the short term and bring them down even lower in the long term through technological change, policymakers will have to harness the power of decentralized markets driven by private incentives.

For some countries, this means using tradable permits (which less-economically-sophisticated environmentalists used to dismiss as permissions to pollute) to achieve national targets. In a trading system, emitters have the choice of meeting targets on their own or paying other emitters to do it for them. Thus, emitters with low costs of control – for example, manufacturers using aging boilers – have an incentive to take on added reductions, so that they can sell their excess permits to emitters that face relatively high control costs. This is the mechanism that was used in the United States in the 1980s to eliminate leaded gasoline from the market, at

PAUL CHESLEY/NATIONAL GEOGRAPHIC/GETTY IMAGES

a savings in refining of more than $250 million dollars per year. It is also the mechanism now being used to cut the sulfur emissions that cause acid rain in the United States by 50 percent, at a savings estimated at about $1 billion annually.

For some countries, carbon taxes that create incentives to switch away from carbon-rich fuels (notably coal) may be a more attractive approach, although there is serious doubt about the political feasibility of such taxes. Norway, for example, introduced a very hefty carbon tax in 1991, but the impact on carbon dioxide emissions was nonetheless modest because major emitters lobbied successfully for tax exemptions.

Another promising market-based approach is a hybrid of tax and tradable-permit systems – that is, an ordinary tradable permit system, plus a government promise to sell additional permits at a fixed price. This guarantees that an adequate number of permits will be available at a reasonable cost, and thus has been labeled a safety-valve system. The approach reduces uncertainty on the cost side without undermining incentives to reduce emissions, and therefore addresses concerns frequently voiced by private industry.

International containment mechanisms are also required, of course. The Kyoto Protocol outlines a system whereby governments can engage in trading their reduction targets. International tradable permits – even a system in which trading was only permitted among the industrialized countries – could reduce compliance costs by as much as half. If trading were expanded to include major developing countries, compliance costs could be lowered to just one-quarter of what they otherwise would be.

An undisputed attraction of an international trading approach is that how and where emissions are reduced is determined by who has the lowest costs of emissions reductions, regardless of how the permits are initially handed out. There is one catch, however: it assumes that no country buying or selling permits has market power. This is a real concern in the context of Kyoto. If, for example, the majority of excess permits (allowable emissions in excess of business-as-usual emissions, or so-called "hot air") were owned by a relatively small number of nations in Eastern Europe and the former Soviet Union, the possibility of collusion among permit sellers would become quite likely.

In any event, the initial allocation of emissions permits can be highly significant in terms of equity. But the alternative, in which countries must meet targets without trading, wouldn't serve anyone's interest.

If international trading is used, it must be integrated with domestic policies that nations use to achieve their respective domestic targets. If all countries use domestic tradable permit systems to meet their national targets – that is, allocate shares from the international permit system to private domestic parties – an international system could certainly be efficient. However, if some countries use non-trading approaches domestically – which seems likely – cost minimization would be more difficult. In such cases, achieving all the potential cost savings of international trading would require some form of international project-by-project credit program that reaches within national economies, such as the Kyoto Protocol's Clean Development Mechanism. But experience with such credit programs suggest that they are less likely to generate major savings because they are difficult to negotiate and implement.

Thus, the way individual nations choose to meet their targets could substantially limit the cost-saving potential of international trading. There's a trade-off between the degree of domestic sovereignty permitted under a global emissions pact and the degree of cost-effectiveness. This is a severe problem for the Kyoto Protocol, which explicitly provides for sovereignty regarding national emissions policies.

Not long ago, most observers would have predicted that few, if any, European countries would employ tradable permit systems internally in light of the European Union's strenuous opposition to such approaches dating

©RAFIQUR RAHMAN/REUTERS/CORBIS

back to the 1990s. But the EU has now launched its own continent-wide trading system. Furthermore, by the time of the 2001 negotiations over the implementation of Kyoto, China and the G-77 (the coalition of less-industrialized economies) had, in effect, dropped their opposition to international emissions trading. Combined with the strong United States preference for trading, this bodes well for making trading part of a viable global climate policy.

Emissions trading is no panacea. Indeed, the more I look to tradable-permits to address climate change, the more I have come to believe that it is the worst possible approach – except for all the others.

THE HARD ROAD AHEAD

Without the United States or the emerging Asian powerhouses on board, and with past concessions to Russia and Canada that undermine the value of the treaty, the Kyoto Protocol is in deep trouble even as it takes on the force of international law. It doesn't much matter whether one labels the next round of a climate control agreement an addition to Kyoto or a substitute. What does matter is that we get the structure right the next time. And, at a minimum, that structure has to include a means for making the pact truly global, a realistic time path for implementation, and a sensible system of market-based incentives for keeping costs under control.

Getting serious about greenhouse emissions won't be easy. But if the state-of-the-science predictions about the consequences of another few decades of inaction are correct, the alternatives are grim, indeed.

[25]

An International Policy Architecture for the Post-Kyoto Era

By SHEILA M. OLMSTEAD AND ROBERT N. STAVINS*

In February 2005, the Kyoto Protocol to the United Nations Framework Convention on Climate Change came into force, but without participation by the United States. Its impacts on emissions of greenhouse gases—including carbon dioxide (CO_2), the primary anthropogenic driver of climate change—will be trivial; but scientific (Robert T. Watson, 2001) and economic (Charles D. Kolstad and Michael A. Toman, 2001) analyses point to the need for a credible international approach.

Because the Kyoto Protocol's ambitious targets apply only to the short term (2008–2012) and only to industrialized nations, the agreement will impose relatively high costs and generate only modest short-term benefits, while failing to provide a real solution (Joseph E. Aldy et al., 2003). For these reasons, most economists see the agreement as deeply flawed (Richard N. Cooper, 1998; David G. Victor, 2001; Warwick J. McKibbin and Peter J. Wilcoxen, 2002), although some see it as an acceptable first step (Axel Michaelowa, 2003). Virtually all agree, however, that the Protocol is not sufficient to the overall challenge.

We describe the basic features of a post-Kyoto international global climate agreement, which addresses three crucial questions: who, when, and how. The respective elements are: first, a means to ensure that key nations—industrialized and developing—are involved; second, an emphasis on an extended time path of action (employing a cost-effective pattern over time); and third, inclusion of market-based policy instruments.

* Olmstead: School of Forestry and Environmental Studies, Yale University, 230 Prospect Street, New Haven, CT 06511 (e-mail: sheila.olmstead@yale.edu); Stavins: John F. Kennedy School of Government, Harvard University, 79 John F. Kennedy Street, Cambridge, MA 02138, and Resources for the Future (e-mail: robert_stavins@harvard.edu). Further details are available in Stavins (2004). We have benefitted from helpful discussions with Jeffrey Frankel and David Victor.

I. Who—Expand Participation to Include All Key Countries

Broad participation, by major industrialized nations and key developing countries, is essential to address this global commons problem effectively and efficiently. Developing countries are likely to account for more than half of global emissions by 2020, possibly sooner (Nebojsa Nakicenovic and Robert Swart, 2000). It can be argued, on an ethical basis, that industrialized countries should take the first steps, since—almost by definition—they are responsible for the bulk of anthropogenic concentrations of greenhouse gases in the atmosphere. But two serious problems remain. First, developing countries currently provide the greatest opportunities for low-cost emissions reductions (Watson, 2001). Second, if developing countries are not included, comparative advantage in the production of carbon-intensive goods and services will shift outside the coalition of participating countries, pushing nonparticipating nations onto more carbon-intensive growth paths, increasing their costs of joining the coalition later.

This poses a policy conundrum. On the one hand, for purposes of environmental effectiveness and economic efficiency, key developing countries should participate. On the other hand, for purposes of distributional equity (and international political pragmatism), they cannot be expected to incur the consequent costs. The answer is a set of "growth targets" that are set initially at business-as-usual (BAU) levels for respective developing countries, but become more stringent as those countries become more wealthy. This would be a natural extension of the allocation pattern in the Kyoto Protocol, where targets for industrialized countries exhibit an income elasticity of reductions of about 0.10 (Jeffrey A. Frankel, 1999).

Initial targets for developing countries could be set at emissions that are above BAU levels. If combined with an international trading program, this would provide significant economic incentives for participation. Developing countries could fully participate without incurring prohibitive (or even any) costs in the short term. Cost-effectiveness and distributional equity would both be addressed.

II. When—Use an Extended Time Path

Global climate change is a long-term problem, because greenhouse gases remain in the atmosphere for decades to centuries, but the Kyoto Protocol fails to reflect this cumulative, stock-pollutant nature of the problem. The Protocol's short-term targets, an average 5-percent reduction from 1990 levels by the 2008–2012 compliance period, may sound modest, but they translate into severe cuts for the United States, 25 to 30 percent from its BAU path, because of the rapid economic growth the country enjoyed during the 1990s, amounting to a 37-percent increase in real gross domestic product (GDP).

Thus, the Kyoto Protocol's targets are too little, too fast. They do little about the problem, but are unreasonable for countries that experienced significant economic growth post-1990. Two elements will ameliorate this problem: moderate targets in the short term to avoid rendering large parts of the capital stock prematurely obsolete, and more stringent targets for the long term to motivate (now and in the future) technological change, in order to bring costs down over time.

Targets ought to begin at BAU levels, then depart gradually, so that emissions increase at first, but at rates below BAU. These targets should reach a maximum level and then decrease—eventually becoming much more severe than the constraints implied by the Kyoto Protocol's targets. This is consistent with estimates of the least-cost time path of emissions for achieving long-term greenhouse-gas concentration goals (Alan S. Manne and Richard G. Richels, 1997). For example, for the frequently discussed goal of stabilizing atmospheric concentrations of CO_2 at twice preindustrial levels, the cost-effective time path of emissions would involve global emissions peaking in 2030 (Thomas M. L. Wigley et al., 1996). Of course, the long-term targets should be flexible, because there are great uncertainties throughout the policy-economics-biophysical system, some of which will be resolved over time (Richels et al., 2004).

We use the word "target" generically to refer not only to emission targets (as in the Kyoto Protocol), but also to "intensity targets"—emissions per unit of GDP. The proposal we offer is also consistent with a time path of targets denominated in financial units, such as carbon prices (taxes on the carbon content of fossil fuels). In any event, this time path of targets—put in place now—would also be consistent with what is often denigrated as "politics as usual." Politicians in representative democracies are frequently condemned because there are incentives to place greater costs on future, not current, voters. It is typically the politically pragmatic strategy. In the case of global climate policy, it may also be the scientifically correct and economically rational approach.

III. How—Employ Market-Based Policy Instruments

All observers agree that conventional regulatory approaches cannot do the job, certainly not at acceptable costs. To keep costs down in the short term and bring them down even lower in the long term through technological change, it is essential to embrace market-based instruments.

On a domestic level, systems of tradable permits might be used to achieve national targets. This approach was used in the United States to phase out leaded gasoline in the 1980s at a savings of more than $250 million per year over an equivalent traditional regulatory approach (Stavins, 2003), and is now used to cut sulfur dioxide (SO_2) emissions from power plants by half, at an annual cost savings of $1 billion (A. Denny Ellerman et al., 2000). The better policy model for climate change is the upstream lead-rights system (analogous to trading on the carbon content of fossil fuels), rather than the downstream SO_2 emissions-trading system.

For some countries, systems of domestic carbon taxes may be more attractive, and a particularly promising approach is a hybrid of tax and tradable-permit systems—an ordinary tradable permit system, plus a government promise to sell additional permits at a stated price (McKibbin and Wilcoxen, 2002; William A. Pizer, 2002). This addresses cost uncertainty by creating a price (and thereby cost) ceiling and has, hence, been labeled a safety-valve approach.

International policy instruments are also required, and the Kyoto Protocol includes a system whereby the parties to the agreement can trade their "assigned amounts"—their national reduction targets—translated into emissions terms. In theory, such a system of international tradable permits—if implemented only for the industrialized countries (as in the Protocol)—could reduce costs by 50 percent. If such a system were to include major developing countries as well, costs could be lowered to 25 percent of what they otherwise would be (James A. Edmonds et al., 1997).

A theoretical attraction of an international trading approach is that the equilibrium allocation of permits, the market-determined permit price, and the aggregate costs of abatement are independent of the initial allocation of permits among countries. This is true only if particularly unlikely types of transaction costs are not prevalent and individual parties—be they nations or firms—are not able to exercise market power. The latter concern is a real one in the climate policy context. If, for example, the majority of excess permits (allowable emissions in excess of BAU emissions) is found in a relatively small number of nations, then the possibility increases of collusion among sellers (Manne and Richels, 2004). Furthermore, an international trading system must be designed to facilitate integration with domestic policies, because individual nations' choices of domestic policy instruments can substantially limit the cost-saving potential of an international system (Robert W. Hahn and Stavins, 1999).

In any event, the initial allocation of permits among nations can imply exceptionally large international wealth transfers. Several analysts have identified this as a major objection to an international carbon trading regime and have endorsed international tax approaches for this and other reasons (Cooper, 1998; McKibbin and Wilcoxen, 2002; Pizer, 2002; Richard G. Newell and Pizer, 2003; William D. Nordhaus, 2005). Many of the objections are well-founded, but it is the wealth-transfer feature of the permit approach that allows cost-effectiveness *and* distributional equity to be addressed simultaneously.

IV. Conclusions

The three-part global climate policy architecture we propose can form the foundation for a successor to the Kyoto Protocol. For such an approach to work, key nations must be involved, including major developing countries, through the use of economic trigger mechanisms such as growth targets. In addition, cost-effective time paths of targets are required: firm but moderate in the short term, and much more stringent yet flexible in the long term. Finally, market-based policy instruments—whether emissions trading, carbon taxes, or hybrids of the two—ought to be part of the package.

A premise of our proposal is that the global commons nature of the climate problem requires a multinational—if not fully global—approach. As long as global marginal benefits exceed every nation's marginal benefits, countries will either want to avoid participating or avoid complying fully if they do participate. Successful international cooperation must change these incentives (Scott Barrett and Stavins, 2003). Can countries credibly commit to the long-term program that is part of this proposed architecture? Our answer is that, once nations have ratified the agreement, implementing legislation within respective nations would translate the agreed-upon long-term targets into domestic policy commitments. And once such implementing legislation was enacted, signals would be sent to private industry, which would begin to take action. Ultimately, such domestic actions provide the signals that other countries need to see. This represents a logical and ultimately feasible chain of credible commitment.

This overall approach can be made to be scientifically sound, economically rational, and politically pragmatic. There is no denying that the challenges facing adoption and successful implementation of this climate policy architecture are significant, but they are no greater than the challenges facing other approaches to the threat of global climate change.

REFERENCES

Aldy, Joseph E.; Barrett, Scott and Stavins, Robert N. "Thirteen Plus One: A Comparison of Global Climate Policy Architectures." *Climate Policy*, 2003, *3*(4), pp. 373–97.

Barrett, Scott and Stavins, Robert N. "Increasing Participation and Compliance in International Climate Change Agreements." *International Environmental Agreements: Politics, Law and Economics*, 2003, *3*(4), pp. 349–76.

Cooper, Richard N. "Toward a Real Global Warming Treaty." *Foreign Affairs*, 1998, *77*(2), pp. 66–79.

Edmonds, J. A.; Kim, S. H.; MacCracken, C. N.; Sands, R. D. and Wise, M. A. *Return to 1990: The cost of mitigating United States carbon emissions in the post–2000 period.* PNNL No. 11819. Washington, DC: Pacific Northwest National Laboratory, 1997.

Ellerman, A. Denny; Joskow, Paul L.; Schmalensee, Richard; Montero, Juan-Pablo and Bailey, Elizabeth M. *Markets for clean air: The U.S. acid rain program.* Cambridge: Cambridge University Press, 2000.

Frankel, Jeffrey A. "Greenhouse Gas Emissions." Brookings Institution, Policy Brief: No. 52, 1999.

Hahn, Robert W. and Stavins, Robert N. *What has the Kyoto Protocol wrought? The real architecture of international tradable permit markets.* Washington, DC: American Enterprise Institute Press, 1999.

Kolstad, Charles D. and Toman, Michael A. "The Economics of Climate Policy," in Karl-Göran Mäler and Jeffrey R. Vincent, eds., *Handbook of environmental economics. Vol. III: Economywide and international environmental issues.* Amsterdam: Elsevier Science, North-Holland, 2005, pp. 1561–1618.

Manne, Alan S. and Richels, Richard G. "On Stabilizing CO_2 Concentrations—Cost-Effective Emission Reduction Strategies." *Environmental Modeling and Assessment*, 1997, *2*(4), pp. 251–65.

Manne, Alan S. and Richels, Richard G. "U.S. Rejection of the Kyoto Protocol: The Impact on Compliance Costs and CO_2 Emissions." *Energy Policy*, 2004, *32*(4), pp. 447–54.

McKibbin, Warwick J. and Wilcoxen, Peter J. "The Role of Economics in Climate Change Policy." *Journal of Economic Perspectives*, 2002, *16*(2), pp. 107–29.

Michaelowa, Axel. "Global Warming Policy: Comment on McKibbin and Wilcoxen." *Journal of Economic Perspectives*, 2003, *17*(3), pp. 204–05.

Nakicenovic, Nebojsa and Swart, Robert, eds. *Special report on emissions scenarios: A special report of working group III of the Intergovernmental Panel on Climate Change.* Cambridge: Cambridge University Press, 2000.

Newell, Richard G. and Pizer, William A. "Regulating Stock Externalities under Uncertainty." *Journal of Environmental Economics and Management*, 2003, *45*(2S), pp. 416–32.

Nordhaus, William D. "Economic Analyses of Kyoto Protocol: Is There Life after Kyoto? Presented at Global Warming: Looking Beyond Kyoto, New Haven, CT, October 21–22, 2005.

Pizer, William A. "Combining Price and Quantity Controls to Mitigate Global Climate Change." *Journal of Public Economics*, 2002, *85*(3), pp. 409–34.

Richels, Richard G.; Manne, Alan S. and Wigley, Thomas M. L. "Moving Beyond Concentrations: The Challenge of Limiting Temperature Change." AEI-Brookings Joint Center for Regulatory Studies, Working Paper: 04-11, 2004.

Stavins, Robert N. "Experience with Market-Based Environmental Policy Instruments," in Karl-Göran Mäler and Jeffrey R. Vincent, eds., *Handbook of environmental economics Vol. 1: Environmental degradation and institutional responses.* Amsterdam: Elsevier Science, North-Holland, 2003, pp. 355–435.

Stavins, Robert N. "Forging a More Effective Global Climate Treaty." *Environment*, 2004, *46*(10), pp. 23–30.

Victor, David G. *The collapse of the Kyoto Protocol and the struggle to slow global warming.* Princeton: Princeton University Press, 2001.

Watson, Robert T. *Climate change 2001: Synthesis report: Third assessment report of the Intergovernmental Panel on Climate Change.* Cambridge: Cambridge University Press, 2001.

Wigley, Thomas M. L.; Richels, Richard G. and Edmonds, Jae. "Economic and Environmental Choices in the Stabilization of Atmospheric CO_2 Concentrations." *Nature*, 1996, *379*(6562), pp. 240–43.

[26]

Linking Tradable Permit Systems: A Key Element of Emerging International Climate Policy Architecture

*Judson Jaffe, Matthew Ranson & Robert N. Stavins**

This Article analyzes the role that linkage between emission trading systems could play in a future international climate policy architecture. Cap-and-trade systems, regional, national, and international in scope, are emerging as a preferred instrument for addressing global climate change throughout the industrialized world, and the Clean Development Mechanism—an emission-reduction-credit system—has also developed a significant constituency. Because links between tradable permit systems can reduce

* Judson Jaffe was a Vice President at Analysis Group; Matthew Ranson is a Ph.D. student in public policy at Harvard University; and Robert N. Stavins is the Albert Pratt Professor of Business and Government at the John F. Kennedy School of Government, Harvard University, University Fellow of Resources for the Future, and Research Associate of the National Bureau of Economic Research. This Article draws, in part, on: JUDSON JAFFE & ROBERT STAVINS, INTERNATIONAL EMISSIONS TRADING ASSOCIATION, LINKING TRADABLE PERMIT SYSTEMS FOR GREENHOUSE GAS EMISSIONS: OPPORTUNITIES, IMPLICATIONS, AND CHALLENGES 1–53 (2007); and Judson Jaffe & Robert Stavins, *Linkage of Tradable Permit Systems in International Climate Policy Architecture*, *in* HARVARD PROJECT ON INTERNATIONAL CLIMATE AGREEMENTS, at 1–20 (Belfer Ctr. for Sci. and Int'l Affairs at Harvard Univ. John F. Kennedy Sch. of Gov't, Discussion Papers 08-07, 2008). It also builds on previous work, including: ERIK HAITES & FIONA MULLINS, MARGAREE CONSULTANTS INC., LINKING DOMESTIC AND INDUSTRY GREENHOUSE GAS EMISSION TRADING SYSTEMS (2001); U.N. Framework Convention on Climate Change, Organization for Economic Co-operation and Development and International Energy Agency, *Towards International Emissions Trading: Design Implications for Linkages*, U.N. DOC. COM/ENV/EPOC/IEA/ SLT(2002)5 (Oct. 2002) (*prepared by* Richard Baron & Stephen Bygrave); U.N. Framework Convention on Climate Change, Organization for Economic Co-operation and Development and International Energy Agency, *Linking Non-EU Domestic Emissions Trading Schemes with the EU Emissions Trading Scheme*, U.N. DOC. COM/ENV/EPOC/IEA/SLT(2004)6 (2004) (*prepared by* William Blyth & Martina Bosi); RICHARD BARON & CÉDRIC PHILIBERT, ORGANIZATION FOR ECONOMIC CO-OPERATION AND DEVELOPMENT AND INTERNATIONAL ENERGY AGENCY, ACT LOCALLY, TRADE GLOBALLY: EMISSIONS TRADING FOR CLIMATE POLICY (2005); U.N. Framework Convention on Climate Change, Organization for Economic Co-operation and Development and International Energy Agency, *Linking GHG Emission Trading Schemes and Markets*, U.N. DOC. COM/ENV/EPOC/IEA/SLT(2006)6 (Oct. 2006) (*prepared by* Jane Ellis & Dennis Tirpak); and Joseph Kruger, Wallace Oates & William Pizer, *Decentralization in the EU Emissions Trading Scheme and Lessons for Global Policy*, 1 REV. OF ENVTL. ECON. & POL'Y 112 (2007). Our research has benefitted from communications with Joseph Aldy, Scott Barrett, Denny Ellerman, Robert Keohane, and David Victor. All errors are our own.

compliance costs and improve market liquidity, the possibility of linking cap-and-trade systems to each other and to emission-reduction-credit systems such as the Clean Development Mechanism has generated considerable interest. We consider whether linkage could pave the way for a future international agreement, play a role as part of a future agreement, or substitute for an agreement. We argue that linkage could promote the near-term goals of participation and cost-effectiveness, while helping to build the foundation for a more comprehensive future agreement to address global climate change.

INTRODUCTION

Tradable permit systems are emerging as a preferred instrument for reducing greenhouse gas (GHG) emissions. Two of the most significant institutions for reducing GHG emissions implemented to date—the European Union Emission Trading Scheme (EU ETS) and the Clean Development Mechanism (CDM)—are tradable permit systems. Further, Australia, Canada, Japan, and the United States, among other countries, are considering tradable permit systems as a primary policy instrument for reducing GHG emissions.[1] As these systems grow in prominence and number, attention has increasingly focused on whether and how to link them.

Linking occurs when a tradable permit system's regulatory authority allows regulated entities to use emission allowances or emission reduction credits from another system in order to meet compliance obligations.[2] Linking thereby allows these entities to take advantage of the cost savings from international trade in allowances or credits.

In this Article, we analyze how linkages between tradable permit systems could influence the form of a future international climate agreement. In particular, we consider how a set of linkages might function in three roles: (1) as a de facto, bottom-up climate architecture; (2) as a step in the evolution of a coherent, top-down architecture; and (3) as a component of a larger climate architecture. We begin our analysis by describing some of the existing and proposed tradable permit systems, and the links that are developing between them. We then explain the major types of linkages and discuss their implications. Finally, we evaluate the various roles that linkages might play in a future climate architecture, and discuss how near-term policy negotiations could facilitate the growth of linkages.

I. TRADABLE PERMIT SYSTEMS

Although there are only a limited number of existing GHG tradable permit systems, as described below, several other systems are likely to emerge in the coming years.[3] The increasing number and prominence of such systems, and the economic and political incentives to link them, provide the motivation for our analysis of linkage as a potential element of the post-2012, post-Kyoto international policy architecture. Because the implications of linking depend on the type of tradable permit systems that are linked, distinguishing between two categories of systems—cap-and-trade systems and emission-reduction-credit systems—is essential.

A. Cap-and-Trade Systems

A cap-and-trade system constrains the aggregate emissions of regulated sources by creating a limited number of tradable emission allowances, which emission sources must secure and surrender in number equal to their emissions.

---

1. *See supra* notes 14–24 and accompanying text.

2. Such linkage of domestic tradable permit systems is different from the state-to-state trading envisaged under Article 17 of the Kyoto Protocol, whereby signatories to the Protocol can trade their "assigned amounts." *See* Kyoto Protocol to the U.N. Framework Convention on Climate Change, Dec. 10, 1997, U.N. DOC. FCCC/CP/1997/7/Add.1, 2303 U.N.T.S. 148, *at* 4-15 (entered into force Feb. 16, 2005), *available at* http://unfccc.int/resource/docs/convkp/kpeng.pdf.

3. *See supra* notes 14–24 and accompanying text; *see also* Matthew Carr, *Seven Carbon Markets May Replace Kyoto, Barclays Says (Update1)*, BLOOMBERG.COM, Oct. 16, 2009, http://www.bloomberg.com/apps/news?pid=20601072&sid=aZSZ4h_uXTzw.

As long as trading costs are low and allowance markets are sufficiently competitive, trading will lead firms to use allowances to cover those emissions that are most costly to reduce, regardless of how allowances are initially distributed.[4] Thus, trading will result in the least-cost combination of emission reduction measures necessary to reach the system-wide cap.

In developing a cap-and-trade system, policy makers must decide on several design elements.[5] Policy makers must define the level of the system's cap by determining how many allowances to issue. They also must determine the scope of the cap's coverage, or what emission sources and types of GHG emissions will be subjected to the overall cap. An associated decision relates to the system's point of regulation. A cap on energy-related carbon dioxide (CO_2) emissions can be enforced by requiring fossil fuel suppliers, such as refineries, to surrender allowances for the carbon content of their fuel sales ("upstream regulation"), or by requiring final emitters, such as factories, to surrender allowances for their emissions ("downstream regulation"). Policy makers also must determine how to distribute allowances. Allowances can be freely distributed or auctioned, or a combination of these approaches can be used. If allowances are freely distributed, methods for determining who receives them and how many allowances each recipient receives are limitless.

Much attention has been given to the possibility of including "cost-containment" measures in cap-and-trade systems, such as offset provisions, allowance banking and borrowing, and safety-valve provisions. An offset provision allows regulated entities to offset some of their emissions with credits from emission reduction measures that are outside the cap-and-trade system's scope of coverage. Banking allows firms to save unused allowances for use in future years. Borrowing allows firms to use allowances that will be issued in future years to demonstrate compliance in an earlier year. A safety valve puts an upper bound on the compliance costs that firms will incur by offering them the option of paying a predetermined fee (the safety-valve "trigger price") to purchase additional allowances. These provisions all limit costs by enhancing the flexibility available to firms in meeting their emission reduction obligations.

The world's largest existing GHG cap-and-trade system is the EU ETS.[6] Phase I of the EU ETS, from 2005 to 2007, capped aggregate CO_2 emissions from more than 11,000 industrial facilities and electricity generators in twenty-

---

4. *See* Robert N. Stavins, *Transaction Costs and Tradeable Permits*, 29 J. ENVTL. ECON. & MGMT 133, 140 (1995); Robert W. Hahn, *Market Power and Transferable Property Rights*, 99 Q.J. ECON. 753, 754 (1984).

5. *See* Robert N. Stavins, *A Meaningful U.S. Cap-and-Trade System to Address Climate Change*, 32 HARV. ENVTL. L. REV. 293, 305 (2008).

6. A. Denny Ellerman & Barbara Buchner, *The European Union Emissions Trading Scheme: Origins, Allocation, and Early Results*, 1 REV. OF ENVTL. ECON. & POL'Y 66 (2007).

five European countries.[7] Those sources collectively emitted approximately two billion metric tons of CO_2 in 2005, about 45 percent of the European Union's CO_2 emissions.[8] The EU ETS cap has been tightened for Phase II, which runs from 2008 to 2012,[9] and its scope has been expanded to cover new sources in countries that participated in Phase I, and to include sources in Bulgaria and Romania, which acceded to the European Union in 2007.[10] Liechtenstein, Iceland, and Norway also joined the EU ETS in 2008, although sources in Iceland are not yet subject to an emissions cap.[11] The European Union plans to extend the EU ETS through at least 2020.[12]

In addition to the EU ETS, the Japanese Voluntary Emissions Trading System has operated since 2006,[13] and Norway operated its own emissions trading system for several years before joining the EU ETS in 2008.[14] Legislation to establish cap-and-trade systems is under debate in Australia[15] and the Canadian province of Ontario,[16] and Japan is also considering a compulsory emission trading system.[17]

In the United States, the Regional Greenhouse Gas Initiative (RGGI), a cap-and-trade system for electricity generators in ten northeastern states, came

7. EUROPEAN COMMISSION, EU ACTION AGAINST CLIMATE CHANGE: EU EMISSIONS TRADING 9 (2005), *available at* http://ec.europa.eu/environment/climat/pdf/emission_trading2_en.pdf.

8. *Id.*; Press Release, European Commission, Emissions Trading: EU-wide Cap for 2008–2012 Set at 2.08 Billion Allowances after Assessment of National Plans for Bulgaria (Oct. 26, 2007), *available at* http://europa.eu/rapid/pressReleasesAction.do?reference=IP/07/1614.

9. *See* EUROPEAN COMMISSION, *supra* note 7, at 9.

10. *See* Denny A. Ellerman, *The EU Emission Trading Scheme: A Prototype Global System?*, *in* HARVARD PROJECT ON INTERNATIONAL CLIMATE AGREEMENTS, at 3 (Belfer Ctr. for Sci. and Int'l Affairs at Harvard Univ. John F. Kennedy Sch. of Gov't, Discussion Papers 08-02, 2008), *available at* http://belfercenter.ksg.harvard.edu/publication/18488/eu_emission_trading_scheme.html. For an analysis of the performance of the EU ETS, see Symposium, *The European Union Emissions Trading Scheme*, 1 REV. ENVTL. ECON. & POL'Y 66 (2007).

11. Press Release, European Commission, Emissions Trading: EU ETS Emissions Fall 3% in 2008 (May 15, 2009), *available at* http://europa.eu/rapid/pressReleasesAction.do?reference=IP/09/794.

12. European Commission, Proposal for a Directive of the European Parliament and of the Council Amending Directive 2003/87/EC so as to Improve and Extend the Greenhouse Gas Emission Allowance Trading System of the Community, at 2, COM (2008) 16 final (Jan. 23, 2008).

13. Kyoto Mechanisms Information Platform, Japan's Voluntary Emissions Trading Scheme (J-VETS), http://www.kyomecha.org/e/info04.html.

14. Norway Ministry of the Environment, Act of 17 December 2004 No.99 relating to Greenhouse Gas Emission Allowance Trading and the Duty to Surrender Emission Allowances (amending 29 June 2007 No.93), *available at* http://www.regjeringen.no/en/doc/Laws/Acts/greenhouse-gas-emission-trading-act.html; *see also* Press Release, *supra* note 11.

15. Australian Climate Change Regulatory Authority Bill, 2009 (Austl.), *available at* http://parlinfo.aph.gov.au/parlInfo/search/display/display.w3p;query=Id:%22legislation/billhome/r4126%22.

16. Legislative Assembly of Ontario, Bill 185, An Act to Amend the Environmental Protection Act with respect to Greenhouse Gas Emissions Trading and Other Economic and Financial Instruments and Market-Based Approaches, 2009, *available at* http://www.ene.gov.on.ca/envision/env_reg/er/documents/2009/010-6467%202.pdf.

17. *Japan Eyes Mandatory Cap-and-Trade in 2011/12: Nikkei*, REUTERS, Sept. 19, 2009, http://www.reuters.com/article/environmentNews/idUSTRE58J03020090920.

into effect on January 1, 2009.[18] In addition to RGGI, other regional and state efforts to limit GHGs in the United States have begun. One of the most prominent is California's enactment of the Global Warming Solutions Act of 2006, which set a statewide GHG emissions limit for 2020 equal to California's 1990 emissions level.[19] In 2008, the California Air Resources Board proposed the use of a cap-and-trade program as a primary policy for achieving this target.[20] The cap initially would cover electric generators and large industrial facilities, and its scope would later be expanded to include smaller facilities and the transportation sector.[21] At the federal level, the U.S. House of Representatives passed the Waxman-Markey American Clean Energy and Security Act on June 26, 2009,[22] and President Obama has expressed support for this legislation.[23] If enacted, this bill would establish an economy-wide cap-and-trade system that would come into force in 2012.[24]

B. *Emission-Reduction-Credit Systems*

Emission-reduction-credit systems are the second major type of tradable permit system. These systems bring about emission reductions by awarding tradable credits to unregulated entities for certified emission reductions. Entities in other cap-and-trade systems can then purchase these credits and use them to meet their own compliance obligations. The credits that are awarded to a particular project are based on an estimate of how much the project reduces emissions from some agreed upon baseline level of what emissions would have been if the project had not been carried out. Thus, it is necessary to estimate what baseline emissions would have been absent the credited action. For example, if an energy company that was planning to build a coal-fired power plant decides instead to build a solar power plant, it can receive and then sell credits equal to the net difference in its emissions—but only if it can prove that

---

18. REGIONAL GREENHOUSE GAS INITIATIVE: AN INITIATIVE OF THE NORTHEAST AND MID-ATLANTIC STATES OF THE U.S. 1 (2009), *available at* http://www.rggi.org/docs/RGGI_Executive%20Summary_4.22.09.pdf; *see also* REGIONAL GREENHOUSE GAS INITIATIVE, MEMORANDUM OF UNDERSTANDING 2, *available at* http://rggi.org/docs/mou_final_12_20_05.pdf.

19. Press Release, Office of the Governor of the State of California, Gov. Schwarzenegger Signs Landmark Legislation to Reduce Greenhouse Gas Emissions (Sept. 27, 2006), *available at* http://gov.ca.gov/press-release/4111. *See also* California Global Warming Solutions Act of 2006 (A.B. 32), CAL. HEALTH & SAFETY CODE § 38500 (West 2008), *available at* http://www.arb.ca.gov/cc/docs/ab32text.pdf.

20. STATE OF CALIFORNIA AIR RESOURCES BOARD, CLIMATE CHANGE SCOPING PLAN 31 (Dec. 2008), *available at* http://www.arb.ca.gov/cc/scopingplan/document/adopted_scoping_plan.pdf.

21. *Id.*

22. American Clean Energy and Security Act of 2009 (Waxman-Markey), H.R. 2454, 111th Cong. (2009), *available at* http://www.govtrack.us/congress/billtext.xpd?bill=h111-2454.

23. *See* The White House Office of the Press Secretary, Weekly Address: President Obama Calls Energy Bill Passage Critical to Stronger American Economy (June 27, 2009), *available at* http://www.whitehouse.gov/the_press_office/UPDATED-and-FINAL-WEEKLY-ADDRESS-President-Obama-Calls-Energy-Bill-Passage-Critical-to-Stronger-American-Economy.

24. American Clean Energy and Security Act of 2009, *supra* note 22.

it would have built the coal-fired plant in the absence of the emission-reduction-credit system.

The most significant GHG emission-reduction-credit system to date is the Kyoto Protocol's CDM.[25] Under the CDM, certified emission reduction credits (CERs) are awarded for voluntary emission reduction projects in developing countries that ratified the Protocol, but are not among the Annex I countries subject to the Protocol's emission limitation commitments—also known as the Annex B countries.[26] While CERs can be used by the Annex I countries to meet their emission commitments, they can also be used for compliance purposes by entities covered by other cap-and-trade systems, including systems in countries that are not Parties to the Protocol, such as the United States.[27]

While emission-reduction-credit systems can be self-standing, as in the case of the CDM, governments can also establish them as elements of domestic cap-and-trade systems. These emission-reduction-credit systems—often referred to as offset programs—serve as a source of credits that can be used by regulated entities to meet compliance obligations under the cap-and-trade system. For example, RGGI includes an offset program that recognizes offsets from activities such as landfill methane capture and destruction, reductions in emissions of sulfur hexafluoride from the electric power sector, and afforestation.[28] Electricity generators covered by RGGI can use these offset credits to cover part of their emissions.[29] Cap-and-trade systems proposed in Australia and at the federal level in the United States also include offset programs.[30]

II. TYPES OF LINKAGES

Not all linkages are created equal: the type of linkage that is established has important implications for its effects. Direct linkages between systems can be one-way (unilateral) or two-way (bilateral or multilateral). Also, while direct links between systems can be established only as a result of explicit decisions to do so, direct links can lead to indirect links between systems even absent

---

25. *See* Kyoto Protocol, *supra* note 2.

26. *Id.* at art. 12. Like the CDM, Joint Implementation (JI) was established as a project-based flexibility mechanism under the Kyoto Protocol. Unlike the CDM, JI applies to emission reduction projects carried out in an Annex I country (the host country) that has a national emissions target under the Protocol. JI projects generate credits, referred to as emission reduction units (ERUs), which can be used to cover increased emissions in other countries. *See id.*

27. The Annex B Parties include thirty-seven industrialized countries and emerging market economies of central and eastern Europe. *See id.* at art. 12.

28. Regional Greenhouse Gas Initiative, Offset Project Categories, http://www.rggi.org/offsets/categories (last visited Oct. 29, 2009).

29. REGIONAL GREENHOUSE GAS INITIATIVE, AN INITIATIVE OF THE NORTHEAST AND MID-ATLANTIC STATES OF THE U.S., *supra* note 18.

30. *See supra* note 24.

explicit decisions to link them.[31] Figure 1 illustrates the various types of linkages.

Figure 1. Types of Linkages between Tradable Permit Systems
(Arrows denote the flow of credits or allowances between systems)

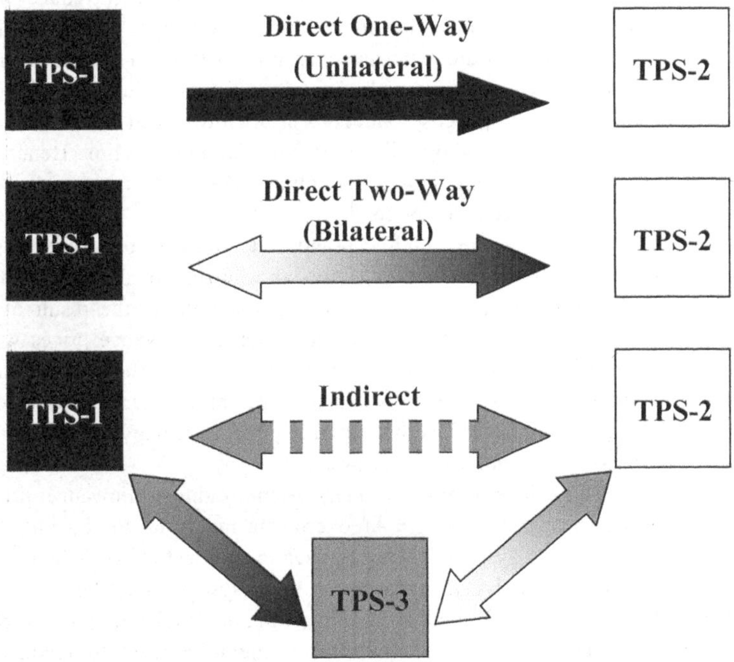

A. Direct Linkages

To establish a direct linkage between two systems, either one or both systems must accept the other's allowances or credits as valid for use in demonstrating compliance in its own system.

In a direct link between a cap-and-trade system and a credit system, the cap-and-trade system chooses to recognize emission reduction credits from the credit system. This linkage is necessarily one-way if the credit system does not place requirements on entities to surrender credits or allowances. If the price of credits is lower than that of emission allowances, then regulated firms in the

31. *See* UNFCCC, *IETA Report on Linking GHG Emissions Trading Systems*, at 13, U.N. Doc. COP13/CMP3 (Nov. 2007) (*prepared by* Judson Jaffe & Robert Stavins).

cap-and-trade system have an incentive to purchase credits. This will reduce the price of allowances in the cap-and-trade system and increase the price of credits in the credit system until the two prices converge.[32]

Direct linkages can also occur between two cap-and-trade systems. The linkage can be one-way or two-way, depending on whether the recognition is mutual. In an unrestricted one-way link in which System *A* recognizes System *B*'s allowances, if *A*'s allowance price is higher than *B*'s, participants in *A* will buy allowances from participants in *B*. These purchases will reduce *A*'s allowance price and increase *B*'s price until the prices converge.[33] Such trading will increase emissions in *A* and decrease emissions in *B* by an equal amount, but will lead to overall cost savings, as higher cost emission reductions in *A* are avoided and replaced by lower cost reductions in *B*. However, if *A*'s allowance price is lower than *B*'s, no trading will result from the one-way link. Hence, a one-way link in which *A* recognizes *B*'s allowances will ensure that *A*'s allowance price never exceeds *B*'s price.[34]

In a two-way direct link, both cap-and-trade systems recognize each other's allowances, making it possible for allowances to flow in either direction. Two-way links can be bilateral or multilateral. As the result of a two-way linkage, any difference between the systems' allowance prices will lead to sales of allowances from the lower price system to the higher price system until the systems' allowance prices converge at an intermediate level, leading to an increase in emissions in the higher price system and an offsetting reduction in emissions in the lower price system.[35]

If governments place limits on inter-system trading, allowance price convergence may not be complete.[36] A government may limit the quantity of allowances from another system that can be used to demonstrate compliance in its own system. Alternatively, participants in a system may be allowed unrestricted use of another system's allowances, but an "exchange rate" might be applied to their use. Such a requirement might be used to reconcile differences in the denomination of different systems' allowances (such as metric tons versus short tons), to reduce inter-system trading, or if there is a

---

32. *See* INTERGOVERNMENTAL PANEL ON CLIMATE CHANGE, IPCC THIRD ASSESSMENT REPORT: CLIMATE CHANGE 2001, WORKING GROUP III: MITIGATION 425 (2001), *available at* http://www.ipcc.ch/ipccreports/tar/wg3/pdf/6.pdf; Peter Bohm, *International Greenhouse Gas Emission Trading—With Special Reference to the Kyoto Protocol*, *in* EFFICIENCY AND EQUITY IN CLIMATE CHANGE POLICY (Carlo Carrero ed., 2000); ALAN S. MANNE & RICHARD G. RICHELS, THE KYOTO PROTOCOL: A COST-EFFECTIVE STRATEGY FOR MEETING ENVIRONMENTAL OBJECTIVES? 1–24 (1998), *available at* http://www.oecd.org/dataoecd/38/53/1923159.pdf.

33. *See* INTERGOVERNMENTAL PANEL ON CLIMATE CHANGE, *supra* note 32; Bohm, *supra* note 32; MANNE & RICHELS, *supra* note 32.

34. *See* INTERGOVERNMENTAL PANEL ON CLIMATE CHANGE, *supra* note 32; Bohm, *supra* note 32; MANNE & RICHELS, *supra* note 32.

35. *See* INTERGOVERNMENTAL PANEL ON CLIMATE CHANGE, *supra* note 32; Bohm, *supra* note 32; MANNE & RICHELS, *supra* note 32.

36. *See* INTERGOVERNMENTAL PANEL ON CLIMATE CHANGE, *supra* note 32; Bohm, *supra* note 32; MANNE & RICHELS, *supra* note 32.

desire, to require that trading with other systems lead to a net reduction in emissions.

B. Indirect Linkages

Even if neither system recognizes the other's allowances, two systems can become indirectly linked through direct links with a common third system. As a result of trading between each of the two systems and the common system, developments in one of the indirectly linked systems can affect the supply and demand for allowances in the other system. Hence, changes in the allowance price and emissions level in one system can affect the allowance price and emissions level in a system with which it is indirectly linked.

Indirect links can be created between two cap-and-trade systems if both have one-way links with a common credit system. As a result of such one-way links, the two indirectly linked systems will compete for credits from the third system. This indirect linkage will reduce the difference between the two cap-and-trade systems' allowance prices, as credits will flow to the system with the higher price. If there is a sufficient supply of credits at a price below the two cap-and-trade systems' pre-link allowance prices, and if there are no constraints on the use of these credits in either system, then prices in the three systems will converge fully.[37]

A series of bilateral links among several systems can also create indirect links among those systems. This kind of indirect linkage is identical in its effects to a direct multilateral link among all of the systems involved. For example, if System *A* has a two-way link with System *B*, which has a two-way link with System *C*, then trading will lead allowance prices to converge across all three systems even though *A* and *C* are not directly linked.[38]

C. Examples of Existing Linkages

Some linkages have already been established among tradable GHG permit systems, reflecting the strong economic incentives that governments face to establish these connections. The EU ETS itself can be viewed as a multilateral linkage among the Member States' own systems, where a central authority enforces the harmonization of certain characteristics of each system, and where allowances issued by any Member State are recognized by all other Member States.[39] Also, through its Linking Directive, the European Commission has allowed EU ETS participants to use CDM CERs to meet compliance

---

37. *See* INTERGOVERNMENTAL PANEL ON CLIMATE CHANGE, *supra* note 32; Bohm, *supra* note 32; MANNE & RICHELS, *supra* note 32.

38. *See* INTERGOVERNMENTAL PANEL ON CLIMATE CHANGE, *supra* note 32; Bohm, *supra* note 32; MANNE & RICHELS, *supra* note 32.

39. *See* Ellerman, *supra* note 10, at 1, 15–16 (making this point, as well as a broader one, namely that the EU ETS provides a range of valuable lessons for the development of the post-2012 international policy architecture).

obligations beginning in 2005, and Joint Implementation Emission Reduction Units (ERUs) beginning in 2008.[40] As a result of these linkages, EU ETS allowance prices are considered a major factor influencing CER prices, and—to a lesser extent—ERU prices.[41] However, the Directive places restrictions on these linkages. CERs and emission reduction units generated from nuclear facilities, land use change, and forestry activities are not recognized,[42] and quantitative limits are placed on the use of CERs and ERUs.[43]

In the United States, the Model Rule governing the implementation of RGGI creates several types of one-way links. Covered sources may use emission reduction credits from qualified domestic offset projects, subject to quantitative limits that depend on the prevailing RGGI allowance price.[44] When the RGGI allowance price exceeds a specific threshold, which increases over time, sources have the additional option to use CERs and allowances from other countries' cap-and-trade systems, such as the EU ETS, in meeting their compliance obligations.[45]

III. IMPLICATIONS OF LINKING

Linking tradable permit systems leads to diverse effects that need to be considered in assessing both the merits of particular linkages and the merits of linkage as a major design element of a post-2012 international policy architecture.[46]

A. Benefits of and Concerns about Linkage

The most significant benefit of linking is the opportunity to lower the costs of achieving emission reduction goals by shifting reductions between linked systems in a manner that minimizes total emission reduction costs.[47] In particular, if systems face very different marginal abatement costs, then linking

40. Council Directive 2004/101/EC, art. 5, 2004 O.J. (L 338) 18.

41. *See, e.g., CER Market Steadies After Price Hit*, CARBONPOSITIVE, Nov. 10, 2008, http://www.carbonpositive.net/viewarticle.aspx?articleID=1335; *CER Prices Ease with EU Carbon*, CARBONPOSITIVE, Feb. 15, 2007, http://www.carbonpositive.net/viewarticle.aspx?articleID=670; POINT CARBON ADVISORY SERVICES, NEW ZEALAND EMISSIONS TRADIGIN GROUP, ISSUES IN THE INTERNATIONAL CARBON MARKET, 2008-2012 AND BEYOND 3–4 (2007), *available at* http://www.mfe.govt.nz/publications/climate/issues-international-carbon-market-oct07/issues-international-carbon-market-07.pdf.

42. EUROPEAN COMMISSION, *supra* note 7, at 17.

43. *Commission Decision of August 31, 2007 Concerning the National Allocation Plan for the Allocation of Greenhouse Gas Emission Allowances Notified by Denmark in Accordance with Directive 2003/87/EC of the European Parliament and of the Council*, at 4, COM (2005) 2515/6 final (Aug. 31, 2007).

44. REGIONAL GREENHOUSE GAS INITIATIVE, MODEL RULE 63 (2007), *available at* http://rggi.org/docs/model_rule_corrected_1_5_07.pdf.

45. *See id.* at 105–106.

46. While these effects are briefly discussed below, see UNFCCC, *supra* note 31, for a more complete discussion of the benefits and concerns regarding linking.

47. *See* INTERGOVERNMENTAL PANEL ON CLIMATE CHANGE, *supra* note 32, at 425.

can substantially improve the overall cost-effectiveness of GHG reduction programs.[48] Additionally, by broadening the market for allowances and credits, linking can improve market liquidity, reduce price volatility,[49] and lessen market power concerns.[50] Finally, through its effects on allowance prices in the linked systems, under certain circumstances, linking can decrease global emissions by reducing emissions leakage.[51] For example, linking can lead to a net reduction in leakage if the link between two systems reduces allowance prices—and hence individual firms' compliance costs—in the system that is more susceptible to leakage, while increasing allowance prices in the system that is less susceptible to leakage.[52]

Although linking can sometimes reduce global emissions, it can also have the opposite effect. For example, any cap-and-trade system that establishes a one-way linkage with a credit system must confront the problem of "additionality": some emission reduction credits offered by a credit system may not represent truly additional emission reductions because of the difficulty of establishing a baseline against which reductions are measured. As a result, to the extent that entities covered by the cap-and-trade system purchase such credits instead of reducing emissions within the cap-and-trade system, the increase in emissions in the cap-and-trade system may not fully be offset by real reductions under the credit system. In addition, although linking can reduce overall emissions leakage under certain circumstances through its effects on prices in each system, these price effects can increase leakage if prices increase instead in the system that is more susceptible to leakage.

The distributional implications of linking can also be a source of concern. Impacts on any firm participating in one of the linked systems depend on linkage's effect on the allowance price that the firm faces, and on whether that firm is a net allowance buyer or seller. For example, net sellers in a system with a low price will be better off after a link to a system with a higher price, because they will be able to sell their allowances at the higher equilibrium

48. Because marginal abatement costs depend on the stringency of a country's targets, the cost savings from linking will also depend on the relative stringency of the targets of the countries involved.

49. Of course, linking also exposes participants to new sources of price volatility from other linked systems. *See* Warwick J. McKibbin & Peter Wilcoxen, *A Credible Foundation for Long Term International Cooperation on Climate Change*, *in* ARCHITECTURES FOR AGREEMENT: ADDRESSING GLOBAL CLIMATE CHANGE IN THE POST-KYOTO WORLD 201–02 (Joseph Aldy & Robert Stavins, eds., 2007).

50. Market liquidity refers to whether there is sufficient depth in a market such that an individual participant can buy and sell without adversely affecting prices. Price volatility is a measure of the degree to which prices fluctuate over time. Market power refers to the ability of large buyers and sellers to influence market prices through their actions in that market.

51. Emissions leakage occurs when regulations to reduce emissions in one region causally lead to increased emissions in another region that is outside the scope of those regulations. For example, a regulation that increases manufacturing costs in a regulated region may cause a shift in manufacturing activity to unregulated regions, thereby leading to increased emissions in those unregulated regions.

52. A system's susceptibility to leakage depends on whether, and how easily, affected economic activity in a regulated region can be shifted to unregulated jurisdictions.

price. However, net buyers in the lower price system will be worse off after linking, because they will have to pay a higher price for allowances. Thus, while yielding overall cost savings, linking can create both winners and losers.

An additional concern related to linkage is that it can reduce national control over the design and impacts of a domestic tradable permit system.[53] Once a system establishes links, the system's allowance price and effect on emissions are influenced by developments in the linked system(s), including decisions made by the government(s) overseeing the linked system(s). A prominent example of this is the fact that a two-way link between two cap-and-trade systems will lead to the propagation of any cost-containment measures—such as banking, borrowing, offsets, or safety-valve provisions—from one system to the other.[54]

A final concern about linkage is that it can alter the incentives that countries face with respect to setting their future caps.[55] In particular, by changing allowance prices in each of the linked systems, a linkage alters the tradeoff that a government faces between the value it can create by issuing additional allowances, and the marginal environmental damage that arises from issuing additional allowances. Moreover, by expanding the scope of the allowance market, linkages reduce the impact that the issuance of additional allowances has on allowance prices, and therefore on the value of existing allowances.[56]

B. *Implications of Different Types of Linkages*

The degree of control that a government can retain over its system depends in part on whether linkage is one-way or two-way. For example, two-way linkages can increase *or* decrease domestic allowance prices.[57] As described above, two-way linkages also lead to complete propagation of cost-containment measures across the linked systems.[58] In contrast, one-way linkages can only decrease the price of allowances in the system that establishes the link. One-way linkages therefore will only lead to propagation

---

53. *See* CHRISTIAN FLACHSLAND ET AL., POTSDAM INSTITUTE FOR CLIMATE IMPACT RESEARCH, DEVELOPING THE INTERNATIONAL CARBON MARKET: LINKING OPTIONS FOR THE EU ETS 16–19 (2008), *available at* http://www.pik-potsdam.de/members/flachs/publikationen/developing-the-international-carbon-market/at_download/file.

54. *See id.* at 16, 18–19.

55. *See* Carsten Helm, *International Emissions Trading with Endogenous Allowance Choices*, 87 J. PUB. ECON. 2737, 2738 (2003); Bjart J. Holtsmark & Dag Einar Sommervoll, *International Emissions Trading in a Non-cooperative Equilibrium*, at 4, (Research Department of Statistics Norway, Discussion Papers No. 542, 2008), *available at* http://www.ssb.no/publikasjoner/DP/pdf/dp542.pdf.

56. By allowing nations with high domestic emission reduction costs to purchase less expensive emission reductions from other systems, however, linking can also reduce the costs that such nations would face in committing to more stringent caps, thereby partially offsetting some of these concerns.

57. *See* INTERGOVERNMENTAL PANEL ON CLIMATE CHANGE, *supra* note 32, at 425.

58. *See* FLACHSLAND ET AL., *supra* note 53, at 13.

of cost-containment measures in one direction—from the system with which a link is established to the system that establishes the link.

The effects of a linkage also depend on whether it connects two cap-and-trade systems or a cap-and-trade system and an emission-reduction-credit system. For example, linkage that involves an emission-reduction-credit system raises the issue of additionality. On the other hand, in a link between two cap-and-trade systems, the increase in allowance prices in one may have more far-reaching economic consequences—such as increasing domestic energy prices—than would the increase in credit prices resulting from a link between a cap-and-trade system and a credit system. In a credit system, entities are not required to meet any emission targets and thus can only benefit from the opportunity to sell credits for higher prices.

IV. THE POTENTIAL ROLE OF LINKAGE IN AN INTERNATIONAL CLIMATE POLICY ARCHITECTURE

The near-term and long-term role that linkage could play in an international policy architecture is limited by political and institutional factors. In particular, establishing direct linkages between cap-and-trade systems may require mutual recognition of emission targets, harmonization of certain design elements, and agreement on procedures for making future adjustments to the linked systems, including setting future emission caps. Therefore, the role of linkages likely will evolve over time as some direct links will initially be less attractive and more difficult to establish than others.

A. Near-Term Role

Links among existing tradable permit systems are already part of the international policy architecture to address global climate change, and new connections among existing and emerging tradable permit systems will undoubtedly be established in the future. Pairs or groups of nations, particularly those that are important trading partners, will likely establish direct two-way links between their respective cap-and-trade systems. However, it may take more time to establish direct links between other cap-and-trade systems, particularly those that use differing cost-containment measures. At the same time, many of these cap-and-trade systems may nonetheless become indirectly linked through direct links with a common credit system, such as the CDM or some alternative future protocol for issuing credits for emission reductions in developing countries. These direct and indirect linkages are currently serving as key operational elements of the de facto global climate policy architecture.

An important feature of such anticipated near-term linkage is that most cap-and-trade systems may be connected with one another only through indirect links via a common emission-reduction-credit system. Although a near-term web of mostly indirect links may not result in a fully cost-effective global market for GHG emission reductions, these indirect links still may yield much of the cost savings and other advantages of a comprehensive system of direct linkages without raising some of the concerns that may impede the development of direct linkages. As a result, indirect linkage via an emission-reduction-credit system such as the CDM could become an important part of the near-term international climate policy architecture. The efficacy of this scenario depends heavily on the widespread acceptance and effectiveness of an international credit system such as the CDM.[59]

B. Long-Term Role

In the long term, linkage could play several different roles. First, it is possible that a comprehensive set of linkages, combined with unilateral emission reduction commitments by many nations, could function as a stand-alone climate architecture. Such a bottom-up architecture could emerge as more countries establish national cap-and-trade systems and begin to seek the gains from linking with other systems. These countries also might use the prospect of linkages as a means of providing incentives to developing countries to participate in an international agreement.

A second long-term possibility is that a collection of bottom-up linkages might serve as a natural starting point in negotiations leading to a top-down agreement. An existing system of linkages may help to develop the experience and mutual trust necessary for global negotiations to succeed. Furthermore, as we discuss below, any future agreement is likely to be heavily influenced by the status quo system of existing linkages and institutional investments.

A third possibility—not mutually exclusive with the second—is that linkage could become an element of a larger global policy architecture. Because the trade-related cost savings available to linked systems may grow as countries adopt increasingly stringent targets, there are strong economic reasons for policy makers to favor linking. Thus, a future global architecture—perhaps based on a negotiated global agreement that specifies timetables and emissions targets—could incorporate a set of direct links among domestic cap-and-trade systems as a key design element.

V. EVALUATION OF THE ROLE OF LINKAGE IN AN INTERNATIONAL POLICY ARCHITECTURE

As described above, linkages are likely to play a long-term role in an international climate architecture. In this Part, we assess three ways in which linkage can contribute to a future climate policy architecture: as an independent, bottom-up architecture; as a transition to a top-down architecture; and as an element of a larger climate architecture.

---

59. For a discussion of the performance of, and concerns about, the CDM to date, see *Options for Reforming the Clean Development Mechanism, in* HARVARD PROJECT ON INTERNATIONAL CLIMATE AGREEMENTS, at 1 (Belfer Ctr. for Sci. and Int'l Affairs at Harvard Univ. John F. Kennedy Sch. of Gov't, Policy Brief 2009-1, 2009), *available at* http://belfercenter.ksg.harvard.edu/files/Issue_Brief_1_Final_4.pdf.

A. Linkage as a Bottom-Up International Policy Architecture

Bilateral linkages are likely to continue to evolve among national and regional cap-and-trade systems and the CDM (or its successor). Could such a set of linkages, established without central coordination, function as an effective, stand-alone, bottom-up international policy architecture?

Although such an architecture would need to include other design elements, including emission reduction commitments and participation incentives, its distinguishing feature would be that it would grow organically from direct and indirect linkages. The degree to which a system of bottom-up linkages could achieve meaningful environmental performance depends on whether participants set sufficient environmental targets, a sufficient number of key countries participate, and participants comply.

With regard to whether participants will set meaningful environmental targets, commitments to reduce emissions in an architecture of bottom-up linkages would result from unilateral decisions by individual nations, or from negotiations among small groups of nations. In developed countries, internal political support would probably be the driving force behind adoption of more stringent emission caps,[60] whereas adoption of emission caps by developing nations may depend upon incentives provided by committed developed countries.[61] To address the possibility that linking may create incentives for some countries to adopt less stringent future caps, countries could negotiate cap trajectories as a condition for linking.[62] On the other hand, a system of linkages may actually allow some countries to adopt more aggressive targets than they otherwise would.

With respect to whether an architecture of bottom-up linkages would generate participation from a sufficient number of key countries, industrialized country participation is likely to be high. Participation by developing countries, on the other hand, will most likely be conditional on participation incentives provided by industrialized countries.[63] Positive participation incentives ("carrots") could take at least three forms: access to demand for emission reduction credits, potential gain from becoming a net seller of allowances, or side payments in the form of technical or development assistance. Negative

60. *See* Robert Keohane & Kal Raustiala, *Toward a Post-Kyoto Climate Change Architecture: A Political Analysis*, *in* HARVARD PROJECT ON INTERNATIONAL CLIMATE AGREEMENTS, at 9 (Belfer Ctr. for Sci. and Int'l Affairs at Harvard Univ. John F. Kennedy Sch. of Gov't, Discussion Papers 08-01, 2008), *available at* http://belfercenter.ksg.harvard.edu/files/KeohaneFinalWebRevised4_09.pdf.

61. *See* Ramgopal Agarwala, Towards a Global Compact for Managing Climate Change, *in* HARVARD PROJECT ON INTERNATIONAL CLIMATE AGREEMENTS, at 1–2 (Belfer Ctr. for Sci. and Int'l Affairs at Harvard Univ. John F. Kennedy Sch. of Gov't, Discussion Papers 08-15, 2008), *available at* http://belfercenter.ksg.harvard.edu/files/AgarwalaWeb.pdf.

62. *See* FLACHSLAND ET AL., *supra* note 53, at 25–26.

63. *See* Daniel S. Hall et al., *Policies for Developing Country Engagement*, *in* HARVARD PROJECT ON INTERNATIONAL CLIMATE AGREEMENTS, at 20–21 (Belfer Ctr. for Sci. and Int'l Affairs at Harvard Univ. John F. Kennedy Sch. of Gov't, Discussion Papers 08-15, 2008), *available at* http://belfercenter.ksg.harvard.edu/publication/18649/policies_for_developing_country_engagement.html.

participation incentives ("sticks") could take at least two forms: industrialized countries could establish border carbon taxes and/or import allowance requirements,[64] or industrialized countries could impose participation and graduation requirements as conditions of allowing continued access to international permit markets.

With respect to whether participants would be likely to comply in an architecture of bottom-up linkage, the picture is no worse—and perhaps better—for some top-down, centralized architectures. This is the case because a bottom-up system only includes industrialized nations where domestic institutions are sufficient to enforce compliance, and only developing nations where the value of the various carrots and sticks employed to encourage participation outweigh the costs of participating.

Links among cap-and-trade systems create gains from trade for the participating countries. Therefore, such an architecture has the potential to be cost-effective if the bottom-up system includes a sufficient set of direct two-way links, or if the system relies primarily on indirect links through a common credit system that has an adequate supply of low-cost credits to bring about allowance price convergence.[65]

A bottom-up system of linkage is already evolving, and could function well in the near-term in the absence of a top-down post-2012 international policy architecture. However, for a bottom-up system to achieve meaningful long-term environmental performance and a high degree of participation, it would require the major emitters—the United States, the European Union, Russia, Japan, China, India, and other key countries—to reach an implicit agreement regarding emissions targets and incentives for participation. Whether this would be possible without centralized negotiations is an open question.

B. Bottom-Up Linkage as a Step towards a Top-Down Architecture

A collection of linkages among cap-and-trade and emission-reduction-credit systems could serve as the foundation for a top-down climate agreement. Indeed, a bottom-up system could evolve into a coherent top-down climate architecture, much as the General Agreement on Tariffs and

64. *See* Jeffrey Frankel, *Global Environmental Policy and Global Trade Policy*, *in* HARVARD PROJECT ON INTERNATIONAL CLIMATE AGREEMENTS, at 12 (Belfer Ctr. for Sci. and Int'l Affairs at Harvard Univ. John F. Kennedy Sch. of Gov't, Discussion Papers 08-14, 2008), *available at* http://belfercenter.ksg.harvard.edu/files/Frankel2Web2.pdf; TREVOR HOUSER ET AL., PETER G. PETERSON INSTITUTE FOR INTERNATIONAL ECONOMICS, LEVELING THE CARBON PLAYING FIELD: INTERNATIONAL COMPETITION AND US CLIMATE POLICY DESIGN 29–30 (2008), *available at* http://pdf.wri.org/leveling_the_carbon_playing_field.pdf.

65. *See* INTERGOVERNMENTAL PANEL ON CLIMATE CHANGE, *supra* note 32; Bohm, *supra* note 32; MANNE & RICHELS, *supra* note 32.

Trade paved the way for the World Trade Organization.[66] Any pre-existing direct or indirect links are likely to influence the evolution of a new top-down international agreement because existing links will function as the status-quo framework. An existing set of links can also foster incentives for countries to participate in a future international agreement. Links will create constituencies within some developing countries who will favor participation if continued demand for emission reduction credits is conditioned on movement toward such an international agreement.

There is another way in which linkage can induce participation in a broader international, if not global, regime. Consider a "leader nation," which can accomplish very little to solve the global problem of climate change were it to act on its own. Linkage provides a mechanism by which that nation's emission reduction activities can be extended more widely. Although shunning, shaming, or punishing non-participant nations is one potential route to trying to broaden the coalition,[67] linkage could offer another route. For example, as the EU ETS has evolved, the European Union (the "leader nation") is using market access—to both allowances markets and to markets for other goods and services—to induce participation and to foster a broader system.[68]

C. Linkage as an Element of an International Architecture

Linkage could also play a significant role as a component of a larger international climate policy architecture—a role that would not conflict with the potential near-term role of linkage as the basis for a future agreement. The overall architecture might be based, for example, on a negotiated global agreement that specifies targets and timetables for emissions reductions from each of the world's nations. Within this architecture, linkage could have distinct near-term and long-term roles. For example, in the near-term, a system of indirect linkages via a common credit system could provide important cost savings while minimizing negative distributional effects and preserving a high degree of national control over allowance markets. In the longer term, the system could transition to negotiated multilateral two-way linkages that would create a single, comprehensive market for allowances and credits. The distinguishing characteristic of this possibility is that the role played by linkage would be specified ahead of time, as part of the overall architecture. In contrast, in Part V.B, we describe how a collection of *independent* linkages could evolve into a broader, top-down agreement.

---

66. *See* Carlo Carraro, *Incentives and Institutions: A Bottom-Up Approach to Climate Policy, in* ARCHITECTURES FOR AGREEMENT: ADDRESSING GLOBAL CLIMATE CHANGE IN THE POST-KYOTO WORLD 165–66 (Joseph Aldy & Robert Stavins, eds., 2007).

67. *See* Keohane & Raustiala, *supra* note 60, at 6–7; Frankel, *supra* 64, at 18–19.

68. *See* Ellerman, *supra* note 10, at 1; Eivind Hoff, *Norway to make money from EU Emissions Trading System*, BELLONA.ORG, Oct. 27, 2007, http://www.bellona.org/articles/articles_2007/norway_emissions_trading_system.

VI. HOW WILL CLIMATE NEGOTIATIONS AFFECT BOTTOM-UP LINKAGES?

Given the potential advantages of linkages in a future international climate architecture, policy makers may want to take steps to facilitate future linkages among tradable permit systems. One way to do this is by choosing design features for a post-2012 international climate agreement that will encourage linkages. First, and perhaps most ambitiously, a post-2012 agreement could establish an agreed trajectory of emissions caps or allowance prices, specify a harmonized set of cost-containment measures, and establish a process for making future adjustments to the level of the emissions cap and other key design elements. Second, it could create an international clearinghouse for transaction records and allowance auctions.[69] Third, it could provide for the ongoing operation of the CDM or a successor, which could play a central role in indirectly linking existing and emerging tradable permit systems. Fourth, a post-2012 international climate agreement could help build capacity in developing countries that would permit and encourage their fuller participation through a system of bottom-up linkages.

If policy makers seek to encourage linkage among tradable permit systems in a post-2012 agreement, some design elements would best be avoided. Any global agreement that encourages strategic behavior in setting domestic emissions caps could impede the development of linkages. For example, an agreement that conditions future commitments on countries' emission levels over the coming years could undermine the ability of linkages to achieve a cost-effective distribution of emission reductions across linked systems. Also, depending on the stringency of such restrictions, an international agreement that imposes "supplementarity" restrictions—in which countries must achieve some specified percentage of emission reductions domestically—may limit the potential benefits of linkage by curtailing the amount of international trading that can occur. Although supplementarity restrictions can force firms in wealthier countries to achieve greater reductions in their own emissions (instead of simply purchasing permits on international markets), these restrictions increase the overall cost of achieving emissions reductions.

CONCLUSION

Cap-and-trade systems are emerging as a preferred domestic instrument for reducing GHG emissions in many parts of the world, and the CDM has developed a substantial constituency despite some concerns about its performance. Because of the considerable political and economic pressure to link these systems, linkage may be expected to play a de facto, if not de jure, role in any future international climate policy architecture.

69. *See* OTTMAR EDENHOFER, CHRISTIAN FLACHSLAND & ROBERT MARSCHINSKI, POTSDAM INSTITUTE FOR CLIMATE IMPACT RESEARCH, TOWARDS A GLOBAL CO2 MARKET: AN ECONOMIC ANALYSIS 29–31 (2007), *available at* http://www.pik-potsdam.de/news/press-releases/files/PIK_Linking_ETS_2007_engl.pdf.

In the short term, linkage will continue to grow in importance as a core element of a bottom-up, de facto international policy architecture. The EU ETS has already established direct links with systems in neighboring countries, and the CDM has emerged as a potential hub for indirect links among cap-and-trade systems worldwide. As new cap-and-trade systems emerge in countries such as Australia, Canada, and the United States, the global network of direct and indirect links will likely continue to spread.

In the longer term, linkage could play a variety of roles. One possibility is that a set of linkages, combined with unilateral emission reduction commitments by many nations, could function as a stand-alone climate architecture. A second long-term possibility is that a collection of bottom-up links may eventually evolve into a comprehensive, top-down agreement. A third long-term possibility is that linkage could play a significant role as a component of a larger international climate policy architecture. Because of the potential benefits of linkage, policy makers may wish to take steps to ensure that a post-2012 agreement includes design features that support the development of linkages.

There is a striking trade-off between direct linkages, which can require a high degree of harmonization and international cooperation, and indirect linkages via a common credit system, which raise concerns about additionality. This tradeoff may suggest a natural progression. In the near-term, indirect linkage of cap-and-trade systems via a common credit system (such as the CDM) could achieve some of the cost savings and risk diversification of direct linkage, but without the need for as much harmonization of emerging and existing cap-and-trade systems. Such indirect linkage would also limit potential distributional concerns and preserve a high degree of national control over allowance markets. In the longer term, international negotiations could establish shared expectations about environmental targets and emission reduction responsibilities that would serve as the basis for a broad set of multilateral, direct links among cap-and-trade systems. This progression could promote the near-term goals of participation and cost-effectiveness while helping to build the foundation for a more comprehensive future agreement.

We welcome responses to this Article. If you are interested in submitting a response for our online companion journal, *Ecology Law Currents*, please contact ecologylawcurrents@ boalt.org. Responses to articles may be viewed at our website, http://www.boalt.org/elq.